Primary and Secondary Metabolism of Plants and Cell Cultures III

Primary and Secondary Metabolism of Plants and Cell Cultures III

Proceedings of the workshop held in Leiden, The Netherlands, 4–7 April 1993

Edited by

J. Schripsema and R. Verpoorte

Reprinted from *Plant Cell, Tissue and Organ Culture* 38: 2/3, 1994

Springer Science+Business Media, B.V.

Library of Congress Cataloging-in-Publication Data

Primary and secondary metabolism of plant cell cultures III / edited by J. Schripsema and R. Verpoorte.
p. cm.
Proceedings from the workshop held in Leiden, Netherlands, April 4th to 7th, 1993.
Includes bibliographical references and index.
ISBN 978-94-010-4106-5 ISBN 978-94-011-0237-7 (eBook)
DOI 10.1007/978-94-011-0237-7
1. Plant cell culture--Congresses. 2. Plants--Metabolism--Congresses. 3. Metabolism, Secondary--Regulation--Congresses.
I. Schripsema, J. II. Verpoorte, R.
QK725.P774 1995
581'.0724--dc20 94-44744

ISBN 978-94-010-4106-5

Printed on acid-free paper

CONTENTS

Editorial

This volume contains the proceedings from the workshop *Primary and Secondary Metabolism of Plants and Plant Cell Cultures III*, which took place in Leiden, April 4th–7th, 1993.

Since the first two meetings on the topic of primary and secondary metabolism of plant cell cultures, in 1984 and 1988, there has been a clear shift in focus of the ongoing research. In the proceedings from the first meeting, particularly, the cell culture itself and the production of secondary metabolites were the dominant themes. In the second proceedings biosynthetic pathways and the activity of enzymes were major topics. In the proceedings of this third meeting these aspects are linked with genes, such that molecular biology becomes more prominent.

These proceedings reflect the state of the art in the field, with contributions on subjects such as fermentation, enzymology of secondary metabolism, catabolism of secondary metabolites, elicitation of pathways and the genetic modification of metabolic pathways.

The book includes contributions on the most recent achievements in the research concerning, among other topics, tropane and indole alkaloids, phenolics, (iso)flavonoids, terpenes and cardenolides. It gives an excellent review of the progress made in the past years and a perspective on future developments in the field.

Leiden, 1994 J. Schripsema and R. Verpoorte

Plant Cell, Tissue and Organ Culture **38**: 85–91, 1994.

Ajmalicine production by cell cultures of *Catharanthus roseus*: from shake flask to bioreactor

Hens J.G. ten Hoopen[1], Walter M. van Gulik[1], Jurriaan E. Schlatmann[1], Paulo R.H. Moreno[2], J.L. Vinke[1], J.J. Heijnen[1] & Robert Verpoorte[2]

Biotechnology Delft Leiden, Project Group Plant Cell Biotechnology, [1]Department of Biochemical Engineering, Delft University of Technology, Julianalaan 67, NL-2628 BC Delft, The Netherlands; [2]Division of Pharmacognosy, Leiden/Amsterdam Center for Drug Research, Leiden University, P.O. Box 9502, NL-2300 RA Leiden, The Netherlands

Key words: Ajmalicine, bioreactor, *Catharanthus roseus*, growth model, scale-up

Abstract

The productivity of a cell culture for the production of a secondary metabolite is defined by three factors: specific growth rate, specific product formation rate, and biomass concentration during production. The effect of scaling-up from shake flask to bioreactor on growth and production and the effect of increasing the biomass concentration were investigated for the production of ajmalicine by *Catharanthus roseus* cell suspensions. Growth of biomass was not affected by the type of culture vessel. Growth, carbohydrate storage, glucose and oxygen consumption, and the carbon dioxide production could be predicted rather well by a structured model with the internal phosphate and the external glucose concentration as the controlling factors. The production of ajmalicine on production medium in a shake flask was not reproduced in a bioreactor. The production could be restored by creating a gas regime in the bioreactor comparable to that in a shake flask. Increasing the biomass concentration both in a shake flask and in a stirred fermenter decreased the ajmalicine production rate. This effect could be removed partly by controlling the oxygen concentration in the more dense culture at 85% air saturation.

Introduction

Commercial application of the production of secondary metabolites by plant cells in suspension culture is mainly hampered by the too low productivity of the cultures. Three factors determine the productivity of the cell culture: the specific growth rate, the specific product formation rate, and the biomass concentration during the production phase. To increase productivity, these three factors have to be optimized. Particularly, the product formation rate of plant cells shows a broad variation. Several techniques are available to increase the product formation rate: screening and selection of cell lines, optimizing culture conditions (medium composition, light, temperature, gas composition, genetic modification) (Verpoorte et al. 1991). Some of these approaches have to be carried out at a small scale in shake flasks, petri dishes or culture tubes, either because the technique demands very small amounts of biological material, or because large numbers of parallel experiments are necessary to optimize a set of conditions (for example, medium optimization through statistical experimental design, Tuominen et al. 1989).

To perform a process on an industrial scale at least two scale-up steps are necessary (Fig. 1). First, the developed system has to be reproduced in a laboratory-scale bioreactor. Secondly, the process has to be scaled-up in one or more steps to the process-size bioreactor. The first step is the most difficult one, because a shake flask and a bioreactor are completely different systems in geometry, mixing and gas regime. Transferring a process to a larger bioreactor of the same type generates also problems, but these problems can be solved with the general scale-up approaches developed in fermentation industry. The type of bioreactor is an essential factor in these studies. In this paper a standard stirred and aerated bioreactor is used. This type is common in fermentation industry and therefore the best option to introduce a new process for the production of a plant product by plant cells in suspension culture. Fur-

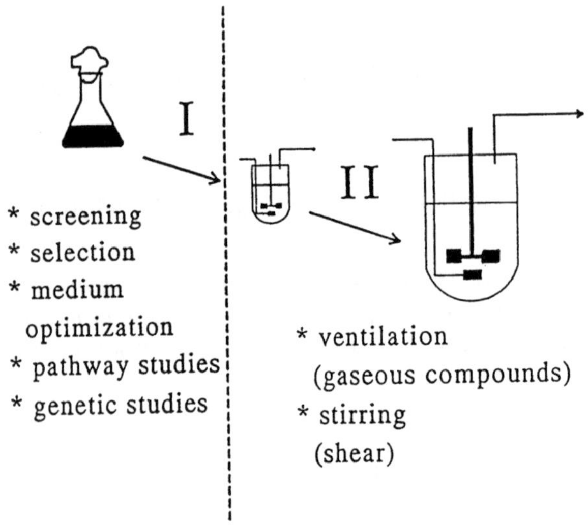

Fig. 1. Scaling-up in plant cell biotechnology.

thermore, scale-up procedures for aerated and stirred bioreactors are well developed.

In this study the first step of the scale-up procedure, from shake flask to bioreactor, will be investigated. The production of ajmalicine by cell cultures of *Catharanthus roseus* was selected as a model system, because a great deal of experience with this system was available from previous studies. Experiments by Meijer (1990) revealed that the cell line was very shear-tolerant. It grew well on LS-medium with either sucrose or glucose as carbon source. The productivity of the cell line used in these studies has been increased to a great deal by Schlatmann et al. (1992, 1993); an optimal production medium for this cell line was selected. In literature, a positive effect on biomass growth of conditioning factors in the medium is mentioned several times (Stuart & Street 1971; Wijnsma et al. 1988). There are conflicting data on the effect of carbon dioxide in the gaseous phase on the growth (Ducos & Pareilleux 1986; Ducos et al. 1988; Hegarty et al. 1986; Maurel & Pareilleux 1985). Taking into consideration these observations, the scale-up effects on the three aspects of productivity: biomass growth, product formation and biomass concentration were investigated.

Materials and methods

Cell material

Cell suspension culture of *Catharanthus roseus* (L.) G. Don MP183D was obtained from the Institute of Plant Molecular Sciences, Leiden University. The culture was initiated from seeds and grown in suspension culture since 1983. Cell lines were subcultured every 14 days (BIX) or every 7 days (BXV) by adding 35 ml of suspension to 165 ml of fresh growth medium. The cultures were grown in 1000-ml Erlenmeyer flasks with silicon stoppers (Shin Etsu, Tokyo, Japan) on a rotary shaker at 100 rpm in the dark.

Media

Growth medium, described by Linsmaier & Skoog (1965) supplemented with 2.0 mg l^{-1} naphthalene acetic acid, 0.2 mg l^{-1} kinetin, and 30 g l^{-1} glucose. The medium was adjusted to pH 5.8 before sterilization (20 min, 121 °C).

Production medium: Growth medium, depleted of nitrate, ammonium, phosphate and hormones, and supplemented with 80 g l^{-1} glucose.

Shake flask experiments

Shake flask experiments were performed in 250 ml Erlenmeyer flasks with silicon stoppers on a rotary shaker (100 rpm) at 25 °C in the dark. The total culture volume was 60 ml; the inoculum size was 10 ml, except for the comparison of low and high biomass concentration. In that case 50 ml of production medium was inoculated with respectively 5 g and 20 g of fresh weight.

Bioreactor experiments

Bioreactor experiments were carried out in a commercially available 3-l turbine stirred tank reactor with a working volume of 1.8–2.1 l (Applikon Dependable Instruments, Schiedam, The Netherlands). The culture was aerated through a sintered steel sparger. The flow was kept at 1/3 vvm with a mass flow control system (Brooks, Veenendaal, The Netherlands). Exhaust gas was led through a glass condenser cooled by a cryostat (Lauda Messgeräte, Lauda, Germany) at 4 °C, in order to minimize evaporation. Two six-bladed turbine impellers (D = 45 mm) were used for mixing; rotation speed was 250 rpm. The bioreactor was equipped with 3 baffles to improve mixing. The temperature was maintained at 25 °C with a thermostated stainless steel pipe in the culture fluid. In experiments on growth medium the pH was maintained at 5.0 by the addition of either 0.25 N NaOH or 0.25 N HCl using ADI 1030 Bio-controllers (Applikon) equipped with sterilisable pH-electrodes (Ingold) and peristaltic pumps for alkali and acid. Dissolved oxygen concentrations were measured with a sterilisable oxygen electrode (Ingold). The formation of foam was prevented by adding, at regular time intervals, a silicon-based antifoaming agent (1% w/w, BDH, Poole, England), or adding 18 mg l^{-1} to the medium in the case of continuous culture experiments. The bioreactor was wrapped in black plastic to keep out light. The bioreactor was inoculated with one part suspension culture and five parts of medium. In the incoming as well as the exhaust air the oxygen (Servomex 1101 paramagnetic O_2 analyzer, Crowborough, UK) and the carbon dioxide concentration (Rosemount analytical 870, La Habra, USA) were measured on line.

Investigations on the effect of high biomass concentrations were performed in 2 identical 15-l turbine stirred bioreactors with a working volume of 11.5 l (Applikon, Schiedam, The Netherlands). One six-bladed turbine impeller (D = 75 mm) was used for mixing; rotation speed was 200 rpm. The aeration rate was 0.3 vvm. The bioreactor was equipped with 3 baffles to improve mixing. In the experiments in which two treatments were compared, two bioreactors were inoculated simultaneously with the same inoculum.

Analytical procedures

The determinations of biomass dry weight, packed cell volume, storage carbohydrates, glucose, phosphate, ajmalicine, tryptamine were performed as described previously (Van Gulik et al. 1992; Schlatmann et al. 1993).

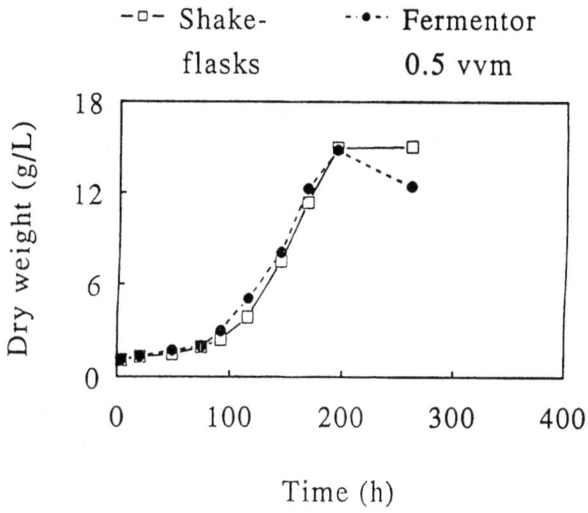

Fig. 2. Comparison of the growth of *C. roseus* cell cultures in shake flasks and in bioreactors at 0.5 vvm aeration rate (Van Gulik et al. 1993b).

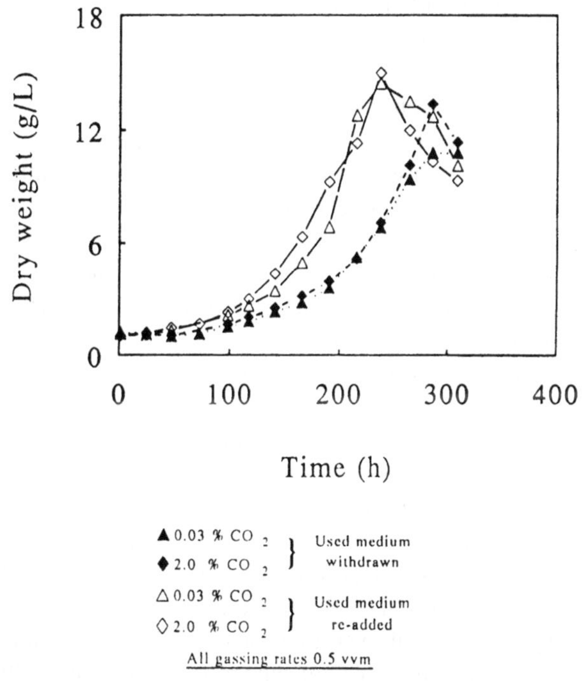

Fig. 3. Effects of CO_2 enrichment of the aeration gas on the growth of *C. roseus* cell suspensions with and without conditioning factors in the medium in stirred fermenters (Van Gulik et al. 1993b).

Results and discussion

Biomass growth

Figure 2 compares the growth in a shake flask and a bioreactor. There is no significant difference during lag or growth phase. Apparently, neither the difference in shear, nor the difference in gas regime between shake flask and stirred fermenter influences the growth of this cell line significantly. If the gas regime does not affect growth, an effect of variations in carbon dioxide concentration in the gaseous phase on the biomass growth is not very likely. This aspect was investigated in detail by comparing growth in stirred bioreactors at different CO_2 concentrations. To exclude an eventual interference with the effect of conditioning factors the experiments were carried out with and without conditioning factors in the medium. The growth curves in Fig. 3 show that there is no significant difference in growth between cultures at 0.03% CO_2 or at 2.0% CO_2 concentration. On the other hand, the positive effect of conditioning factors in the inoculum is clearly proven for both CO_2 concentrations.

For scale-up to larger fermenters, the availability of a mathematical growth model is very useful. A biological parameter in itself is scale independent, but productivity is influenced by chemical and physical parameters, which are production scale dependent. If these parameters are included in a model, the production can be predicted at larger scale. Van Gulik et al. (1992) concluded that a growth model in which the biomass is considered to be a single compound with an average chemical composition is of limited value because large changes in the composition of the biomass may occur. They developed a structured model describing the growth as a function of two limiting substrates: glucose and phosphate (Van Gulik et al. 1993a). All other nutrients essential for cell growth, including oxygen, were assumed to be non-limiting. In the biomass four compartments are assumed:

1. free intracellular phosphate (P);
2. low molecular weight phosphorylated compounds (E); e.g. nucleotides and phosphorylated sugars;
3. storage carbohydrates (C);
4. structural biomass (B).

The P-pool has an important control function. Under non-limiting conditions, the formation of phosphorylated precursors is stimulated. These compounds serve as energy carriers and/or precursors for structural-biomass formation. On the other hand under phosphate limitation the conversion of C-source in starch-like storage products is stimulated. Besides, the formation of extra-cellular polysaccharides is included in the model (Fig. 4). This model was translated in a mathematical model by formulation of balance equations for the various compartments and rate equations for the conversions among the compartments. Most model parameters were derived from experimental results or literature data, lacking parameters were found by optimization of the fit of the model with the experimental data. Simulations of this model were compared with experimental results with the initial concentration of phosphate as the variation factor. In Fig. 5 the results of these comparisons are shown on LS medium with 20%, 100% and 500% of the standard amount of phosphate in the medium. All experimental results are rather well described by the model. At a low phosphate concentration biomass growth is linear during the whole growth period and there is a high production of storage carbohydrates caused by the low internal phosphate concentration. With the standard amount of phosphate in the medium, exponential growth during a few days is followed by linear growth. The growth limiting factor in the LS medium appears to be phosphate. After consumption of the external C-source, structural biomass and storage products are consumed for maintenance

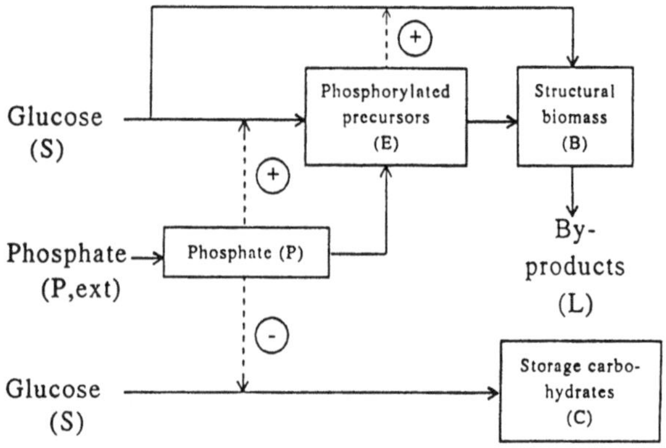

Fig. 4. Scheme of the structured growth model.

of the cell mass. At a higher phosphate concentration exponential growth is sustained until the medium glucose is completely consumed. There is hardly any storage of starch-like products, because there is no limitation by internal phosphate. The model describes the internal phosphate pools and continuous culture experiments also very well (results not shown).

The model will be extended with ajmalicine production in relation to glucose and oxygen concentration to make it useful for scaling up ajmalicine production.

Ajmalicine production

Figure 6 compares the growth and product formation of *C. roseus* cell suspension in respectively a shake flask and a bioreactor on production medium. In the shake flask, there is an increase of biomass, which must be caused mostly by storage of starch-like products on this phosphate-free medium (see preceding section). The production of ajmalicine starts after about 10 days and proceeds during the next two weeks. During the production phase part of the produced ajmalicine is excreted. The behaviour of the cell culture in the bioreactor with production medium is completely different: hardly any biomass increase and no ajmalicine production. Furthermore, the culture turns brown after 10 days. The important differences between the shake flask and the bioreactor are the stirring by a Rushton turbine and the forced aeration through a sparger underneath the stirrer. Therefore, either shear forces generated by the stirrer, or differences in concentration of gaseous compounds caused by the aeration can be responsible for the observed differences. Schlatmann et al. (1993) developed an experimental set-up in which these aspects can be studied separately keeping all other conditions unchanged. To achieve this,

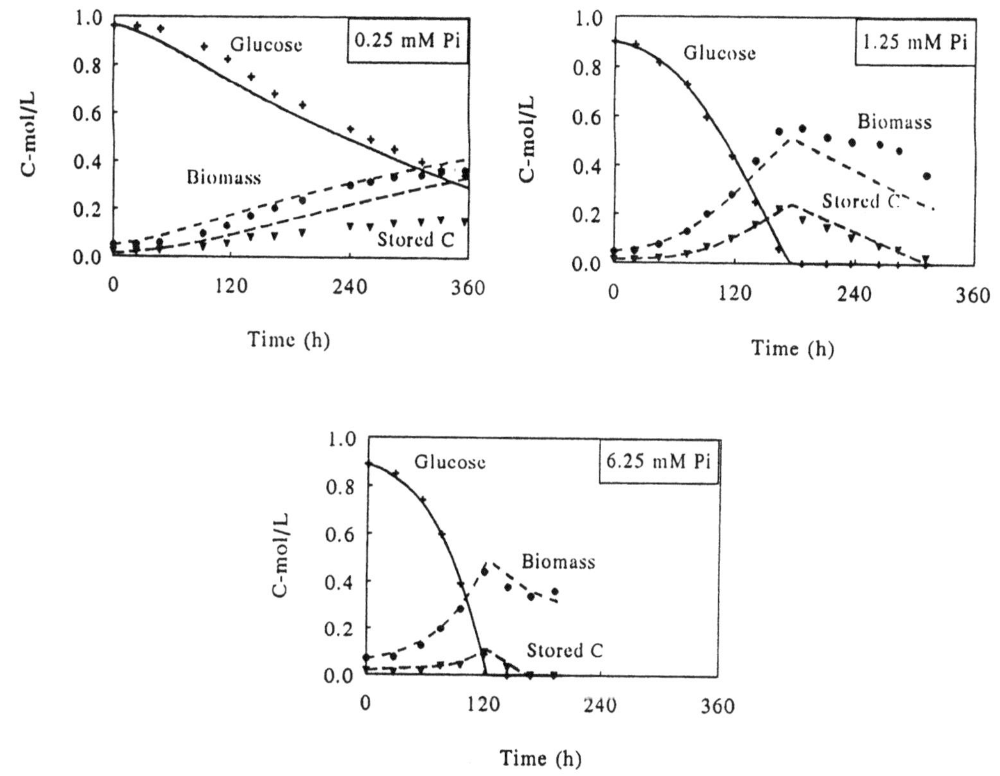

Fig. 5. Comparison of experimental growth data (symbols) and model predictions (lines) for three initial phosphate concentrations in LS medium (Van Gulik et al. 1993a).

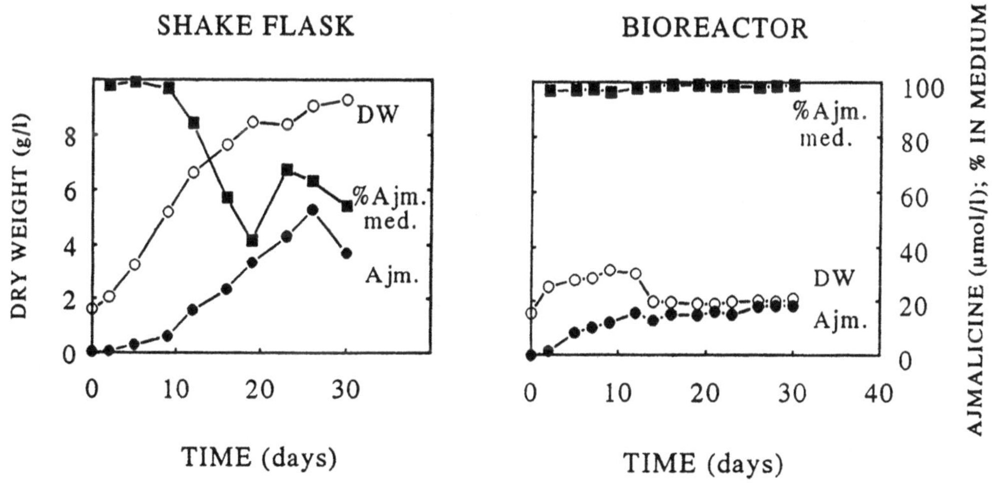

Fig. 6. Comparison of the biomass and ajmalicine production of *Catharanthus roseus* cell cultures on production medium in respectively a shake flask and a bioreactor (Schlatmann et al. 1993).

they modified the bioreactor, either into a bubble column avoiding shear by the stirrer, or into a recirculation system keeping about 90% of the gaseous phase in the bioreactor with an unchanged total gas flow, simulating the gaseous conditions in a shake flask and keeping the hydrodynamic conditions unchanged (Fig. 7). The experimental results from the bubble column and the recirculation reactor are shown in Fig. 8. Although the performance of the bubble column is better than of the standard bioreactor, ajmalicine production is still very low. Browning of the culture was not observed, suggesting that there was shear damage of the biomass in

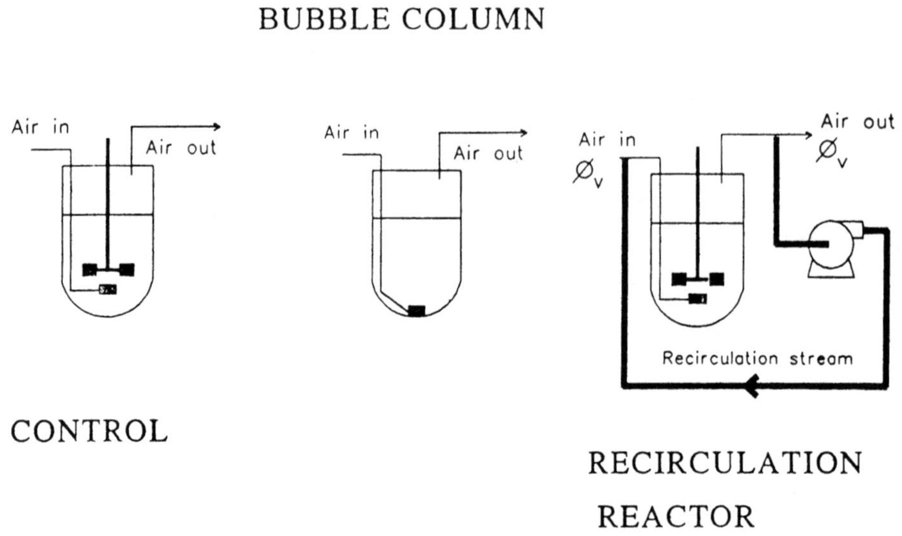

Fig. 7. Experimental set-up for the separate investigation of shear effects and ventilation gas composition effects on ajmalicine production in a bioreactor (Schlatmann et al. 1993).

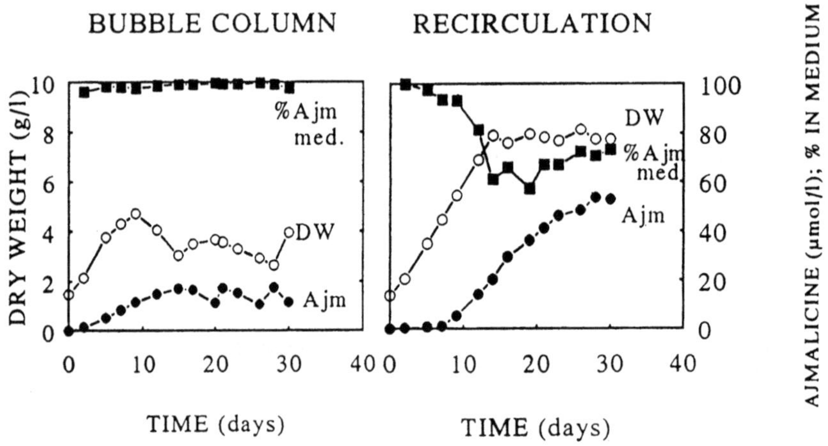

Fig. 8. Comparison of the biomass and ajmalicine production of *Catharanthus roseus* cell cultures on production medium respectively in a bubble column and a recirculation bioreactor (Schlatmann et al. 1993).

the stirred fermenter, but this effect was not responsible for the low product formation. In the recirculation reactor the performance of the shake flasks was completely reproduced, clearly showing that the gas regime in the production system is essential. Attempts to pinpoint a specific component of the gas phase as responsible for this observation were not successful until now.

Comparing the results of experiments in growth and in production medium, the effects of gas regime in production medium appear much more pronounced. However, one should realize that there was no ajmalicine production in growth medium, hence the effects of gas regime on production could not be established in that case. Biomass increase in growth medium (with phosphate) is mainly structural biomass, in production medium (without phosphate) it is mainly storage carbohydrate (see Fig. 4). This implies that different pathways are involved, which may be differently influenced by the gas regime. Nevertheless, obviously the production medium is not a very favourable environment for plant cells: no phosphate and nitrate and high osmotic pressure. This may sensitize the culture.

Biomass concentration

Increasing the biomass concentration had a detrimental effect on the ajmalicine production in both shake flask and bioreactor (Moreno et al. 1993). The high biomass concentration changes the gas regime in the culture system, because of a higher volumetric conversion of nutrients and decreased transport of oxygen, carbon dioxide and other volatile compounds. The dissolved oxygen concentration will be lower for example and appeared a possible candidate to be responsible for the observed effect. Therefore, a fermenter experiment was set up in which the dissolved oxygen concentration could be kept at respectively 15% and 85% by controlled addition of oxygen or nitrogen to a constant total gas flow. The lower oxygen concentration is representative for the uncontrolled dissolved oxygen regime at high biomass concentration. Although this concentration is not harmful for biomass production in growth medium it might effect the ajmalicine production in production medium. The experiments were carried out in 15-l bioreactors and ajmalicine, tryptamine and biomass were determined. The preliminary results show that the dissolved oxygen concentration has indeed a great effect on the product formation. At low oxygen concentration an intermediate from the tryptophan pathway, tryptamine, is accumulated, at high oxygen ajmalicine is formed. To get a better insight in this control on the ajmalicine pathway, the levels of some key-enzymes of the pathway will be studied at different oxygen levels in the future.

Conclusions

Growth of *C. roseus* on growth medium in an aerated and stirred bioreactor is comparable with the growth in shake flasks. The growth can be predicted with a structured mathematical model. The biomass production rate can be influenced by conditioning factors in the medium. The carbon dioxide concentration in the gas phase appeared to have no effect upon the growth rate between 0.03 and 2% CO_2.

Ajmalicine production on production medium is inhibited in an aerated and stirred bioreactor. Gaseous metabolites appear to be responsible for this effect. Recirculation of the ventilation gas restores the production. At low dissolved oxygen concentrations ajmalicine production appeared to be replaced by tryptamine production.

References

Ducos JP & Pareilleux A (1986) Effect of aeration rate and influence of pCO2 in large scale cultures of *Catharanthus roseus* cells. Appl. Microbiol. Biotechnol. 25: 101–105

Ducos JP, Feron G & Pareilleux A (1988) Growth and activities of enzymes of primary metabolism in batch cultures of *Catharanthus roseus* cell suspension under different pCO_2 conditions. Plant Cell Tissue Organ Cult. 13: 167–177

Hegarty PK, Smart NJ, Scragg AH & Fowler MW (1986) The aeration of *Catharanthus roseus* L. G. Don suspension cultures in airlift bioreactors: The inhibitory effect at high aeration rates on culture growth. J. Exp. Bot. 37: 1911–1920

Linsmaier EM & Skoog F (1965) Organic growth factor requirements of tobacco tissue cultures. Physiol. Plant. 18: 100–127

Maurel B & Pareilleux A (1985) Effect of carbon dioxide on the growth of cell suspensions of *Catharanthus roseus*. Biotechnol. Lett. 7: 313–318

Meijer JJ (1990) Effects of hydrodynamic and chemical/osmotic stress on plant cells in a bioreactor. Thesis, Delft, The Netherlands

Moreno PRH, Schlatmann JE, Van der Heijden R, Van Gulik WM, Ten Hoopen HJG, Verpoorte R & Heijnen JJ (1993) Induction of ajmalicine formation and related enzyme activities in *Catharanthus roseus* cells: effect of inoculum density. Appl. Microbiol. Biotechnol. 39: 42–47

Schlatmann JE, Ten Hoopen HJG & Heijnen JJ (1992) Optimization of the medium composition for alkaloid production by *Catharanthus roseus* using statistical experimental designs. Med. Fac. Landbouww. Univ. Gent 57: 1567–1569

Schlatmann JE, Nuutila AM, Van Gulik WM, Ten Hoopen HJG, Verpoorte R & Heijnen JJ (1993) Scale-up of ajmalicine production by plant cell cultures of *Catharanthus roseus*. Biotechnol. Bioeng. 41: 253–262.

Stuart B & Street HE (1971) Studies on the growth in culture of plant cells X. Further studies on the conditioning of culture media by suspensions of *Acer Pseudoplatanus* cells. J. Exp. Bot. 22: 96–106

Tuominen U, Toivonen L, Kauppinen V, Markkanen P & Björk L (1989) Studies on growth and cardenolide production of *Digitalis lanata* tissue cultures. Biotechnol. Bioengin. 33: 558–562

Van Gulik WM, Ten Hoopen HJG & Heijnen JJ (1992) Kinetics and stoichiometry of growth of plant cell cultures of *Catharanthus roseus* and *Nicotiana tabacum* in batch and continuous fermenters. Biotechnol. Bioeng. 40: 863–874

Van Gulik WM, Ten Hoopen HJG & Heijnen JJ (1993a) A structured model describing carbon and phosphate limited growth of *Catharanthus roseus* plant cell suspensions in batch and chemostat culture. Biotechnol. Bioeng. 41: 771–780

Van Gulik WM, Nuutila AM, Vinke JL, Ten Hoopen HJG & Heijnen JJ (1993b) Effects of carbon dioxide, air flow rate and inoculation density on the batch growth of *Catharanthus roseus* cell suspensions in stirred fermenters. (submitted for publication)

Verpoorte R, Van der Heijden R, Van Gulik WM & Ten Hoopen HJG (1991) Plant biotechnology for the production of alkaloids: Present status and prospects. In: Brossi A (Ed) The Alkaloids, Vol 40 (pp 1–187). Academic Press, Orlando

Wijnsma R, Verpoorte R, Harkes PAA, Van Iren F & Ten Hoopen HJG (1988) Conditioning of media: An elaborate method of optimizing initial growth hormone concentration. In: Pais MSS et al. (Eds), Plant Cell Biotechnology. Nato ASI Series H18, Springer Verlag, Berlin/Heidelberg

Plant Cell, Tissue and Organ Culture **38**: 93–102, 1994.

Production of steroidal alkaloids by hairy roots of *Solanum aviculare* and the effect of gibberellic acid

M. Ahkam Subroto & Pauline M. Doran*
Department of Biotechnology, University of New South Wales, Sydney NSW 2052, Australia
(requests for offprints)*

Key words: *Agrobacterium rhizogenes*, gibberellic acid, hairy root, *Solanum aviculare*, solasodine, steroidal alkaloids

Abstract

Cultures of *Solanum aviculare* hairy roots were established after transformation with *Agrobacterium rhizogenes* A4. High levels of steroidal alkaloids measured as solasodine equivalents were produced in shake-flasks and bioreactor, even though relatively low concentrations are found in roots in vivo. In shake flasks the maximum alkaloid yield was 32 mg g^{-1} dry weight; in a 3–l air-driven bioreactor the yield was 29 mg g^{-1}. These yields represent a 5-fold increase over previous reports for in vitro production, and are comparable with levels found in the aerial parts of intact *S. aviculare* plants. Production of steroidal alkaloids was growth-associated. High sugar levels at stationary phase and insensitivity to increased levels of medium components suggest that root cultures were limited by oxygen mass-transfer. In Petri-dish culture with and without exogenous gibberellic acid, root length and number of root tips increased exponentially; growth proceeded with a constant length per root tip of about 35 mm. Addition of gibberellic acid enhanced growth but reduced the specific steroidal-alkaloid level. Taking into account both growth and alkaloid yield, accumulation of steroidal alkaloids was improved by about 40% at gibberellic-acid concentrations of 10 and 100 μg l^{-1}.

Introduction

Solasodine is a precursor for commercial production of steroidal hormones. It is synthesised by plants of the Solanaceae family, and is found in highest quantity in *Solanum aviculare*, a native species of Australia and New Zealand. In vitro production by plant tissue culture is an alternative method for supply of this compound to the pharmaceutical industry, and overcomes many of the problems associated with field production (Macek 1989).

In plants, solasodine is found in glycosidic form with sugar residues attached through the C-3 hydroxyl group. Several glycosides are reported to occur; the most common are solasonine and solamargine. On hydrolysis, the glycosides yield solasodine in the aglycone form. Solasodiene is formed by dehydration of solasodine under conditions of acid hydrolysis. Colorimetric analyses, such as the methyl-orange method of Birner (1969) and the bromothymol-blue method developed by Lancaster & Mann (1975), do not distinguish between several related steroidal alkaloids. For example, identical bromothymol-blue dye-complexing properties have been reported for solasodine, solasodiene, soladulcidine and tomatidine (Ehmke & Eilert 1986), although diosgenin and cholesterol do not bind (Chandler & Dodds 1983). Varying responses for solasodine, solanidine, solasodiene, tomatidine, demissidine and other compounds have been found using methyl-orange (Bradley et al. 1978). Results published by different authors are therefore difficult to compare, as different methods of extraction, hydrolysis and analysis have been used. Many reported solasodine levels are not absolute, but represent 'solasodine equivalents'.

Undifferentiated callus and suspension cultures of *S. aviculare* contain low levels of solasodine (0.145–0.9 mg g^{-1} dry weight) as analysed using TLC and colorimetric procedures (Khanna et al. 1976, 1977). These yields have been improved to ca. 5 mg g^{-1} by addition of cholesterol to the medium (Khanna et al. 1977), or strain selection and manipulation of culture conditions

(Macek 1989). In this laboratory, self-immobilisation of *S. aviculare* cells into aggregates was used to promote cell-cell contact and tissue organisation (Tsoulpha & Doran 1991); solasodine equivalents up to 3 mg g^{-1} were obtained from this culture system without media additives. So far, solasodine yields produced in vitro have been considerably lower than those found in intact plants; levels of solasodine in plants analysed using methyl-orange colorimetry have been reported as ca. 12 mg g^{-1} in the leaves and 30 mg g^{-1} in green fruit (Bradley et al. 1978). An average solasodine content of 15.4 mg g^{-1} has been reported from radioimmunoassay of *S. aviculare* leaves (Weiler et al. 1980).

As with most plant species, secondary-product synthesis in *Solanum* cultures can be improved by allowing the plant tissue to differentiate morphologically; shoots of *S. laciniatum* regenerated from callus contain significant levels of solasodine (Chandler & Dodds 1983; Conner 1987). Organ culture is therefore a promising approach to gaining improvements in solasodine production. Increasing numbers of medicinal plant species are being transformed with *Agrobacterium rhizogenes* to develop hairy-root cultures for enhanced phytochemical synthesis without the need for exogenous growth-regulators (Flores et al. 1987; Hamill et al. 1987; Hamill & Rhodes 1993). Previous studies have shown that the pattern of product accumulation in hairy roots closely follows that of roots in vivo (Parr & Hamill 1987; Parr et al. 1988). Until now, the suitability of hairy-root technology for production of steroidal alkaloids such as solasodine has been unclear; roots are not the primary commercial source of solasodine and there is uncertainty about the site of synthesis in the whole plant.

The effect of growth regulators such as auxin and cytokinin on root growth and morphology has been studied extensively; the influence of these substances on hairy-root development has also been examined (Bercetche et al. 1987; Zhan et al. 1990). However, there is little information about the effect of exogenously supplied gibberellins (GA_3) on hairy roots. In excised untransformed roots, the effect of gibberellic acid appears variable; root elongation and stimulation of cell division can occur depending on parameters such as sucrose concentration, light quality and intensity, and interactions with other growth regulators (Torrey 1976). Concentrations of gibberellic-acid between 10 ng l^{-1} and 1 mg l^{-1} have been found to accelerate the fresh weight increase of *Datura innoxia* hairy roots and enhance elongation and lateral branching (Ohkawa et al. 1989); however the effect on secondary-product synthesis was not recorded. It was also suggested that *A. rhizogenes* T-DNA may produce factors which interact with gibberellins. As there is evidence that gibberellic acid increases the solasodine content in fruit of producing plants (Chaudhuri & Chatterjee 1979), the combined effects of enhanced growth and enhanced product accumulation would be advantageous for in vitro solasodine synthesis.

In the present work, *S. aviculare* was transformed by *Agrobacterium rhizogenes* to produce hairy roots. Kinetics of growth, sugar consumption and steroidal alkaloid production in shake-flask and reactor cultures were investigated. The effect on growth and alkaloid synthesis of varying concentrations of exogenous gibberellic acid was also studied.

Materials and methods

Hairy-root cultures

Solanum aviculare seeds were obtained from the Botanic Gardens of Adelaide, South Australia. *Agrobacterium rhizogenes* A4 was kindly donated by Prof. A. Kerr, Waite Agricultural Institute, Adelaide, Australia, and maintained as described previously (Sharp & Doran 1990).

Hairy roots were obtained by wounding *S. aviculare* plantlets with a syringe needle containing *A. rhizogenes*. Root clones from different wound sites were cultured separately. Excised roots were cleared of bacteria with cefotaxime (Roussel Pharmaceuticals Pty Ltd) as described elsewhere (Sharp & Doran 1990), and maintained in Murashige & Skoog (MS) medium (Flow Laboratories, Scotland) containing either 3% or 6% sucrose without plant growth-regulators. After 6 months, root material from a single hairy-root clone was used as inocula for shake-flask and reactor experiments.

Shake-flask experiments

Thirty-six shake flasks (250 ml) containing 100 ml MS medium with 6% sucrose were each inoculated with the same fresh weight of hairy roots. The flasks were incubated for 54 d at 25 °C with shaking at 110 rpm. Cultures were illuminated using Osram and Philips 18W and 20W fluorescent lights at an intensity of ca. 3000 lux. Samples were taken once a week by harvesting triplicate flasks for fresh weight, dry weight and medium sugar analyses, and extraction of steroidal alkaloids from both liquid medium and roots.

Additional shake-flask experiments were carried out over a period of 49 d to determine the effect of high concentrations of MS salts and vitamins on growth of roots. Flasks containing 1.0, 1.3 or 1.5 times the normal level of MS powder and 6% sucrose were inoculated with roots; duplicate flasks were harvested every 7 d for analysis of dry weight.

Measurement of root extension and branching

Kinetics of hairy-root extension and branching were examined using Petri-dish culture. Approx. 30 ml liquid MS medium containing 6% sucrose was added to triplicate Petri dishes; each dish was then inoculated with a single root-tip of initial length 2.3–2.5 cm. The plates were shaken gently at 25 °C under 1000 lux for a period of 26 d; photographs of the roots were taken every 2–6 d for measurement of root length and number of root tips. The root growth-unit was calculated as defined by the equation:

$$\text{Root growth-unit (mm)} = \frac{\text{Total root length (mm)}}{\text{Number of root tips}} \qquad (1)$$

Experiments with gibberellic acid

To assess the effect of gibberellic acid on root growth and steroidal-alkaloid synthesis, shake flasks containing MS medium with 0, 10, 100 or 1000 $\mu g\ l^{-1}$ gibberellic acid (Sigma) and 6% sucrose were inoculated with ca. 0.5 g fresh weight hairy roots. All flasks were incubated at 25 °C and shaken at 110 rpm under ca. 1500 lux. After 56 d, the roots and culture liquid were separated for measurement of dry weight and alkaloid content. This experiment was performed in duplicate.

The effect of gibberellic acid on hairy-root morphology was examined using liquid Petri-dish culture as described above. Gibberellic acid was added to the plates at a final concentration of 1000 $\mu g\ l^{-1}$; plates without gibberelic acid were used as control. This experiment was performed in triplicate.

Bioreactor experiments

The bioreactor used was a converted conical-flask air-driven reactor as described by Tsoulpha & Doran (1991). Temperature was controlled at 25 °C. Dissolved oxygen in the medium was monitored with a galvanic sensor (locally made) and dissolved-oxygen meter (LH Fermentation, UK). Water at 4 °C was recirculated through a condenser to control evaporation losses.

The reactor was inoculated with 3.0 l of growth-regulator-free MS medium containing 3% sucrose and 5.6 g fresh weight *S. aviculare* hairy roots. The reactor was run batch-wise under ca. 3000 lux. Air was supplied at a constant rate of 60 ml min^{-1}. The reactor vessel was periodically drained of liquid and weighed on a Mettler PJ12 balance to determine increase in fresh biomass. Liquid samples were taken for sugar analysis. At the end of the culture period the roots were harvested, weighed and freeze-dried for alkaloid determination.

Analyses

Fresh weight, dry weight and medium sugar concentrations were analysed as described previously (Tsoulpha & Doran 1991). Steroidal alkaloids in roots and medium were analysed using HPLC, TLC and chemical-analysis techniques.

For HPLC, 100 mg hairy root was refluxed in 1N HCl in methanol at 90–92 °C for 2 h. The reaction mixture was then neutralised with concentrated ammonia solution before filtering. The filtrate was evaporated under reduced pressure to yield a solid residue. To purify the sample, the solid residue was dissolved in 75:25 (v/v) methanol: 0.01M Tris buffer (pH 7.0) and passed through a Sep-Pak C18 cartridge (Waters). Purified samples of solasonine and solamargine kindly provided by Dr JD Mann, DSIR Lincoln, New Zealand, and solasodine (Sigma) were hydrolysed with 1N HCl in methanol using the same procedure. Samples were injected into a 30 cm × 3.9 mm HPLC column containing 10 μm Bondex C18 packing (Phenomenex) operated at ambient temperature. The mobile phase was 75:25 (v/v) methanol:0.01M Tris buffer (pH 7.0) at a flow rate of 1.4 ml min^{-1}. Steroidal alkaloids were detected at 205 nm. Hydrolysed hairy-root extraact gave a single broad peak at a retention time of 22 min. Hydrolysed solasodine added to these samples appeared within the peak at 18.5 min. Identification of compounds likely to be contained in the sample peak was carried out using TLC.

Both hydrolysed and unhydrolysed hairy-root samples were analysed using TLC. Hydrolysed hairy-root residue after evaporation of methanol was dissolved in a mixture of benzene and 0.1M NH_4Cl; samples for TLC were taken from the benzene layer. Unhydrolysed hairy-root extract was obtained by macer-

Table 1. Overall yields for *S. aviculare* hairy roots grown on MS medium containing 6% sucrose (shake flasks) and 3% sucrose (bioreactor).

	Yield of biomass from substrate (Y_{XS}, g g^{-1})	Yield of steroidal alkaloids from substrate (Y_{PS}, mg g^{-1})	Yield of steroidal alkaloids from biomass (Y_{PX}, mg g^{-1})
Shake flasks Day 54	0.43	14	32
Bioreactor Day 82	0.32	9.1	29

Table 2. Average specific growth rates and length of root growth-unit from triplicate cultures.

	Specific growth rate* (d^{-1})		Root growth-unit (mm)
	Root length	Number of root tips	
Without gibberellic acid	0.09 ± 0.002	0.10 ± 0.006	37 ± 5.3
With gibberellic acid	0.18 ± 0.02	0.14 ± 0.03	32 ± 3.3

* Values are calculated for the period of exponential growth.

ation in 19:1 (v/v) ethanol: 28% ammonia solution. Three TLC separations were tested in this work: (1) silica gel 60 plates (Merck) with mobile phase 95:5 (v/v) chloroform:methanol and Dragendorff's reagent (Sigma) as spraying agent; (2) 15% $AgNO_3$/silica gel (Alltech) with mobile phase 95:5 (v/v) chloroform:methanol and anisaldehyde (Sigma)/H_2SO_4 as spraying agent; and (3) silica gel 60 F_{254} (Merck) with mobile phase 9:1 (v/v) dichloromethane:methanol and anisaldehyde/H_2SO_4 as spraying agent. Using (1), hydrolysed hairy-root samples formed a single spot with R_f value corresponding to authentic solasodine. R_f values for the standards were solasodine 0.29, tomatidine 0.49, diosgenin 0.85. In (2), hydrolysed hairy-root produced a spot at R_f 29 corresponding to solasodine. Other spots appeared at R_f 64, 68 and 78. From (3), hydrolysed hairy root produced spots at R_f 47, 53, 78, 81, 83 and 96. Compounds at R_f 47 and 53 were identified by co-chromatography as solasodine and a product of solasodine hydrolysis (presumably solasodiene), respectively. Tomatidine (R_f 62 in (3)) was not present in hairy root; the colour reaction for soladulcidine indicated absence of this compound also (Hunter et al. 1976). These results identify solasodine as a component of hairy roots; other compounds less polar than solasodine were also present.

Quantitative analysis of steroidal alkaloids was carried out using the colorimetric method based on complexing with bromothymol blue developed by Lancaster & Mann (1975) and modified by Chandler & Dodds (1983). Samples were prepared as described previously (Tsoulpha & Doran 1991). Solasodine was used as calibration material. Total steroidal-alkaloid concentrations in the medium plus roots are expressed as mg solasodine equivalents l^{-1} culture fluid.

Results

Shake-flask culture

Figure 1 shows the results from batch culture of *S. aviculare* hairy roots in shake flasks. The final biomass achieved was 8.2 g l^{-1} dry weight, an increase of about 40-fold from the inoculum. The maximum concentration of steroidal alkaloids was 260 mg solasodine equivalents l^{-1}. The culture reached stationary phase after about 42 d, even though a very high total sugar level (ca. 40 g l^{-1}) still remained in the medium. Figure 2 shows the pattern of sugar utilisation in shake-flask culture. Sucrose was completely inverted to fructose and glucose by day 35; after this time uptake of reducing sugars was slow.

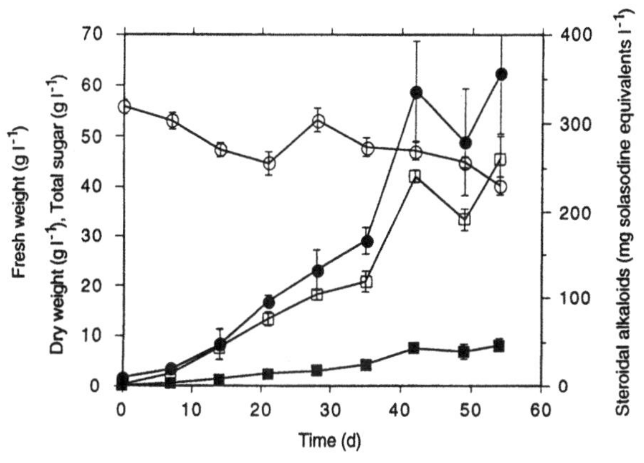

Fig. 1. Fresh weight (●), dry weight (■), total sugar (○) and steroidal-alkaloid (□) concentrations in shake-flask cultures of *S. aviculare* hairy roots.

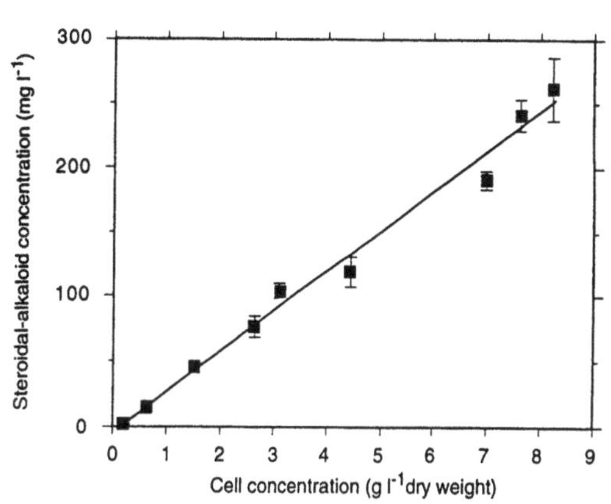

Fig. 3. Growth-associated production of steroidal alkaloids in shake-flask culture of *S. aviculare* hairy roots.

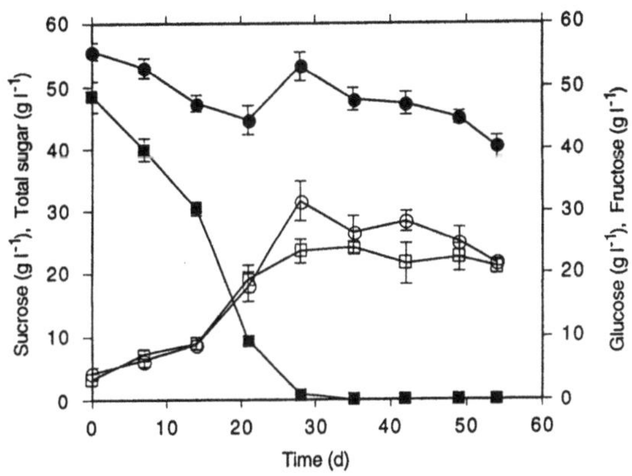

Fig. 2. Sugar utilisation in shake-flask culture of *S. aviculare* hairy roots. (■) sucrose, (○) fructose, (□) glucose, (●) total sugar.

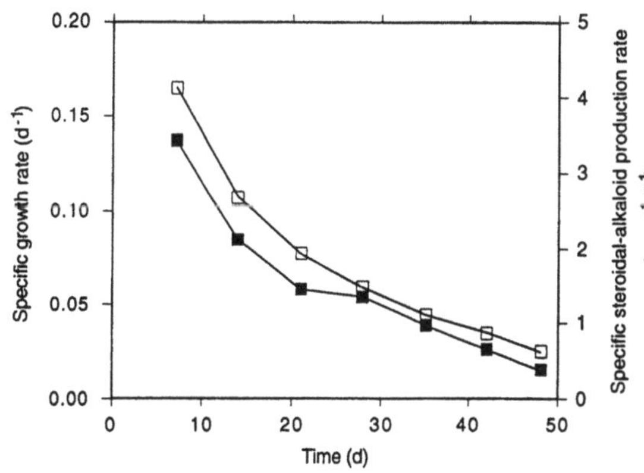

Fig. 4. Specific rates of growth (■) and steroidal-alkaloid production (□) during shake-flask culture. Both specific rates declined continuously throughout the experiment. Specific rates are based on dry cell weight.

Production of steroidal alkaloids was growth-associated, as shown in Fig. 3. Product yield from biomass (Y_{PX}; Table 1) was approximately constant at 32 mg g^{-1} dry weight. Calculated values for yield of biomass from substrate (Y_{XS}) and yield of steroidal alkaloids from substrate (Y_{PS}) are given in Table 1. The proportion of alkaloid present in the medium varied considerably during the culture period. At day 7, 50% of the total product was found in the liquid; this level decreased steadily to less than 10% by day 49.

Figure 4 shows the results of kinetic analysis of growth and steroidal-alkaloid production in shake-flask culture. Overall specific rates of growth and production of solasodine equivalents were calculated from the data shown in Fig. 1 and are based on cell dry weight. The observed specific growth rate decreased continuously throughout the culture period from an initial value of about 0.14 d^{-1} (doubling time: 5.0 d). Specific steroidal alkaloid production also decreased throughout the culture period from an initial value of about 4 mg g^{-1} d^{-1}.

The effect of increasing the concentration of MS salts and vitamins is shown in Fig. 5. When concentrations of all media components except sugar were elevated by up to 50%, growth was not increased relative to the control.

Kinetics of root extension and branching

Figure 6 shows typical results for increase in root length and number of root tips for a single *S. aviculare* hairy root in liquid culture. After formation of the first branch, total root length increased exponentially. Production of root tips is a discontinuous process; however after 3–4 branches had been formed, the number of root tips also increased exponentially with approxi-

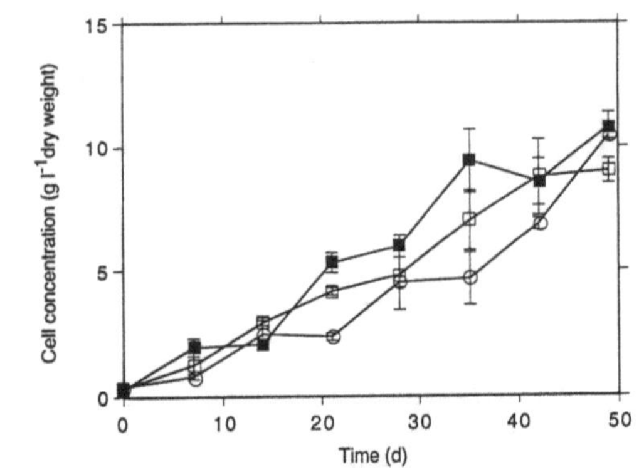

Fig. 5. Growth of *S. aviculare* hairy-roots in shake flasks containing different initial concentrations of MS salts and vitamins. (■) MS medium, (□) MS medium with 30% higher concentration of all components except sucrose, (○) MS medium with 50% higher concentration of all components except sucrose. Increase in medium components did not increase the extent or rate of growth.

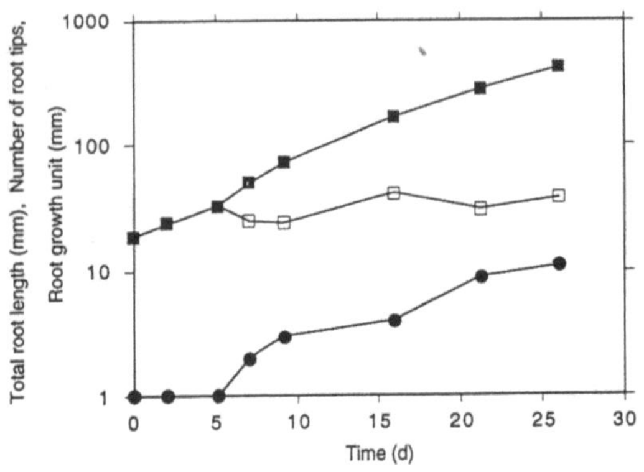

Fig. 6. Growth of *S. aviculare* hairy root in liquid Petri-dish culture at 25 °C. (■) Total root length, (●) number of root tips, (□) length of root growth-unit.

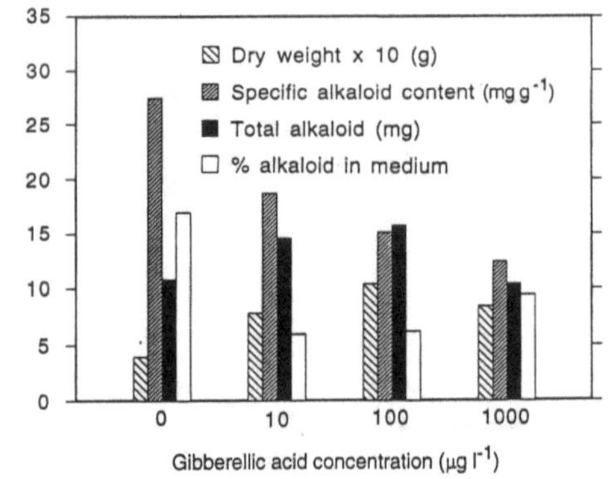

Fig. 7. Effect of gibberellic acid on growth, steroidal-alkaloid production and product excretion in *S. aviculare* hairy roots. Samples were analysed after 56 d culture in shake flasks.

mately the same specific rate as root length (Table 2). Length of the root growth-unit oscillated during the culture. Directly after inoculation the growth unit is equal to the total root length; when the first branch is produced the length of the growth unit decreases sharply. As growth proceeded, the root growth-unit attained a roughly constant value of about 37 mm.

Effect of gibberellic acid

The effect of gibberellic acid on growth and steroidal alkaloid production is summarised in Fig. 7. Gibberellic acid had a favourable effect on root growth; over a period of 56 d, about 2.3 times more biomass was produced in media containing 100 and 1000 μg l^{-1} gibberellic acid compared with the control culture. Steroidal-alkaloid production also increased. Of the gibberellic-acid concentrations tested, 10 and 100 μg l^{-1} produced the highest total amount of alkaloid expressed as solasodine equivalents, representing an increase of about 40% compared with the control culture. Figure 7 also shows that specific steroidal alkaloid levels (mg g^{-1} dry weight) decreased as the concentration of gibberellic acid increased. However, at 10 and 100 μg l^{-1}, increase in biomass more than compensates for the reduced level of alkaloid per g dry weight, so that gibberellic acid has an overall beneficial effect on product accumulation. At 1000 μg l^{-1} gibberellic acid, the total alkaloid level is similar to the control even though the biomass obtained is considerably higher.

The effect of gibberellic acid on root morphology was examined using liquid Petri-dish culture from single root tips. As shown in Fig. 8, addition of 1000 μg l^{-1} gibberellic acid had a considerable influence on total root length and lateral branching. Results for total length and number of root tips are shown in Fig. 9. The pattern of growth is very similar to that shown in Fig. 6 without exogenous gibberellic acid. From triplicate measurements, the average specific growth rate based on root length was twice that found in the control cultures (Table 2). With gibberellic acid there was considerable scatter in the replicate data for the number of branches, with significant variation between individual roots. Values for specific rate of growth based on root tips ranged close to those based on root length. The average root growth-unit was about 32 mm; the standard deviation from triplicate cultures indicates that

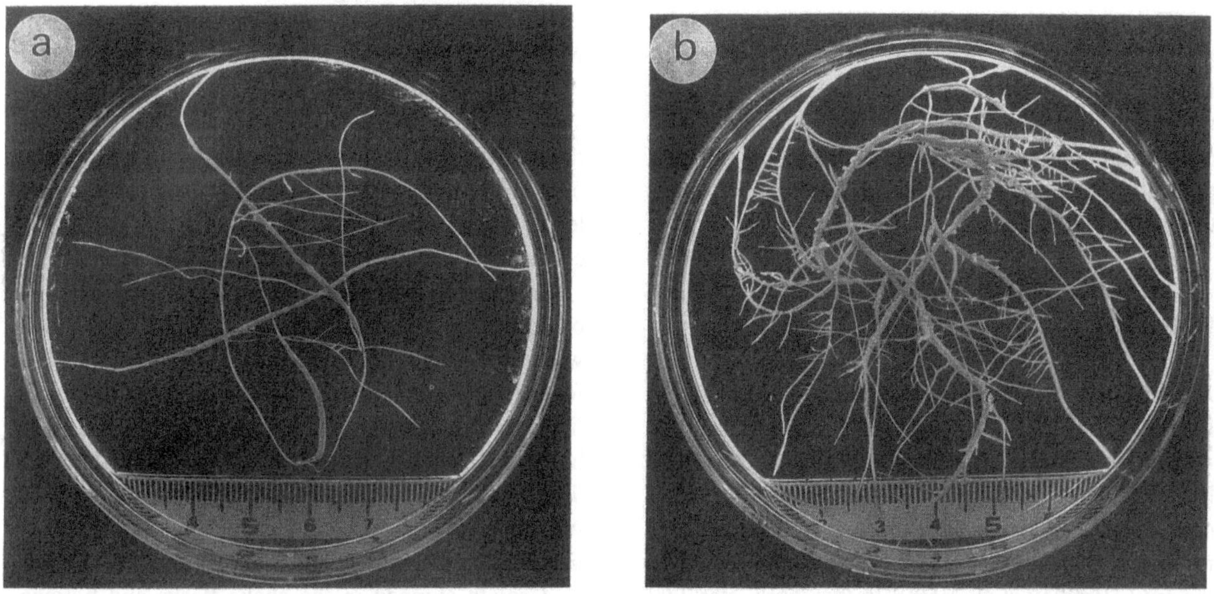

Fig. 8. Hairy roots in liquid medium in Petri dishes after 26 d culture. *(a)* control MS medium; *(b)* MS medium with 1000 μg l^{-1} gibberellic acid. The linear scale shows length in cm. Cultures were initiated from a single root tip.

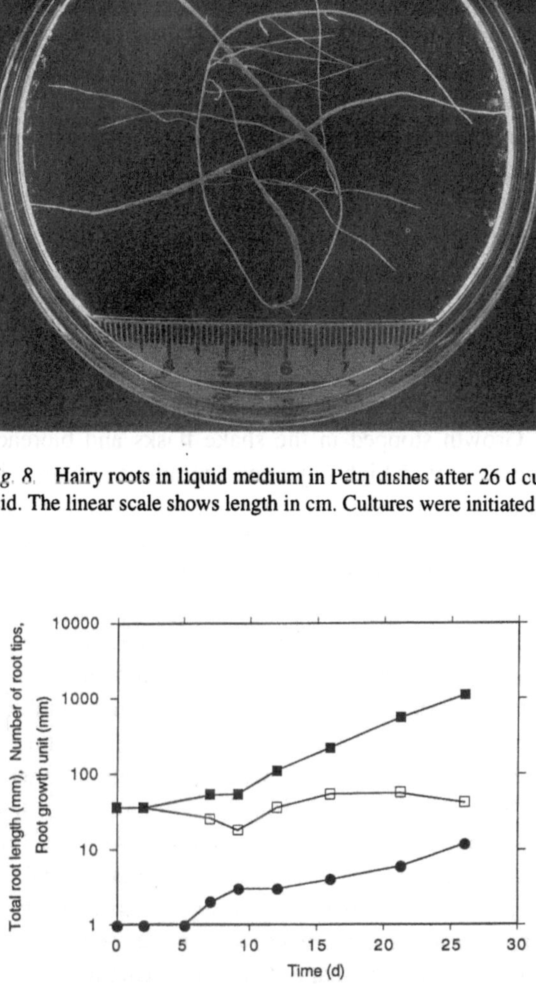

Fig. 9. Growth of *S. aviculare* hairy root in liquid Petri-dish with 1000 μg l^{-1} gibberellic acid. (■) Total root length, (●) number of root tips, (□) length of root growth-unit.

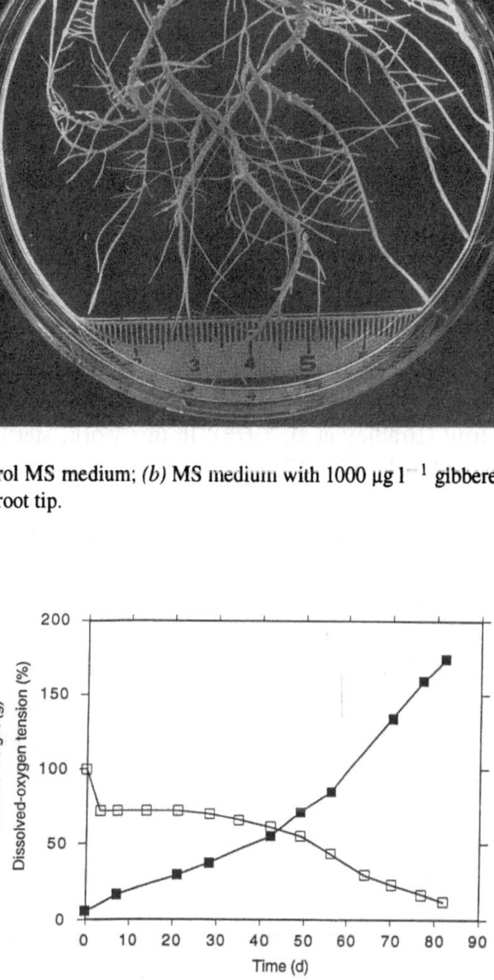

Fig. 10. Bioreactor culture of *S. aviculare* hairy roots. Biomass measured as drained weight (■); dissolved oxygen in the medium (□).

this value is approximately the same as that determined without gibberellic acid.

Bioreactor culture

Growth in the reactor was monitored in terms of drained weight, as sampling of roots during the experiments was not possible. The results are shown in Fig. 10. The ratio of drained weight:dry weight varied from 10.7 at the beginning to 16.9 at the end of the experiment. Because this variation is substantial, drained weight measurements during the culture could not be reliably converted to dry weight values. About 10 g dry biomass (3.8 g l^{-1}) was obtained at the end of the 82-d culture period, an increase of about 20-fold. As shown in Fig. 10, dissolved-oxygen content in the medium decreased progressively during biorector culture from 100% saturation at inoculation to 12% after 82 d. From an initial 30 g l^{-1}, about 20 g l^{-1} total sugar remained in the medium at the end of the experiment. Sucrose was completely hydrolysed to fructose and glucose by day 49.

The quantity of steroidal alkaloid increased by about 33-fold during bioreactor culture; 29 mg solasodine equivalents was produced per g dry weight. At the end of the culture, steroidal alkaloid released into the medium accounted for only about 7% of the total present. Overall yields calculated for the reactor culture are listed in Table 1.

Discussion

A survey of Australian *S. aviculare* plants sampled from different locations has shown that the average solasodine contents of leaf and green fruit determined by colorimetric analysis are 12 and 30 mg solasodine equivalents g^{-1} dry weight respectively, with maximum values of 31 mg g^{-1} in the leaf and 35 mg g^{-1} in the fruit (Bradley et al. 1978). In this work, steroidal alkaloid levels of 29–32 mg g^{-1} were achieved using *S. aviculare* hairy-root cultures. These yields are about 5 times higher than the maximum values reported previously with undifferentiated or aggregated cells (Macek 1989), and are comparable with levels in field-grown plants measured using similar analytical procedures. The results indicate that in vitro culture of *S. aviculare* hairy roots is a promising technique for commercial production of solasodine. Production was not greatly affected by reduced sucrose level or scale-up to the bioreactor, although lower biomass and steroidal-alkaloid yields from sugar were measured in the reactor experiment.

Genetic transformation of plants to form hairy roots has been accomplished with several commercial species (Flores et al. 1987; Hamill et al. 1987; Hamill & Rhodes 1993). In previous studies, the profile and extent of secondary product accumulation in hairy roots has been found very similar to roots of the whole plant (Parr & Hamill 1987; Parr et al. 1988). In contrast, in this work high levels of solasodine equivalents have been obtained from hairy roots even though similar analysis of *Solanum* species has shown relatively low solasodine levels in roots in vivo. For example, in *S. laciniatum*, 16–20 mg g^{-1} solasodine is found in green fruit, while the roots accumulate only 0.3–3 mg g^{-1} (Waclaw-Rozkrutowa 1966; Lancaster & Mann 1975). From studies of the variation of alkaloid level in different organs of *S. laciniatum*, it has been suggested that immature leaves are the site of solasodine synthesis (Lancaster & Mann 1975). However, because steroidal alkaloids are present in the roots as well as in the leaves shortly after germination, the role of the various organs in synthesis cannot be easily resolved in the whole plant. The results of this work show that roots in vitro possess the ability to synthesise and store significant quantities of solasodine. Other work in this laboratory with genetically-transformed shooty teratomas of *S. aviculare* has shown that, without roots, very low levels of steroidal alkaloids (ca. 0.85 mg solasodine equivalents g^{-1} dry weight) are produced.

In shake-flask culture, the initial root doubling-time based on dry weight was approx. 5.0 d. This value is within the range 2–7 d given by Wilson et al. (1987) for hairy roots of several plant species. As this is the first report of growth characteristics for *S. aviculare* hairy roots, no data from other studies are available for direct comparison. The initial growth rate of the roots was about 1.5 times that measured previously with *S. aviculare* aggregates (Tsoulpha & Doran 1991).

Growth stopped in the shake flasks and bioreactor even though high sugar concentrations remained in the medium, suggesting that some nutrient other than sugar was growth limiting. The continual reduction in specific growth rate during the shake-flask experiments supports this conclusion. Experiments with increased levels of medium components were conducted to determine whether any of the salts or vitamins in the MS formula were growth-limiting in shake-flask culture. The results with medium containing 1.0, 1.3 and 1.5 times these compounds showed that growth rates were not enhanced with added medium components. It is most likely therefore that growth in this work was limited by oxygen. Experiments carried out in shallow Petri dishes showed that when oxygen availability is improved and roots are grown outside of the dense clumps formed in shake flasks and reactors, specific growth rate remains roughly constant with respect to both total length and number of root tips. In the reactor there was direct evidence of oxygen limitation. Dissolved-oxygen tension measured at the top of the vessel declined significantly throughout the culture period to reach 12% air saturation at the end of the experiment. As the root concentration was most dense at the bottom of the vessel, the actual dissolved-oxygen concentration experienced by the roots was most likely significantly lower than that recorded. Since the critical dissolved-oxygen tension for plant cells is 16–20% saturation under average culture conditions (Kessell & Carr 1972; Payne et al. 1987), severe oxygen limitations would be expected in the root ball. In design of reactors for hairy roots, a balance must be found between the need for high-density culture to enhance volumetric productivity and the need for adequate oxygen transfer within the root clumps. Because oxygen limitation in dense hairy-root culture is difficult to avoid, growth data obtained from shake flasks and reactors are likely to be significantly affected by mass transfer.

Production of steroidal alkaloids in *S. aviculare* hairy roots occurred during growth. This is in contrast with the results of previous work with aggregated cells (Tsoulpha & Doran 1991), in which production accelerated with the onset of stationary phase. Whether or not product synthesis accompanies growth in *S. aviculare* appears to depend on the state of differentiation of the plant tissue. Exogenous auxin levels have been found previously to alter growth-production relationships in tissue culture (Davies 1972; Sahai & Shuler 1984). Hairy-root medium does not contain auxin; however it is possible that alterations in the timing of secondary-product formation are a consequence of *A. rhizogenes* infection. T-DNA from the Ri-plasmid has a direct effect on auxin synthesis and sensitivity to auxin in plants (Schmülling et al. 1988).

The pattern of growth found in this work showed that hairy roots grew with a constant 'root growth-unit', or mean length per root tip, of about 35 mm. This result suggests that branch initiation is regulated by changes in root length or volume; when the total length of a root increases by a critical value a new branch is initiated. Similar patterns of growth have been observed for branching fungal mycelia (Trinci 1974; Reichl et al. 1990). Gibberellic acid at a concentration of 1000 $\mu g\ l^{-1}$ had a marked effect on the rate of root elongation and lateral branching, but did not affect the critical length between branchings.

Acknowledgements

We thank Kian Kwok for initiating the hairy-root cultures, and Wandee Yanpaisan, Malcolm Noble and Ian McFarlane for TLC and HPLC analyses. We are also grateful to J.D. Mann for providing samples of solasodine, solamargine and solasonine, and to J.D. Steve for identifying the glycoside sugars. This work was funded by the Australian Research Council (ARC) and the CSIRO/University of NSW Collaborative Research Scheme. P.M. Doran conducted this work as a Queen Elizabeth II Research Fellow.

References

Bercetche J, Chriqui D, Adam S & David C (1987) Morphogenetic and cellular reorientations induced by *Agrobacterium rhizogenes* (strains 1855, 2659 and 8196) on carrot, pea and tobacco. Plant Sci. 52: 195–210

Birner J (1969) Determination of total steroid bases in *Solanum* species. J. Pharm. Sci. 58: 258–259

Bradley V, Collins DJ, Crabbe PG, Eastwood FW, Irvine MC, Swan JM & Symon DE (1978) A survey of Australian *Solanum* plants for potentially useful sources of solasodine. Aust. J. Bot. 26: 723–754

Chandler S & Dodds J (1983) Solasodine production in rapidly proliferating tissue cultures of *Solanum laciniatum* Ait. Plant Cell Rep. 2: 69–72

Chaudhuri RK & Chatterjee SK (1979) Effect of altitude and gibberellin-feeding on solasodine contents in *Solanum khasianum* Clarke, fruits of Darjeeling Hills. Ind. J. Exp. Biol. 17: 611–612

Conner AJ (1987) Differential solasodine accumulation in photoautotrophic and heterotrophic tissue cultures of *Solanum laciniatum*. Phytochem. 26: 2749–2750

Davies ME (1972) Polyphenol synthesis in cell suspension cultures of Paul's Scarlet rose. Planta 104: 50–65

Ehmke A & Eilert U (1986) Steroidal alkaloids in tissue cultures and regenerated plants of *Solanum dulcamara*. Plant Cell Rep. 5: 31–34

Flores HE, Hoy MW & Pickard JJ (1987) Secondary metabolites from root cultures. Trends in Biotechnol. 5: 64–69

Hamill JD, Parr AJ, Rhodes MJC, Robins RJ & Walton NJ (1987) New routes to plant secondary products. Bio/Technol. 5: 800–804

Hamill JD & Rhodes MJC (1993) Manipulating secondary metabolism in culture. In: Grierson D (Ed) Plant Biotechnology, Vol 3: Biosynthesis and Manipulation of Plant Products (pp 178–209). Blackie and Son, Glasgow

Hunter IR, Walden MK, Wagner JR & Heftmann E (1976) Thin-layer chromatography of steroidal alkaloids. J. Chromatogr. 118: 259–262

Kessell RHJ & Carr AH (1972) The effect of dissolved oxygen concentration on growth and differentiation of carrot (*Daucus carota*) tissue. J. Exp. Bot. 23: 996–1007

Khanna P, Uddin A, Sharma GL, Manot SK & Rathore AK (1976) Isolation and characterization of sapogenin and solasodine from in vitro tissue cultures of some Solanaceous plants. Ind. J. Exp. Biol. 14: 694–696

Khanna P, Sharma GL, Rathore AK & Manot SK (1977) Effect of cholesterol on in vitro suspension tissue cultures of *Costus speciosus* (Koen) Sm., *Dioscorea floribunda* Mart. & Gal., *Solanum aviculare* Forst. and *Solanum xanthocarpum* Schard & Wendl. Ind. J. Exp. Biol. 15: 1025–1027

Lancaster JE & Mann JD (1975) Changes in solasodine content during the development of *Solanum laciniatum* Ait. NZ J. Agric. Res. 18: 139–144

Macek TE (1989) *Solanum aviculare* Forst., *Solanum laciniatum* Ait. (poroporo): in vitro culture and the production of solasodine. In: Bajaj YPS (Ed) Biotechnology in Agriculture and Forestry 7: Medicinal and Aromatic Plants II (pp 443–467). Springer-Verlag, Berlin

Ohkawa H, Kamada H, Sudo H & Harada H (1989) Effects of gibberellic acid on hairy root growth in *Datura innoxia*. J. Plant Physiol. 134: 633–636

Parr AJ & Hamill JD (1987) Relationship between *Agrobacterium rhizogenes* transformed hairy roots and intact, uninfected *Nicotiana* plants. Phytochem: 26: 3241–3245

Parr AJ, Peerless ACJ, Hamill JD, Walton NJ, Robins RJ & Rhodes MJC (1988) Alkaloid production by transformed root cultures of *Catharanthus roseus*. Plant Cell Rep. 7: 309–312

Payne GF, Shuler ML & Brodelius P (1987) Large scale plant cell culture. In: Lydersen BK (Ed) Large Scale Cell Culture Technology (pp 193–229). Hanser, Munich

Reichl U, Buschulte TK & Gilles ED (1990) Study of the early growth and branching of *Streptomyces tendae* by means of an image processing system. J. Microscopy 158: 55–62

Sahai OP & Shuler ML (1984) Environmental parameters influencing phenolics production by batch cultures of *Nicotiana tabacum*. Biotechnol. Bioeng. 26: 111–120

Schmülling T, Schell J & Spena A (1988) Single genes from *Agrobacterium rhizogenes* influence plant development. EMBO J. 7: 2621–2629

Sharp JM & Doran PM (1990) Characteristics of growth and tropane alkaloid synthesis in *Atropa belladonna* roots transformed by *Agrobacterium rhizogenes*. J. Biotechnol. 16: 171–186

Torrey JG (1976) Root hormones and plant growth. Ann. Rev. Plant Physiol. 27: 435–459

Trinci APJ (1974) A study of the kinetics of hyphal extension and branch initiation of fungal mycelia. J. Gen. Microbiol. 81: 225–236

Tsoulpha P & Doran PM (1991) Solasodine production from self-immobilised *Solanum aviculare* cells. J. Biotechnol. 19: 99–110

Waclaw-Rozkrutowa B (1966) Content of solasodine during the growth period in *Solanum laciniatum*. Diss. Pharm. Pharmacol. 18(6): 595–599; abstracted in Chem. Abstr. 67: 29848 (1967)

Weiler EW, Krüger H & Zenk MH (1980) Radioimmunoassay for the determination of the steroidal alkaloid solasodine and related compounds in living plants and herbarium specimen. Planta Med. 39: 112–124

Wilson PDG, Hilton MG, Robins RJ & Rhodes MJC (1987) Fermentation studies of transformed root cultures. In: Moody GW & Baker PB (Eds) Bioreactors and Biotransformations (pp 38–51). Elsevier, London

Zhan X, Jones DA & Kerr A (1990) The pTiC58 *tzs* gene promotes high-efficiency root induction by agropine strain 1855 of *Agrobacterium rhizogenes*. Plant Molec. Biol. 14: 785–792

Plant Cell, Tissue and Organ Culture **38**: 103–113, 1994.

Cyclodextrins as a useful tool for bioconversions in plant cell biotechnology

Wim van Uden, Herman J. Woerdenbag & Niesko Pras
University Centre for Pharmacy, Laboratory for Pharmaceutical Biology, University of Groningen, A. Deusinglaan 2, NL-9713 AW Groningen, The Netherlands

Key words: Bioconversion, catechols, cyclodextrins, lignans, plant cell cultures, solubility enhancement

Abstract

The application of cyclodextrins as precursor solubilizers in biotechnological processes, in which plant cells are involved, is new. In this paper the possibilities for cyclodextrin facilitated bioconversions by freely suspended and/or immobilized plant cells or plant enzymes are demonstrated. After complexation with β-cyclodextrin, the phenolic steroid 17β-estradiol could be ortho-hydroxylated into a catechol, mainly 4-hydroxyestradiol, by a phenoloxidase from in vitro grown cells of *Mucuna pruriens*. By complexation with β-cyclodextrin the solubility of the steroid increased from almost insoluble to 660 μM. In addition, by complexation with β-cyclodextrin, a solution of 3 mM coniferyl alcohol could be fed to cell cultures of *Podophyllum hexandrum* in order to enhance the accumulation of podophyllotoxin. Finally, the glucosylation of podophyllotoxin by cell cultures derived from *Linum flavum* was investigated. Four cyclodextrins: β-cyclodextrin, γ-cyclodextrin, hydroxypropyl-β-cyclodextrin and dimethyl-β-cyclodextrin were used to improve the solubility of podophyllotoxin. Dimethyl-β-cyclodextrin met our needs the best and the solubility of podophyllotoxin could be enhanced from 0.15 to 1.92 mM. Podophyllotoxin-β-D-glucoside was formed at a rate of 0.51 mmol l^{-1} suspension per day by the *L. flavum* cells growing in the presence of 1.35 mM podophyllotoxin, complexed with dimethyl-β-cyclodextrin.

Abbreviations: DW – dry weight, E2 – 17β-estradiol, FW – fresh weight, PCV – packed cell volume

Introduction

Cyclodextrins are cyclic oligosaccharides, consisting of 6, 7 or 8 glucose units, designated by the Greek letter α, β or γ, respectively. They can be regarded as torus-like rings (Fig. 1) (Uekama & Irie 1987). Cyclodextrins with fewer than six glucose units are not known to exist.

The glucose units are linked by α-1,4-glucosidic bonds. As a consequence of the chair formation of the sugar units, all secondary hydroxyl groups (at C_2, C_3) are located at one side of the torus, while all the primary hydroxyl groups at C_6 are situated on the other side. As a result, the external faces are hydrophilic, making the cyclodextrins water-soluble (Fig. 2A, B).

In contrast, the cavities of the cyclodextrins are hydrophobic, since they are lined by hydrogen atoms of C_3 and C_5, and by ether-like oxygens. These matrices allow complexation with a variety of relatively hydrophobic compounds, including drugs, flavours and aromas, sweeteners, plant extracts, oils and fats, surfactants, anti-oxidants and pesticides. The complexation takes place by Van der Waals interactions and by hydrogen bond formation. The molecules have to fit either entirely, or at least partially in the cavities of 0.49, 0.62 or 0.79 nm (Fig. 1).

The stability of the complex formed depends both on the size and polarity of the guest molecule as well as on the particular cyclodextrin used (Saenger 1984; Le Bas & Rysanek 1987; Duchêne & Wouessidjewe 1990a).

Cyclodextrins are prepared from potato starch by a two-step process. In the first step, cyclodextrin-glycosyltransferase, an enzyme that is produced by several *Bacillus*, *Micrococcus* and *Klebsiella* strains, converts the pre-hydrolysed starch into a mixture of cyclodextrins and linear dextrins. For industrial production, specially selected strains of *Bacillus* are to be preferred (Sicard & Saniez 1987). In the second step the mixture of degradation products is separated into its pure constituents. The methods used here, are crystal-

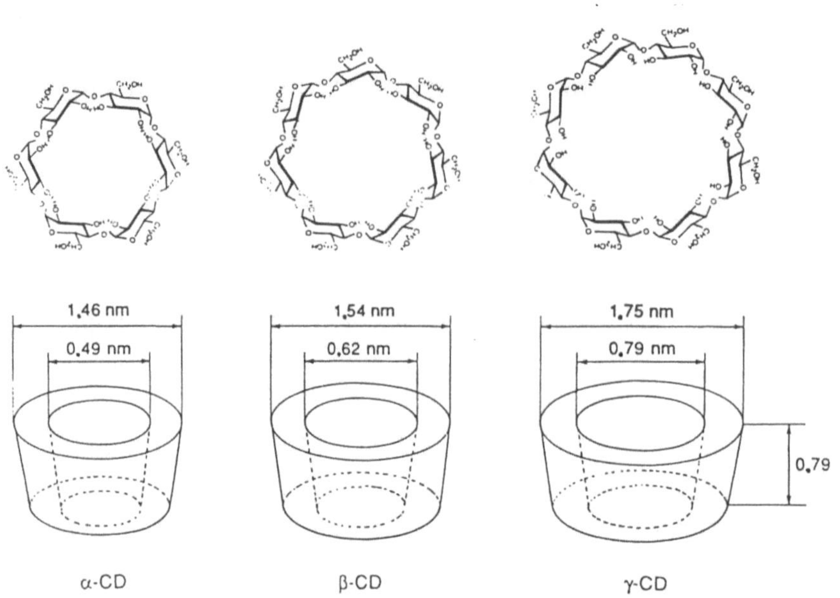

Fig. 1. Structures of α-, β- and γ-cyclodextrin and their molecular dimensions (adapted from Saenger 1984).

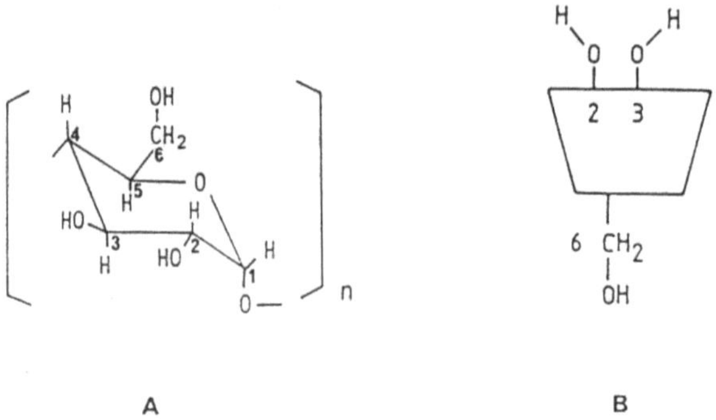

Fig. 2. Structure of the glucose unit as present in cyclodextrins, n = 6, 7 or 8 (*A*) (adapted from Hirayama & Uekama 1987). Due to the position of the hydroxyl groups in the chair formation, the external faces of cyclodextrins are hydrophilic (*B*) (adapted from Sébille 1987).

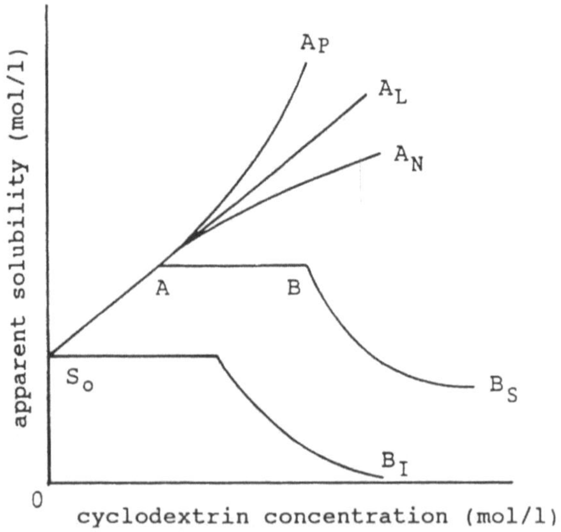

Fig. 3. The different types of phase-solubility diagrams (adapted from Hirayama & Uekama 1987).

lization with the help of organic solvents, ultrafiltration or chromatography.

Next to α-, β- and γ-cyclodextrin, several derivatives have been synthesized. The 18 to 24 hydroxyl groups of the respective cyclodextrin molecules are the starting points for the synthesis of such derivatives. The differences in reactivity between the alcohol groups are sufficiently large to induce some selectivity (Sébille 1987). However, because of the polyfunctional character of the cyclodextrin molecule this often results in mixtures of derivatives, which are separated by chromatographic processes. Methyl-, ethyl-, hydroxyethyl- and hydroxypropyl-substituted cyclodextrins have become available so far. Particularly the methylated cyclodextrins and hydroxypropyl-β-cyclodextrin seem to possess promising properties for pharmaceutical applications (Duchêne & Wouessidjewe 1990b). The physico-chemical properties of the cyclodextrin derivatives depend strongly on the kind and the degree of substitution. As a common feature, they are all water-soluble and many derivatives display a high surface activity. In addition, they are soluble in many organic solvents (Müller & Brauns 1986; Yamamoto et al. 1989).

To detect the formation of inclusion complexes of cyclodextrins with guest molecules in solution, the phase-solubility method, as described by Higuchi & Connors (1965), is most commonly used in pharmaceutical sciences (Szejtli & Pagington 1989). To aqueous suspensions, containing an excess amount of guest molecules, increasing amounts of cyclodextrin are added. The mixtures are shaken at a constant temperature until equilibrium is reached. The solid particles are then removed and the solution is analysed to determine the total concentration of guest dissolved. The solubility of the guest is now plotted against the cyclodextrin concentration, yielding a so called phase-solubility diagram.

Two major types of phase-solubility diagrams can be obtained, the A-type and B-type (Fig. 3). In the A-type diagrams the solubility of the guest increases with increasing cyclodextrin concentration, sometimes to a certain extent. The solubility of the complex is higher than that of pure cyclodextrin. When a 1:1 inclusion complex is formed, a diagram from the A_l-type is obtained, whereas the A_p-type diagram is indicative for the formation of higher order complexes. The A_n-type diagram is difficult to analyse quantitatively, because the system is rather complex due to solute/solvent and solute/solute interactions. In the case of a B-type diagram, the solubility of the complex is lower than that of the pure cyclodextrin but higher than the uncomplexed guest molecule. This results in a more complicated diagram. In the B_s-type diagram, the solubility of the guest increases in the first part of the curve. At point A the solubility limit of the inclusion complex is reached. Further addition of cyclodextrin results in the precipitation of inclusion complex. Therefore, the concentration of dissolved guest remains constant. The subsequent decrease in solubility upon further addition of cyclodextrin is to be ascribed to the precipitation of complex with the amount of freely dissolved guest (S_0), still present at point B. The decrease in the last part of the diagram is often more than S_0 due to the formation of higher order complexes.

From these diagrams the complex stability constant (K_c) as well as the stoichiometry of the complex can be calculated. In the investigation of inclusion complexes in solution, spectroscopic methods are used, including circular dichroism spectroscopy, fluorescence spectroscopy and nuclear magnetic resonance spectroscopy (Hirayama & Uekama 1987).

When the complex stability constant (K_c) and stoichiometry are known, soluble complexes can be prepared by simply adding appropriate amounts of guest molecule and cyclodextrin to an aqueous solution. Then the mixture is either shaken at a constant temperature (Hirayama & Uekama 1987) or is heated during a certain period, in which complexation is completed. For tissue culture purposes, complexation may be obtained in a convenient way, by simply autoclaving the medium together with cyclodextrin and guest molecule (Woerdenbag et al. 1990a). Thermostability of the guest is of course a prerequisite in this case. Solid complexes can be prepared analogously, in that case an excess of guest molecules as well as cyclodextrin are used. Finally, the precipitated complex is isolated by filtration followed by drying (Uekama et al. 1983).

The application of cyclodextrins as precursor solubilizers in biotechnological processes, in which microorganisms as well as plant cells are involved, is relatively new.

Recently, cyclodextrins have been applied successfully to enhance the bioconversion rate of steroidal precursors by microorganisms. In the case of cholesterol degradation into androstene-like compounds by *Mycobacterium* species, this resulted in an increased production and reduced side product formation (Hesselink et al. 1989).

Next to microorganisms, plant cells are able to carry out interesting bioconversions. Especially regioselective hydroxylations and glycosylations seem very promising (Pras 1992).

In this paper possibilities for cyclodextrin-facilitated bioconversions by plant cells or plant enzymes are demonstrated by the *ortho*-hydroxylation of 17β-estradiol, complexed with β-cyclodextrin, into 2- and 4-hydroxyestradiol by cells of *Mucuna pruriens* (Woerdenbag et al. 1990a), by the bioconversion of β-cyclodextrin-complexed coniferyl alcohol into podophyllotoxin by cells of *Podophyllum hexandrum* (Woerdenbag et al. 1990b), by the glucosylation of podophyllotoxin complexed with several cyclodextrins into its β-D-glucoside (Van Uden et al. 1993) and of dimethyl-β-cyclodextrin-complexed coniferyl alcohol into coniferin using cells of *Linum flavum*.

Bioconversion of 17β-estradiol by a phenoloxidase from *Mucuna pruriens* cell suspensions

Cells of *Mucuna pruriens* L. (Fabaceae) contain a phenoloxidase (EC 1.14.18.1) that is able to ortho-

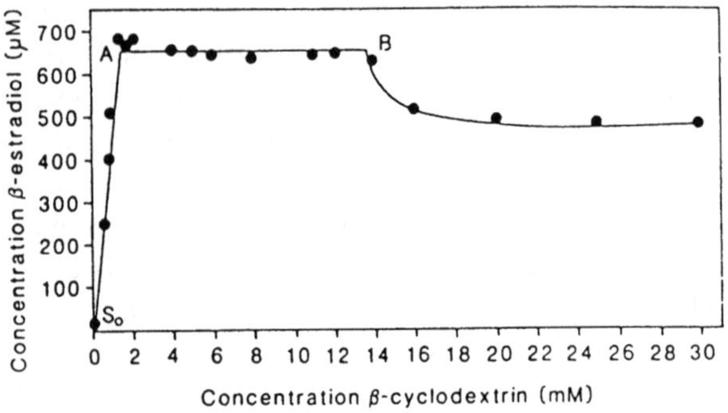

Fig. 4. Phase-solubility diagram of the E2/β-cyclodextrin system (from Woerdenbag et al. 1990a).

Table 1. The bioconversion of E2 into 2OH-E2 and 4OH-E2 in the different biocatalytic systems, expressed as nmol product formed per mg protein (mean ± s.d.; n = 3) after 72 h (Woerdenbag et al. 1990a).

Biocatalytic system	2OH-E2	4OH-E2
freely suspended cells	0	0
immobilized cells	6.5 ± 1.2	37.6 ± 6.9
cell homogenate	12.6 ± 3.2	62.1 ± 9.4
phenoloxidase preparation*	60.8 ± 8.7	688 ± 83.5

* purification ca. 185-fold.

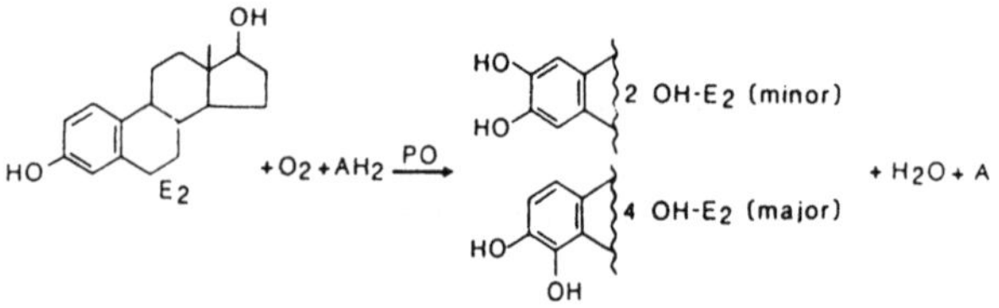

Fig. 5. Reaction scheme for the bioconversion of E2 into 2OH-E2 and 4OH-E2 by *Mucuna*-phenoloxidase (PO). Sodium ascorbate (AH_2) serves as a co-factor and an anti-oxidant, to prevent quinone formation (from Woerdenbag et al. 1990a).

hydroxylate phenolic substrates. L-tyrosine could be converted into the anti-Parkinson drug L-DOPA (Pras et al. 1988). Also the chemically more complex and cell-foreign aminotetralines could serve as substrates (Pras 1989; Pras et al. 1989).

To investigate whether a poorly water-soluble compound, like the naturally occurring steroid hormone 17β-estradiol (E2), could serve as a substrate for the phenoloxidase, it was complexed with β-cyclodextrin.

Freely suspended and alginate entrapped cells, a cell homogenate and a phenoloxidase preparation of *M. pruriens* were compared for their capability and efficiency to convert E2 into a catechol (Woerdenbag et al. 1990a).

Complexation of E2 with β-cyclodextrin enhanced the solubility of E2 from 12 μM to 66o μM, as determined by the method described by Higuchi & Connors (1965). The phase-solubility diagram of this system is shown in Fig. 4. The E2/β-cyclodextrin system showed a solubility curve of the so-called B-type, the K_c value of the complex was calculated to be 20,000 M^{-1}. This relatively high K_c points to a very stable complex of E2 with β-cyclodextrin. On a weight basis the complex, used in the experiments described, contained 10% E2; the molar ratio between E2 and β-cyclodextrin was 1:2. With this knowledge, solid complex was prepared by method of heating.

The reaction scheme for the bioconversion of E2 into 2-hydroxy-estradiol (2OH-E2) and/or 4-hydroxy-estradiol (4OH-E2) is given in Fig. 5. When E2 was fed as a complex with β-cyclodextrin, the efficiency of the bioconversion, after a 72 h incubation period, increased in the following order: free cells (0%), immobilized cells (1% 4OH-E2), cell homogenate (6% 4OH-E2, 1% 2OH-E2), phenoloxidase preparation (40% 4OH-E2, 3% 2OH-E2).

The bioconversions of E2 into 2OH-E2 and 4OH-E2 are compared for the different biocatalytic systems in Table 1, in terms of nmol product formed per mg protein after 72 h.

As the phenoloxidase occurs intracellularly (Mayer 1987), the substrate has to penetrate through the plant cell wall and cell membrane, being an apparent barrier for the steroidal structure. This is reflected in the finding that no substrate could be bioconverted by freely suspended cells. The calcium-alginate matrix itself was proven not to be a barrier for the E2/β-cyclodextrin complex. The diffusion behaviour of E2 was equal to earlier tested phenolic substrates, which could diffuse freely into and out of the alginate matrix (Pras et al. 1989). However, a rather poor bioconversion efficiency was seen, especially when compared to the phenoloxidase preparation.

During the growth cycle the β-cyclodextrin concentration in the culture medium remained unchanged, indicating that it was not broken down or used as a carbon source by the cells.

This study represented the first report that dealt with the bioconversion of a steroid by a plant enzyme. More importantly, cyclodextrins were used succesfully for the first time in plant cell biotechnology, because of their solubilizing action.

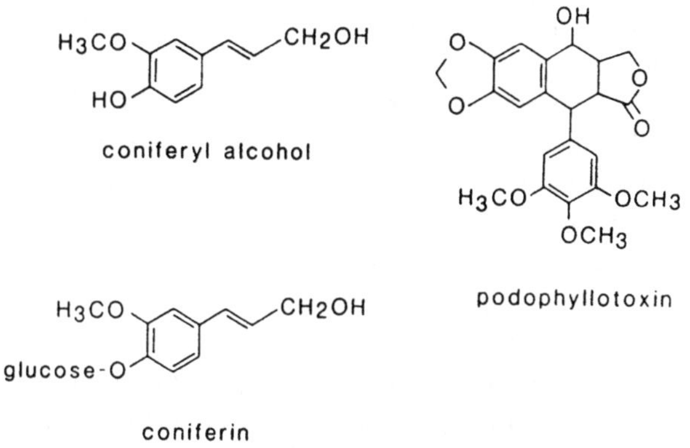

Fig. 6. Chemical structures of podophyllotoxin and its biosynthetic precursors coniferyl alcohol and coniferin (from Woerdenbag et al. 1990b).

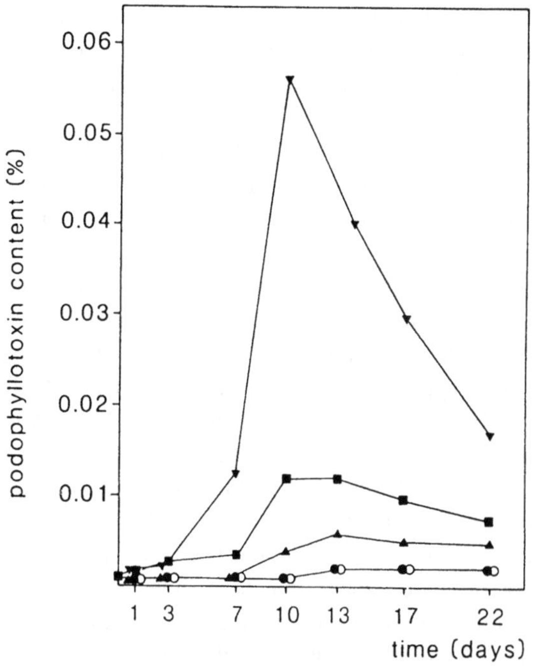

Fig. 7. Time course of the podophyllotoxin content, expressed as percentage of dry weight, in suspension-grown cultures of *P. hexandrum*. Each point represents the mean value of at least duplicate experiments. Control (●); 3 mM β-cyclodextrin (○); 3 mM coniferyl alcohol/β-cyclodextrin complex (■); 3 mM coniferyl alcohol (▲); 3 mM coniferin (▼) (from Woerdenbag et al. 1990b).

Bioconversion of coniferyl alcohol by cell suspension cultures of *Podophyllum hexandrum*

Van Uden et al. (1989) described the initiation of cell cultures from roots of *P. hexandrum* that accumulated podophyllotoxin (Fig. 6). The production of this lignan could be improved by feeding biosynthetic precursors (Van Uden et al. 1990b). This method was restricted to water-soluble compounds and it has been found that coniferin (Fig. 6), the water-soluble β-D-glucoside of coniferyl alcohol, was the substrate of choice to increase the production of podophyllotoxin.

In a parallel study by Woerdenbag et al. (1990b), coniferyl alcohol, being a key precursor in the biosynthesis of podophyllotoxin (Jackson & Dewick 1984), was chosen as a model substrate because of its very poor water-solubility. Podophyllotoxin contents were investigated in *P. hexandrum* cell cultures after feeding coniferyl alcohol, complexed with β-cyclodextrin. For comparison, the cultures were fed with non-complexed, suspended, coniferyl alcohol as well as with coniferin. In addition, the influence of β-cyclodextrin on cell viability was monitored.

At 26 °C the water-solubility of coniferyl alcohol increased from 0.15 mM without, to a maximum of 3.4 mM with β-cyclodextrin, in a molar ratio of 1:1. A stable complex was formed with a K_c of 1,360 M^{-1}. Cells of *P. hexandrum*, endogenously accumulated podophyllotoxin in concentrations ranging from 0.001 to 0.002%, calculated on dry weight, during their growth cycle (Fig. 7). Feeding of 3 mM coniferyl alcohol, dissolved in the culture as a β-cyclodextrin complex, resulted in enhanced podophyllotoxin accumulation, with a maximum of 0.012% on day 10 of the growth cycle. Non-complexed coniferyl alcohol, suspended in the medium in a concentration of 3mM, also enhanced the lignan production, but only to a maximum of 0.006% on day 13. Feeding 3 mM coniferin caused a giant increase, with a maximum of 0.056% on day 10. This was in agreement with previous results, obtained with water-soluble precursors, including coniferin (Van Uden et al. 1990b).

The accumulation pattern of podophyllotoxin, after adding only β-cyclodextrin to the culture medium, did not differ from the control conditions. The culture characteristics, i.e. cell growth in terms of dry weight, pH and conductivity, were not altered by any of the additions. The β-cyclodextrin concentration in the culture medium remained unchanged during the whole growth cycle and no β-cyclodextrin could be detected intracellularly. This indicates that the oligosaccharide, as was the case with *M. pruriens* cells, was not metabolized by *P. hexandrum* cells and did not penetrate the plant cells.

No podophyllotoxin could be detected in the culture medium.

Under control conditions, standard-grown cell suspensions contained 2 μg coniferyl alcohol per g dry weight. After feeding the β-cyclodextrin/substrate complex, an increased coniferyl alcohol content was found, 13 μg/g dry weight, until day 3. In later stages of the growth cycle the low control levels were measured again. From the culture medium, coniferyl alcohol had disappeared within 1 day. Thus, despite this rapid disappearance, no direct podophyllotoxin formation was measured.

From the increase in podophyllotoxin accumulation, a bioconversion percentage of maximal 1% could be calculated. The excess of coniferyl alcohol may have disappeared via other pathways, such as the formation of lignin, which is a well-known constituent of plant cell walls (Luckner 1986).

The direct benefit, with respect to a higher podophyllotoxin content, was probably obtained from the fact that coniferyl alcohol is delivered to the cells in a dissolved state, due to complexation with β-cyclodextrin. This carrier function of β-cyclodextrin may lead to an increased supply rate of the precursor in the cells, finally resulting in an increased conversion, as compared with the non-complexed precursor. At present, a lack of knowledge exists on the underlying mass transfer processes that take place at the cell envelope.

Glucosylation of podophyllotoxin by a cell suspension of *Linum flavum*

At present, glucosylation seems one of the most interesting bioconversions. It occurs readily in plant cells, but only with difficulty in microorganisms and the organic chemist has problems with this reaction as well (Pras 1992). This makes the glucosylation of cytotoxic lignans by plant cells a scientifically and biotechnologically interesting bioconversion.

Glucosyltransferases are able to perform the glucosylation reaction under strict stereochemical control (Hösel 1981).

Cell cultures of *Linum flavum* (Linaceae), yellow flax, are able to synthesize several podophyllotoxin-related lignans including 5-methoxypodophyllotoxin and its glucoside (Berlin et al. 1986, 1988; Van Uden et al. 1990a, 1991a,b, 1992; Wichers et al. 1990, 1991).

In a recent study, the cytotoxic lignan podophyllotoxin, which is poorly water-soluble, was chosen as the substrate to be glucosylated by cell suspensions of *L. flavum* (Van Uden et al. 1993). The glucosylation reaction scheme is depicted in Fig. 8.

Four types of cyclodextrins were applied to prepare water-soluble podophyllotoxin/cyclodextrin complexes. Before bioconversion experiments directed to the glucosylation of podophyllotoxin by *L. flavum* cell suspensions were started, the possible toxicity of cyclodextrins to the plant cells was investigated first. It appeared that the growth characteristics of *L. flavum* cell suspensions were not affected at all, when using these clathrating agents at a concentration of 2 mM. For the control as well as for the cyclodextrin-containing cultures, the packed cell volume (PCV) increased from 20 to 70%, the dry weight (DW) from 6 to 21 g l^{-1}, the fresh weight (FW) from 125 to 350 g l^{-1}, while the conductance decreased from 4 to 0.7 mS within one growth cycle.

The preparation of the podophyllotoxin/cyclodextrin complex was achieved by autoclaving or by shaking the suspensions. The high temperature during the sterilisation process appeared to be disadvantageous because this method resulted in ca. 15% decomposition of podophyllotoxin. Therefore, it was concluded that the most convenient method to prepare cyclodextrin-complexed podophyllotoxin is by shaking, where no decomposition was found.

The solubility of podophyllotoxin could be increased by the complexation with all four types of applied cyclodextrins. Using guest and host molecule in a ratio of 1:1 and at a concentration of 2 mM, γ-cyclodextrin yielded a maximal solubility of 0.20 mM, hydroxypropyl-β-cyclodextrin of 0.68 mM, β-cyclodextrin of 0.91 mM, dimethyl-β-cyclodextrin of 1.92 mM, whereas the control (podophyllotoxin without cyclodextrin) only yielded a maximum of 0.15 mM dissolved podophyllotoxin. From these results it was concluded that dimethyl-β-cyclodextrin met our needs the best in terms of substrate availability. The solubility of podophyllotoxin could thus be increased by a factor 12.8 using this cyclodextrin. When changing the ratio of podophyllotoxin/cyclodextrin, using 2 mM and 5 mM concentrations of podophyllotoxin, it was found that even a concentration of 5 mM dissolved podophyllotoxin could be reached.

The solubility of podophyllotoxin in water supplemented with dimethyl-β-cyclodextrin is depicted in Fig. 9.

In the bioconversion experiments five podophyllotoxin-containing media (without cyclodextrin, with 2 mM β-cyclodextrin, γ-cyclodextrin, dimethyl-β-cyclodextrin or hydroxypropyl-β-cyclodextrin) were incubated with *L. flavum* cell suspensions. A standard-grown culture was used as a control.

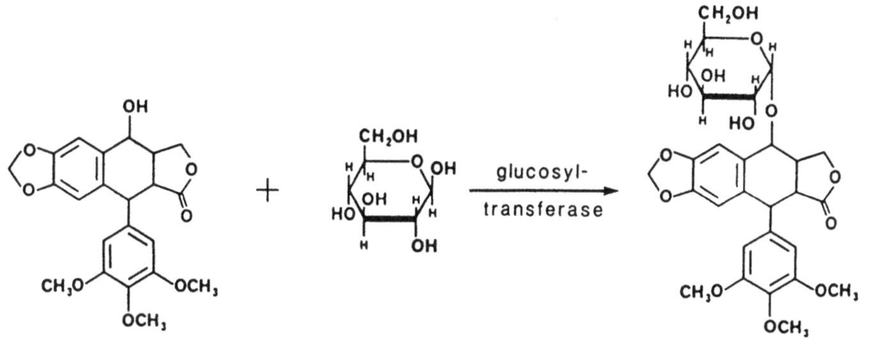

1 2 3

Fig. 8. Glucosylation reaction; podophyllotoxin (*1*) and (UDP)-glucose (*2*) are coupled by a glucosyltransferase to yield podophyllotoxin-β-D-glucoside (*3*) (from Van Uden et al. 1993).

Table 2. Solubility of podophyllotoxin in culture medium, non-complexed and as a cyclodextrin complex, and the corresponding bioconversion percentages (from Van Uden et al. 1993).

Cyclodextrin	PT[1] in suspension (mM)	Bioconversion after 1 day (mM)	(%)
none	0.11	0.11	100
β-CD[2]	0.64	0.18	28
β-DMCD[3]	1.34	0.51	38
β-HPCD[4]	0.48	0.23	47
γ-CD[5]	0.14	0.10	72

[1] podophyllotoxin.
[2] β-cyclodextrin.
[3] dimethyl-β-cyclodextrin.
[4] hydroxypropyl-β-cyclodextrin.
[5] γ-cyclodextrin.

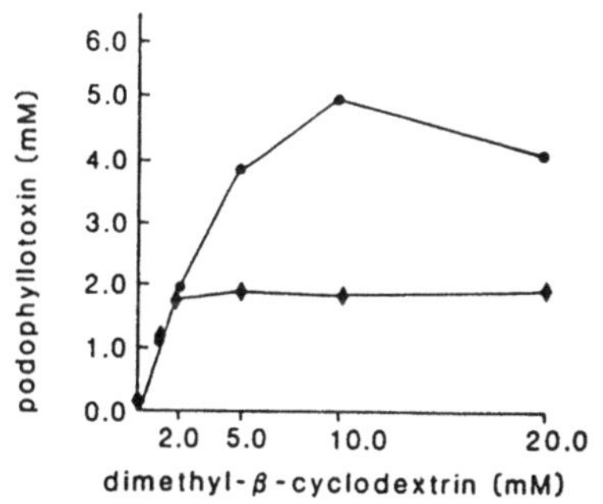

Fig. 9. Solubility of podophyllotoxin at a concentration of 2 mM (♦) or 5 mM (●) in medium supplemented with different amounts of dimethyl-β-cyclodextrin (from Van Uden et al. 1993).

Compared with the control, growth characteristics were hardly affected, although the cells grew in the presence of a rather high concentration of the cytotoxic lignan. Generally, the parameters PCV, FW and DW reached the same values as found for cultures growing without podophyllotoxin. In the bioconversion experiments however, maximal values of these parameters were reached two days later. Podophyllotoxin very rapidly vanished from the culture media (Fig. 10). Already one day after incubation, podophyllotoxin could not be detected in the culture media any more, except in the case when podophyllotoxin was complexed with dimethyl-β-cyclodextrin. Under these conditions still ca. 0.025 mg ml^{-1}, being only 4% of added compound, was present after one day. After three days podophyllotoxin was also undetectable (< 0.001 mg ml^{-1}) in this culture medium. Quite remarkably, endogenously only small amounts were found. Between 0 and 8 h after the incubation, ca. 0.2% podophyllotoxin (DW), corresponding with 16 mg podophyllotoxin l^{-1} suspension, was found in the cells, for all tested cyclodextrin/lignan-containing media. In the case of the dimethyl-β-cyclodextrin-complexed podophyllotoxin this implicated that only

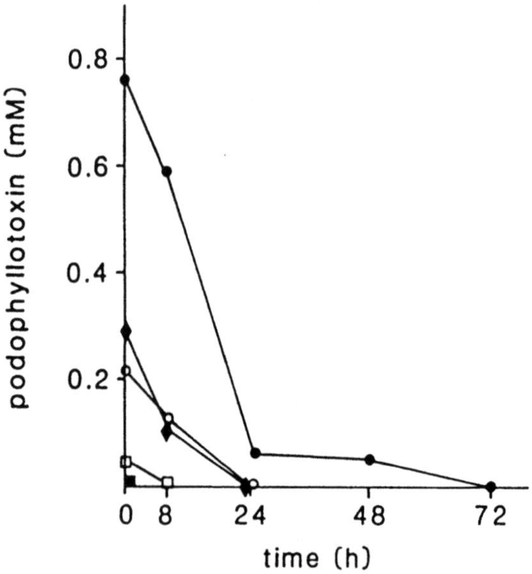

Fig. 10. Time course of the podophyllotoxin concentration in the medium of a *L. flavum* cell suspension fed with podophyllotoxin (■) solely, or with β- (♦), γ- (□), dimethyl-β- (●) and hydroxypropyl-β-cyclodextrin- (○) complexed podophyllotoxin (from Van Uden et al. 1993).

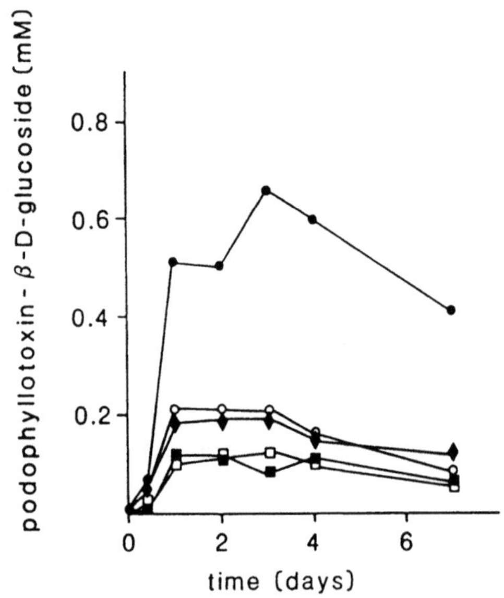

Fig. 11. Time course of the accumulation of podophyllotoxin-β-D-glucoside in *L. flavum* cells after feeding of podophyllotoxin, non-complexed (■), or complexed with β- (♦), γ- (□), dimethyl-β-(●) and hydroxypropyl-β-cyclodextrin (○) (from Van Uden et al. 1993).

3% of the added podophyllotoxin was detectable at that time.

The glucosylation of podophyllotoxin started quickly. Already 8 h after incubation podophyllotoxin-β-D-glucoside was detectable (> 0.001% DW) in the cells.

The addition of the highly cytotoxic podophyllotoxin to the growth medium of *L. flavum* cell suspensions, probably stimulated the cells to detoxify the lignan. Bioconversion to the more polar podophyllotoxin-β-D-glucoside may enable storage in vacuoles and in that way toxic effects of the aglucone are prevented.

The time course of endogenously accumulated podophyllotoxin-β-D-glucoside is depicted in Fig. 11. It can be seen that podophyllotoxin was bioconverted rapidly to podophyllotoxin-β-D-glucoside. Nearly under all conditions maximal levels of podophyllotoxin-β-D-glucoside were found after one day. Only with the medium containing the highest concentration of podophyllotoxin, i.e. complexed with dimethyl-β-cyclodextrin, the highest level of glucoside was found after three days. No lignan glucoside could be detected in the spent culture medium.

The highest bioconversion rate was found for this latter culture condition and was calculated to be 6.6 μmol g^{-1} FW day^{-1}, corresponding with 0.51 mmol l^{-1} suspension day^{-1} (see Table 2). Non-complexed

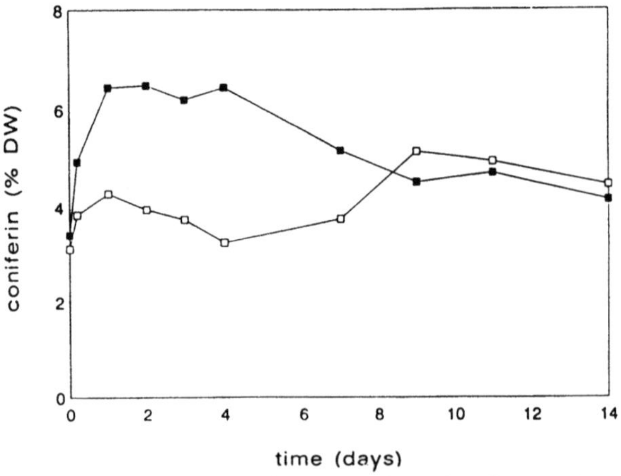

Fig. 12. Time course of coniferin accumulation in cell suspensions of *L. flavum* under standard conditions (□) and after feeding 3 mM coniferyl alcohol complexed with dimethyl-β-cyclodextrin (■).

podophyllotoxin was converted at a rate of 0.11 mmol l^{-1} suspension day^{-1}. If the solubility of podophyllotoxin was not the limiting factor, more could have been bioconverted into its glucoside.

From Table 2 it can be seen that the highest bioconversion percentage after one day was found for the culture growing with uncomplexed podophyllotoxin. However, the absolute amount of converted podophyllotoxin was maximal (0.51 mM) in the medium that contained dimethyl-β-cyclodextrin.

After bioconversion, the cell cultures used in this study accumulated maximally 2.9% (DW) podophyllotoxin as its glucoside when the substrate was complexed with dimethyl-β-cyclodextrin. Free podophyllotoxin, or podophyllotoxin complexed with β-, γ-, or hydroxypropyl-β-cyclodextrin yielded maximally 0.6%, 0.9%, 0.5% and 1.2% podophyllotoxin (present as -β-D-glucoside), respectively.

To our knowledge, this is the first example of an exogenously supplied lignan, containing a non-phenolic hydroxyl moiety, that could be bioconverted into its corresponding glucoside by a plant cell culture.

Glucosylation of coniferyl alcohol by a cell suspension of *Linum flavum*

Cell cultures of *L. flavum* accumulate coniferin, the β-D-glucoside of coniferyl alcohol, in large amounts (up to 12% on a dry weight basis) (Van Uden et al. 1991a). This gives evidence for the presence of a coniferyl alcohol glucosyl transferase.

The application of cyclodextrins as solubilizing agents in the glucosylation of coniferyl alcohol into coniferin by suspension-grown cells of *L. flavum* has been investigated. In bioconversion experiments, 3 mM dimethyl-β-cyclodextrin-complexed coniferyl alcohol (obtained by method of shaking) was added to *L. flavum* cell suspensions. The substrate disappeared completely from the culture medium within two days. Coniferin accumulation occurred rapidly, within one day after the addition of coniferyl alcohol-con-taining fresh medium (Fig. 12). A maximal content of 6.5% on a dry weight basis was found at day four. For comparison, the untreated cultures contained 4.0% coniferin at this point of time. A highest bioconversion rate of 5.7 μmol g^{-1} fresh weight day^{-1} was calculated, corresponding with 33.7% bioconversion of added coniferyl alcohol.

Conclusions

It may be concluded that cyclodextrins can be applied successfully in plant cell biotechnology, because very smooth bioconversion conditions are created with respect to cell vitality and maintenance of enzyme activity. We have shown that the feeding of poorly water-soluble, otherwise ineffective precursors, such as β-estradiol with *Mucuna pruriens* cells, coniferyl alcohol with *Podophyllum hexandrum* and *Linum flavum* cells, and podophyllotoxin with *Linum flavum* cells resulted in significant increases in bioconversion rates and percentages. Generally, the products can easily be extracted from the cyclodextrin-containing reaction mixture by using a suitable organic solvent. In addition, the complexing agent can be recovered for reuse. As was demonstrated for the glucosylation of podophyllotoxin by *L. flavum* cells, different cyclodextrin derivatives can be applied without toxicity for the cells. This opens the possibility to choose the most suitable cyclodextrin for an individual substrate.

Furthermore, cyclodextrins enable the feeding of lipophilic compounds, which are regarded as possible intermediates in a biosynthetic routing, to plant cell cultures and plants as well. This may render more fundamental knowledge on the formation of certain secondary metabolites.

Cyclodextrins act as precursor solubilizers, but they probably also play a role as product protecting agents and reactant carriers. This has been suggested for the steroid side chain cleavage by *Mycobacterium* species (Hesselink et al. 1989). If product protection indeed occurs, this also implies that product inhibition can basically be prevented during a bioconversion process, thereby improving the product yield.

The role of cyclodextrins as reactant carriers seems likely, since the cyclodextrin concentration in the media remained unchanged during product formation by the plant cells. Moreover, we measured only intracellularly formed products when using freely suspended and immobilized plant cells, i.e. 2- and 4-hydroxy-estradiol, podophyllotoxin, podophyllotoxin-β-D-glucoside and coniferin.

With respect to mass transfer, plant cell systems are rather complicated and limitations in precursor supply can easily occur. One of the most interesting questions is whether the inclusion complexes are able to penetrate the cell wall and cell membrane, and even enter the living plant cell. The multi-functional behaviour of cyclodextrins can only be revealed by detailed kinetic studies. A suggestion is to follow cyclodextrin-complexed labelled precursors (and products) during a bioconversion process.

The concept of using cyclodextrins as a means of introducing relatively insoluble precursors for bioconversions by plant cells or enzymes, possibly yielding new or improved drugs, certainly merits further attention.

Acknowledgements

The authors like to thank Ms. A.M.A. van Dijken who performed the experiments concerning the glucosylation of coniferyl alcohol by a cell suspension of *L. flavum*.

Kluwer Academic Publishers, Dordrecht, The Netherlands, Pergamon Press, Oxford, UK, and Springer-Verlag, Heidelberg, Germany, are thanked for providing permission to use earlier published material.

References

Berlin J, Wray V, Mollenschott C & Sasse F (1986) Formation of β-peltatin-A-methylether and coniferin by root cultures of *Linum flavum*. J. Nat. Prod. 49: 435–439

Berlin J, Bedorf N, Mollenschott C, Wray V, Sasse F & Höfle G (1988) On the podophyllotoxins of root cultures of *Linum flavum*. Planta Med. 54: 204–206

Duchêne D & Wouessidjewe D (1990a) Physicochemical characteristics and pharmaceutical uses of cyclodextrin derivatives, Part I. Pharm. Technol. 14 (6): 26–34

Duchêne D & Wouessidjewe D (1990b) Physicochemical characteristics and pharmaceutical uses of cyclodextrin derivatives, Part II. Pharm. Technol. 14 (8): 22–30

Hesselink PGM, Van Vliet S, De Vries H & Witholt B (1989) Optimization of *Mycobacterium* steroid side chain cleavage in the presence of cyclodextrins. Enzyme Microbiol. Technol. 11: 398–404

Higuchi T & Connors K (1965) Phase solubility techniques. Adv. Anal. Chem. Industr. 4: 117–212

Hirayama F & Uekama K (1987) Methods of investigating and preparing inclusion compounds. In: Duchêne D (Ed) Cyclodextrins and Their Industrial Uses (pp 131–172). Editions de Santé, Paris

Hösel W (1981) Glycosylation and glycosidases. In: Conn EE, Stumpf PK (Ed) The Biochemistry of Plants, Vol 7 (pp 725–753). Academic Press, New York

Jackson DE & Dewick PM (1984) Biosynthesis of *Podophyllum* lignans – II. Interconversions of aryltetralin lignans in *Podophyllum hexandrum*. Phytochemistry 23: 1037–1042

Le Bas G & Rysanek N (1987) Structural aspects of cyclodextrins. In: Duchêne D (Ed) Cyclodextrins and Their Industrial Uses (pp 105–130). Editions de Santé, Paris

Luckner M (1986) In: Secondary Metabolism in Microorganisms, Plants and Animals, 2nd edition. Springer-Verlag, Heidelberg/New York/Tokyo

Mayer AM (1987) Polyphenol oxidases in plants – recent progress. Phytochemistry 26: 11–20

Müller BW & Brauns U (1986) Hydroxypropyl-β-cyclodextrin derivatives: influence of average degree of substitution on complexing ability and surface activity. J. Pharm. Sci. 75 (6): 571–572

Pras N (1989) Biotechnological production of catechols: bioconversion spectrum and related kinetic aspects of entrapped cells of *Mucuna pruriens* L. Pharm. Weekbl. Sci. Ed. 11: 30–31

Pras N (1992) Bioconversion of naturally occurring precursors and related synthetic compounds using plant cell cultures: a review. J. Biotechnol. 26: 29–62

Pras N, Wichers HJ, Bruins AP & Malingré ThM (1988) Bioconversion of para-substituted monophenolic compounds into corresponding catechols by alginate-entrapped cells of *Mucuna pruriens*. Plant Cell Tiss. Org. Cult. 13: 15–26

Pras N, Hesselink PGM, Guikema WM & Malingré ThM (1989) Further kinetic characterization of alginate-entrapped cells of *Mucuna pruriens* L. Biotechnol. Bioeng. 33: 1461–1468

Saenger W (1984) Structural aspects of cyclodextrins and their inclusion complexes. In: Atwood JL, Davies JED & MacNicol DD (Eds) Inclusion Compounds (pp 231–259). Academic Press, London

Sébille B (1987) Cyclodextrin derivatives. In: Duchêne D (Ed) Cyclodextrins and Their Industrial Uses (pp 353–392). Editions de Santé, Paris

Sicard PJ & Saniez MH (1987) Biosynthesis of cycloglucosyltransferase and obtention of its enzymatic reaction products. In: Duchêne D (Ed) Cyclodextrins and Their Industrial Uses (pp 77–103). Editions de Santé, Paris

Szejtli J & Pagington J (1989) Solubility isotherms. Cyclodextrin News 3: 77–79

Uekama K, Narisawa S, Hirayama F & Otagiri M (1983) Improvement of dissolution and absorption characteristics of benzodiazepines by cyclodextrin complexation. Int. J. Pharm. 16: 327–338

Uekama K & Irie T (1987) Pharmaceutical applications of methylated cyclodextrin derivatives. In: Duchêne D (Ed) Cyclodextrins and Their Industrial Uses (pp 395–439). Editions de Santé, Paris

Van Uden W, Pras N, Visser JF & Malingré ThM (1989) Detection and identification of podophyllotoxin produced by cell cultures derived from *Podophyllum hexandrum* Royle. Plant Cell Rep. 8: 165–168

Van Uden W, Pras N, Vossebeld EM, Mol JNM & Malingré ThM (1990a) Production of 5-methoxypodophyllotoxin in cell suspension cultures of *Linum flavum* L. Plant Cell Tiss. Org. Cult. 20: 81–87

Van Uden W, Pras N & Malingré ThM (1990b) On the improvement of the podophyllotoxin production by phenylpropanoid precursor feeding to cell cultures of *Podophyllum hexandrum* Royle. Plant Cell Tiss. Org. Cult. 23: 217–224

Van Uden W, Pras N, Batterman S, Visser JF & Malingré ThM (1991a) The accumulation and isolation of coniferin from a high-producing cell suspension of *Linum flavum* L. Planta 183: 25–30

Van Uden W, Pras N, Homan B & Malingré ThM (1991b) Improvement of the production of 5-methoxypodophyllotoxin using a new selected root culture of *Linum flavum* L. Plant Cell Tiss. Org. Cult. 27: 115–121

Van Uden W, Homan B, Woerdenbag HJ, Pras N & Malingré ThM, Wichers HJ & Harkes M (1992) Isolation, purification, and cytotoxicity of 5-methoxypodophyllotoxin, a lignan from a root culture of *Linum flavum*. J. Nat. Prod. 55: 102–110

Van Uden W, Oeij H, Woerdenbag HJ & Pras N (1993) Glucosylation of cyclodextrin-complexed podophyllotoxin by cell cultures of *Linum flavum* L. Plant Cell Tiss. Org. Cult. (in press)

Wichers HJ, Harkes MP & Arroo RJ (1990) Occurrence of 5-methoxypodophyllotoxin in plants, cell cultures and regenerated plants of *Linum flavum*. Plant Cell Tiss. Org. Cult. 23: 93–100

Wichers HJ, Versluis-De Haan GG, Marsman JW & Harkes MP (1991) Podophyllotoxins in plants and cell cultures of *Linum flavum*. Phytochemistry 30: 3601–3604

Woerdenbag HJ, Pras N, Frijlink HW, Lerk CF & Malingré ThM (1990a) Cyclodextrin-facilitated bioconversion of 17β-estradiol by a phenoloxidase from *Mucuna pruriens* cell cultures. Phytochemistry 29: 1551–1554

Woerdenbag HJ, Van Uden W, Frijlink HW, Lerk CF, Pras N & Malingré ThM (1990b) Increased podophyllotoxin production in *Podophyllum hexandrum* cell suspension cultures after feeding coniferyl alcohol as a β-cyclodextrin complex. Plant Cell Rep. 9: 97–100

Yamamoto M, Yoshida A, Hirayama F & Uekama K (1989) Some physicochemical properties of branched β-cyclodextrins and their inclusion characteristics. Int. J. Pharm. 49: 163–171

Plant Cell, Tissue and Organ Culture **38**: 115–122, 1994.

Embryogenesis of photoautotrophic cell cultures of *Daucus carota* L.

Bärbel Grieb[1], Ulrich Groß[2], Eva Pleschka[1,3], Birgit Arnholdt-Schmitt[1] &
Karl-Hermann Neumann[1]
[1]*Institut für Pflanzenernährung, Abt. Gewebekultur, Justus Liebig-Universität Gießen, Südanlage 6, 35390 Gießen, Germany;* [2]*Abteilung für Klinische Biochemie der Philipps-Universität Marburg, Deutschhausstr. 17 1/2, 35037 Marburg, Germany;* ([3]*This paper contains part of the PhD thesis of E. Pleschka)*

Key words: Daucus carota, elevated CO_2 concentration, photoautotrophy, somatic embryogenesis, sucrose

Abstract

In this paper photoautotrophic carrot (*Daucus carota* L.) suspension cultures are described which are able to produce somatic embryos. The development of somatic embryos, however, requires a sucrose supplement. Although an elevation of the CO_2 concentration up to 2.3% results in the same level of dry weight production as with sucrose in the medium, somatic embryos could not be observed.

Results on the influence of sucrose on some aspects of the photosynthetic apparatus of cultured cells are discussed.

Abbreviations: 2,4-D – 2,4-dichlorophenoxyacetic acid, DW – dry weight, ELISA – enzyme-linked-immunosorbent-assay, FW – fresh weight, IAA – indole-3-acetic acid, NAA -naphthaleneacetic acid, PEPCase – phosphoenol-pyruvate carboxylase, Rubisco – Ribulose- 1,5-bisphosphate carboxylase/oxygenase, se – somatic embryogenesis

Introduction

Although zygotic embryogenesis is the most common process of propagation of higher plants, some other means of embryogenesis exist, e.g. polyembryony for *Citrus* ssp., often summarized as apomixis. One of these detours to zygotic embryogenesis is also somatic embryogenesis as described originally for carrot cell cultures and later also for cultured cells of other species of higher plants.

As summarized in Fig. 1 somatic embryogenesis can be subdivided into several phases. First ontogenetic processes lead to competent cells with the capacity to be inducible to embryogenesis (directive induction, Christianson 1985). In vitro, these cells can be determined to embryogenic competent cells in the presence of an auxin (induction phase; permissive induction, Christianson 1985). In the following realization phase the program of embryogenesis unfolds, leading to mature embryos (Grieb 1992; Neumann & Grieb 1992). Between the latter two phases often an intervening culture period that can last for many years is practiced during which embryogenic cell material can be propagated by many subculture passages. This is called a propagation phase. Most investigations (see e.g. Aleith & Richter 1990; Choi & Sung 1984; Komamine et al. 1990; Sengupta & Raghavan 1980; Sung & Okimoto 1981) concerned with somatic embryogenesis of cultured plant cells used cell material kept in an embryogenic state during the propagation phase which requires an auxin supplement to the nutrient medium to prevent the realization of embryo development. Upon transfer of such cultures into an auxin-free medium the development of somatic embryos is initiated. Evaluating the results of experiments with such material, it has to be kept in mind that the induction of competence to somatic embryogenesis took place maybe many years before at explantation and subsequent cultivation of the original explants in an auxin containing nutrient medium. Such experimental systems describe the realization of the embryogenic program only, but not the induction for competence to somatic embryogenesis in the cultured cells.

From all what we know till now, the induction of embryogenic competence requires an auxin in the medium, generally either IAA, NAA or 2,4-D. Some variation in hormone requirement, however, exists

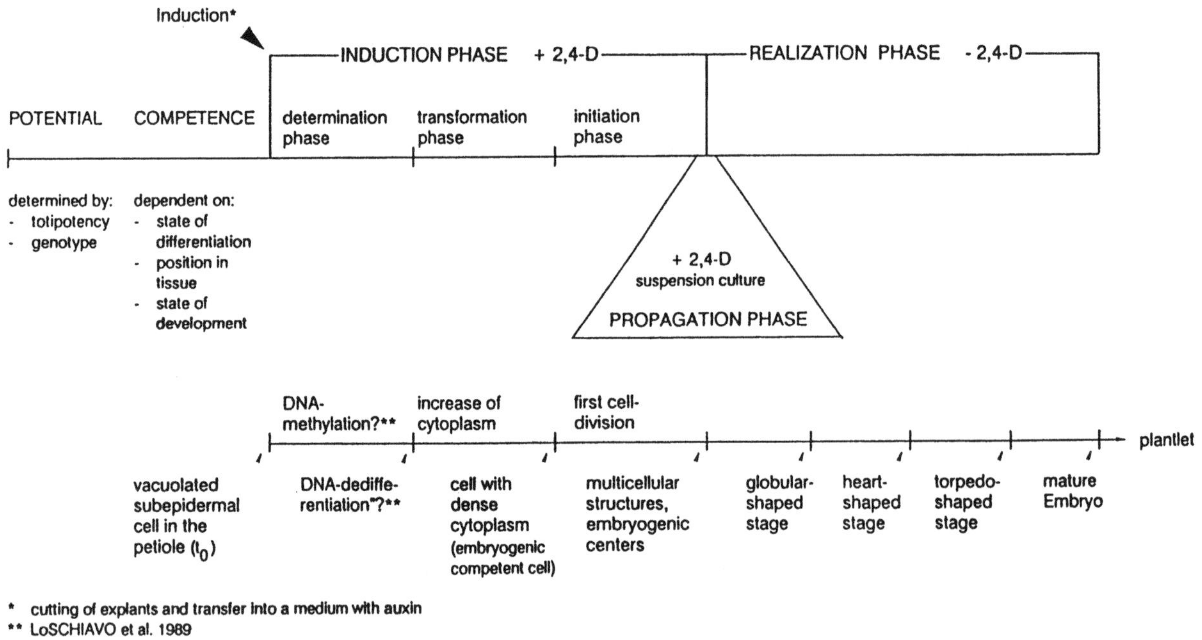

Fig. 1. Hypothetical scheme of somatic embryogenesis in petiole explants of *Daucus carota* at the cellular level (distances in the scheme are formal and no indication of the duration of time intervals between individual steps).

for gymnosperms (e.g. Norgaard & Krogstrup 1991). Here, cytokinins play a more important role and an auxin may be even characterized by negative effects. To induce the development of embryos in competent cells in angiosperms, i.e. the unfolding of the program of embryogenesis in such cultures, a transfer into an auxin-free medium is necessary. If IAA is used as an auxin a transfer is not required due to photooxydative decomposition of this phytohormone if the cultures are kept in the light (Bender & Neumann 1978).

In many investigations a characterization of events leading to embryogenesis out of competent cells at the macromolecular level concerning the organization of DNA, the occurrence of special RNAs or proteins was attempted during the last decade (for general reference, see Nijkamp et al. 1990). Almost no experiments, however, were performed to describe processes during the induction of somatic embryogenesis, the propagation of embryogenic cells and the realization phase of somatic embryogenesis at the level of primary metabolism.

At the macromolecular level somatic embryogenesis seems to be at least to a certain extent comparable to zygotic embryogenesis (Crouch 1982; Stuart et al. 1988) and by comparing several plant species the developmental program leading to the differentiation of embryos seems to be phylogenetically highly conserved within the plant kingdom. Some variation may exist in primary metabolism. Therefore, investigations on somatic embryogenesis will also contribute to the understanding of this key program within the development of higher plants.

For a start to describe primary metabolism going along with somatic embryogenesis, an embryogenic photoautotrophic strain of cultured carrot cells was used to study nutritional requirements of somatic embryogenesis. These studies are by far not yet completed. Results using callus cultures with autotrophic or mixotrophic nutrition and petiole cultures will be included to interpret the results obtained with this system.

Results and discussion

Somatic embryogenesis of photoautotrophic carrot cell cultures

The photoautotrophic strain of carrot cell suspensions (var. 'Vosgeses') used in the experiments described in this paper was originally isolated from a system of cultured petiole explants in a modified B5-medium supplemented with 2% sucrose and 0.1 ppm 2,4-D (Gamborg et al. 1968; Schäfer et al. 1988). During

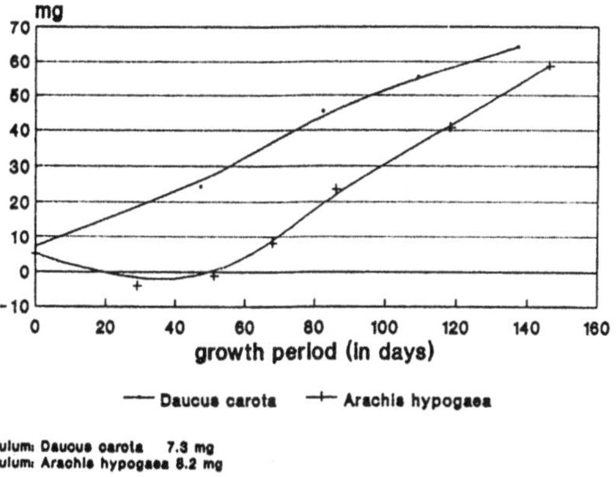

Fig. 2. Photoautotrophic growth of strains from *Daucus carota* and *Arachis hypogaea* (dry weight increment in mg). Culture conditions: 28°C; continuous light, 6.4 W m^{-2}, Osram day light.

Table 1. Influence of various carbohydrates (0.06 mol l^{-1}) on induction phase and realization phase of somatic embryogenesis in carrot petiole explants.

	Realization phase ($B5^-$)				
Induction phase ($B5^+$)	O*	Sucrose	Glucose	Fructose	Maltose
O*	-	e	-	-	-
Sucrose	-	se	se	se	se
Glucose			se	-	
Fructose	-	se	se	(se)	(se)
Maltose	-	se	-		se

culture conditions: liquid medium, 28°C, continuous light, modified B5-medium (Schäfer et al. 1988).
$B5^+$: 14 days in B5-medium with 0.5 ppm 2,4-D
$B5^-$: 18 days in B5-medium without 2,4-D
O*: medium without carbohydrates
se: somatic embryogenesis
e: somatic embryogenesis was induced in auxin-free medium (later induction)
-: no somatic embryos
(sc): minimal formation of somatic embryos

culture callus material developed distinguished by a much higher chlorophyll concentration judged by visual comparison than in other similar experiments. The concentration of total chlorophyll in these cultures is still only 10–15% of that of carrot leaves. The culture was initiated 6 years ago and kept at ambient CO_2 since. During the first seven month subcultures were set up every 4 weeks using the same nutrient medium as described above. At each subculture the most green cell material was selected. After this period for 20 month the subculturing was carried out into a B5-medium without sugar at ambient CO_2. The only organic constituents were the vitamins (0.5 ppm nicotinic acid, 0.1 ppm thiamine, 0.1 ppm pyridoxine), caseinhydrolysates (250 ppm), myo-inositol (5 ppm) and 2,4-D (0.1 ppm). The subculture intervals were extended to 5 month. The growth performance of these cultures during a five month culture period (without subculture) can be seen in Fig. 2 in which for comparison also data on an autotrophic, but non- embryogenic *Arachis hypogaea* strain are included. The latter cultures were initiated some twenty years ago and a detailed description of this culture system is given elsewhere (Groß et al. 1993).

To check the embryogenic competence and its dependence on a sugar supply these autotrophic cultures were transferred into a medium either free of carbohydrates or supplemented with 2% sucrose for 4 weeks, both containing 0.1 ppm 2,4-D. After this treatment the cultures were transferred into a medium of the same composition from which, however, 2,4-D was omitted to induce the realization phase of embryogenesis. At inspection of these cultures about 3 weeks later it was obvious that only those cultures kept in a sucrose containing medium were able to produce adventitious roots and somatic embryos (Figs 3, 4).

The experimental program just described is summarized in Fig. 3. The induction of embryogenic competence was achieved during the cultivation of petiole explants in the medium supplemented with both, sucrose and 2,4-D.Whereas the maintenance of embryogenic competence is possible in autotrophic conditions, apparently a sucrose supplement is a prerequisite to realize the program of somatic embryogenesis in induced cells.

In Table 1 the influences of some carbohydrates on somatic embryogenesis supplied to the nutrient medium during the induction and the realization phase of cultured carrot petiole explants are compared to the influence of sucrose. As shown in Fig. 3, the same system was used to initiate the photoautotrophic cultures. The data in Table 1 show that of the carbohydrates tested sucrose seems to be most effective because its supplement was efficient during the induction as well as the realization phase of somatic embryogenesis. Although not all combinations were tested, from the data available it can be derived, that sucrose can be replaced by glucose, fructose and maltose. If sucrose is supplied during the induction phase, somatic embryos will be produced if it is replaced by glucose, fructose or maltose in the medium used in the realization phase. Following a substitution of sucrose by glucose

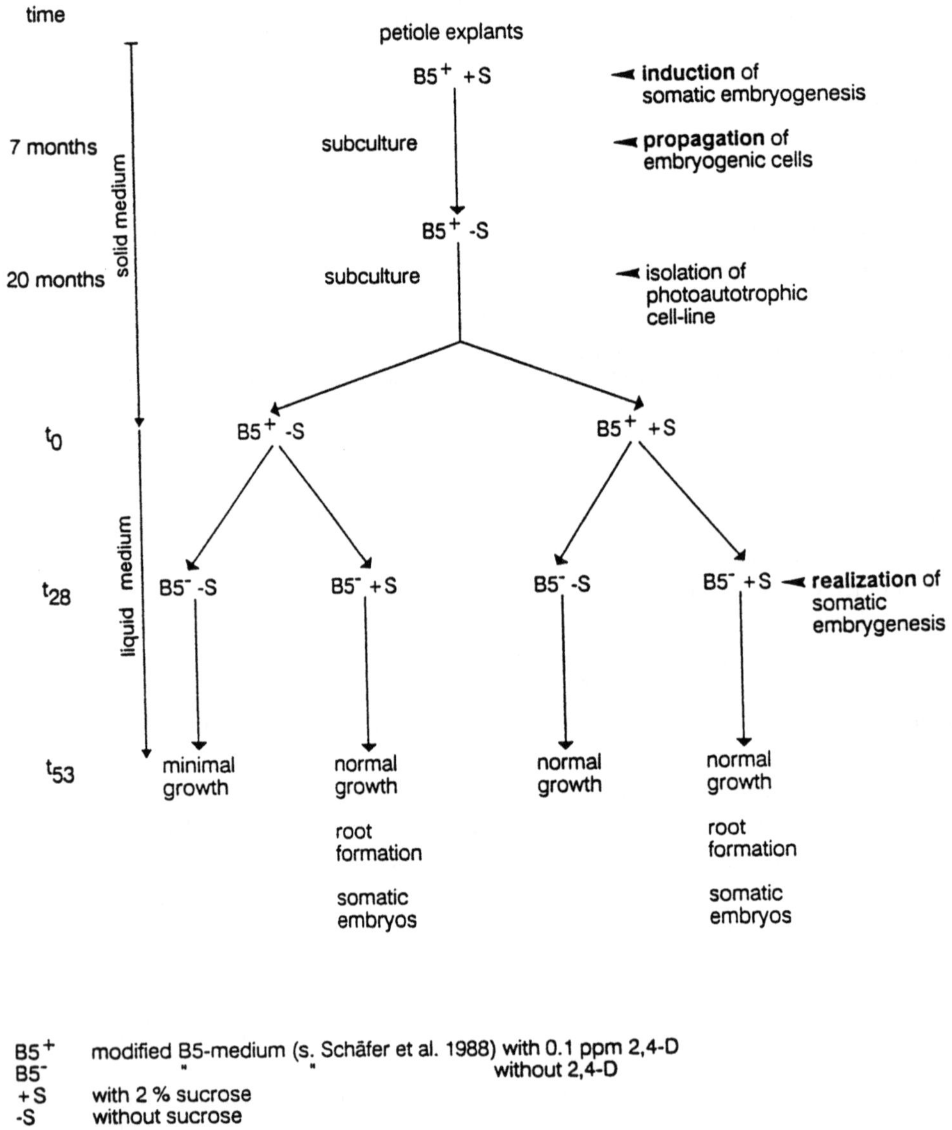

Fig. 3. Protocol of somatic embryogenesis in a strain of photoautotrophic carrot culture (28°C, continuous light, Osram lumilux day light).

during induction, in the nutrient medium containing fructose during the realization phase no embryos could be observed. Also, if maltose substitutes sucrose during induction, then only with sucrose or maltose in the auxin free medium (realization phase) embryos are produced, but not with glucose. Although not given in the table, also if mannose or galactose are supplied consecutively to both phases somatic embryos are produced, but not with ribose or mannitol.

These results indicate that both phases of somatic embryogenesis require some carbohydrate in the nutrient medium to produce mature somatic embryos of

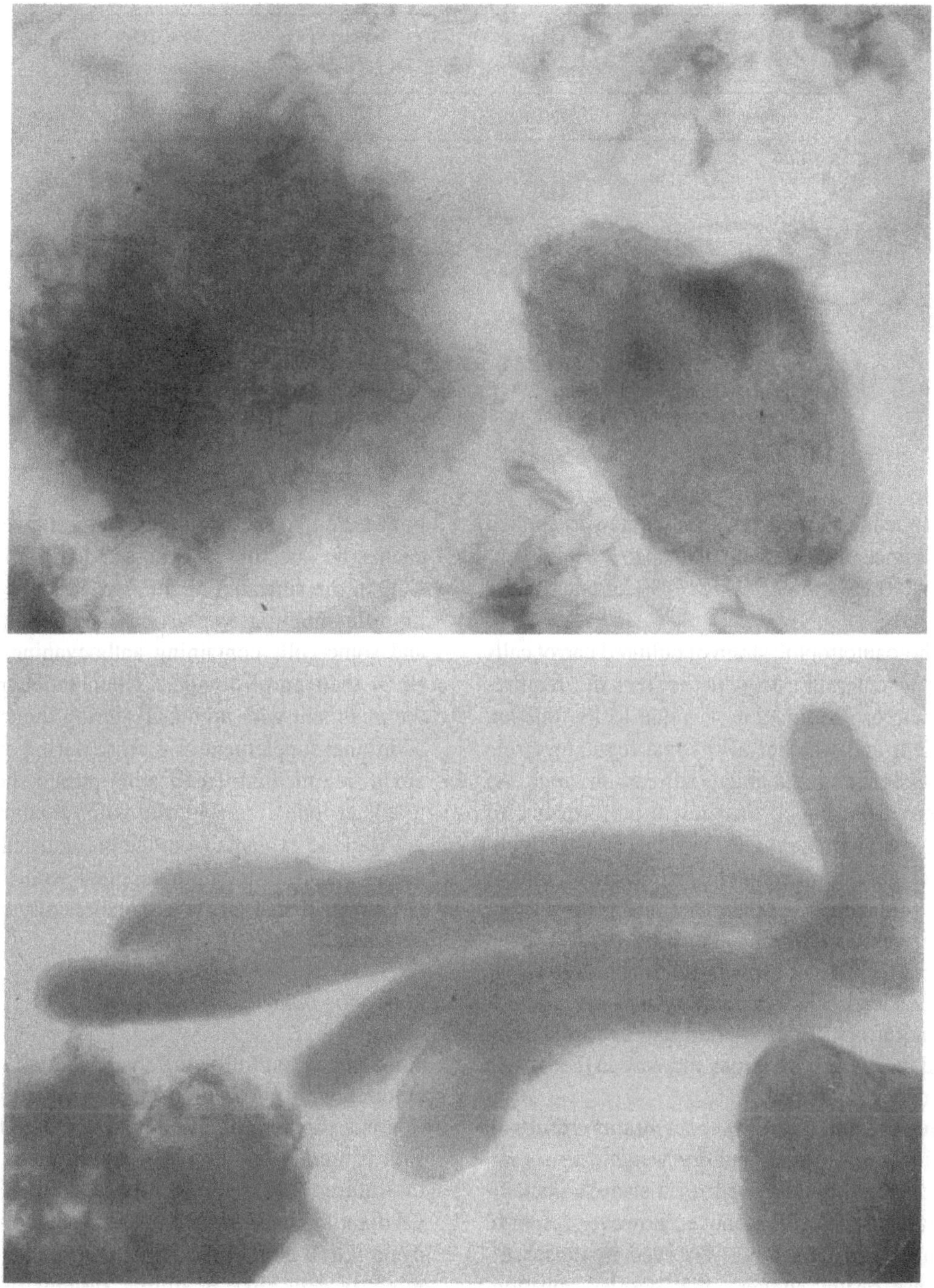

Fig. 4. Somatic embryos of a photoautotrophic strain from *Daucus carota* in an auxin-free B5-medium with sucrose.

which sucrose is the most efficient. Some qualitative variation however, seems to exist in the efficiency of the others tested during the realization phase. Since neither ribose nor mannitol applied at the same osmolarity as the effective carbohydrates were efficient to induce and to realize the program of somatic embryogenesis osmotic effects of the other carbohydrates with promotive influences can be discarded.

To determine the minimal requirement of carbohydrates during both phases of somatic embryogenesis,

Table 2. Influence of elevated CO_2 (2.34%) and sucrose (2%) on growth of a photoautotrophic strain from *Daucus carota*.

	ambient CO_2 without sucrose	ambient CO_2 2% sucrose	2.34% CO_2 without sucrose
Inoculum (FW)	150 mg	150 mg	600 mg
Increase of fresh weight			
mg/42 days	433	3512	4523
mg/day	10.3	83.6	107.7
Increase of dry weight			
mg/42 days	11.4	167	293
mg/day	0.27	3.98	6.98

culture conditions: liquid medium, 28°C, continuous light, Osram day light, 6.4 W m^{-2}. modified B5-medium (Schäfer et al. 1988) without 2,4-D and without sucrose.

sucrose as the most efficient carbohydrate was applied at various concentrations from 0.05 up to 2.0%. A concentration of 0.1% suffices, which is 5% of that usually supplied.

The photoautotrophic strain of cultured carrot cells was used to determine to which extent the requirement for sucrose application was due to its function as a nutrient and whether also some regulatory role could be assigned to this ubiquitous carbohydrate. As can be seen from Fig. 2 the growth performance of these cultures at ambient CO_2 is rather low. As published elsewhere in photoautotrophic *Arachis* cultures a hundred fold increase of the CO_2 concentration using double-tier vessels (Hüsemann & Barz 1977) a seven to eight fold increase of growth can be achieved (Neumann & Bender 1987). The same system was used for the photoautotrophic carrot strain and as can be seen from the data in Table 2 a strong increase in the growth performance was induced.

Although raising the CO_2 concentration results in a strong increase in fresh and dry weight, no somatic embryos could be observed. In a sucrose containing medium as described above, however, embryo development can be readily observed in those cultures. These results suggest, that besides nutritional functions as a source for carbon and energy the requirement for exogenous sucrose to induce somatic embryos may serve also some regulatory aspects of the process. Again primary petiole explants were used to check influences of the elevated CO_2 concentration on the induction of embryogenic competence. The results summarized in Table 3 clearly indicate that also the induction of embryogenic competence requires a carbohydrate like sucrose in the nutrient medium, which cannot be substituted by a hundredfold increase of CO_2 in the surrounding atmosphere. The only cellular differentiations were some tracheid-like structures and some cells containing anthocyanine, as indicated by their purplish color. The plantlets produced in the treatment with high CO_2 during induction and an additional supplement of sucrose during culture in the auxin-free medium (realization phase) should be due to a later induction achieved with remaining traces of 2,4-D in the explants from the previous culture period in the induction medium supplied with this auxin. It is known, that 2,4-D is a metabolically rather stable compound.

Metabolic influences of sucrose

A main reaction following an application of sucrose to cultures with photoautotrophic or mixotrophic nutrition is a suppression of photosynthesis. This suppression is mainly due to a decrease in the concentration of Rubisco and consequently in the performance of C3 fixation of CO_2, i.e. the activity of the Calvin cycle (Groß et al. 1993). In cultured photoautotrophic carrot cells, like in similar systems of other plant species, besides Rubisco also PEPCase operates as part of a second carboxylating enzyme system. At ambient CO_2 specific activities of these two enzymes in the sugar-free medium were determined as 0.66 nkat for Rubisco and 0.06 nkat mg^{-1} soluble protein for PEPCase respectively (Groß 1990). The concentration of PEPCase in these cultures is about twice as high as in carrot leaves. This agrees also with data obtained

Table 3. Influence of elevated CO_2 (2.34%) on the induction phase and realization phase of somatic embryogenesis in carrot petiole explants.

	Realization phase ($B5^-$)			
Induction phase ($B5^+$)	O*	S	CO_2	S + CO_2
O*	-	e	-	e
S	-	se	-	se
CO_2	-	e	n.d.	n.d.

Culture conditions: liquid medium, 28°C, continuous light, modified B5-medium (Schäfer et al. 1988).
$B5^+$: 14days in B5-medium with 0.5ppm 2,4-D
$B5^-$: 14days in B5-medium without 2,4-D
O*: medium without sucrose
S: 2% sucrose
se: somatic embryogenesis
e: somatic embryogenesis was induced in auxin-free medium (later induction)
-: no somatic embryos
n.d.: not determined

Table 4. The PEPCase concentration (relative units mg^{-1} total soluble protein) of leaves and photoautotrophic cell cultures of *Daucus carota* and *Arachis hypogaea* as determined by ELISA.

Species		PEPCase
Daucus carota	leaves	9.84 ± 3.70
	cell cultures	18.07 ± 8.80
Arachis hypogaea	leaves	7.20 ± 1.90
	cell cultures	11.70 ± 3.40

for autotrophic peanut cell cultures and leaves (Table 4). Whereas a sucrose supply to the medium has no influences on the PEPCase concentration Rubisco is strongly suppressed at 4% of sucrose in the medium. Only PEPCase remains whereas Rubisco can be detected only as traces. This can, however, be reversed upon a transfer into a sugar free medium (Groß et al. 1993). As a consequence of these changes in the situation of primary metabolism following a sucrose supplement many other changes in metabolism of low molecular weight components like amino acids or carbohydrates occur as described earlier for *Daucus* or *Arachis* cultures (Neumann & Bender 1987, Neumann et al. 1989). It remains to be seen of which significance such changes are for the induction and the realization of the developmental program of somatic embryogenesis. The maintenance of embryogenic competence, however, is achieved without carbohydrates. Basic changes in the protein moiety of cultured cells describe mainly the capacity for alterations in the developmental competence, however, its realization is to a great extent dependent on the biochemical and physiological situation of the cell material concerned.

Conclusions

All the data reported indicate that an exogenous carbohydrate like sucrose seems to play also a regulatory role in the induction and the performance of somatic embryogenesis. This requirement exists for the induction as well as for the realization phase. However, embryogenic competence can be maintained during a propagation phase of about two years of autotrophic nutrition at ambient CO_2. Possibly, the regulatory function of exogenous sucrose (or any of the other carbohydrates here discussed) in the light consists just in the induction of a mixotrophic nutritional regime and since all the phases of somatic embryogenesis can be performed also in the dark (Grieb, 1992), only a requirement for autotrophic nutrition besides other factors seems to exist for the induction and the realization of this developmental program.

These photoautotrophic cell cultures growing at ambient CO_2 could also be used as model systems to study metabolic problems of photosynthesis without interference of physiological factors like stomatal aperture etc. These cultures should be more representative for cells of intact higher plants in natural CO_2 concentration environment than those requiring an elevated CO_2 concentration up to an unnatural level. Since these cultures are embryogenic, this material could also be included in breeding programs.

Acknowledgement

We thank Dr. A. Nato for the specific antibody against PEPCase from tobacco.

References

Aleith F & Richter G (1990) Gene expression during induction of somatic embryogenesis in carrot cell suspension. Planta 183:17–24

Bender L & Neumann KH (1978) Investigations on the influence of pre-culture in IAA- and kinetin containing media on subsequent growth of cultured carrot explants. Z. Pflanzenphysiol. 88. 201–208

Choi JH & Sung ZR (1984) Two-dimensional gel analysis of carrot somatic embryogenic proteins. Plant Mol. Biol. Rep. 2:19–25

Christianson ML (1985) An embryogenic culture of soybean: towards a general theory of somatic embryogenesis. In: Henke RR, Hughes KW, Constantin MJ & Hollaender A (Eds) Tissue Culture in Forestry and Agriculture (pp 83–103). Plenum Press, New York

Crouch ML (1982) Non-zygotic embryos of *Brassica napus* L. contain embryospecific storage proteins. Planta 156: 520–524

Gamborg OL, Miller RA & Ojima K (1968) Nutrient requirements of suspension cultures of soybean root cells. Exp. Cell Res. 50:151–158

Grieb B (1992) Untersuchungen zur Induktion der Kompetenz zur somatischen Embryogenese in Karotten-Petiolenexplantaten (*Daucus carota* L.) - Histologie und Proteinsynthesemuster -. Wissenschafts-Verlag Dr. W. Maraun, Frankfurt/M

Groß U (1990) Der Einfluß von Saccharose und der CO_2-Konzentration auf die Aktivität von Ribulose-1,5-bisphosphat Carboxylase/Oxygenase und Phosphoenolpyruvat Carboxylase und die Konzentration der Ribulose-1,5-bisphosphat Carboxylase/Oxygenase in photoautotrophen Zellkulturen von *Arachis hypogaea* L. und *Daucus carota* L. Dissertation, Giessen

Groß U, Gilles F, Bender L, Berghöfer P & Neumann KH (1993) The influence of sucrose and an elevated CO_2 concentration on photosynthesis of photoautotrophic peanut (*Arachis hypogaea* L.) cell cultures. Plant Cell Tiss. Org. Cult. 33:143–150

Hüsemann W & Barz W (1977) Phototautotrophic growth and photosynthesis in cell suspension cultures of *Chenopodium rubrum*. Physiol. Plant. 40: 77–81

Komamine A, Matsumoto M, Tsukahara M, Fujiwara A, Kawahara R, Ito M, Smith J, Nomura K & Fujimura T (1990) Mechanism of somatic embryogenesis in cell cultures - physiology, biochemistry and molecular biology. In: Nijkamp HJJ, van der Plas LHW & van Aartrijk J (Eds) Progress in Plant Cellular and Molecular Biology, Current Plant Science and Biotechnology in Agriculture, Vol 9 (pp 307–313). Proc. Vllth Int. Congr. Plant Tissue and Cell Culture, Amsterdam. Kluwer Academic Publ., Dordrecht

LoSchiavo F, Pitto L, Giuliano G, Torti G, Nuti-Ronchi V, Marazziti D, Vergara R, Orselli S & Terzi M (1989) DNA methylation of embryogenic carrot cell cultures and its variations as caused by mutation, differentiation, hormones and hypomethylating drugs. Theor. Appl. Genet.77: 325–331

Neumann KH & Bender L (1987) Photosynthesis in cell and tissue culture systems. In: Green CE, Somers DA, Hackett WP & Biesboer DD (Eds) Plant Tissue and Cell Culture (pp 151–165). Alan R. Liss, Inc., New York

Neumann KH & Grieb B (1992) Somatische Embryogenese bei höheren Pflanzen: Grundlagen und praktische Anwendung. Wiss. Zt. der Humboldt-Univ. zu Berlin R. Mathematik/Naturwiss. 41: 63–80

Neumann KH, Groß U & Bender L (1989) Regulation of photosynthesis in *Daucus carota* and *Arachis hypogaea* cell cultures by exogenous sucrose. In: Kurz WGW (Ed) Primary and Secondary Metabolism of Plant Cell Cultures II. Springer Verlag, Berlin, Heidelberg, New York 281–291

Nijkamp HJJ, van der Plas LHW & van Aartrijk J (Eds) (1990) Progress in Plant Cellular and Molecular Biology, Current Plant Science and Biotechnology in Agriculture Vol 9, Proc. Vllth Int. Congr. Plant Tissue and Cell Culture, Amsterdam. Kluwer Academic Publ., Dordrecht

Norgaard JV & Krogstrup AA (1991) Cytokinin induced somatic embryogenesis from immature embryos of *Abies nordmanniana* Lk. Plant Cell Rep. 9: 509–513

Schäfer F, Grieb B & Neumann KH (1988) Morphogenetic and histological events during somatic embryogenesis in intact carrot plantlets (*Daucus carota* L.) in various nutrient media Bot. Acta 101: 362–365

Sengupta C & Raghavan V (1980) Somatic embryogenesis in carrot cell suspension. I. Pattern of protein and nucleic acid synthesis. J. Exp. Bot. 31: 247–258

Stuart DA, Nelsen J & Nichol JW (1988) Expression of 7S and 11S alfalfa seed storage proteins in somatic embryos. J. Plant Physiol.132: 134–139

Sung ZR & Okimoto R (1981) Embryonic proteins in somatic embryos of carrot. Proc. Nat. Acad. Sci. USA 78: 3683–3687

Plant Cell, Tissue and Organ Culture **38**: 123–134, 1994.

Dedicated to Prof. F.-C. Czygan on the Occasion of his 60th Birthday

Semicontinuous cultivation of photoautotrophic cell suspension cultures in a 20 l airlift-reactor

Uwe Fischer[1], Uwe J. Santore[1], Wolfgang Hüsemann[2], Wolfgang Barz[2] & A. Wilhelm Alfermann[1,*]

[1]*Institut für Entwicklungs- und Molekularbiologie der Pflanzen, Heinrich-Heine-Universität Düsseldorf, Universitätsstraße 1, 40225 Düsseldorf 1, Germany;* [2]*Institut für Biochemie und Biotechnologie der Pflanzen, Westfälische-Wilhelms-Universität Münster, Hindenburgplatz 55, 48143 Münster 1, Germany (* request for offprints)*

Key words: Airlift-reactor, cell culture, *Chenopodium rubrum*, growth characteristics, photoautotroph, semicontinuous cultivation

Abstract

An airlift-bioreactor system was established for semicontinuous growth of photosynthetically active plant cell suspension cultures in a controlled environment. The bioreactor unit was constructed as a conventional, internal draught tube airlift-reactor, which is characterized by a $H D^{-1}$ ratio of 2.9, a ratio of the cross-sectional area of the riser to the cross-sectional area of the downcomer of 0.25 and a surface area of 0.435 m^2 for illumination. Cultivation experiments could be scaled up to working volumes of maximal 20 l. Sixteen fluorescent tubes were fixed around the outer glass cylinder to provide cells continuously with light. An external cooling device was used to keep the temperature constantly at 27 °C. Agitation as well as supply with CO_2 was performed by injecting air enriched with CO_2 through a ring-shaped sparger at the bottom of the vessel. A first set of experiments was carried out with a photoautotrophic culture of *Chenopodium rubrum* L. Cell material adapted to large scale culture conditions was used to inoculate a modified MS medium (Murashige & Skoog 1962) without any organic constituents. Under these conditions a biomass increase of 1870% was achieved in 18 days. Several physiological parameters (e.g. pigmentation, photosynthetic O_2 evolution, carbohydrate content) were measured routinely to elucidate the growth characteristics of large-scale grown *Chenopodium* cells. Electron microscopic photographs from different phases of culture growth clearly demonstrate the pattern of cellular development. Special emphasis was placed upon the differentiation of chloroplast ultrastructure. The presented data confirm the feasibility of large-scale culture techniques with photosynthetic active plant cell cultures.

Abbreviations: D – diameter, DW – dry weight, FW – fresh weight, H – height, K_La – volumetric oxygen transfer coefficient (h^{-1}), MES – 2-(N-morpholino)-ethanesulfonic acid, μ – specific growth rate (d^{-1}), PAR – photosynthetically active radiation (400–700 nm), Pepcase – phosphoenolpyruvate carboxylase, Rubisco – ribulose-1,5-biphosphate carboxylase/oxygenase, t_d – doubling time (d), vvm – (aeration volume) (medium volume)$^{-1}$ min^{-1}

Introduction

Photoautotrophic cultures represent a special type of green pigmented plant cell cultures. They are characterized by their ability to grow in a mineral salt solution in the presence of a CO_2 enriched atmosphere and a sufficient supply of light. This contrasts with heterotrophic cell culture systems which grow by dissimilation and respiration of carbohydrates. Recently, photoautotrophic growth in ambient air was demonstrated in a number of cell cultures (Bender et al. 1980; Blair et al. 1988) but most cultures are dependent on 1–5% CO_2 (vv^{-1}) in the atmosphere for biomass accumulation (Hüsemann 1985). In the last 25 years, photoautotrophic cultures have been established from only a limited number of plant species. Because it is not possible to cultivate isolated mesophyll cells of higher plants over long periods of time, photoautotrophic cell cultures can serve as suitable tools for investigations of those physiological and molecular events which require intact chloroplasts. This is especially true for investigations of photosynthetic CO_2-assimilation (Roeske et al. 1989; Hüsemann et al. 1984), electron transport mechanisms (Xu et al. 1989) or photorespiration (McHale et al. 1987, 1989; Carrier et al. 1989; Avelange et al. 1991). Several groups took advantage of the physiological similarities between mesophyll cells and autotrophic cell cultures for studies of herbicide effects on photosynthesis (Ashton & Ziegler 1987; Sato et. al. 1987, 1991; Thiemann et al. 1989). Some enzymes and metabolic pathways are located exclusively in plastids. Therefore, the unique metabolism of photoautotrophic cells can offer a new potential for investigations of secondary metabolites. Wink & Hartmann (1980) discovered that the synthesis of lupanine, a quinolizidine alkaloid, is located in chloroplasts of intact plants and photomixotrophic cultures of *Lupinus polyphyllus*. Igbavboa et al. (1985) described that the production of lipoquinones is restricted to photoautotrophic cultures of *Morinda lucida*, whereas heterotrophic cultures synthesize root-like anthraquinones. Ikemeyer & Barz (1989) found trigonelline (N-methyl-nicotinic acid) to accumulate predominantly in photoautotrophic cell cultures of *Nicotiana tabacum*, whereas heterotrophic cultures produced no trigonelline. Due to the restricted number of photoautotrophic systems, only few data dealing with the production of secondary metabolites are available to date.

Because of the close relationship between the production of useful metabolites and the elaboration of large scale culture units, we investigated bioreactor systems for photoautotrophic cells. While heterotrophic cell cultures were already cultivated in volumes up to 75,000 l (Rittershaus et al. 1989), the propagation of photoautotrophic suspension cultures is limited usually to small lab scale reactors with volumes not exceeding 2.0 l (Dalton 1980; Hüsemann 1982, 1983; Peel 1982; Hardy et al. 1987). Bender et al. (1980) and Yamada et al. (1981) used mechanically agitated bioreactors for the cultivation of photoautotrophic cell suspensions on the 5 l scale. As far as we know, no experiments exceeding the capacity of these bioreactor systems were performed with this type of cells. Therefore, a 20 l system for the growth of photoautotrophic suspension cultures has been established. The bioreactor has been running in a long-term semicontinuous mode, because the use of shake-flask grown cell material for inoculation could have provoked at least transient negative scale-up effects. Diluting a bioreactor culture with fresh medium in regular time intervals resulted in a 20 l standard culture system which delivered sufficient cell material adapted to this volume. This method enabled semicontinuous growth of a photoautotrophic suspension culture of *Chenopodium rubrum* L. in the 20 l airlift reactor for 22 months (37 growth cycles).

Material and methods

Design of the airlift bioreactor

The experiments were performed in a simply designed, low cost, concentric draught tube airlift fermenter with a maximum working volume of 20 l. The construction which is based on the model designed by Wahl (1977), is depicted in Fig. 1. Top and bottom of the cylindrical glass vessel (4) are covered by metal plates (1, 12), which are tightly sealed (13, 17) to maintain aseptic conditions. Several functional parts of the bioreactor are inserted through the lid. Air is provided through the air inlet tube (8) and liberated at the ring-shaped sparger section (3) through 15 openings of 0.8 mm diameter. A draught-tube (6), fixed between the cooling tubes (5, 7), is located close to the sparger. The draught-tube is used to minimize turbulence in liquid circulation by separating a rising region in the centre from a downcomer region on the periphery of the suspension. To prevent cell sedimentation in areas of poor mixing, a teflon-ring of concave shape (2) is located at the bottom. A sampling tube (15) enables removal of cell material under aseptic conditions. Probes (Ingold, Urdorf, Switzerland) for measuring the content of dissolved O_2 as well as CO_2 (10, 16) are located in the downcomer section. Additional ports (14) in the lid offer the possibility to supply the culture with fresh medium or to connect a second bioreactor, which allows transfer of cell material for inoculation. A water-filled pipe (9), closed at its lower end, is fitted with a thermometer for controlling the temperature of the cell suspension. Any excess of air is removed through the air outlet (11). The airlift bioreactor is characterized by a height to diameter ratio (H D^{-1}) of 2.9 and a ratio of the cross-sectional area of the riser to the cross-sectional area of the downcomer

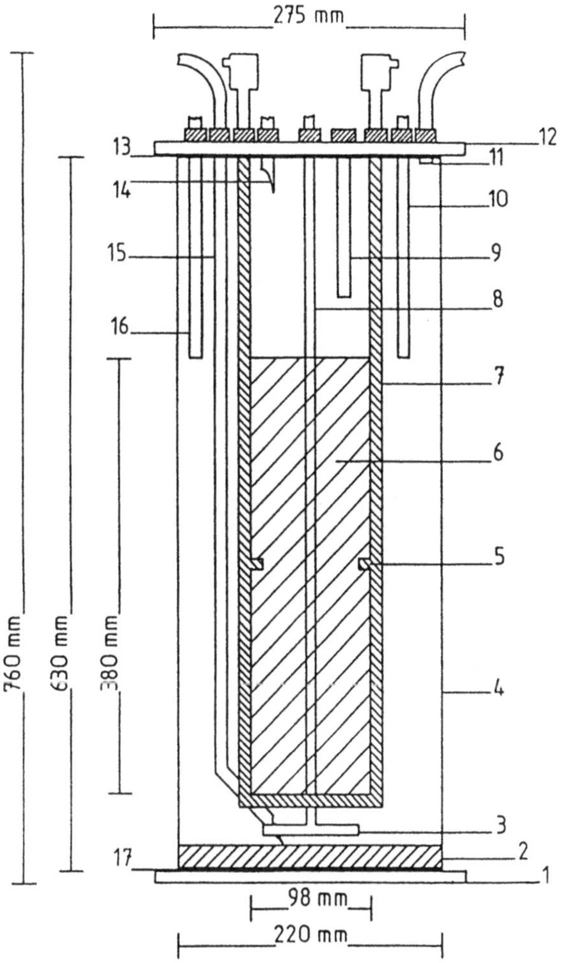

Fig. 1. Design of the airlift bioreactor. (1) metal plate, (2) teflon-ring, (3) sparger, (4) glass vessel, (5) holding device for the draught-tube, (6) draught-tube, (7) cooling tube, (8) air inlet tube, (9) water-filled pipe, (10) probe for dissolved O_2 content, (11) air outlet, (12) lid, (13) sealing, (14) additional port, e.g. for inoculation, (15) sampling tube, (16) probe for dissolved CO_2 content, (17) sealing.

of 0.25. The bioreactor has a surface area of 0.435 m^2 for illumination.

Construction of the bioreactor system

Figure 2 provides a general survey of the bioreactor system. For agitation and gassing a mixture of compressed air and CO_2 is applied to the suspension. Before use, care is taken to remove possible oil contamination from the air (1) with an oil separator and to adjust gas pressure to 100 kPa with pressure reducers (2). Gas flow rates are controlled with flow meters (3, 17) and the gas stream is continuously sterilized with a sterile filter device (5). To measure the CO_2 content of the fresh and the consumed air mixture, an infrared gas analyzer (7, Maihak Unor 6N, Hamburg, Germany) is connected (4) with the air inlet (10) and outlet (9), respectively. Disturbing water vapour is removed from the samples prior to analysis with a cooling trap (6). Excess of gas can leave the bioreactor through the air outlet and is bubbled through a solution of $CuSO_4$ (14). This procedure should minimize the risk of contaminations. Probes to measure the dissolved content of O_2 and CO_2 (11, 12) as well as suitable amplifiers (18, Infors, Basel, Switzerland) are included. Light is supplied by up to 16 fluorescent tubes (8, Philips TLD 18W, light colour 83 + 84, Eindhoven, Netherlands), which are fixed around the outer glass cylinder. To avoid an uncontrolled heating of the suspension, water is circulated (13) through the tubes from a cooled water bath (15), which itself is temperature controlled by an additional cooling circuit (16). Different storage tanks (19, 20), for example, to provide fresh medium to the suspension can be connected to the bioreactor. Table 1 summarizes the main operational features of the bioreactor system.

Analytical procedures

Biomass accumulation was recorded as increase in fresh and dry mass. Fresh weights were obtained from cells collected by vacuum filtration and dry weights of samples were measured after freeze-drying (Breda Scientific Freeze Dryer LY-3-TT, Breda, Netherlands). The specific growth rate μ as well as t_d referring to the DW accumulation were calculated according to Einsele et al. (1985). The specific conductivity of the suspension culture was measured with a conductometer (WTW, Weinheim, Germany) and probes (Ingold, Urdorf, Switzerland) were used to determine the concentration of dissolved O_2 and CO_2, respectively. Chlorophyll was estimated as described by Ziegler & Egle (1965) and total carotenoids according to Röbbelen (1957). Net-photosynthetic O_2 evolution as well as dark respiration rates were measured with a clark-type electrode (Bachofer DW-1, Reutlingen, Germany). Freshly collected cell material was resuspended in 2 ml buffer (50 mM MES, pH 5.5) and transferred to the reaction chamber. To induce O_2 evolution, a $NaHCO_3$-solution was injected to give a final concentration of 0.1 mM. Photosynthetic O_2 evolution was measured under strong illumination (420 $\mu E\ m^{-2}\ s^{-1}$) at 25 °C. Respiration activities were measured as O_2 consumption under dark conditions. Carbohydrate extraction and analysis were performed

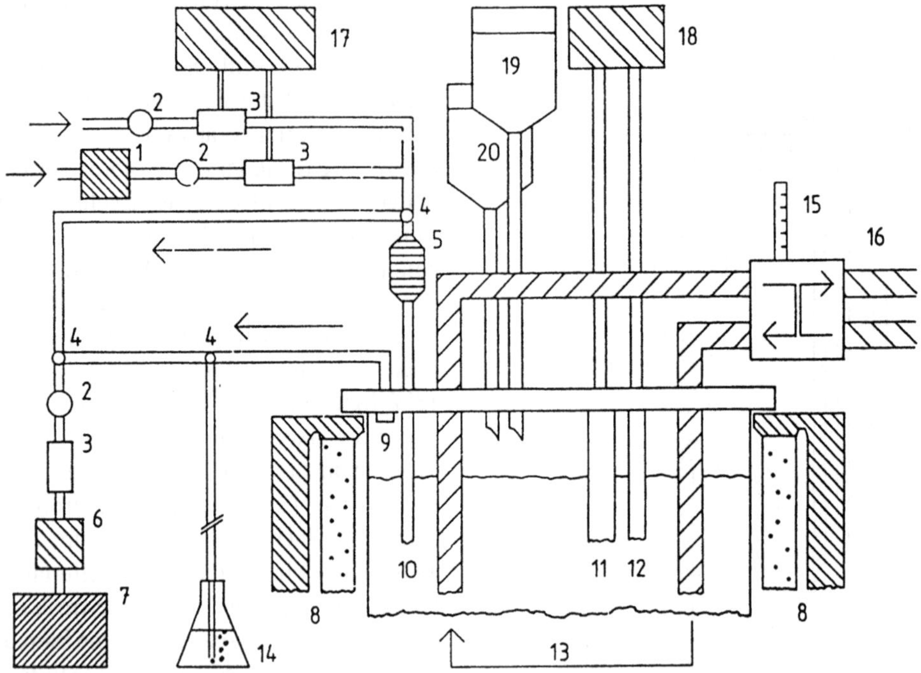

Fig. 2. Construction of the bioreactor system. (1) oil separator, (2) pressure reducers, (3) flow meter, (4) branching point, (5) sterile filter device, (6) cooling trap, (7) infrared gas analyzer, (8) fluorescent tubes, (9) air outlet, (10) air inlet, (11) CO_2-probe, (12) O_2-probe, (13) cooling circuit, (14) solution of $CuSO_4$, (15) water bath, (16) additional cooling circuit, (17) gas flow control, (18) amplifiers for O_2 and CO_2 probes, (19, 20) storage tanks.

Table 1. Standard parameters for semicontinuous growth of photoautotrophic cell suspension cultures of *Chenopodium rubrum* L.

Bioreactor	airlift bioreactor with internal loop, working volume 20.0 l
Culture medium	MS without organic constituents
Light	16 fluorescent tubes, Philips TLD 18W/83 and 84 continuous light; 420 $\mu E\ m^{-2}\ s^{-1}$; 24,000 lux
Gassing and agitation	180 $l\ h^{-1}$ air + 6 $l\ h^{-1}$ CO_2, gas pressure 100 kPa ring-shaped sparger, $K_La = 14.2\ h^{-1}$, vvm = 0.155
Temperature	27 °C ± 1.5
Inoculum	2.0–3.0 $g\ l^{-1}$ FW
Growth cycle	18 days
Operating mode	semicontinuous, dilution factor 1:10

according to Rebeille (1988) with slight modifications. Cell material (1.0 g FW) was extracted in 10 N NaOH with an Ultraturrax (Janke und Kunkel, Staufen, Germany) for 1 min and incubated for 30 min on ice. This procedure was repeated 2 times. After disruption of cells and plastids in a Potter-Elvehjem homogenizer, extracts were neutralized with 10 N HCl. Hydrolysis of sucrose and starch with invertase and amyloglucosidase, respectively, as well as determination of liberated monosaccharides was performed according to Rebeille (1988). The concentration of inorganic phosphate in the culture medium was determined using the method of Barthlen (1983). To measure the concentrations of calcium, magnesium and iron in the medium, an atomic absorption spectrometer (Perkin Elmer Type 2280, Überlingen, Germany) was used. Prior to analysis, samples were supplemented with 0.1% (vv^{-1}) $La(NO_3)_3$. The following wavelengths were used for quantification: magnesium 202.6 nm, calcium 422.7 nm and iron 372.0 nm. An isocratic HPLC-System with a conductivity detector (Conducto monitor III, Therma Separation Products, Darmstadt, Germany) was used to measure the concentrations of nitrate, ammonium and potassium in the culture medium by means of single column ion chromatography. Cations were separated on a PRP X-200 column (250 × 4.1

mm I.D.,) with 4.0 mM HNO_3 in 30% MeOH, pH 2.4 as solvent. A PRP X-100 column (250 × 4.1 mm I.D., Hamilton, Darmstadt, Germany) and 4.0 mM p-hydroxybenzoic acid in 2.5% MeOH, pH 8.5 as eluent was used for determination of nitrate. In both cases, samples of 100 μl were injected and flow rates were adjusted to 3.0 ml min^{-1}. Crude enzyme extracts were prepared essentially according to Hüsemann (1981). After centrifugation (Sorvall RC-5B, Dupont, USA, SS 34, 48,000 g, 4 °C, 15 min) the clear supernatant was collected and low molecular weight compounds were removed by gelfiltration with Sephadex G 25, medium (PD 10, Pharmacia-LKB, Freiburg, Germany). All enzyme activities were determined spectrophotometrically (Kontron Uvikon 930, Echingen, Germany). Enzyme assays were performed according to Hüsemann (1981): ribulose-1,5-biphosphate carboxylase, E.C. 4.1.1.39; phosphoenolpyruvate carboxylase, E.C. 4.1.1.3; Holtum & Winter (1982): glucose-6-phosphate dehydrogenase, E.C. 1.1.1.49 and Blackwell et al. (1990): hydroxypyruvate reductase, E.C. 1.1.1.29; catalase, E.C. 1.11.16. Protein concentrations were determined according to Bradford (1976) with bovine serum albumine as reference protein. For electron microscopy cell material from different stages of the growth cycle was transferred into the fixation medium of 2.5% glutaraldehyde in 200 mM sodium cacodylate buffer at pH 7.0. After washing in buffer and postfixation in 1% osmium tetroxide in 200 mM cacodylate buffer for 1 h, samples were dehydrated in a graded ethanol series and embedded in Agar 100 epoxy resin with the aid of epoxy propane as a linking agent. The resin mixture was equivalent to the 3A:1B formula of Luft (1961) with catalyst present at all stages of embedding. Sections were double-stained with 2% aqueous uranyl acetate followed by Reynold's lead citrate (1963). The preparations were examined in a Philips 301 electron microscope.

K_La was determined with a dynamic gassing out technique as described by Spieler (1985).

Results

Photoautotrophic suspension cultures of *Chenopodium rubrum* L. were able to grow in a 20 l airlift bioreactor. Cell material from the linear growth phase was used for inoculation, either by transferring suspension of the stock culture into fresh medium or by diluting cells with medium. The data describing the growth of *Chenopodium* cells represent the mean values of at least 3 bioreactor runs which were performed under semicontinuous culture conditions in the 20 l scale. Growth kinetics are shown in Fig. 3. An initial cell density of about 2.5 g l^{-1} FW (0.2 g l^{-1} DW) was found to be beneficial for growth induction. No initial lag-phase could be observed and after a growth period of 18 days the dry matter increase was 2.95 gl^{-1} or 1870%, respectively. The specific growth rate μ was calculated to be 0.223 d^{-1}, corresponding to t_d of 3.11 days. The concentration of total soluble protein showed only little variation during a culture cycle.

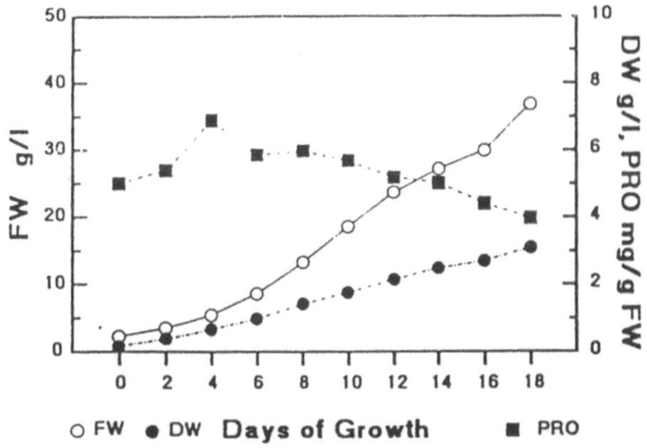

Fig. 3. Growth characteristics of photoautotrophic *Chenopodium rubrum* cell suspension cultures which were grown in a 20 l airlift bioreactor. Biomass accumulation is presented as FW and DW accumulation. Additionally, the protein content (PRO) of cells is shown.

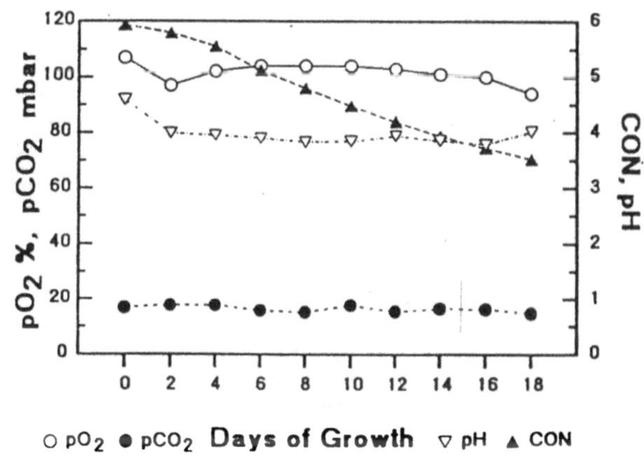

Fig. 4. Changes in the amounts of dissolved O_2 (dO_2) and CO_2 (dCO_2) in the suspension during semicontinuous growth of photoautotrophic *Chenopodium rubrum*. Agititation as well as gassing is performed with CO_2 enriched air. Modifications of medium pH and specific conductivity (CON) are also shown.

The amounts of dissolved O_2 and CO_2, the specific conductivity and the pH-value of the suspension are shown in Fig. 4. The dissolved gaseous compounds

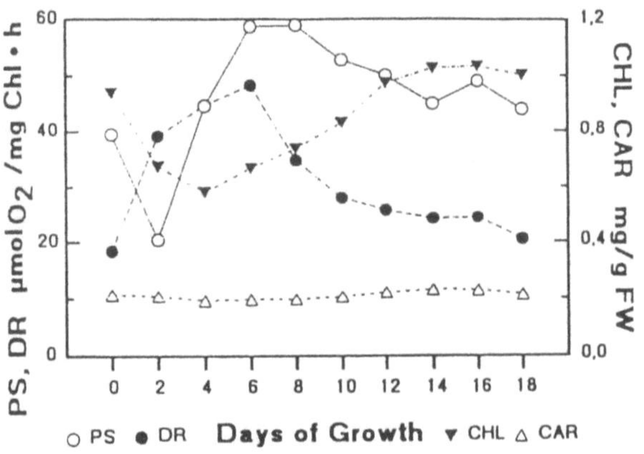

Fig. 5. Changes in chlorophyll content (CHL), carotenoid content (CAR), rates of net-photosynthetic O2 evolution (PS) and dark respiration (DR) in autotrophic bioreactor cultures of *Chenopodium rubrum* during a 18 days growth period.

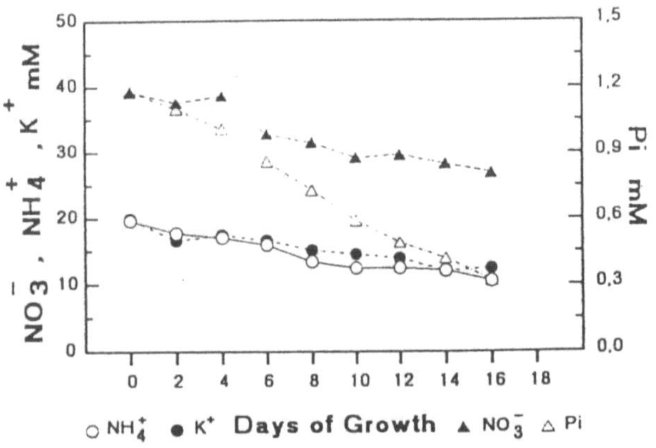

Fig. 7. Change in medium concentration of nitrate (NO_3^-), ammonium (NH_4^+), potassium (K^+) and inorganic phosphate (P_i) during a growth cycle of 16 days of photoautotrophic cultures of *Chenopodium rubrum*. The cell cultures were grown in MS medium without any organic constituents.

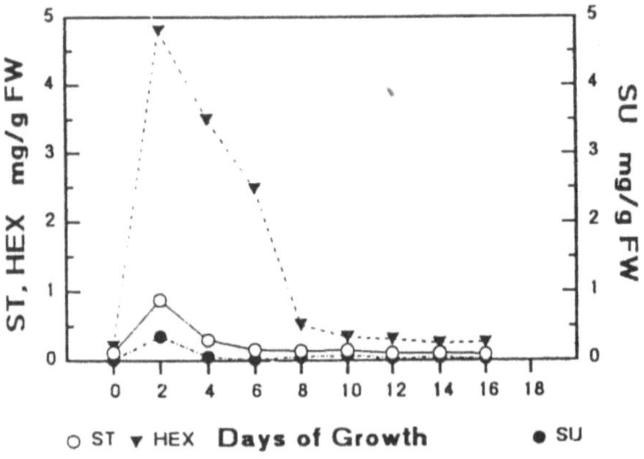

Fig. 6. Starch (ST), sucrose (SU) and monosaccharide (HEX) content of large scale cultures of *Chenopodium rubrum* at different stages of growth. The values for monosaccharides are the sum of glucose and fructose content, respectively.

during the growth cycle were constant, with values for dO_2 near saturation (100%) and 17 mbar for dCO_2. The pH of the suspension decreased from an initial value of 4.5 to 3.8 during the first days and then remained stable for the rest of the experiment. The reduction of the specific conductivity of the culture from 5.9 to 3.5 mS cm^{-1} indicated an incomplete overall uptake of ionic compounds.

The characteristic pattern of pigmentation, photosynthetic O_2 evolution as well as dark respiration is presented in Fig. 5. The concentration of total carotenoids seemed to remain unchanged during the growth cycle, whereas the chlorophyll content (Chl a + b) started to decline immediately after inoculation. After day 4, when only 60% of the initial concentration of 1 mg g^{-1} FW was detectable, an increased chlorophyll formation led to a restoration of the initial value in the following days. The ratio of Chl a and Chl b was fixed at values between 3.1–3.2 throughout the growth cycle. The time course of photosynthetic O_2 evolution was also characterised by a transient decrease in connection with the start of the growth cycle. The initial activity was recovered already after 4 days and maximum activities (60 µmol O_2 mg^{-1} Chl h^{-1}) occurred between days 6 and 8. In contrast to these findings, dark respiration was pronounced only in early growth phases and after day 6 a declining activity could be observed. The cellular content of carbohydrates (starch, sucrose and monosaccharides) is presented in Fig. 6. Starch accumulated to significant amounts only in the first few days of the culture interval. The measured values were below 1% on a dry weight basis and they were decreasing very soon to 0.1%. A similar result was obtained for sucrose, which could be detected in even minor concentrations. The cellular content of monosaccharides (glucose + fructose) increased significantly during the first 2 days and high cellular amounts were found until day 6. After that time the concentrations decreased to the level of starch and sucrose.

The characteristic uptake-pattern of different constituents of the Murashige and Skoog medium are presented in Fig. 7. The consumption of ammonium, potassium and inorganic phosphate started immediately after inoculation. Inorganic phosphate was taken up most rapidly, but at the end of the culture time only 77% of the initial concentration was absorbed. Incom-

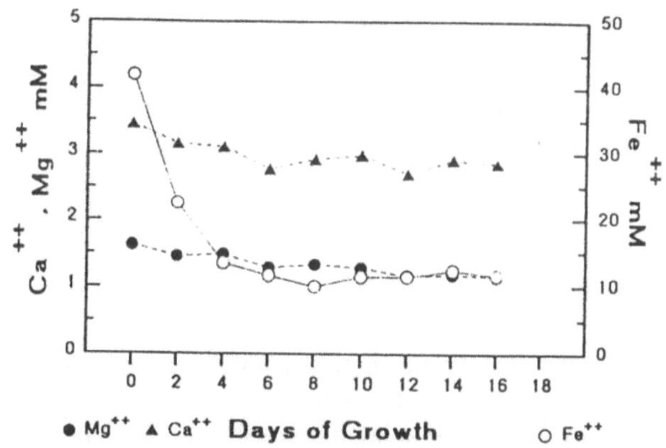

Fig. 8. Change in medium concentration of calcium (Ca), magnesium (Mg) and iron (Fe) during a growth cycle of *Chenopodium rubrum* under photoautotrophic conditions.

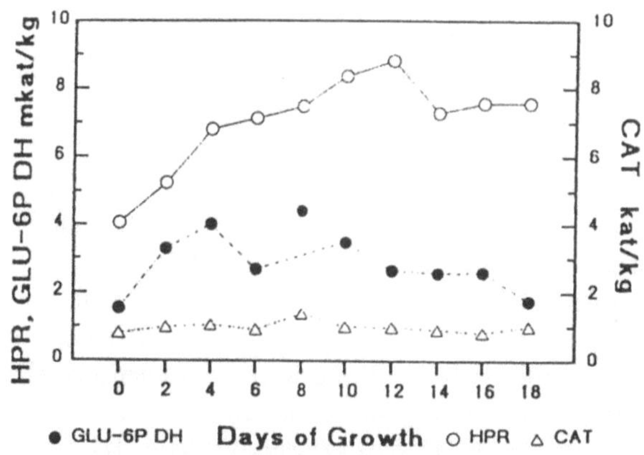

Fig. 10. Change in the in vitro activities of glucose-6-phosphate dehydrogenase, E.C. 4.1.1.31, (GLU-6P DH), catalase, E.C. 1.11.16 (CAT), and hydroxypyruvate reductase, E.C. 1.1.1.29 (HPR) during semicontinuous growth of photoautotrophic *Chenopodium rubrum* cells under large scale conditions.

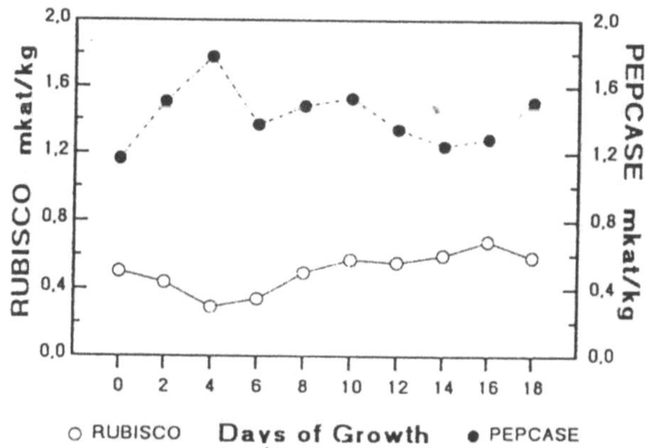

Fig. 9. Change in the in vitro activities of ribulose-1,5-biphosphate carboxylase/oxygenase E.C. 4.1.1.39 (RUBISCO) and phosphoenolpyruvate carboxylase, E.C. 4.1.1.31 (PEPCASE) during a 18 days growth period of *Chenopodium rubrum*.

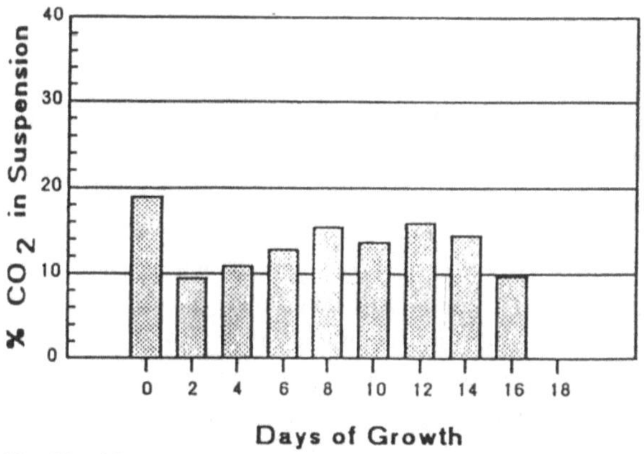

Fig. 11. Measurements of CO_2 content of the fresh and the consumed gas mixture. The results are expressed as the percentage of CO_2 that remained in the suspension.

plete consumption was also determined for ammonium (49.7%), potassium (39.7%) and nitrate (31.5 %).

Figure 8 provides additional information about nutrient uptake. Calcium (17%) and magnesium (28%) were used only in minor amounts. In contrast, only 40% of the initial iron concentration of 0.1 mM were found even at the beginning of the growth cycle. The difference to the original iron content was most likely caused by precipitation of iron-phosphate during sterilization (Dalton et al. 1983). The remaining iron was reduced to 0.01 mM within 4 days and this level remained constant for the rest of the culture interval.

Figure 9 presents data on the in vitro activities of enzymes of primary metabolism expressed by the large scale grown *Chenopodium* cells. The time course of Rubisco and Pepcase confirm the findings of several authors, that fast growing autotrophic cells are characterized by high Pepcase and low Rubisco activities. After inoculation, the specific activity of Rubisco decreased, whereas Pepcase activity increased. In the following days, a restoration of the initial Rubisco activity was achieved, whereas the Pepcase activity remained constant after a decline between day 4 and 6. A high ratio of these activities was found in all phases of growth.

The in vitro activities of two further enzymes which are important for primary metabolism are shown in Fig. 10. An increase in glucose-6P dehydrogenase activity could be observed in the first half of the growth cycle, but the original level was restored in the following days. In contrast to these findings, the increase in the in vitro activity of hydroxypyruvate reductase, an enzyme of photorespiration, was extended for 12 days and a high level of enzyme activity was conserved during the following days. Obviously, a high photorespiratory activity was developed as a response to a high O_2 pressure caused by the mode of aeration. This is also stressed by the high catalase activity, which could be detected throughout the growth cycle.

An infrared gas analyzer was connected to the gas inlet and outlet in order to evaluate the difference in CO_2 content between the fresh and consumed gas. The percentage of CO_2 which remained in the suspension, is shown in Fig. 11. It is evident, that only 10–15% of the initial CO_2 supply provided a potential source for autotrophic carbon assimilation. These values showed only small fluctuations, with a time course reflecting the kinetics of photosynthetic O_2 evolution.

The ultrastructure of photoautotrophically cultured *Chenopodium rubrum* cells from different phases of semicontinuous growth was examined with the electron microscope. Figure 12a gives a survey of a small group of cells of spherical shape 3 days after inoculation. The shape of these cells is very characteristic for bioreactor cultivated suspensions, because cylindrical-formed cells are more sensitive to the mechanical stress produced by the mode of agitation. Cells from this phase of growth (Fig. 12b) were characterized by large chloroplasts, containing numerous starch grains and well developed stroma but only few grana lamellae. A typical plastid of this developmental stage is shown in Fig. 12c. Mitochondria were small in size in comparison with plastids, but their large number (Fig. 12d) well agrees with the increased dark respiration activity at the beginning of the growth cycle. The occurrence of numerous dictyosomes confirmed the metabolic active state of these cells (Fig. 12e).

Figure 13 provides detailed information concerning the developmental pattern of plastidial ultrastructure throughout a growth cycle. After 7 days of cultivation, the formation of grana stacks was induced in different areas of the plastids, but membrane stacking was still limited to 4–6 layers (Fig. 13a). Almost no starch grains were visible at this state of growth. The number of grana lamellae as well as the height of membrane stacking (Fig. 13b) was increased after 5 additional days of growth. In spite of starch grains, which were characteristic for cells of the beginning of the growth cycle, an increasing number of plastoglobuli occurred within the plastids. At the end of the culture interval (Fig. 13c), the plastids contained grana stacks with maximal 20 membrane layers and numerous plastoglobuli. Differentiation processes were not restricted to chloroplasts. Cell walls were changed in structure and size during the growth cycle of 18 days. They became thicker and they seemed to have a more lipophilic appearance (Fig. 13d) in comparison with cells from early growth phases.

Discussion

The data presented in this communication clearly demonstrate that photoautotrophic cell suspensions are able to grow under semicontinuous culture conditions in a 20 l airlift bioreactor system.

Airlift reactors are generally thought to be suitable for the cultivation of plant cell cultures. Agitation as well as gassing is performed by a stream of air, which is liberated at the bottom of the culture vessel. The introduction of a concentric draught tube enables the differentiation between a rising section in the central and a downcomer region in the peripheral parts of the bioreactor which reduces turbulence in liquid circulation. A sufficient supply with light and CO_2 is essential for growth of photoautotrophic *Chenopodium* cells. A mixture of air/CO_2 was found to be suitable for gassing, although cells were forced to grow under conditions of a high O_2 partial pressure and therefore negative effects on photosynthesis cannot be excluded.

Because of the large cross-section area of the glass vessel and the continuously increasing cell density it is difficult to ensure a sufficient supply of light during a growth cycle. Peel (1982) established a turbidostat system for photoautotrophic cells of *Asparagus officinalis* and found higher specific growth rates for low density cultures (1.8 g l^{-1} DW, $\mu = 0.36\ d^{-1}$) in comparison with cultures of high density (4.8 g l^{-1} DW, $\mu = 0.12\ d^{-1}$), which, in fact, was caused by an improved utilization of light. For that reason, it was decided to run the bioreactor under conditions of strong illumination (PAR 420 $\mu E\ m^{-2}\ s^{-1}$) and reduced cell density. The conditions for CO_2 and light supply enabled a biomass increase of 2.95 g l^{-1} DW in 18 days, which corresponded to a relative increase of 1870%.

The ratio of the in vitro activities of the carboxylating enzymes Rubisco and Pepcase was characteristic for fast growing photoautotrophic cell suspension cultures with high Pepcase and low Rubisco activities. Although this situation has been found for large scale grown cell material, the data are in contrast to Hüsemann (1985) because this relationship was conserved throughout the growth cycle and not inversed

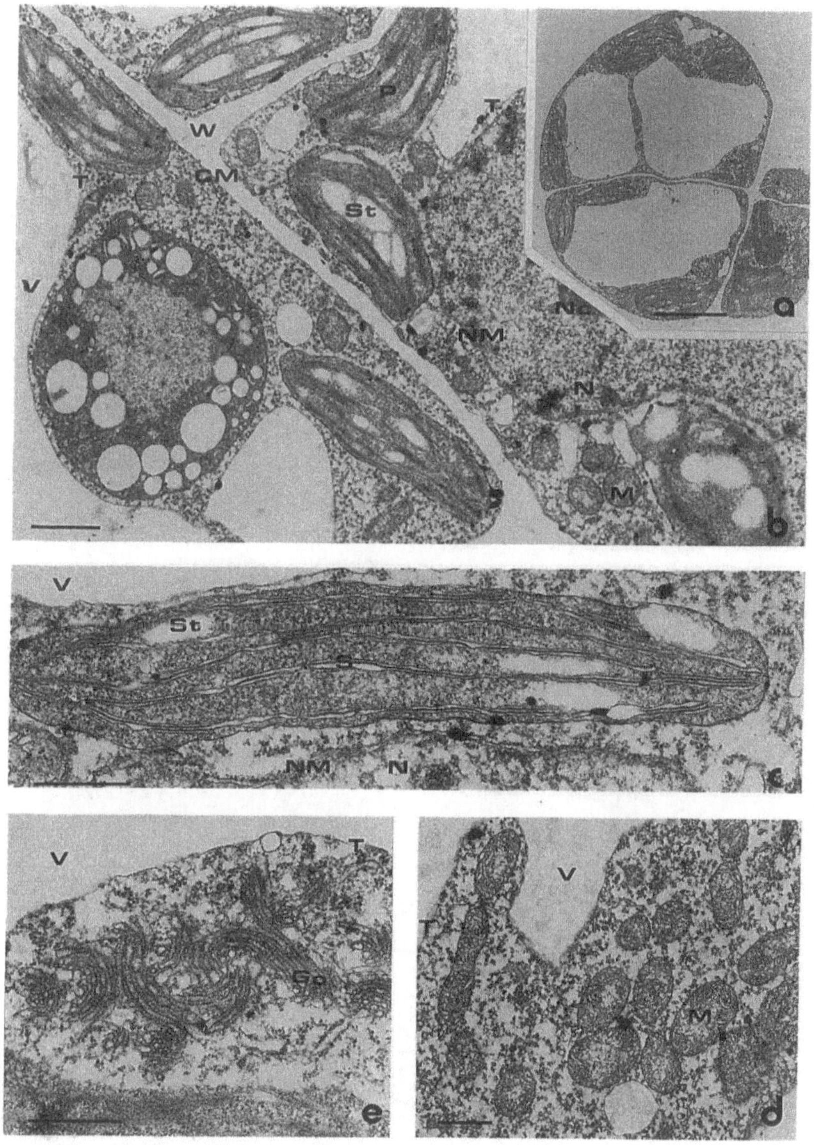

Fig. 12. Ultrastructure of photoautotrophic *Chenopodium rubrum* cells after 3 days of culture, (Fig. 12a) survey of a small group of cells, (Fig. 12b) cell region with several starch containing plastids, (Fig. 12c) single plastid with stroma thylacoids, (Fig. 12d) numerous dictyosomes, (Fig. 12e) mitochondria rich region. Abbreviations: cytoplasma membrane (CM), grana stacks (G), golgi apparatus (Go), plastoglobuli (L), mitochondrium (M), nucleus (N), nucleolus (Nc), nuclear membrane (NM), chloroplast (P), starch grains (St), tonoplast (T), vacuole (V), cell wall (W). Bars are representing 0.5 μm, except Fig. 12a and Fig. 13d with a scale of 5.0 μm.

in the late growth phase. This was probably caused by the high O_2 concentration in the suspension, which resulted in a pronounced photorespiratory activity. The time courses of hydroxypyruvate reductase and catalase activities are in agreement with this hypothesis. The time course of glucose-6-P dehydrogenase showed only a small increase in the first half of the growth cycle, thus indicating that the semicontinuous mode of cultivation arrested photoautotrophic cell material in a state of high physiological activity.

Supply with CO_2 was performed by injecting CO_2 enriched air to the suspension by means of a ring-shaped sparger, but measurements of the CO_2 concentrations in fresh and consumed gas mixtures revealed that only 10–15% of the delivered CO_2 provide the source for autotrophic carbon assimilation. Most prob-

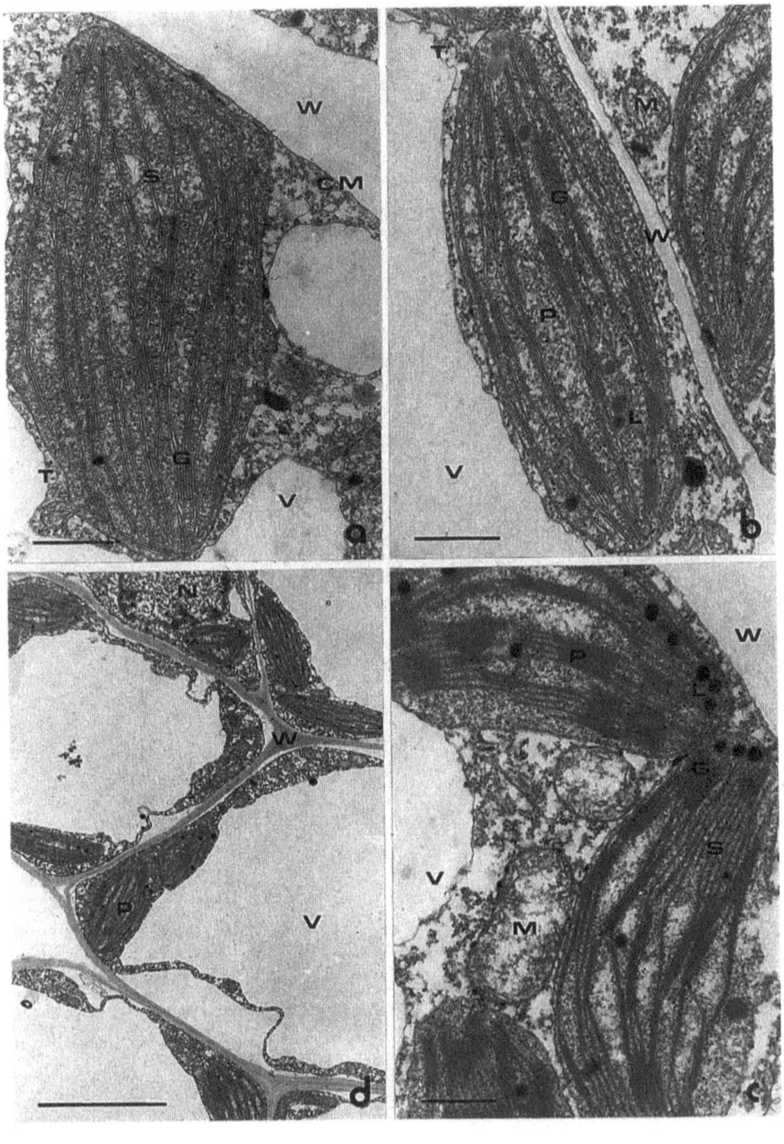

Fig. 13. Developmental pattern of plastidial ultrastructure of photoautotrophic *Chenopodium rubrum* cells. Cells after 7 (Fig. 13a), 12 (Fig. 13b) and 18 days (Fig. 13c) of growth are shown. Fig. 13d represents a group of cells at the end of the culture interval.

ably, a more evolved system for gas supply (e.g. porous metal plates or ceramics) or the use of gas mixtures with significantly enhanced CO_2 concentrations would improve gas exchange and biomass accumulation, respectively. The assimilation of CO_2 was probably used solely to build up biomass, because cell material containing starch in significant amounts could only be observed immediately after the inoculation.

Large scale cultivated cells of *Chenopodium rubrum* undergo continuous physiological modifications. This was especially true for chlorophyll formation, photosynthetic O_2 evolution and dark respiration. It is evident, that in early growth phases respiratory activities were more pronounced, whereas photosynthetic activities were developed only after a lag-phase of several days. These physiological data were confirmed by ultrastructural observations, which documented the conversion of high light plastids (mainly stroma lamellae, large starch grains) to low light plastids (large grana stacks, no starch). Differentiation processes were not restricted to the structure of plastids. Cell walls became thicker and developed a more lipophilic appearance in comparison with cells from early growth phases. During the growth cycle, an increasing number of lipophilic bodies occurred within plastids. At the beginning of the culture interval single cells, small cell groups and few aggregates of 20–50 cells were commonly observed in the suspension, whereas in later growth phases the number as well as the size of cell aggregates were significantly increased.

Very often growth of cell cultures is restricted because one major nutrient, e.g. inorganic phosphate, was depleted during growth, while other constituents of the medium were still present in sufficient amounts. Our data on nutrient uptake clearly indicated that this was most probably not the case with semicontinuously grown *Chenopodium rubrum* cells. Incomplete uptake was characteristic for several major constituents of the MS-medium. Uptake rates below 50% for almost all compounds investigated (except inorganic phosphate) indicated that a reduction in the overall concentration of nutrients may be beneficial for further improvement of growth.

The data presented in this communication demonstrate for the first time, that growth of photoautotrophic cell suspension cultures of *Chenopodium rubrum* L. in a bioreactor system with a working volume of 20 l is possible and that high growth rates can be achieved under these conditions. The results are promising and should be followed by further investigations on large scale cultivation of other photosynthetically active plant cell cultures.

Acknowledgements

Financial support by the Federal Minister for Research and Technology, Bonn, (Projects Photosynthetically active plant cell cultures, no. 0318983A and 0318984B) and the Fonds of Chemical Industry, Frankfurt, is gratefully acknowledged.

References

Ashton AR & Ziegler P (1987) Lack of effect of the photosystem II-based herbicides diurone and atrazine on growth of photoheterotrophic *Chenopodium rubrum* cells at concentrations inhibiting photoautotrophic growth of these cells. Plant Sci. 51: 269–276

Avelange MH, Thiery JM, Sarry F, Gans P & Rebeille F (1991) Mass-spectrometric determination of O_2 and CO_2 gas exchange in illuminated higher-plant cells. Planta 183: 150–157

Barthlen U (1983) 12β-Hydroxylierung von β-Methyldigitoxin in Fermenter- und Suspensionskulturen von *Digitalis lanata*. Doctoral Thesis, University of Tübingen

Bender L, Kumar A & Neumann KH (1980) Photoautotrophe pflanzliche Gewebekulturen in Laborfermentern. In: Lafferty RM (Ed) Fermentation II Rotenburger Symposium (pp 193–203). Springer Verlag, Wien, New York

Blackwell RD, Murray AJS & Lea PJ (1990) Enzymes of the photorespiratory carbon pathway. In: Lea PJ (Ed) Methods in Plant Biochemistry, Vol 3 (pp 130–144). Academic Press, London

Blair LC, Chartain CJ & Widholm JM (1988) Initiation and characterization of a cotton (*Gossypium hirsutum* L.) photoautotrophic cell suspension culture. Plant Cell Rep. 7: 266–269

Bradford MM (1976) A rapid and sensitive method for the quantitation of microgram quantities of protein utilizing the principle of protein-dye binding. Anal. Biochem. 72: 248–254

Carrier P, Chagvardieff P & Tapie P (1989) Comparison of the oxygen exchange between photosynthetic cell suspensions and detached leaves of *Euphorbia characias* L. Plant Physiol. 91: 1075–1079

Dalton CC, Iqbal K & Turner DA (1983) Iron phosphate precipitation in Murashige and Skoog media. Physiol. Plant. 57: 472–476

Dalton CC (1980) Photoautotrophy of spinach cells in continuous culture: photosynthetic development and sustained photoautotrophic growth. J. Exp. Bot. 122: 791–804

Einsele A, Finn RK & Samhaber W (1985) Mikrobiologische und biochemische Verfahrenstechnik, VCH Verlagsgesellschaft, Weinheim

Hardy T, Chaumont D, Brunel L & Gudin C (1987) Photoautotrophic suspension cultures I – obtention of photoautotrophic cultures from *Euphorbia characias* L. J. Plant Physiol. 128: 11–19

Holtum JAM & Winter K (1982) Activity of enzymes of carbon metabolism during the induction of crassulacean acid metabolism in *Mesembryanthemum crystallinum* L. Planta 155: 8–16

Hüsemann W (1981) Growth characteristics of hormone and vitamine independent photoautotrophic cell suspension cultures from *Chenopodium rubrum*. Protoplasma 109: 415–431

Hüsemann W (1982) Photoautotrophic growth of cell suspension cultures from *Chenopodium rubrum* in an airlift fermenter. Protoplasma 113: 214–220

Hüsemann W (1983) Continuous culture growth of photoautotrophic cell suspensions from *Chenopodium rubrum*. Plant Cell Rep. 2: 59–62

Hüsemann W (1985) Photoautotrophic growth of cells in culture. In: Vasil IK (Ed) Cell Culture and Somatic Cell Genetics of Plants, Vol 2 (pp 213–252). Academic Press, New York

Hüsemann W, Herzbeck H & Robenek H (1984) Photosynthesis and carbon metabolism in photoautotrophic cell suspensions of *Chenopodium rubrum* from different phases of batch growth. Physiol. Plant. 62: 349–355

Igbavboa U, Sieweke HJ, Leistner E, Röwer I, Hüsemann W & Barz W (1985) Alternative formation of anthraquinones and lipoquinones in heterotrophic and photoautotrophic cell suspension cultures of *Morinda lucida* Benth.. Planta 166: 537–544

Ikemeyer D & Barz W (1989) Comparison of secondary product accumulation in photoautotrophic, photomixotrophic and heterotrophic *Nicotiana tabacum* cell suspension cultures. Plant Cell Rep. 8: 479–482

Luft JH (1961) Improvements in epoxy resin embedding methods. J. Biophys. Biochem. Cytol. 9: 409–414

McHale NA, Zelitch I & Peterson RB (1987) Effects of CO_2 and O_2 on photosynthesis and growth of autotrophic tobacco callus. Plant Physiol. 84: 1055–1058

McHale NA, Havir EA & Zelitch I (1989) Photorespiratory toxicity in autotrophic cell cultures of a mutant of *Nicotiana sylvestris* lacking serine:glyoxylate aminotransferase activity. Planta 179: 67–72

Murashige T & Skoog F (1962) A revised medium for rapid growth and bioassays with tobacco tissue cultures. Physiol. Plant. 15: 473–497

Peel E (1982) Photoautotrophic growth of suspension cultures of *Asparagus officinalis* L. cells in turbidostats. Plant Sci. Lett. 24: 147–155

Rebeille F (1988) Photosynthesis and respiration in air-grown and CO_2-grown photoautotrophic cell suspension cultures of carnation. Plant Science 54:11–21

Reynolds ES (1963) The use of lead citrate at high pH as an electron opaque stain in electron microscopy. J. Cell Bio. 17: 208–212

Rittershaus E, Ulrich J, Weiss A & Westphal K (1989) Großtechnische Fermentation von pflanzlichen Zellkulturen: Betriebserfahrungen bei der Kultivierung von Pflanzenzellen in einer Fermentatioskaskade bis zu 75000 l. BioEng. 2: 28–34

Roeske CA, Widholm JM & Ogren WL (1989) Photosynthetic carbon metabolism in photoautotrophic cell suspension cultures grown at low and high CO_2. Plant Physiol. 91: 1512–1519

Röbbelen G (1957) Bestimmung des Pigmentgehalts. Z. indukt. Abstamm. u. Vererb.-L. 88: 189

Sato F, Yamada Y, Kwak SS, Ichinose K, Kishida M, Takahashi N & Yoshida S (1991) Photoautotrophic cultured plant cells: a novel system to survey new photosynthetic electron transport inhibitors. Z. Naturforsch. 46c, 563–568

Sato F, Takeda S & Yamada Y (1987) A comparison of effects of several herbicides on photoautotrophic, photomixotrophic and heterotrophic cultured tobacco cells and seedlings. Plant Cell Rep. 6: 401–404

Spieler H (1985) Untersuchungen zur Sauerstoff-Versorgung von *Digitalis lanata* Fermenterkulturen. Auswirkungen auf Wachstum und 12β-Hydroxylierung Doctoral Thesis, University of Tübingen

Thiemann J, Nieswandt A & Barz W (1989) A microtest system for the serial assay of phytotoxic compounds using photoautotrophic cell suspension cultures of *Chenopodium rubrum*. Plant Cell Rep. 8: 399–402

Wahl J (1977) Fermentation von Pflanzlichen Zellkulturen und 12-β-Hydroxylierung von β-Methyldigitoxin durch Zellkulturen von *Digitalis lanata*. Doctoral Thesis, University of Tübingen

Wink M & Hartmann T (1980) Production of quinolizidine alkaloids by photomixotrophic cell suspension cultures: biochemical and biogenetic aspects. Planta Med. 40: 149–155

Xu C, Rogers SMD, Goldstein C, Widholm JM & Govindjee (1989) Fluorescence charcteristics of photoautotrophic soybean cells. Photosyn. Res. 21: 93–106

Yamada Y, Imaizumi K, Sato F & Yasuda T (1981) Photoautotrophic and photomixotrophic culture of green tobacco cells in a jar fermenter. Plant Cell Physiol. 22: 917–922

Ziegler P & Egle K (1965) Zur quantitativen Analyse der Chloroplastenpigmente. Beitr. Biol. Pflanzen 41: 11–37

Plant Cell, Tissue and Organ Culture **38:** 135–141, 1994.

Studies on the relationship between ploidy level, morphology, the concentration of some phytohormones and the nicotine concentration of haploid and doubled haploid tobacco (*Nicotiana tabacum* L.) and NICA plants

B. Zeppernick, F. Schäfer, K. Paasch, B. Arnholdt-Schmitt & K.-H. Neumann
Institut für Pflanzenernährung, Justus Liebig Universität, 35390 Giessen, Germany

Key words: Gibberellin activity, *Nicotiana tabacum*, nicotine concentration, NICA plants, ploidy level

Abstract

As compared to doubled haploid plants of the same origin, haploid tobacco plants are characterized by narrow leaves and in these leaves the endogenous concentration of gibberellins was considerably higher than in doubled haploids. This higher GA activity is almost entirely due to elevated levels of polar gibberellins. The same leaf shape as in haploids could be induced by GA_3 sprays to doubled haploids. A similar leaf shape was also observed on tissue culture derived so called NICA plants displaying the morphology of tobacco plants as described by Dudits et al. (1987) from whom the plant material was obtained as a gift. Here, in the leaves of a special strain with narrow lamina again a much higher gibberellin activity was detected than in the leaves of plants of the original tobacco strain. Histochemical determination of the relative DNA content indicated that leaves of NICA were chimaeras containing 1C cells besides cells with higher C values. Obviously, haploidy is somehow related to the endogenous gibberellin activity in tobacco plant material with consequences on the morphological appearance of 1n plants. Comparing some haploid and doubled haploid strains in tissue culture and pot and field experiments in several years apparently the genotype of the plant material is more significant for nicotine concentration than the ploidy level.

Abbreviations: DW – dry weight, FW – fresh weight, LSI – leaf shape index

Introduction

Although a vast literature exists on the induction of haploid plants by anther and microspore culture, rarely investigations were carried out to characterize the morphophysiological background of these plants in detail. Some papers describing the haploid material were either concerned with its performance at the cell culture level or with the production of secondary metabolites in cell cultures or intact plants (e.g. Kibler & Neumann 1979; Mechler & Kohlenbach 1978; Burk & Matzinger 1976; Schiltz et al. 1980). In order to elucidate influences of the ploidy level on the development of intact plants, in the present paper some growth parameters, in particular of the leaves, the gibberellin activity and the nicotine concentration of haploid and doubled haploid plants derived from the same haploid strain were compared. The ploidy level of the plants was routinely checked and it remained stable during the investigation period. The doubled haploids were expected to be homozygous.

For comparison, a strain of NICA plants originally derived from fusion experiments of *Nicotiana* with *Daucus* protoplasts (Dudits et al. 1987) was included into the study. In this material besides gibberellin activity also the concentrations of IAA and ABA were determined.

Table 1. Some morphological features of haploid, doubled haploid (2 x n) and diploid tobacco plants (*Nicotiana tabacum*, var. *Xanthi*) at the flowering stage.

	Stem height (cm)	Number of leaves	Number of floral parts[a]	g FW of stem	g FW of leaves	g FW of floral parts	Total FW of shoot	Leaf shape index (LSI)[b]
n	76	28	146	84	86	31	201	0.47
2n	112	35	52	173	185	8	366	0.63
2 × n	77	28	66	99	134	9	242	0.62

[a] flower buds and flowers
[b] LSI= (cm leaf width)(cm leaf length)$^{-1}$

Fig. 1. The morphological habitus of haploid (left) and double haploid (right) tobacco plants (strain 8/1).

Results and discussion

Morphology and gibberellin activity of haploid and doubled haploid tobacco plants

The method to obtain haploid plants of *Nicotiana tabacum* var. *Xanthi* by anther culture was described by Forche & Neumann (1977). On some of the haploid plants spontaneous diploidization of axillary buds was observed and these were propagated by stem cuttings since. Out of the population of doubled haploid plants three strains indicating special morphological features were selected together with their haploid mother strains for these investigations. A detailed description of these strains and a hybrid derived thereof was given earlier (Zeppernick 1988). The plants were raised in soil in pot or field experiments supplemented with the usual fertilizers.

A comparison of haploids with diploid material from which the anthers for androgenesis were obtained indicate a quantitative reduction of most parameters for haploids as summarized in Table 1. Only the number and the weight of floral parts were higher in haploids, mostly due to the initiation of flowering 8 to 10 days earlier than in diploids. Doubled haploid plants, however, were similar to haploids except for total leaf weight which was higher in doubled haploids. Again the number and weight of floral parts was increased due to earlier flowering in haploids. The leaf shape index (LSI) indicates, that the reduction of total leaf weight of the former was at least partly due to a reduction in the width of the leaf lamina (Fig. 1). Additionally leaf

Table 2. The gibberellin activity in leaves of haploid and doubled haploid tobacco plants (*Nicotiana tabacum*, var. *Xanthi*) at flower emergence (ng g^{-1} FW in GA_3 equivalents, average of 3 strains and 4 replicates each).

	Polar gibberellins	s.e.[a]	Unpolar gibberellins	s.e.[a]	Total
n	1.34	0.2	0.53	0.05	1.87
2 × n	0.45	0.06	0.40	0.03	0.85

[a] s.e.= standard error

Table 3. The influence of GA_3 spraying (60 ppm at 10 ml twice a week from the 6th leaf emergence to flowering) on leaf growth of doubled haploids (4 replicates, 65 days after the first application, 15th leaf, *Nicotiana tabacum*, var. *Xanthi*).

	Leaf length (cm)	Leaf width (cm)	Leaf area (cm^2)	LSI[a]
Control	26	14	182	0.54
+ GA_3	25	11	138	0.44

[a] LSI= (cm leaf width)(cm leaf length)$^{-1}$

weight was influenced by a decrease of leaf thickness mostly due to a shortening of palisade cells by about 30% (Zeppernick 1988). Also, the epidermal cells of leaves of haploids are smaller (about 40%) as compared to the doubled haploids. Comparing the data given in Table 1 for haploids and diploid or doubled haploids, it is evident, that the differences in morphology of haploid and diploid plants are only partly due to the ploidy level, and some other factors have to be considered as well. One of these factors could be heterozygoty of the diploid plants, resulting in some kind of "heterosis".

Leaf growth and development is influenced by the hormonal system of the plant, notably the gibberellins (e.g. Schwab & Neumann 1975; Barlow 1987), and to correlate it with the ploidy level gibberellin activity was determined in comparable leaves (10 to 12 cm in length of the two genotypes at flower bud appearance of the haploid plants; dwarf rice test, Murakami 1968). As can be seen from Table 2, total gibberellin activity in haploids exceeds that of doubled haploids by about a 100%. A comparison of the contribution of the two gibberellin groups (polar and nonpolar compounds as separated by thin layer chromatography) to total gibberellin activity indicates, that the enhanced GA-activity is almost entirely due to an increase of polar compounds.

To check the significance of gibberellins for leaf growth, GA_3 was applied to doubled haploid plants. As the data in Table 3 indicate, following this treatment a leaf shape similar to that observed for haploid plants developed, with a reduced growth in lamina width (see LSI). For GA_3 treated plants, however, a postponement of flowering of about 10–14 days was observed.

The higher gibberellin activity in haploids than in doubled haploids agrees with an increased response of cultured shoot explants of the latter to a GA_3 supplement to the nutrient medium. Whereas in the control the fresh weight of cultured explants of doubled haploids was less than 50% of that of haploids, after a GA_3 supplement growth of both ploidy levels was increased. The increase of growth of doubled haploids is much higher and the fresh weight eventually is the same in the GA_3 treatments of n and 2 × n cultures (Table 4).

As the number of cells per explant shows, the increased growth of the GA_3 treatments apparently was due to a stimulation of cell division activity. This indicates that the influence of ploidy on the gibberellin system could be executed at the cellular level. Possibly, metabolic inactivation of gibberellins was somehow disturbed in haploids or their synthesis was increased, the former seems to be more likely.

Some characteristics of NICA-plants

NICA plants were originally obtained by Dudits et al. (1987) by protoplasts fusion of *Nicotiana tabacum*, cv. *Petit Havanna* SR1 (Maliga et al. 1973) and *Daucus carota* H47 (Sung 1979) (The plant material used in our study was a gift from Prof. Dudits). The carrot protoplasts were irradiated with X-rays prior to protoplast

Fig. 2. The morphological appearance of NICA (left) and plants of the tobacco parent line (right).

Table 4. Fresh weight and number of cells per explant of cultured pith tissue of haploid and doubled haploid tobacco plants (*Nicotiana tabacum* var. *Xanthi*) as influenced by a GA_3 supplement (10^{-6} M, stationary cultures on 0.8% agar for two weeks, 27°C, ca. 5000 lux, developmental stage at explantation: 6 leaves per plant).

	mg FW per explant	Number of cells per explant X 10^3
Control		
n	121	710
2 × n	50	348
+ GA_3		
n	263	1382
2 × n	261	1522

fusion. Our investigations concentrated on the tobacco parent material and on the line NICA 401 described as resistant to methotrexate and to 5-methyl-tryptophan. During our investigations only resistance to the latter was checked and assured. The NICA plants had a tobacco habitus and its most obvious morphological characteristics as distinguished from the tobacco parent line was stunted growth with short internodes and very narrow leaf blades (Fig. 2). Both lines were propagated by cuttings since 1986 and the morphological features of the plants remained stable. Subcultures were obtained in about 2 month intervals on Agar with a modified MS-Medium containing inositol, but no other growth substances. Following an application of 5-methyl-tryptophan as a selection agent at each subculture most cuttings die off, indicating instability of the resistance. Many of those remaining alive had vital leaves only at the upper part of the shoot from which side branches out of axillary buds developed. These were used again to obtain cuttings for the next subculture.

To characterize the hormonal system of the NICA plants the concentrations of IAA, ABA and the gibberellin activity were determined in leaves with the tobacco parent line as the control. For IAA and for ABA the ELISA technique (Weiler 1982) with monoclonal antibodies was used and gibberellin activity was determined by the dwarf rice test (Murakami 1968). As can be seen from the results summarized in Table 5. the concentration of IAA was more than doubled in the NICA plants as compared to the control and that of ABA is about half. The higher concentration of IAA is in agreement with the function of 5-methyl-tryptophan as an inhibitor to IAA synthesis. The disturbed IAA synthesis is apparently compensated by the cells by an overproduction of this phytohormone (Sung 1979). As the data show, gibberellin activity was dramatically increased in the NICA plants which agrees with

Table 5. The concentration of IAA and ABA and the gibberellin activity[a] in leaves of NICA plants (product of fusion experiments with protoplasts of *Nicotiana tabacum* and *Daucus carota*) and of tobacco parent plants.

	IAA ng g^{-1} FW	ABA ng g^{-1} FW	Gibberellin activity ng g^{-1} FW
Nicotiana parent	14.0	22.9	1.4
NICA	32.8	12.6	23.5

[a]calculated on the basis of responses to GA_3 by the dwarf rice bioassay.

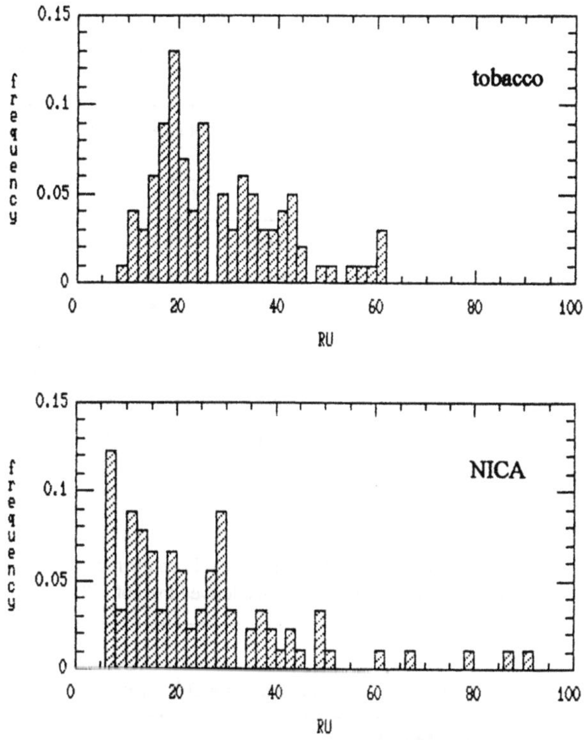

Fig. 3. The frequency of DNA content (as relative units, RU) in nuclei of leaves of NICA (product of protoplast fusion experiments with *Nicotiana* and *Daucus*) and the tobacco parent plant. The DNA content of tobacco microspores was measured as 8.6 RU.

the drastic reduction of width of the leaf lamina of this line. In disaccord with the strongly increased gibberellin activity, however, are the shorter internodes of the NICA plants.

As a first characterization of the genetic system of the NICA material the relative DNA content in nuclei was determined with a microscope photometer following staining with Bisbenzimid (Forche & Blaschke 1978). In Fig. 3 the frequency of nuclei with a given value for DNA content as relative units (RU) is plotted viz. the relative units measured. As can be seen from the data, in the tobacco parent strain the majority of nuclei inspected concentrates at the DNA value of 20, which is somewhat more than doubled as compared to the values obtained for nuclei of tobacco microspores (8.6 RU), by definition haploids. The DNA value of about 40 should represent diploid cells in the G2-Phase of the cell cycle. Obviously, the cells of the tobacco parent plant are mostly diploid, interspersed with some cells of a higher DNA-value.

A completely different picture emerges for the DNA values of the NICA plants. NICA 401 leaf cells displayed a much broader spectrum of DNA values. This is in accordance with the report of Dudits et al. (1987) on variations of chromosome number of NICA plants, however, no haploid cells were described as occurring in the meristems inspected. Nevertheless, our investigations demonstrate that a fairly good share of cells in NICA 401 leaves possessed nuclei with DNA-values comparable to those measured in microspores of diploid tobacco. It remains to be seen whether the occurrence of cells with such a low DNA content in NICA 401 will be limited to the leaves. At any rate, the occurrence of haploid cells would be in agreement with the enhanced gibberellin activity and the narrow width of the leaf lamina of the NICA plants as described for haploids derived by anther cultures.

Since the resistance to 5-methyl-tryptophan in the NICA plants originating from carrot cells used for protoplast fusion is still maintained, some carrot DNA could still be part of the genome of the NICA strain used. In a preliminary characterization of the DNA of these plants by digestion with restriction nucleases like BspN-I with genomic DNA of *Daucus* and *Nicotiana* as controls, no obvious differences between NICA and *Nicotiana* genomic DNA were detected. At least for carrots it could be shown that by culturing on a medium containing 5-methyl-tryptophan cells resistant to this drug could be selected out of wild type cell suspensions within 7 months (Chinachit 1991; s.a. Hauptmann et al. 1983). The resistance of NICA plants could be of a similar origin without components of the carrot genome. A resistance check to methotrexate was not attempted.

Nicotine concentrations of haploid and dihaploid tobacco plants

The same strains of haploid and doubled haploid tobacco plants as used for morphological studies and

Table 6. The nicotine concentration in leaves of haploid lines of *Nicotiana tabacum* var. *Xanthi*, and doubled haploids, derived thereof (Hauptgut, mg nicotine g^{-1} DW).

Line	Pot experiments			
	n		2xn	
	1st year	2nd year	1st year	2nd year
2/2	1.7	2.4	2.1	2.1
4/3	1.9	2.3	2.4	1.9
8/1	0.8	0.8	3.6	3.4
	Field experiments			
2/2	5.2		3.0	
8/1	3.5		10.6	

Table 7. The influence of IAA and of 2,4-D on the nicotine concentrations of cell suspensions of two haploid lines of *Nicotiana tabacum* var. *Xanthi* and of doubled haploids derived thereof (28 days of culture, NL + 0.1 ppm Kinetin + 50 ppm m-inositol).

Strain	4/3			8/1		
	g DW l^{-1}	μg Nicotine per g DW	μg Nicotine per vessel	g DW l^{-1}	μg Nicotine per g DW	μg Nicotine per vessel
1.14×10^{-5}M IAA						
n	5.6	29.4	336.6	5.1	1.8	25.2
2xn	2.9	11.8	130.2	8.9	7.2	80.1
9×10^{-6}M 2,4-D						
n	4.4	3.4	85.0	8.6	22.9	237
2xn	5.6	14.9	167.4	7.3	11.1	97

investigations on gibberellins were also employed for determinations concerning nicotine accumulations. Nicotine concentrations were compared in leaves (Hauptgut) of plants raised in pot and field experiments, and in cell suspensions. The data in Table 6 show the nicotine concentrations in leaves in two successive years. Whereas for the strains 2/2 and 4/3 no significant differences in the concentration of nicotine were found comparing both ploidy levels, a strong increase was observed in doubled haploids of strain 8/1 in both years. An increase in nicotine concentration in doubled haploids of strain 8/1 was also determined in field experiments with again no clear variation due to the ploidy level for the strain 2/2.

As these few data for tobacco show apparently the genotype of plants derived from androgenesis is more important for nicotine concentrations in leaves than the ploidy level. This contradicts data for *Datura* alkaloids (Kibler et al. 1980; Kibler & Neumann 1979; Mechler & Kohlenbach 1978; Burk & Matzinger 1976; Schiltz et al. 1980). Another deviation of *Datura innoxia* haploids was a reduced leaf area with about the same LSI as diploids.

Nicotine concentrations of haploid and doubled haploid tobacco cell cultures

Using a nutrient medium with inorganic nitrogen only in the form of nitrate cell suspensions of two of the strains were checked for nicotine concentration in cell material and in the nutrient medium after a culture period of 4 weeks. The results are summarized in Table 7. In the nutrient medium with IAA as the auxin, again as for leaves the doubled haploids of strain 8/1 contained a significantly higher nicotine concentration and a higher production rate per culture vessel (including nicotine determined in the nutrient medium) than the haploids. In those of strain 4/3, however, nicotine concentration was only about 30% of haploids. A completely different picture emerges comparing these strains if IAA is substituted by 2,4-D. Here the nicotine concentration is significantly higher in the doubled haploid cells of 4/3 then in haploids and the reverse was found for strain 8/1. This confusing result in cultured cells may only indicate that at the cellular level again genomic differences of plants derived by the haploid technique seem to be more important for nicotine concentrations than either the ploidy level or the kind of auxin supplied to the culture medium. The latter factors, however, modify strongly and statistically significant the genomic potential.

Conclusions

The data reported in this paper indicate that two components of *Nicotiana tabacum*, i.e. gibberellins and nicotine, belonging to various areas of metabolism are differently influenced by the ploidy level as a *quantitative* factor of the genome. Whereas gibberellin activity is strongly increased in leaves of haploid plants as compared to doubled haploids of the same strain with clear consequences on leaf shape, a direct correlation between the ploidy level and the nicotine concentration seems not to exist. Here *qualitative* genomic differences between strains seem to be more important. Similar conclusions were drawn by Jacquin-Dubreuil et al. (1991) in a comparable investigation using *Nicotiana plumbaginifolia* as plant material. The studies reported in this paper also show again the difficulties to compare results obtained on secondary metabolism as represented by the nicotine concentration of intact plants with those from cultured cells of the same genetic strain or vice versa.

References

Barlow PW (1987) Requirements for hormone involvement in development at different levels of organization. In: Hoad GV, Lenton JR, Jackson MB & Atkin RK (Eds) Hormone action in plant development – A critical appraisal. Butterworths, London

Burk LGg & Matzinger DF (1976) Variation among anther derived doubled haploids from an inbred line of tobacco. J. Hered. 67: 381–384

Chinachit W (1991) Die somatische Hybridisierung von hemmstoffresistenten Wildkarotten und Kulturkarotten (*Daucus carota* L.) und die Charakterisierung der Hybriden. Dissertation Justus-Liebig Universität Giessen.

Dudits D, Maroy E, Praznovsky T, Olah Z, Gyorgyey J & Ella R (1987) Transfer of resistance traits from carrot into tobacco by asymmetric somatic hybridization: Regeneration of fertile plants. Proc. Natl. Acad. Sci USA 84: 8434–8438

Forche E & Blaschke JR (1978) Der Einsatz des Mikroskop-Photometers MPV-1 bei der flourometrischen Bestimmung der Ploidie pflanzlicher Zellen. Leitz-Mitt. Wiss. u. techn. Bd. VII: 90–93

Forche E & Neumann KH (1977) Der Einfluß verschiedener Kulturfaktoren auf die Gewinnung haploider Pflanzen aus Antheren von *Datura innoxia* und *Nicotiana tabacum* ssp. Z. f. Pflanzenzüchtung 79: 250–255

Hauptmann R, Kumar P & Widholm JM (1983) Carrot (x) tobacco somatic cell hybrids selected by amino acid analog resistance complementation. 6th Intern. Protoplast Symposium: 92–93

Jacquin-Dubreuil A, Chetrit S, Fliniaux MA, Trinh TH, Cosson L & Tran Thanh Van K (1991) Study of the alkaloid content of four "hypohaploids" of *Nicotiana plumbaginifolia* grown in vitro. Plant Cell Tissue Organ Cult. 27: 1–6

Kibler R & Neumann KH (1979) Alkaloidgehalte in Blättern und Zellsuspensionen von *Datura innoxia*. Planta Med. 35: 354–359

Kibler R & Neumann KH (1980) On cytogenetic stability of cultured tissue and cell suspensions of haploid and diploid origin. In: Sala F, Parisi B, Cella & Ciferri O (Eds). Plant cell cultures. Results and Perspectives. Elsevier North-Holl, Biomedical Press 59–65

Kibler R, Blasche JR, Forche E & Neumann KH (1980) Investigations on ploidy levels of haploid and diploid callus and cell suspension cultures of *Datura innoxia* Mill. J. Cell Sci. 44: 365–373

Maliga P, Sz.-Bredsnovits L & Marton L (1973). Streptomycin-resistant plants from callus culture of haploid tobacco. Nature (London/New Biol.) 244: 29–30

Mechler E & HW Kohlenbach (1978) Alkaloid content in leaves of diploid and haploid *Datura* species. Planta Med. 33: 350–355

Murakami Y (1968) A new rice seedling test for gibberellins, "microdrop method", and its use for testing extracts of rice and morning glory. Bot. Mag. Tokyo 81: 33–43

Schiltz P, Coussirat JC, Cazamajour F, Hitier G, Albo JP & Delon R (1980) Modifications induites chez *Nicotiana tabacum* après doublement chromosomique de types haploides obtenus par culture in vitro d' antheres. Ann. Tabac. sect. 2, 16: 5–12

Schwab B & Neumann KH (1975) Der Einfluß von Gibberellinsäurespritzungen auf das Blühverhalten von Karotten. Z. Pflanzenernähr. Bodenk. 1: 13–18

Sung ZR (1979) Relationship of indole-3-acetic acid and tryptophan concentration in normal and 5-methyl-tryptophan resistant cell lines of wild carrots. Planta 145: 339–345

Weiler EW (1982) Plant hormone immunoassay. Physiol. Plant. 54: 230–234

Zeppernick B (1988) Untersuchungen zur Bedeutung des Ploidieniveaus (n, 2 × n) für den Entwicklungsverlauf, das Gibberellinsystem und die Nikotinkonzentration von Blättern und Zellkulturen von *Nicotiana tabacum*, var. *Xanthi*. Dissertation, Justus-Liebig Universität

Plant Cell, Tissue and Organ Culture **38**: 143–151, 1994.

Influence of exogenous hormones on the growth and secondary metabolite formation in transformed root cultures

M. J. C. Rhodes, A. J. Parr, A. Giulietti & E. L. H. Aird
Genetics and Microbiology Department, Institute of Food Research, Norwich Research Park, Colney, Norwich NR4 7UA, UK

Key words: Alkaloid formation, ascorbate oxidase, auxin, *Brugmansia candida*, *Cucumis sativus*, gibberellic acid, *Nicotiana rustica*, transformed organ cultures

Abstract

Transformed organ cultures formed following transformation of plant tissues with *Agrobacterium* species owe their phenotypes to alterations in hormone metabolism. Exogenously supplied hormones have been used to probe the relationship between the growth and morphology of transformed root cultures of a number of species and their ability to accumulate secondary products. Auxins in the presence of low levels of kinetin induce the rapid disorganisation of transformed roots of *Nicotiana rustica* ultimately to form suspension cultures of transformed cells and this process is associated with a decrease in nicotine content of the cells. This is related to cells in the culture losing competence in alkaloid biosynthesis. In contrast, exogenously supplied GA_3 enhanced branching in two transformed root clones of the tropane-alkaloid producing species, *Brugmansia candida* and so enhanced their typical "hairy root" phenotype. This growth substance had the effect of reducing the overall alkaloid accumulation but in one case significantly altered the relative concentrations of different tropine esters.

In transformed roots of *Cucumis sativus*, the phenotype of the roots is influenced by the expression of auxin synthesis genes on T_R-DNA resulting in roots with two distinct morphologies. The pattern of expression of the enzyme ascorbate oxidase in populations of control roots of different morphologies is described. The significance of these phenotypic variations on the utility of transformed root cultures for the study of secondary metabolic pathways will be discussed.

Abbreviations: AO – ascorbate oxidase, DW – dry weight, FW – fresh weight, GA_3 – gibberellic acid

Introduction

Transformed plant organ cultures have proved valuable in the study of aspects of secondary metabolism. Their advantages over conventional cell suspension cultures lie in their genetic and biochemical stability over long periods in culture and the potential for introducing novel genes to modify growth and secondary metabolism. Transformed cultures of both roots and shoots have been developed. They are derived following infection with the plant pathogens, *Agrobacterium rhizogenes* and *A. tumefaciens*. Their common mode of action is to transfer sections of plasmid DNA (T-DNA) into plant cells and to insert this T-DNA into the plant genome where it is expressed, transforming the infected plant cell. The T-DNA encodes genes which modify hormone metabolism of the transformed cell and diverts it into novel routes of cellular and organ differentiation. In the case of *A. rhizogenes*, the transformed cell is induced to initiate rhizogenesis. Sections of T-DNA bearing three genes *rolA, B* and *C* are capable of inducing transformed ("hairy") root formation (Spena et al. 1987). The precise roles of the individual rol genes in the initiation and maintenance of such transformed roots are unclear and the phenotypic responses to the rol genes individually expressed varies to some extent between plant species. Vilaine et al. (1987) found that *rolA* was primarily responsible for development of hairy roots in tobacco whereas in other work *rolB* appears to be the main factor in hairy root formation (Cardarelli et al. 1987, Spena et al. 1987). *RolB* encodes an enzyme which hydrolyses indoxyl glucosides (Estruch et al. 1991a), while *rolC* encodes one which hydrolyses cytokinin conjugates to hormonally active forms (Estruch et al. 1991b). The product of the *rolA* gene is unknown.

A. tumefaciens normally induces the initially transformed cell(s) to grow in an uncontrolled manner to form callus tissue. This induction of tumour formation depends largely on the expression of two auxin biosynthetic genes *iaaM* and *iaaH* and of a cytokinin biosynthesis gene *ipt* and the over-accumulation of both auxin and cytokinin within the tumourous tissue (Hooykaas & Schilperoort 1992). Other genes such as gene5 which encodes an enzyme involved in the accumulation of an auxin-antagonist, indolelactic acid (Körber et al. 1991) and gene 6b which is also thought to regulate auxin or cytokinin effects (Tinland et al. 1989) may also play a role in induction of the tumourous state. In tobacco, expression of the *ipt* alone behind either the endogenous or stronger promoters (Smigocki & Owens 1988) induces the formation of shooty teratomas rather than galls. In such cultures, high levels of cytokinins accumulate in the shooty tissue (Akiyoshi et al. 1983)) and they show a loss of apical dominance, increased chlorophyll content, delayed senescence and marked inhibition of root formation and growth (Li et al. 1992). We have used constructs which over-express the *ipt* gene to induce formation of transformed shoot cultures in *Mentha* (Spencer et al. 1990, Rhodes et al. 1992). These shoot cultures maintain a stable phenotype over several years and have expressed the pathway to form monoterpenes stably during this period. These cultures can be used to study the expression of genes relevant to metabolic pathways specific to green tissue.

This paper will consider two related aspects of the properties of transformed root cultures as experimental systems in which to study secondary metabolism. Both relate to the extent to which changes in hormonal metabolism can influence secondary metabolism. Transformed root cultures grow independently in the absence of externally supplied hormones. However, exogenous hormones do affect transformed roots and perturbing roots with hormones enables the relationship between hormonal status and effects on root morphology, growth and secondary metabolism to be probed. This will be exploited in studies using either auxin/cytokinin mixtures or gibberellic acid in two solanaceous species. In addition, variations in morphology are found in transformed roots of some species due to expression of T-DNA genes not essential for root formation. The influence of such modifying genes and their morphological effects will be considered in transformed roots of cucumber and the pattern of changes in ascorbate oxidase activity will be discussed.

Materials and methods

The methods used for the establishment and maintenance of the transformed root cultures were essentially as described previously (Hamill et al. 1986). All were based on infections with the agropine strain of *A. rhizogenes* LBA9402. The properties of the transformed root cultures of *N. rustica* (Hamill et al. 1986), and *Brugmansia candida* (Giulietti et al. 1993) are as described elsewhere. The cloned cultures of cucumber (*Cucumis sativus*) were derived by infection of sterile seedlings of the variety Perfection with *A. rhizogenes* LBA9402. Individual roots emerging from the wound site were excised, cleared of contaminating bacteria with antibiotics and grown as separate lines. All root lines were maintained in B50 medium (Gamborg's B5 salts, 3% sucrose). Two media, B5NK (Gamborg's B5 salts, 3% sucrose, naphthalene acetic acid 2 mg l^{-1}, kinetin 0.1 mg l^{-1}) or MS6 (Murashige and Skoog salts, 3% sucrose, 2,4-dichlorophenoxyacetic acid 1 mg l^{-1} and kinetin 0.1 mg l^{-1}) were used to induce disorganisation of *Nicotiana* roots. The degree of disorganisation of the cultures was assessed on a scale (Aird 1988) of 0 (typical hairy root phenotype) to 5 (fully dispersed cultures). The methods for assay of nicotine in *N. rustica* and of tropane alkaloids in *Brugmansia candida* are as described elsewhere (Hamill et al. 1986, Giulietti et al. 1993). Ascorbate oxidase was assayed by standard methods (Esaka et al. 1988).

Results and discussion

Effect of exogenous hormones

Auxin/cytokinin effects

The effects of exogenously supplied auxins and cytokinins on growth of transformed root cultures tend to vary between species. In some species such as *Tagetes* (Croes et al. 1989) and *Panax* (Yoshikawa & Furuya 1987) auxins either alone or in combination with cytokinins stimulate growth by increasing branching. In others such as *Beta vulgaris* such treatments are without effect. However, in some solanaceous species these hormones cause significant inhibi-

Table 1. The effect of transfer of a low inoculum (0.15 g FW) of transformed roots of *Nicotiana rustica* into 250 ml flasks containing 50 ml of media which either promote disorganisation (MS6 or B5NK) or into a hormone-free medium (B50) which maintains root integrity.

B50 medium			
Time (days)	Growth (g DW flask^{-1})	Disorganisation Index	Nicotine (mg g^{-1} DW)
0	0.01	0	2.1
3	0.025		
6	0.04	0	3.9
9		0–1	
12			
15	0.31	0–1	3.2
18			
21	0.42	0–1	2.9
B5NK medium			
Time (days)	Growth (g DW flask^{-1})	Disorganisation Index	Nicotine (mg g^{-1} DW)
0	0.01	0	2.1
3	0.01	1–2	0.8
6	0.01	2	0.6
9	0.01	2–3	0.5
12	0.015	2–3	
15	0.09	3–4	0.4
18	0.075	4	0.2
21	0.10	4	0.4
MS6 medium			
Time (days)	Growth (g DW flask^{-1})	Disorganisation Index	Nicotine (mg g^{-1} DW)
0	0.01	0	2.1
3	0.01	1–2	1.6
6	0.02	2	1.0
9	0.025	3	0.1
12	0.040	3	0.1
15	0.065	3–4	0.1
18	0.055	3–4	0.1
21	0.075	4	0.1

tion of root growth and indeed stimulate disorganisation of the root structure.

Table 1 shows the effect of transferring small inocula (0.15 g) of transformed roots of *N. rustica* previously maintained in hormone-free media to either of two media (B5NK or MS6) containing high levels of auxin and lower levels of cytokinin (Aird 1988). These hormonal treatments promote rapid disorganization of the root structure so that by 21 days the culture consisted of masses of callus formed around the original roots. The onset of such disorganization is evident within 3–6 days of the transfer and is largely complete by 18–21 days (see Fig. 1). Table 1 shows that during the initial period before significant disorganization was evident, the nicotine content of the roots fell in either medium by 50–70% and then continued to fall so that by 21 days the nicotine content of the roots had fallen by over 81% in B5NK and by over 95% in MS6 medium. However, in both cases the growth of the cultures undergoing disorganization was severely restricted compared with control transformed roots incubated in hormone-free medium (B50).

In order to minimise this effect on growth, a further experiment (see Table 2) using a much higher inoculum (1.5 g FW) was performed. Under these conditions the effects on growth are relatively small. Indeed, roots

A B C

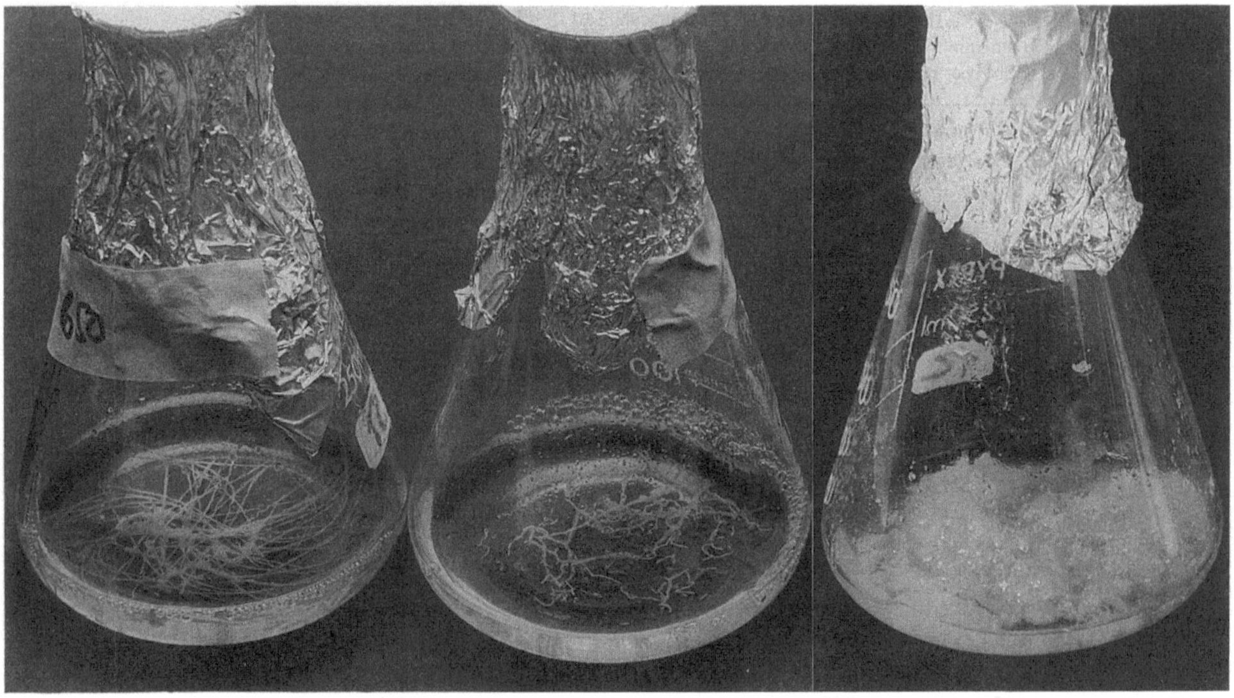

Fig. 1. Photograph of cultures of transformed roots of *Nicotiana rustica* (A): 0 days. (B): 5 days and (C): 28 days after transfer into B5NK medium.

dispersed in B5NK grew faster than the controls while those in MS6 grew at only a slightly lower rate. Disorganization, however, was slower to develop and was not apparent until day 8 in the B5NK medium and not until days 8–10 in MS6. In the periods where there were no obvious morphological changes (days 0–6) the nicotine content fell by 80% in B5NK and by 67% in MS6.

In the case of dispersion of roots at the higher inoculum level in B5NK medium, growth and the loss of nicotine were correlated ($R^2 = 0.85$). Thus the apparent loss of the alkaloid could be due to dilution with callus cells in which nicotine biosynthesis is repressed. However, in the other three situations viz. high inoculum in MS6 or low inoculum in either MS6 or B5NK, there is no such correlation. In these cases most of the loss of nicotine occurs before significant growth of callus is observed. The simplest explanation is that the hormone treatment rapidly induces in both residual root tissue and in the induced callus either an increase in degradation or an inhibition of synthesis under conditions in which turnover of nicotine is rapid. Both *Nicotiana* and *Datura* (Robins et al. 1991) show similar patterns of dispersion in auxin/cytokinin media. In both species, during the period of rapid depletion of alkaloid on exposure of roots to the hormone containing media, there is a rapid loss of activity in two enzymes of the early stages of alkaloid biosynthesis. Within 24 h of onset of the hormone treatment there is a 90% loss in putrescine methyltransferase (PMT) activity and a smaller decrease in N-methylputrescine oxidase (MPO) (Aird 1988, Robins et al. 1991). This suggests that an effect on the biosynthesis, rather than degradation, of alkaloids may be important in the response of roots on exposure to the hormone medium.

Gibberellin effects

The effects of gibberellic acid on transformed root cultures were studied in the solanaceous species, *Brugmansia candida*. Several clones derived from transformation of this species with *A. rhizogenes* were established and characterised (Giulietti et al. 1993). The effects of GA_3 on growth of 15 of these clones were found to be variable. Some clones show significant stimulation of growth at low concentrations of GA_3 while others show inhibition at all concentrations. Figure 2 shows two clones (clones 1 and 40) in which GA_3 concentrations between 0.1 and 10 μM stimulated growth measured by increase in FW. This stimulation was most marked at the time of fastest growth

Table 2. The effect of transfer of a high inoculum (1.5 g FW) of transformed roots of *Nicotiana rustica* into 250 ml flasks containing 50 ml of media which either promote disorganisation (MS6 or B5NK) or into a hormone-free medium (B50) which maintains root integrity.

B50 medium			
Time (days)	Growth (g DW flask^{-1})	Disorganisation Index	Nicotine (mg g^{-1} DW)
0	0.075	0	3.05
1			
2			
3			
4			
6	0.26	0–1	2.9
8			
10			
12			
14	0.43	0–1	2.9
B5NK medium			
Time (days)	Growth (g DW flask^{-1})	Disorganisation Index	Nicotine (mg g^{-1} DW)
0	0.075	0	3.05
1	0.09	0–1	2.65
2	0.10	0–1	1.75
3	0.18	0–1	1.50
4	0.20	1	1.20
6	0.28	1–2	0.60
8	0.40	2	0.60
10	0.40	3	0.90
12	0.70	3–4	0.90
14	0.63	4	0.70
MS6 medium			
Time (days)	Growth (g DW flask^{-1})	Disorganisation Index	Nicotine (mg g^{-1} DW)
0	0.075	0	3.05
1	0.08	0–1	2.0
2	0.08	0–1	1.1
3	0.09	1	1.1
4	0.11	1	1.0
6	0.11	1	1.0
8	0.15	1–2	0.75
10	0.23	2–3	0.35
12	0.21	3	0.35
14	0.30	3	0.30

rate (day15) but there is also an increase in the final biomass achieved as the GA_3 is raised. This stimulation in growth is characterised (Fig. 3) by a small effect on the rate of elongation (up to 30% increase in clone 40) but a much more substantial effect on the rate of branching. At day 20, there was a 5-fold increase in the number of side branches in clone 1 and a 3-fold increase in clone 40.

Figure 4 shows the levels of the main alkaloids of clone 1 in roots treated with levels of GA_3 between 0.1 and 10 μM and harvested and analysed after 14 days. In clone 1 all levels of GA_3 are inhibitory to total alkaloid accumulation; at 10 μM the inhibition is 70% compared to roots grown in the absence of GA_3. However, this disguises an alteration in alkaloid composition. Compared to the controls, 0.1 μM

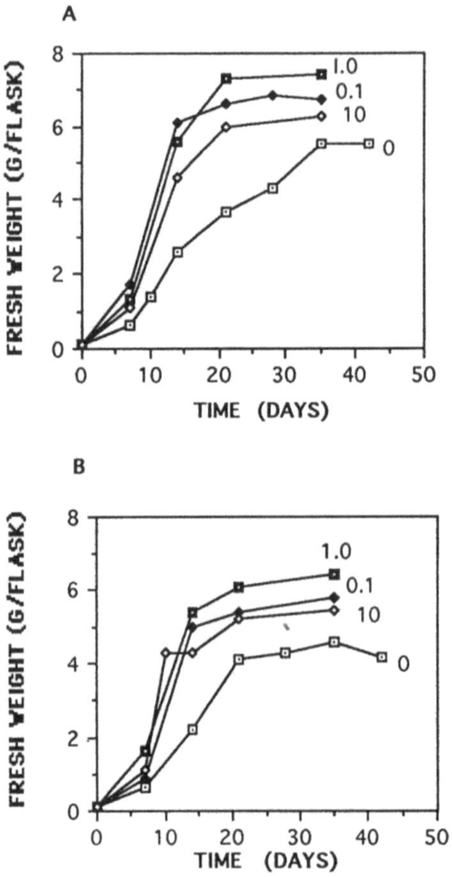

Fig. 2. Time course of growth of clone 1 (A) and clone 40 (B) of transformed root of *Brugmansia candida* on B50 medium (□) or B50 medium containing 0.1 (♦), 1 (□) or 10 (◊) μM GA_3.

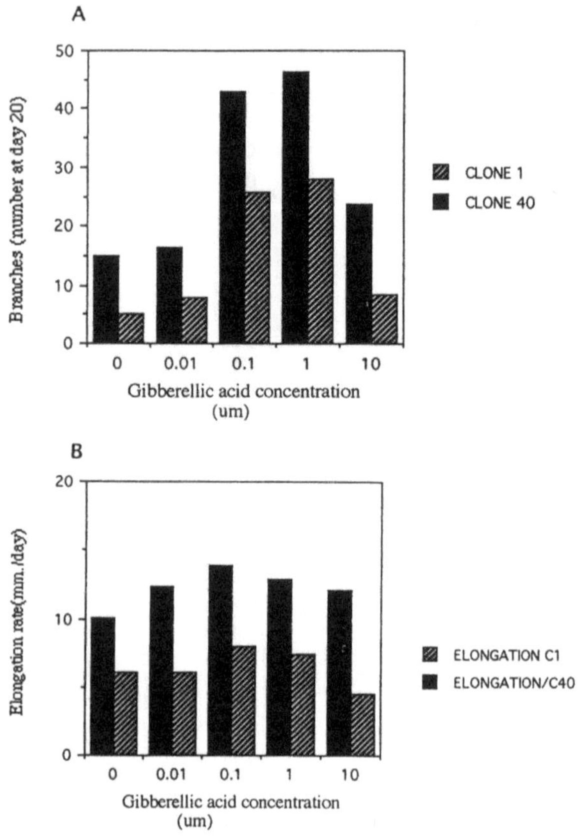

Fig. 3. Effect of GA_3 on the branching (A) and elongation rates (B) of clones 1 and 40 of transformed root of *Brugmansia candida*.

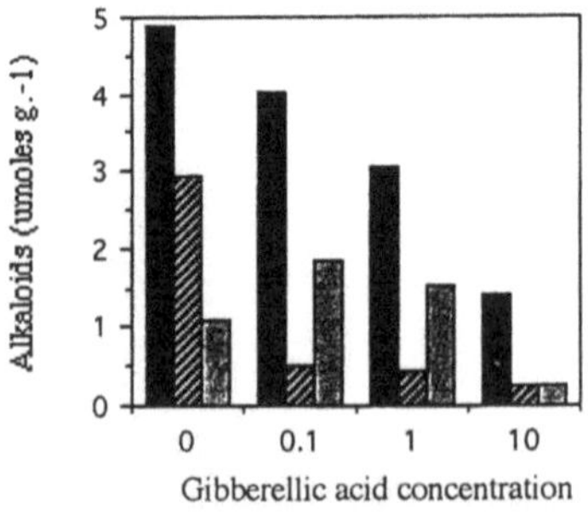

Fig. 4. Histogram showing effect of a range of GA_3 concentrations (0.1–10 μM) on the alkaloid content of 14-day old transformed root cultures of *Brugmansia candida* clone 1. Order of histograms 1. Total tropane alkaloids, 2. 3α-acetyloxytropane and 3. hyoscyamine.

GA_3 stimulates hyoscyamine accumulation by about 40% but this is accompanied by an 82% decrease in the level of the other major alkaloid of this line, 3α-acetyloxytropane. In line 40 (see Fig. 5) a small increase in total alkaloid and in hyoscyamine occurs between 0 and 0.1 μM GA_3 but higher concentrations are inhibitory. 3α-Acetyloxytropane and scopolamine are minor alkaloid constituents in this line but their levels show a small decrease at all GA_3 levels tested. In the transformed clone of the related species, *Datura stramonium* DS15/5, GA_3 was also shown to be inhibitory to alkaloid accumulation.

Under none of these conditions was there any evidence of loss of root structure and integrity. Indeed GA_3 treatment tends to reinforce the typical highly branched phenotype of transformed roots. The causes of their inhibitory effects on alkaloid accumulation are thus quite different to those observed using auxin/cytokinin containing media in these and other solanaceous species and it is interesting that qualitative as well as quantitative effects are involved.

Much emphasis is placed on the roles of auxins and cytokinins in transformed tissues (Hooykaas & Schilperoort 1992) but there are cases where other plant growth substances may influence both the mor-

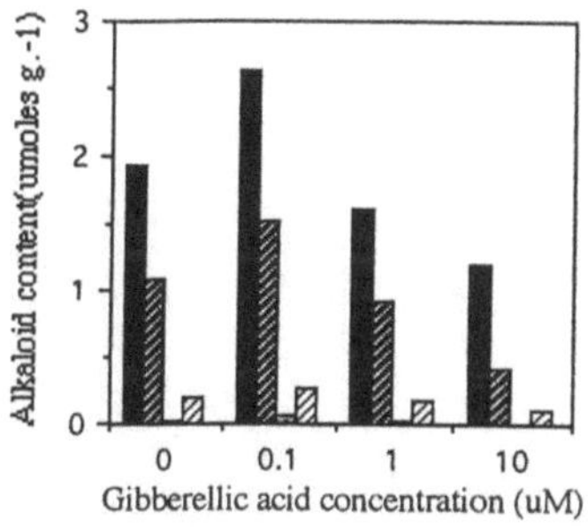

Fig. 5. Histogram showing the effect of a range of GA_3 concentrations on the alkaloid content of 14-day old transformed root cultures of *Brugmansia candida* clone 40. Order of histograms 1. total tropane alkaloids, 2. hyoscyamine, 3. 3α-acetyloxytropane and 4. scopolamine.

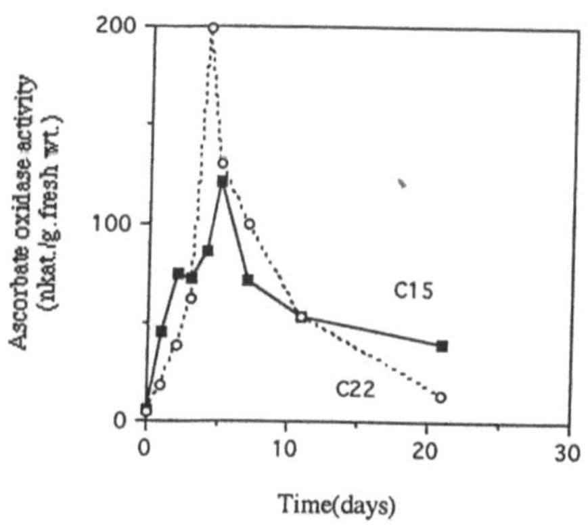

Fig. 6. Time course of changes in AO activity in two control hairy root lines, one C15 showing type A morphology and the other C22 showing type B morphology.

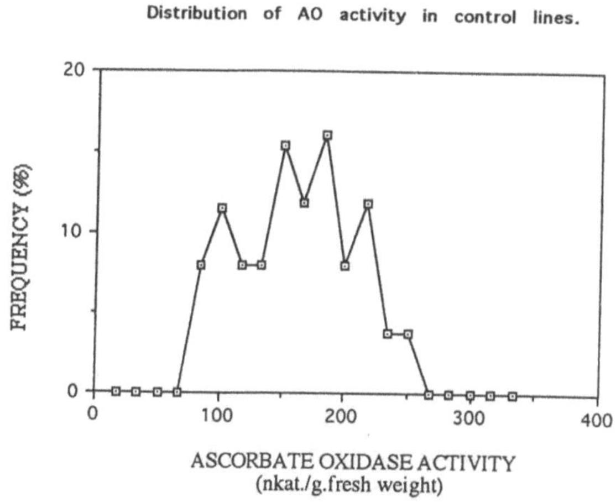

Fig. 7. Distribution of AO activities among a population of 26 cloned transformed control root cultures of *Cucumis sativus* formed following infection with the wild type strain of *Agrobacterium rhizogenes*, LBA9402.

phology and metabolism of hairy root cultures. GA_3 can, in some species, influence branching and to a smaller extent elongation rates. These effects tend to be variable between lines and may indicate variations in the internal hormonal metabolism which renders only some root lines susceptible to manipulation by exogenous regulators. Hairy roots are generally considered to be independent of exogenous growth substances (Spano et al. 1981, Tepfer & Tempé 1981) and while this is generally true there is a growing list of examples in which exogenous supply of auxins and cytokinins will stimulate growth in transformed roots of certain species (Croes et al. 1989, Yoshikawa & Furuya 1987). There are other examples where added gibberellins will influence the growth of transformed root cultures (Kamada et al. 1989, Ohkawa et al. 1989) but as yet there are no indications that transformation *per se* influences gibberellin metabolism. In our experiments auxin and cytokinin induce disorganisation in transformed roots of *Datura* and *Nicotiana* to produce non-alkaloid producing cell suspension cultures but our data shows that in addition there are effects on nicotine formation which precede and are largely independent of effects on growth or growth pattern. The treatment with GA_3 leads to suppression of tropane alkaloid formation which is associated with enhancement of the hairy root structure. These effects on alkaloid formation could result from changes in the proportions of different cell types in the cultures but effects not mediated through changing growth patterns cannot be ruled out.

Morphology and expression of genes in transformed root cultures

Transformed root cultures have been used for the study of expression of ascorbate oxidase in cucumber. This enzyme is present in transformed roots of this species at relatively high levels (up to 200 nkat g^{-1} FW) and during a typical growth cycle (see Fig. 6) shows a characteristic pattern of changes in activity, rising steeply to a peak at day 4 and then falling back to low levels.

Sterile cucumber seedlings were infected with *A. rhizogenes* strain LBA9402 and individual roots emerging from the wound site were freed of living bacteria using ampicillin and grown as separate axenic clones. Figure 7 shows the variation in AO activities in 4 day-old roots from 26 such cultures. This shows the expected normal distribution of activities around a mean of 159.6 nkat g^{-1} FW and a standard deviation of

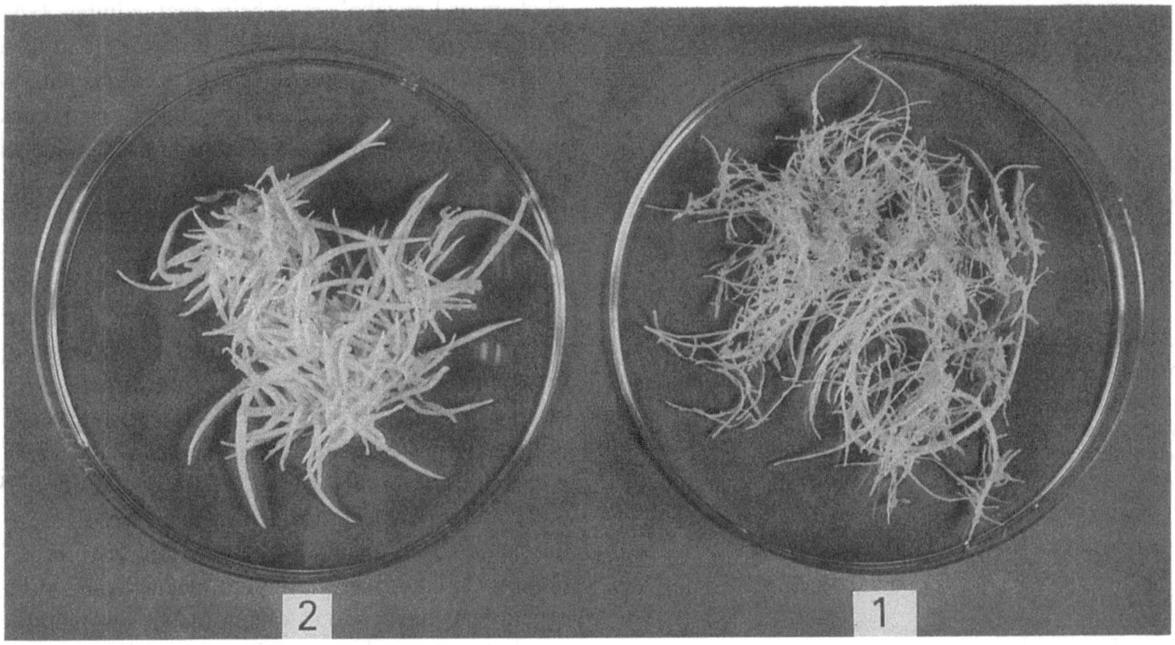

Fig. 8. Photograph of transformed root cultures of *Cucumis sativus* formed following infection with *Agrobacterium rhizogenes* LBA9402 showing either type A (line C15, 1) or type B morphology (line C22, 2).

this sample of 59 nkat g^{-1}. Overall within this sample a 2.5-fold range of activity was observed.

However, this is underlaid by a further source of variation. Within this control population we observed two morphologically distinct populations of transformed roots. Roots of one group (type A) were fine and highly branched showing typical hairy root structure but were relatively slow growing while those of the other group (type B) were coarse, less well branched, yet were generally faster growing. The morphology of these two classes of roots is illustrated in Fig. 8.

Recently an explanation for these distinct morphologies has come from the work of Amselem & Tepfer (1992) who showed that the roots with the coarse morphology expressed T_L-DNA bearing the rol genes together with T_R-DNA which carries two genes for auxin biosynthesis homologous to the *iaaH* and *iaaM* genes of *A. tumefaciens*. However, those showing the fine morphology expressed only T_L-DNA. T_R-DNA is not essential to the maintenance of transformed roots (Spano et al. 1981, Tepfer & Tempé 1981) but expression of the auxin genes leads to over-production of auxin which influences their patterns of growth. There is essentially no difference in the pattern of AO activity in the two classes of transformed cucumber roots. Indeed if the population of control roots is subdivided on the basis of morphological type there is no significant difference between the means of the two populations (Type A 171 ± 40 nkat g^{-1}, n = 6 ; Type B 156 ± 38 nkat g^{-1}, n = 14).

The hormonal regimes established within transformed organ cultures influence the pattern of growth of the culture and may directly or indirectly affect the operation of secondary pathways. Indirect effects may follow changes in branching pattern leading to changed cell age and cell type distributions. Hormonal levels within transformed tissues are influenced by the locus, copy number and integrity of the T-DNA. In cucumber, the integration of T_R-DNA has a marked effect on root morphology but in the case of AO this does not appear to be a major influence on the pattern of changes in the activity of this enzyme.

Transformed organ cultures are useful models in which to study secondary metabolic pathways. Such pathways are likely to reflect the *in planta* situation in the route and enzymology of the pathway. However, the unique features of transformed roots cultures, their altered hormone metabolism, their production of opines and their isolation from the rest of the plant preventing transport of metabolites may significantly alter the regulation of such pathways. Therefore caution should be applied in extrapolating of results on the regulation of such pathways from studies in organ culture to the whole plant.

References

Aird ELH (1988) The study of the relationship between organisation, genetic stability and secondary metabolite production in plant cell and organ cultures. Ph.D. Thesis, University of East Anglia.

Akiyoshi DE, Morris RO, Hinz R, Mischake SB, Kosuge T, Garfunkel DJ, Gordon MP & Nester EW (1983) Cytokinin/auxin balance in crown gall tumors is regulated by specific loci in the T-DNA. Proc. Natl. Acad. Sci. USA 80: 407–411

Amselem J & Tepfer M (1992) Molecular basis for novel root phenotypes induced by *Agrobacterium rhizogenes* A4 on cucumber. Plant Mol. Biol. 19: 421–432

Cardarelli M, Marriotti D, Pomponi M, Spano L, Capone I & Constantino P (1987) *Agrobacterium rhizogenes* T-DNA genes capable of inducing the hairy root phenotype. Mol. Gen. Genet. 209: 475–480

Croes AF, van den Berg AJR, Bosveld M, Breteler H & Wullems GJ (1989) Thiophene accumulation in relation to morphology in roots of *Tagetes patula*. Effects of auxin and transformation by *Agrobacterium*. Planta 179: 43–50

Esaka M, Uchida M, Fukui H, Kubota K & Suzaki K (1988) Marked increase in ascorbate oxidase protein in pumpkin callus by adding copper. Plant Physiol. 88: 656–660

Estruch J, Chriqui D, Grossmann K, Schell J & Spena A (1991a) The plant oncogene rolC is responsible for the release of cytokinins from glucoside conjugates. EMBO J. *10* 2889–2895

Estruch J, Schell J & Spena A (1991b) The protein encoded by the rol B plant oncogene hydrolyses indole glucosides. EMBO J. 10: 3125–3128

Giulietti AM, Parr AJ & Rhodes MJC (1993) Tropane alkaloid production in transformed roots of *Brugmansia candida*. Planta Med. (in press)

Hamill JD, Parr AJ, Robins RJ & Rhodes MJC (1986) Secondary product formation by cultures of *Beta vulgaris* and *Nicotiana rustica* transformed with *Agrobacterium rhizogenes*. Plant Cell Rep. 5: 111–114

Hooykaas PJJ & Schilperoort RA (1992) *Agrobacterium* and plant genetic engineering. Plant Mol. Biol. 19: 15–38.

Kamada H, Ohkawa H, Harada H & Shimomura K (1989) Effects of GA_3 on growth and alkaloid production by hairy roots of *Datura innoxia*. Proceedings of joint meeting Plant Growth Regulator Society of America and Japanese Society for the Chemical Regulation of Plants pp 227–232

Körber H, Strizhov N, Staiger D, Feldwisch J, Olsson O, Sandberg G, Palme K, Schell J & Koncz C (1991) T-DNA gene 5 of *Agrobacterium* modulates auxin response by autoregulated synthesis of a growth hormone antagonist in plants. EMBO J. 10: 3983–3991

Li Y, Hagen G & Guilfoyle TJ (1992) Altered Morphology in Transgenic tobacco Plants That Overproduce Cytokinins in Specific tissues and Organs. Developm. Biol. 153: 386–395

Ohkawa H, Kamada H, Suodo H & Harada H (1989) Effects of GA_3 on hairy root growth in *Datura innoxia*. J. Plant Physiol. 134: 633–636

Rhodes MJC, Spencer AJ, Hamill JD & Robins RJ (1992) Flavour improvement through plant cell Culture. In: Patterson RIS, Charlwood BV, MacLeod G & Williams AA (Eds) Bioformation of Flavours (pp 42–65) Royal Society of Chemistry, London

Robins RJ, Bent ES & Rhodes MJC (1991) Studies on the biosynthesis of tropane alkaloids by *Datura stramonium* L. transformed root cultures. 3. The relationship between morphological integrity and alkaloid biosynthesis. Planta 185: 385–390

Smigocki AC & Owens LD (1988) Cytokinin gene fused with a strong promoter enhances shoot organogenesis and zeatin levels in transformed plant cells. Proc. Natl. Acad. Sci. USA 85: 5131–5135

Spano A, Wullens GJ, Schilperoort RA & Constantino P (1981) Hairy root: *in vitro* growth properties of tissues induced by *Agrobacterium rhizogenes* in Tobacco. Plant Sci. Lett. 23: 299–305

Spena A, Schmülling T, Koncz C & Schell J (1987) Independent and synergistic activity of rolA, B and C loci in stimulating abnormal growth in plants. EMBO J. 6: 3891–3899

Spencer AJ, Hamill JD & Rhodes MJC (1990) Production of terpenes by differentiated shoot cultures of *Mentha citrata* transformed with *Agrobacterium tumefaciens* T37. Plant Cell Rep. 8: 601–604

Tepfer D & Tempé J (1981) Production d'agropine par des racine formées sous l'action d'*Agrobacterium rhizogenes*. Comptes Rendus, Académie des Sciences, Paris 292: 153–156

Tinland B, Huss B, Paulus F, Bonnard G & Otten L (1989) *Agrobacterium tumefaciens* 6b genes are strain specific and affect the activity of auxin as well as cytokinin genes. Mol. Gen Genet. 219: 217–224

Vilaine F, Charbonnier C & Casse-Delbart F (1987) Further insight concerning the TL-DNA region of the Ri plasmid of *Agrobacterium rhizogenes* strain A4: Transfer of a 1.9kb fragment is sufficient to induce transformed roots on tobacco leaf fragments. Mol. Gen. Genet. 210: 111–115

Yoshikawa T & Furuya T (1987) Saponin production by cultures of *Panax ginseng* transformed with *Agrobacterium rhizogenes*. Plant Cell Rep. 6: 449–453

Plant Cell, Tissue and Organ Culture **38**: 153–158, 1994.

Molecular cloning and expression of key enzymes for biosynthesis of cysteine and related secondary non-protein amino acids

Kazuki Saito, Naoko Miura, Mami Yamazaki, Kazuyo Tatsuguchi, Makoto Kurosawa, Reiko Kanda, Masaaki Noji & Isamu Murakoshi
Faculty of Pharmaceutical Sciences, Chiba University, Yayoi-cho 1–33, Inage-ku, Chiba 263, Japan

Key words: O-acetylserine (thiol) lyase, cDNA cloning, cysteine synthase, genetic complementation, *Spinacia oleracea*, transgenic plants

Abstract

Cysteine synthase plays a key role in the sulfur assimilation pathway in plant cells. The cDNA clones encoding two isoforms of this enzyme were isolated from spinach by synthetic oligonucleotide probes. The modes of expression of these two genes differed in tissues of spinach. A heterologous expression system in *Escherichia coli* and transgenic tobacco was made. The application of heterologous expression to modify sulfur metabolism and to produce non-protein amino acids is discussed.

Abbreviation: CSase – cysteine synthase

Introduction

The concentrations of essential elements differ in plant cells and in soil (Table 1). Sulfur needs to be concentrated in plant cells relative to soil by a factor 4.9, nitrogen 30-fold and carbon 22.7-fold. Most inorganic sulfate assimilated by plants appears ultimately in cysteine and methionine in proteins. Cysteine is the principal starting metabolite for the synthesis of other sulfur-containing metabolites such as methionine, glutathione and S-alkenylcysteine sulfoxides (alliin), flavor precursors in species of *Allium*.

Animals require a dietary source of methionine for sulfur metabolism to inorganic sulfate, and plants in turn assimilate inorganic sulfate back to cysteine (Fig. 1). Therefore, the step of cysteine biosynthesis is one of the key reactions in biology, being comparable in importance to CO_2 assimilation in photosynthesis and nitrogen assimilation by rhizobacteria. The crucial fixation step of inorganic sulfide to cysteine, the first sulfur containing organic compound, is catalyzed by cysteine synthase (CSase) [O-acetyl-L-serine (thiol)-lyase, O-acetyl-L-serine acetate-lyase (adding hydrogen sulfide), EC 4.2.99.8]. This pyridoxal phosphate-dependent enzyme catalyzes the formation of cysteine and acetic acid from O-acetylserine and hydrogen sulfide (Anderson 1980; Lea et al. 1985).

Non-protein amino acids in higher plants form a large group of secondary plant products (Rosenthal 1982). Some of them are toxic to humans and livestocks. Recently, our study has indicated that some non-protein β-substituted alanines, *e.g.*, L-mimosine, are biosynthesized by isozymes of CSase (Fig. 2) (Murakoshi et al. 1972; Ikegami et al. 1990).

In plant cells, there are three isoforms of CSase of different subcellular localization. In spinach (*Spinacia oleracea* L.) green leaves, the two major activities are localized in cytoplasm and chloroplasts as CSase A and B, respectively. A minor activity is localized in mitochondria (Lunn et al. 1990). In the present study, we present our recent data on the molecular cloning of CSase, a comparison of deduced amino acid sequences and expression in prokaryotic and eukaryotic transgenic cells.

Table 1. Concentration of essential elements in plant and soil (Masuda 1988).

Element	Concentration(ppm)		Ratio (plant/soil)
	Plant	Soil	
1. C	454,000	20,000	22.7
2. O	410,000	490,000	0.84
3. H	55,000	–	–
4. N	30,000	1,000	30
5. Ca	18,000	13,700	1.3
6. K	14,000	14,000	1.0
7. S	3,400	700	4.9
8. Mg	3,200	5,000	0.64
9. P	2,300	650	3.5

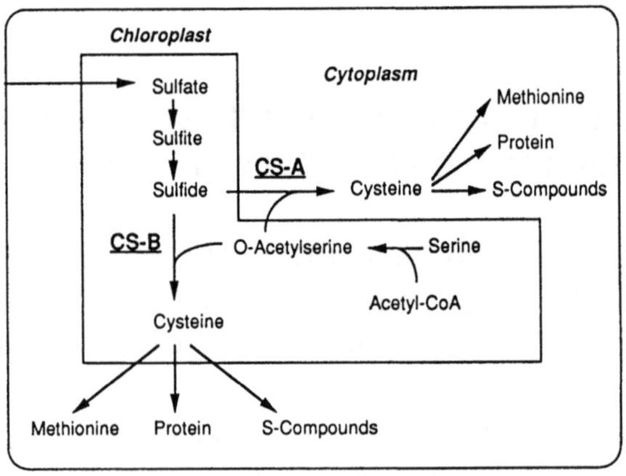

Fig. 1. Pathway for sulfur assimilation in plant cells.

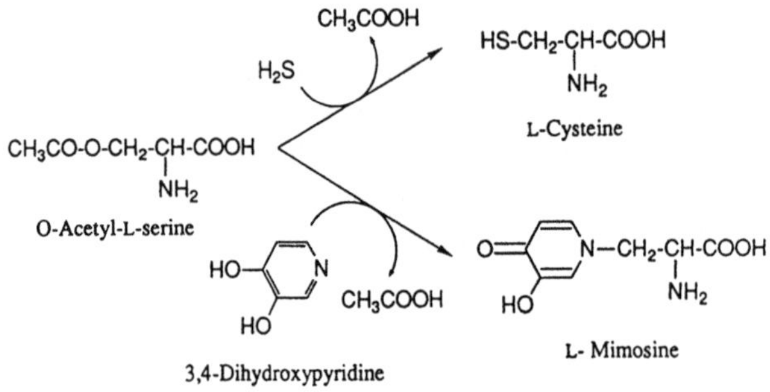

Fig. 2. Biosynthesis of cysteine and mimosine from O-acetylserine by cysteine synthase.

Results and discussion

Purification, microsequencing of digested peptide fragments and cDNA cloning of plant CSases (Saito et al. 1992a, b)

In green leaves of spinach, we detected the activity of, at least, two forms of CSases, isozymes A and B, which are different in sub-cellular localization. CSase A has a molecular weight around 35kDa on SDS-polyacrylamide gel electrophoresis (SDS-PAGE) and occurs primarily in the cytoplasm; whereas CSase B is localized in chloroplasts. Two isozymes were purified from the green leaves of spinach by steps involving heat treatment, fractional precipitation with ammonium sulfate, DEAE-Sephadex chromatography, gel filtration, EAH-Sepharose chromatography and preparative PAGE. The N-terminal amino acid of the purified CSase A was blocked and thus gave no information on the amino acid sequence, even after an acid treatment to remove a N-formyl group or a pyroglutamyl peptidase treatment. Thus, internal amino acid sequences were determined after digestion with *Staphylococcus aureus* V8 protease.

Poly $(A)^+$ RNA was isolated from young green leaves of *S. oleracea* cv Parade and the cDNA library in λgt10 was made. The non-amplified cDNA library was comprised of 2×10^5 independent phage clones and was screened with two synthetic DNA probes designed from partially determined amino acid sequences of CSase A. The probe V812 was a 50 mer in length and had 16 base degeneracy with 11 inosines at ambiguous positions. The probe V822 was a "guessmer" of 56 mer length and 8 base degeneracy containing 9 inosines. Among 19 doubly positive clones for both V812 and V822, two clones of ca 1.3 kb-length insert were selected for further analysis, because this length of cDNA matched to that deduced from the determined molecular mass (35 kDa) of purified CSase A. By a similar approach, 16 clones for CSaseB were also isolated by using the specific oligonucleotides designed for the digested fragments of CSaseB as probes.

Nucleotide and deduced amino acid sequences of CSases

Sequence determination of cDNA clones for CSaseA revealed an open reading frame of 975 bp encoding 325 amino acids (Saito et al. 1992a, 1992b). The calculated molecular mass of the encoded peptide was 34,185 Da, which was coincident with that of purified CSase A determined by SDS-PAGE. All four determined peptide sequences were identical with the sequences predicted from the cDNA sequences except for 3 equivocal residues in V8–5. The sequence around the ini-

```
                                                                                                   ↓
1. Consensus                  1 : MAS..NN............E.......R.........N...KV......CKAVS.............I.......IG.TP.V.L............K.
2. CS B (Sp. oleracea)        1 : MASLVNNAYAAIRTSKLELREVKNLANFRVGPPSSLSCNNFKKVSSSPITCKAVSL--SPPSTIEGLNIAEDVSQLIGKTPMVYLNNVSKGSVANIAAKL
3. CS B (C. annuum)           1 : MASIINNPFTSL-CCNTNKCEPNRICSLRSQQSLVFD-NVNRKVGFPSVVCKAVSVQTKSPTEIEGLNIAEDVTQLIGNTPMVYLNTIVKGCVANIAAKL
4. CS A (T. aestivum)         1 :                                                             MGEASSPAIAKDVTELIGNTPLVYLNKVTDGCVGRVAAKL
5. CS A (Sp. oleracea)        1 :                                                              MVEEKAFIAKDVTELIGKTPLVYLNTVADGCVARVAAKL
6. cysK (E. coli)             1 :                                                                  MSKIFEDNSLTIGHTPLVRLNR--IGNGRIL-AKV
7. cysK (Sa. typhimurium)     1 :                                                                  MSKIYEDNSLTIGHTPLVRLN--RIGNGRIL-AKV
8. cysM (E. coli)             1 :                                                                      MSTLEQTIGNTPLVKLQRMGPDNGSEVWLKL

                            101 : E...P..SVK.RI...MI..AE..G...PG...L.E.T.GNTGI.LA..AA..GY.....MP..MS.ERR....A.GA.L.L.....GM.GA...A.E..
                            101 : ESMEPCCSVKDRIGYSMIDDAEQKGVITPGKTTLVEPTSGNTGIGLAFIAAARGYKITLTMPASMSMERRVILKAFGAELVLTDPAKGMKGAVEKAEEIL
                            101 : EIMEPCCSVKDRIGFSMISDAEEKGLISPGKTVLVEPTSGNTGIGLAFIAASRGYKLILTMPASMSLERRVILKAFGAELVLTDPAKGMKGAVSKAEEIL
                             41 : ESMEPCSSVKDRIGYSMITDAEEKGFIVPGKSVLIEPTSGNTGIGLAFMAAAKGYRLVLTMPASMSMERRIILKAFGAELILTDPLLGMKGAVQKAEELA
                             40 : EGMEPCSSVKDRIGFSMITDAEKSGLITPGESVLIEPTSGNTGIGLAFIAAAKGYKLIITMPASMSLERRTILRAFGAELILTDPAKGMKGAVQKAEEIR
                             36 : ESRNPSFSVKCRIGANMIWDAEKRGVLKPG-VELVEPTSGNTGIALAYVAAARGYKLTLTMPETMSIERRKLLKALGANLVLTEGAKGMKGAIQKAEEIV
                             36 : ESRNPSFSVKCRIGANMIWDAEKRGVLKPG-VELVEPTNGNTGIALAYVAAARGYKLTLTMPETMSIERRKLLKALGANLVLTEGAKGMKGAIQKAEEIV
                             32 : EGNNPAGSVKDRAALSMIVEAEKRGEIKPG-DVLIEATSGNTGIALAMIAALKGYRMKLLMPDNMSQERRAAMRAYGAELILVTKEQGMEGARDLALEMA

                            201 : .........L.QF.N..NP..H..TTGPEIW..T.G.........GT.GT..G............................L...............
                            201 : KKTPDSY-MLQQFDNPANPKIHYETTGPEIWEDTKGKVDIFVAGIGTGGTISGVGRYL--KERNPGVQVIGIEPTES----NILSG--GKPGPHKIQGLG
                            201 : NNTPDAY-ILQQFDNPANPKIHYETTGPEIWEDTKGKIDILVAGIGTGGTISGTGRYL--KEKNPNIKIIGVEPTES----NVLSG--GKPG--------
                            141 : AKTPNSY-ILQQFENAANPKIHYETTGPEIWKGTGGKIDGLVSGIGTGGTITGTGKYL--QEQNPNIKLYGVEPTES----AILNG--GKPGPHKIQGIG
                            140 : DKTPNSY-ILQQFENPANPKVHYETTGPEIWKGTGGKIDIFVSGIGTGGTITGAGKYL--KEQNPDVKLIGLEPVES----AVLSG--GKPGPHKIQGLG
                            136 : ASNPEKYLLLQQFSNPANPEIHEKTTGPEIWEDTDGQVDVFIAGVGTGGTWTGVTPYIKGTKGKTDLISVAVEPTDSPVIAQALAGEEIKPGPHKIQGIG
                            136 : ASDPQKYLLLQQFSNPANPEIHEKTTGPEIWEDTDGQVDVFISGVGTGGTLTGVTRYIKGTKGKTDLITVAVEPTDSPVIAQALAGEEIKPGPHKIQGIG
                            132 : NRGEGK--LLDQFNNPDNPYAHYTTTGPEIWQQTGGRITHFVSSMGTTGTITGVSRFMR-------------EQSKPVTIVGLQPEEGSSIP-GIRRWP

                            301 : ....P.......D................L...E....G.SSG.A...A....................G.RYLS...F.............
                            301 : AGFVPSNLDLGVMDEVIEVSSEEAVEMAKQLAMKEGLLVGISSGAAAAAAVRIGKRPENAGKLIAVVFPSFGERYLSSILFQSIREECENMKPE
                            301 : --FIPGNLDQDVMDEVIEISSDEAVETAKQLALQEGLLVGISSGAAALAAIQVAKRPENAGKLIAVVFPSFGERYLSSILFQSIREECEKMKPEL
                            241 : AGFIPGVLDVDIIDETIQVSSDESIEMAKSLALKEGLLVGISSGAAAAAAIKVAQRPENAGKLFVVVFPSFGERYLSSVLFHSIKKEAESMVVE
                            240 : AGFIPGVLDVNIIDEVVQISSEESIEMAKLLALKEGLLVGISSGAAAAAAIKVAKRPENAGKLIVAVFPSFGERYLSSVLFDSVRKEAESMVIES
                            236 : AGFIPANLDLKLVDKVIGITNEEAISTARRLMEEEGILAGISSGAAVAAALKLQEDESFTNKNIVVILPSSGERYLSTALFADLFTEKELQQ
                            236 : AGFIPGNLDLKLIDKVVGITNEEAISTARRLMEEEVFLAGISSGAAVAAALKLQEDESFTNKNIVVILPSSGERYLSTALFADLFTEKELQQ
                            232 : TEYLPGIFNASLVDEVLDIHQRDAENTMRELAVREGIFCGVSSGGAVAGALRVAKANPDAV--VVAIICDRGDRYLSTGVFGEEHFSQGAGI
```

Fig. 3. Comparison of deduced amino acid sequences of cysteine synthases from various sources and consensus sequence. The *Arrow* indicates the cleavage site of transit peptide. *Dashes* indicate gaps in sequence for the best alignment.

tiation ATG codon, CAAAATGGT, closely matched the proposed consensus sequence for plant gene initiation codons, AACAATGGC (Lütcke et al. 1987), indicating this ATG to be the proper start methionine codon.

Sequence analysis of CSaseB cDNAs showed the calculated molecular mass of pre-CSaseB containing a 52 amino acid transit peptide as 40.636Da and that of mature CSaseB as 34.998Da.

The multiple alignment of predicted amino acid sequences of CSases indicated that the sequences after the cleavage position at 52–53 of chloroplastic isoforms, CSaseB, are homologous in all seven proteins (Fig. 3).

Hybridization analysis

Southern blot analysis of genomic DNA indicated the presence of at least 2–3 copies *cysA* and *cysB* genes encoding CSase A and B, respectively, in spinach. Presumably *cysA* and *cysB* comprise independent small multi-gene families and are expressed differently during the developments as postulated in the case of glutamine synthase genes for ammonia fixation (Tingey et al. 1988).

Expression of *cysA* and *cysB* genes was analyzed by RNA blot hybridization. *CysB* gene was expressed primarily in green tissues of spinach. Small but substantial amounts of transcripts were also accumulated in roots of the plants, presumably due to the expression in non-green plastids. On the contrary, the expression of *cysA* gene was constitutive in leaves and roots of 4- and 10-week-old spinach. These suggested that CSase B localized in chloroplasts is functionally related to sulfate reduction coupled with photosynthesis and utilizes sulfide immediately after the reduction of sulfate; whereas CSase A localized in cytoplasm has the general role for assimilation and detoxication of sulfide in the cells.

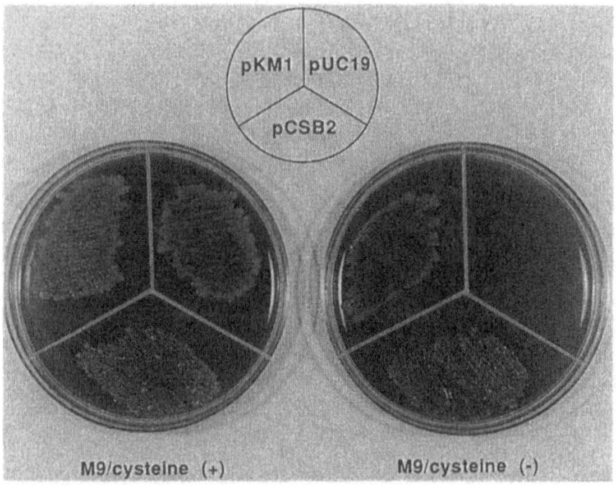

Fig. 4. Genetic complementation of Cys⁻ *E. coli* NK3 by transformation with expression vectors, pCSB2 and pKM1. Transformed bacteria were spread on M9 minimal agar plates supplemented with 0.02% leucine and tryptophan plus 0.5mM cysteine (*left* plate) or without cysteine (*right* plate).

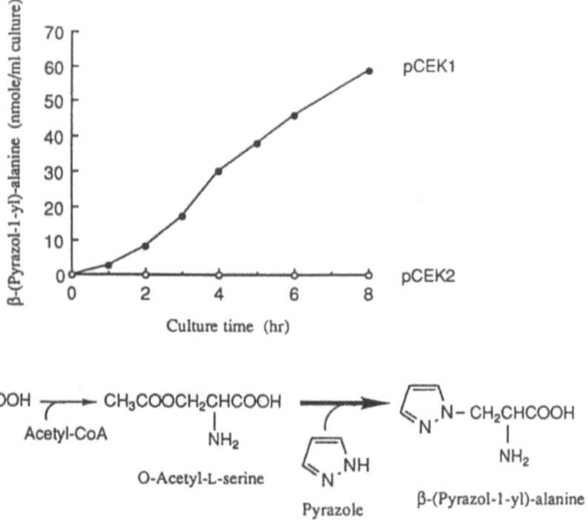

Fig. 5. Production of β-(pyrazol-1-yl)-alanine by the over-expression system of a plant CSase cDNA in *E. coli*. cDNA sequence of spinach CSase A is ligated under the transcriptional control of T7 promoter in pCEK1 and pCEK2 with sense and antisense orientation, respectively. The plasmids were used for the transformation of *E. coli* BL21 for over-expression of the chimeric genes. After pre-culture for 1.5 hr at 37 °C in LB medium, IPTG (1 mM), serine (50 mM) and pyrazole (50 mM) were added in the culture medium. The concentration of β-(pyrazol-1-yl)-alanine in the *E. coli* cells was determined by an amino acid analyzer.

Expression of spinach CSase cDNA and genetic complementation in Cys⁻ Escherichia coli

To confirm the identity of the isolated clones encoding CSase, a cysteine-auxotroph mutant, *E. coli* NK3 (Kredich 1987), was complemented genetically by the expression of spinach CSase cDNA under the control of the *lacZ* promoter. *E. coli* NK3 transformed with expression vectors, pKM1 and pCSB2, were able to grow in the minimal medium without cysteine; whereas *E. coli* transformed with a control vector, pUC19, was unable to grow in minimal medium (Fig. 4). The expression of functionally active CSases in *E. coli* NK3 transformed with pKM1 and pCSB2 was also demonstrated by Western blot analysis using an antibody against CSase A and by the assay of enzymatic activity of CSases. These results confirmed that those cDNA clones encode CSases which are functionally active in *E. coli*.

Over-expression in E. coli and production of β-(pyrazol-1-yl)-alanine

The over-expression system using a pET vector with the strong T7 transcriptional promoter was established for CSaseA in *E. coli*. The spinach CSaseA protein was accumulated maximally up to 45% in the total soluble protein of *E. coli* extract expressing the cDNA. This system was successfully used for the production of a plant specific metabolite, β-(pyrazol-1-yl)-alanine. This non-protein secondary amino acid is distributed in the plants of Cucurbitaceae, i.e., watermelon and cucumber. The *in vivo* production of β-(pyrazol-1-yl)-alanine was only achieved by use of the over-expression vector, pCEK1, but not by the control plasmid, pCEK2 (Fig. 5). This is one of the first successful examples for production of a plant secondary metabolite by expression of a plant gene in prokaryotic cells.

Expression of the chimeric genes in transgenic tobacco plants

We made three expression vectors for plants containing cDNA sequence of CSaseA. Vector pCSK3F contained the CaMV 35S promoter fused to the cDNA of CSaseA in a sense orientation. A second construct pCSK3R contained CaMV35S promoter fused to the cDNA in an antisense orientation. A third vector pCSK4F was made by chimeric fusion of CaMV35S promoter to the transit peptide (TP) of pea in the sense orientation to enable targeting to chloroplasts. Transgenic tobacco containing these constructs was obtained using the transformation system of *Agrobacterium tumefaciens*-pGV2260 as described previously (Saito et al. 1991).

Analysis of transgenic plants showed a four-fold enhanced CSase activity in the plants with the TP fused construct pCSK4F and a 2.5-fold enhanced

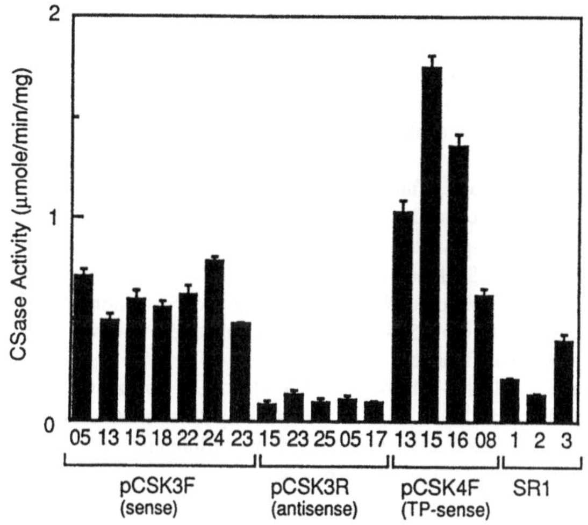

Fig. 6. Enzymatic activity of CSase in transgenic tobacco plants transformed with pCSK3F (35S promoter – sense cDNA), pCSK3R (35S promoter – antisense cDNA) and pCSK4F (35S promoter – TP for chloroplast transport – sense cDNA). *SR1* indicates the activity in non-transformed plants. The soluble protein extracts of independent plants were used for the assay of CSase activity as described (Saito 1992a).

activity in the plants with pCSK3F. The activity in the plants containing the antisense pCSK3R construct were decreased to ca 50% of those of the control plants (Fig. 6). The cellular concentrations of cysteine were also changed in transgenic plants as expected, although the changes of cysteine contents in the leaves were not directly correlated with those of CSase activity. Further detailed analyses will be necessary on the effects of modified CSase activities on the metabolic flux of sulfur-containing amino acids and secondary products.

Cloning strategy for cDNA encoding cysteine biosynthetic enzymes by genetic complementation of E. coli Cys⁻ auxotroph

The biosynthetic pathways for cysteine are similar in plants and bacteria. A number of cysteine-auxotrophic *E. coli* mutants lacking an enzyme for each step are available (Kredich 1987). Thus, a cloning strategy using genetic complementation of *E. coli* mutants by screening with an expression cDNA library can certainly work for isolation of cDNA clones for each step in the reaction. *E. coli* NK3 Cys^- auxotroph lacking *cysK* and *cysM* for CSases was used for isolation of cDNA in watermelon (*Citrullus vulgaris*) for CSase responsible for biosynthesis of β-(pyrazol-1-yl)-alanine. We have obtained a number of positive clones which could complement the lack of endogenous CSases in *E. coli* NK3. The cDNAs for other enzymes involved in the assimilation of inorganic sulfur into cysteine are being cloned by this strategy using mutants lacking corresponding step of sulfur assimilation.

Conclusions

The present study showed the first successful isolation and characterization of cDNA clones encoding different forms of CSase, a key enzyme of sulfur assimilation, from plants. The expression of the plant CSase in *E. coli* could functionally complement the lack of endogenous CSases. The production of a plant secondary metabolite, β-(pyrazol-1-yl)-alanine, was achieved by over-expression of the plant cDNA in *E. coli*. Transgenic plants containing the cDNA in sense and antisense orientation were made and analyzed for CSase activity and cellular cysteine contents. We are also currently isolating the cDNA encoding the specific isozymes responsible for biosynthesis of other secondary non-protein amino acids, *i.e.*, mimosine in *Leucaena leucocephala*. It would be very interesting to clarify the molecular relations of biosynthesis of cysteine and the secondary non-protein amino acids and their molecular evolutionary relations.

Acknowledgements

We thank Dr. N. M. Kredich (Duke University, Medical Center) for supplying *E. coli* NK3; Dr. S. Youssefian (Akita Prefectural College of Agriculture, Akita) for sharing his results before publication. This research was supported, in part, by Grants-in-Aid for Scientific Research from the Ministry of Education, Science and Culture, Japan; from the Research Foundation for Pharmaceutical Sciences, Japan; from Shourai Foundation; from Naito Foundation.

References

Anderson JW (1980) Assimilation of inorganic sulfate into cysteine. In: Miflin BJ (Ed) The biochemistry of plants, Vol 5 (pp 203–225) Academic Press, New York

Ikegami F, Mizuno M, Kihara M & Murakoshi I (1990) Enzymatic synthesis of the thyrotoxic amino acid mimosine by cysteine synthase. Phytochemistry 29: 3461–3465

Kredich NM (1987) Biosynthesis of cysteine. In: Neidhardt FC, Ingraham JL, Low KB, Magasanik B, Schaechter CM & Umbarger HE (Eds) *Escherichia coli* and *Salmonella typhimurium*: Cellular and Molecular Biology, Vol 1 (pp 419–428) American Society for Microbiology, Washington DC

Lea PJ, Wallsgrove RM & Miflin BJ (1985) The biosynthesis of amino acids in plants. In: Barret GC (Ed) Chemistry and Biochemistry of the Amino Acids. (pp 197–226) Chapman and Hall, London

Lunn JE, Droux M, Martin J & Douce R (1990) Localization of ATP sulfurylase and O-acetylserine (thiol) lyase in spinach leaves. Plant Physiol. 94: 1345–1352

Lütcke HA, Chow KC, Mickel FS, Moss KA, Kern HF & Scheele GA (1987) Selection of AUG initiation codons differs in plants and animals. EMBO J. 6: 43–48

Masuda Y (1988) Plant Physiology, revised edition (in Japanese), Baifu-kan, Tokyo.

Murakoshi I, Kuramoto H, Haginiwa J & Fowden L (1972) The enzymatic synthesis of β-substituted alanines. Phytochemistry 11: 177–182

Rosenthal GA (1982) Plant Nonprotein Amino and Imino Acids. Academic Press, New York

Saito K, Noji M, Ohmori S, Imai Y & Murakoshi I (1991) Integration and expression of a rabbit liver cytochrome P-450 gene in transgenic *Nicotiana tabacum*. Proc. Natl. Acad. Sci. USA 88: 7041–7045

Saito K, Miura N, Yamazaki M, Hirano H & Murakoshi I (1992a) Molecular cloning and bacterial expression of cDNA encoding a plant cysteine synthase. Proc. Natl. Acad. Sci. USA 89: 8078–8082

Saito K, Miura N, Yamazaki M & Murakoshi I (1992b) Isolation and sequence analysis of cDNA clones encoding cysteine synthase from spinach (*Spinacia oleracea*). In: Oono K, Hirabayashi T, Kikuchi S, Handa H & Kajiwara K (Eds) Plant Tissue Culture and Gene Manipulation for Breeding and Formation of Phytochemicals, (pp 157–164) National Institute of Agrobiological Resources, Tsukuba

Tingey SV, Tsai F-Y, Edwards JW, Walker EL & Coruzzi GM (1988) Chloroplast and cytosolic glutamine synthetase are encoded by homologous nuclear genes which are differentially expressed *in vivo*. J. Biol. Chem. 263: 9651–9657

Plant Cell, Tissue and Organ Culture **38**: 159–165, 1994.

Thiophene biosynthesis in *Tagetes* roots: molecular versus metabolic regulation

Anton F. Croes, John J.M.R. Jacobs, Randy R.J. Arroo & George J. Wullems
NOVAPLANT Cell Biotechnology Group, Department of Experimental Botany, University of Nijmegen, Toernooiveld, 6525 ED Nijmegen, the Netherlands

Key words: Polyacetylenes, secondary metabolism, sulfur partitioning, *Tagetes*, thiophenes, transformed roots

Abstract

The biosynthesis of thiophenes from polyacetylenes and its regulation were studied in seedlings and transformed roots of *Tagetes erecta* and *T. patula*. The key steps in the conversion are the closure of the first and second heterocyclic ring by addition of reduced sulfur to two acetylenic groups of tridecapentaynene. Two presumptive intermediates in the sequence of reactions were isolated from a mutant of *T. erecta* with an altered thiophene spectrum. The compounds were purified by HPLC and identified by GC-MS and ^{1}H-NMR. One of them proved to be a monothiophene, the other a methylated bithienyl. The position of these thiophenes in the biosynthetic route was clarified by studying their conversion by *Tagetes* tissues. After formation of the monothiophene, the pathway branches. One branch leads to the major bithienyls in these species whereas the end product of the other branch are methylated forms which are only minor components in *Tagetes*. Substitutions at the vinyl end of the molecule presumably are the last reactions of both routes.

Because of the importance of sulfur addition in thiophene biosynthesis, its regulation was studied by varying the sulfur supply to isolated *Tagetes patula* roots. A forty-fold reduction in sulfate concentration in the medium relative to the standard level had no effect on root growth and development but severely slowed down thiophene accumulation. When roots cultured at low sulfate were returned to standard conditions, the rate of thiophene biosynthesis slowly increased. The rise was completely blocked by the transcription inhibitor cordycepin. It was inferred that the incorporation of sulfur is mainly regulated at the molecular level.

Abbreviations: $AcOCH_2BBT$ – 5′-(acetoxymethyl)-5-(3-buten-1-ynyl)-2,2′-bithienyl, BBTOH – 5-(4-hydroxy-1-butenyl)-2,2′-bithienyl, BBTOAc – 5-(4-acetoxy-1-butenyl)-2,2′-bithienyl, BBT – 5-(3-buten-1-enyl)-2,2′-bithienyl, BPT – 2-(but-3-en-1-ynyl)-5-(penta-1,3-diynyl)-thiophene, MeBBT – 5′-(methyl)-5-(3-buten-1-ynyl)-2,2′-bithienyl

Introduction

All secondary metabolites that accumulate in plants, are ultimately derived from primary metabolism. This notion implies that the metabolic routes leading to the biosynthesis of secondary compounds branch off primary pathways. At the branching points, the cell makes decisions as to how much of a precursor will be converted to fulfill the needs for essential primary compounds, and how much will channeled away into secondary pathways. In other words, primary and secondary metabolism compete for the same substrates at points where the secondary routes deviate from the main pathway.

Any regulated channeling of common precursors requires the operation of a control mechanism. One possibility is that secondary metabolism only proceeds at a substantial rate when the precursor concentration is high. In this model the precursor is preferentially channeled into primary metabolism. Only when the primary pathway is saturated, an 'overflow' of precursor into the secondary route occurs. This mechanism operates whenever two enzymes compete with different K_m values for the same substrate. The observation

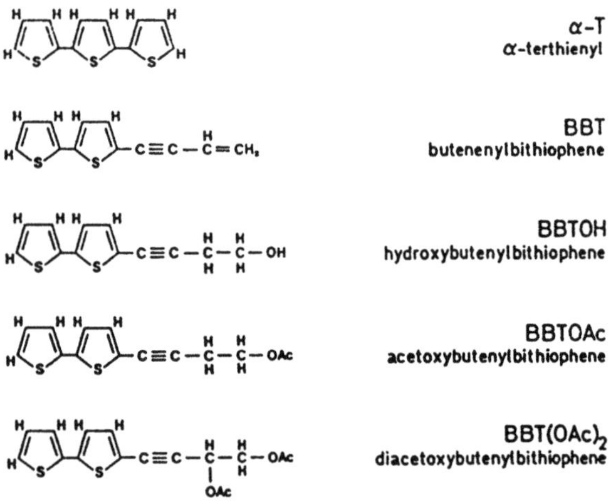

Fig. 1. Structural formulas of the five main thiophenes accumulated in *Tagetes erecta* and *T. patula*.

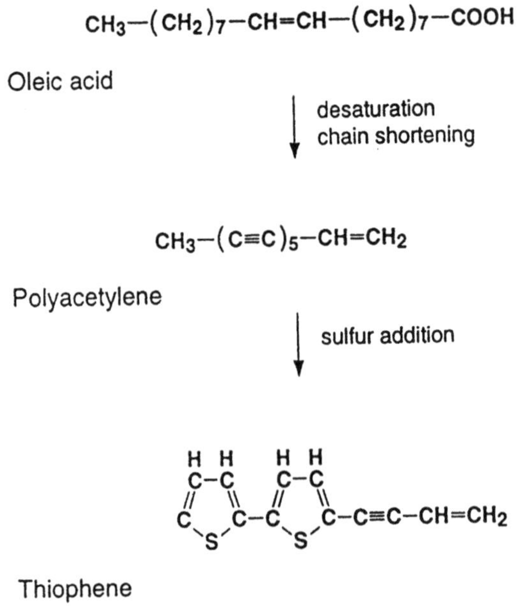

Fig. 2. General scheme of the biosynthesis of thiophenes from polyacetylenes.

that enzymes in secondary pathways usually do not work under V_{max} conditions due to substrate limitation (Luckner 1990) fits into this model of metabolic regulation.

A regulatory mechanism operating at the molecular level modulates the amount of catalyst either by synthesis and breakdown of enzyme molecules or by activation of pre-existing protein. Elicitation is an example of secondary product accumulation as a result of induced enzyme synthesis involving transcription and translation. The regulated enzyme may or may not be the first enzyme of the secondary branch (Galneder & Zenk 1990).

The synthesis of thiophenes (Fig. 1) , nematicidal compounds that accumulate in the roots of many Asteraceous plant species (Bohlmann et al. 1973), involves the partitioning of sulfur between primary and secondary metabolism. The carbon skeleton of the thiophenes is thought to be derived from oleic acid (Fig. 2). A largely unknown process of chain shortening and desaturation leads to the formation of polyacetylenes which are considered as the immediate thiophene precursors (Bohlmann & Hinz 1965). The introduction of the sulfur moiety in these molecules leads to the formation of one or more of the characteristic heterocyclic rings (Fig. 1). The sulfur comes from sulfate which is reduced to become the end-standing group of cysteine. This amino acid thus is at the branching point between primary and secondary sulfur metabolism. It may be a precursor in the synthesis of methionine and proteins but it may also act as the carrier of reduced sulfur which is added to the acetylenic groups of a polyacetylene to form the thiophene ring.

The predominant thiophenes in plants of the genus *Tagetes* are the bithienyls BBT and BBTOAc (Fig. 1) (Downum & Towers 1983). They are formed from the C_{13}-polyace-tylene tridecapentaynene (Bohlmann et al. 1966). The conversion of this compound to BBTOAc involves, apart from the formation of the two rings, chain shortening by demethylation and substitution of an acetyl group at the vinyl end. The order in which these processes occur is not known yet. There are some indications that the position of the first ring is not fixed (Bohlmann et al. 1964, 1965; Jente et al. 1981) and that formation of both rings precedes demethylation (Jente et al. 1981).

The purpose of the present study was to investigate the regulation of the synthesis of thiophenes from polyacetylenes. Isolated *Tagetes* roots transformed by *Agrobacterium rhizogenes* served as the model system. Insight in the pathway was obtained by identification of intermediates and by precursor feeding to wild-type tissues and to a mutant with an altered thiophene composition. The regulation of the process was studied by manipulating the sulfate concentration in the medium. The effects of sulfur limitation on primary and secondary sulfur metabolism were quantified. Root growth and development were used as parameters for the essential primary processes whereas thiophene synthesis and accumulation served as measures for the flow of sulfur through the secondary route.

Table 1. Thiophene content of six-days-old seedlings of *Tagetes patula* grown in vermiculite on half-strength Hoagland solution.

Organ	Thiophene content (μmol g FW^{-1})
Cotelydon	0.16 ± 0.035
Hypocotyl	0.39 ± 0.075
Root	0.77 ± 0.20

Data are means of four determinations ±S.E.

Table 2. Sulfur incorporation in thiophenes in the aerial parts of three-weeks-old plants of *Tagetes erecta*.

Organ	Sulfur in organ (kBq g FW^{-1})	Sulfur in thiophenes (kBq g FW^{-1})
Hypocotyl	588 ± 110	29.6 ± 5.3 (5.1)
Leaf	218 ± 22	2.6 ± 0.6 (1.2)
Apex	503 ± 154	6.2 ± 0.6 (1.8)

Intact plants were labeled for one day with carrier-free [^{35}S]sulfate supplied to sulfate-free half-strength Hoagland solution. The plants were then transferred to medium with sulfate and analyzed after another day. The percentage of label in thiophenes relative to the total radioactivity in the organ is given in parentheses. Data are means of four determinations ±S.E.

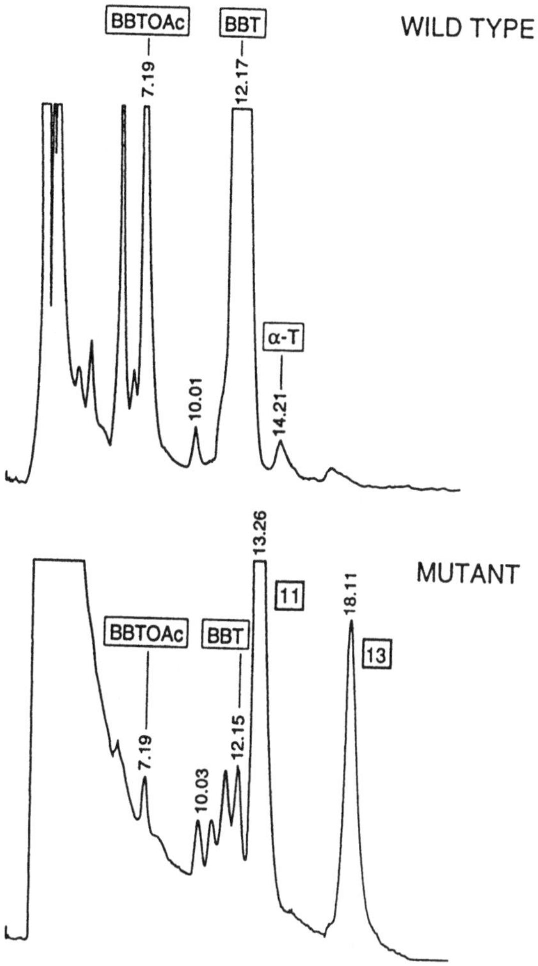

Fig. 3. HPLC profiles of hexane extracts from wild-type and mutant plants of *Tagetes erecta*. The thiophenes were monitored by their UV-absorption at 340 nm.

Results

Thiophene distribution

Thiophenes are mainly accumulated in the lower portion of the *Tagetes* plant (Sütfeld 1982). In the young seedling the total thiophene content of the plant is concentrated in the roots (Table 1). The concentration is lowest in the leaves. This distribution might reflect the rate of synthesis in the different organs but could also result from processes occurring after synthesis, such as transport or local breakdown of the metabolites. Thiophene synthesis in the aerial parts was quantified by feeding young plants [^{35}S]sulfate in the liquid culture medium and measuring the radioactivity in the thiophenes of the various organs after two days (Table 2). Much less sulfur was incorporated in the thiophenes in the leaves than in the other parts. This low incorporation is due to two factors:

- the radioactivity in the thiophenes expressed as a percentage of the total counts in the organ is lowest in the leaves, and
- the radioactive concentration which is indicative of the availability of sulfurous precursors, is lower in the leaf than in the hypocotyl and the apex.

It may be concluded that the low thiophene content of the leaf is due to a low rate of synthesis rather than to transport from the leaves or preferential localized breakdown.

The five main thiophenes accumulated in *Tagetes* tissues (Croes et al. 1989) are shown in Fig. 1. Two of these, BBT and BBTOAc, contribute for over 80% to the total amount (Fig. 3). Therefore, the study on thiophene biosynthesis was focused on these compounds.

Thiophene biosynthesis

From a biochemical point of view, the formation of BBTOAc requires the most elaborate modification of the acetylenic precursor tridecapentaynene, including closure of two rings, demethylation, and substitution (Fig. 2). The only possible intermediate that accumulates in a small amount in *Tagetes* tissue, is the acetoxy-form of MeBBT eluting from the HPLC at 10 min (Figs 3, 5). Feeding of ^{35}S-labeled $AcOCH_2BBT$ to roots did not lead to the appearance of label in any other thiophene (data not shown). This indicates that $AcOCH_2BBT$ is an end product rather than an intermediate in biosynthesis.

Other potential intermediates were obtained from a mutant of *Tagetes erecta* with an aberrant thiophene spectrum (Fig. 3). The deviant originally arose as a sector mutant after treatment of seeds with ethylmethanesulfonate. The phenotype was retained after three selfings. Transformed root cultures were incited on third-generation plants by transformation with *Agrobacterium rhizogenes.* Two new thiophenes marked '11' and '13' in Fig. 3 were isolated and identified by GC-MS and ^{1}H-NMR. The peak at 13.3 min in the HPLC chromatogram proved to be the monothiophene BPT whereas the compound with a retention time of 18.1 min is MeBBT (Fig. 5).

The question of whether BPT and MeBBT are intermediates in thiophene biosynthesis in non-mutant plants was addressed by feeding normal and mutant *Tagetes* tissues radiolabeled BPT and MeBBT. The compounds were isolated from transformed root cultures of the mutant grown in the presence of [^{35}S]sulfate, and purified by HPLC. BPT was converted by wild-type tissue into four thiophenes, methylated (MeBBT, $AcOCH_2$-BBT) as well as demethylated (BBT, BBTOAc) (Fig. 4). In contrast, little BBTOAc and no BBT was formed from BPT by the mutant whereas conversion to the methylated bithienyls occurred as in the wild type. The results indicate that BPT is an intermediate in the synthesis of all bithienyls and that the mutation affects the demethylation step. MeBBT was only converted to its acetoxy-form (Fig. 4). This result almost rules out the possibility that MeBBT is an intermediate in the biosynthesis of BBT and BBTOAc. An immediate consequence of this is that the pathway branches after BPT (Fig. 5) and that in the route leading to BBT and BBTOAc closure of the second ring occurs after or coincides with demethylation. The scheme would also explain why BPT accumulates in a mutant impaired in the demethylation step.

In normal root tissue the concentration of the methylated bithienyls is negligible in comparison to the levels of BBT and BBTOAc (Fig. 3). This would indicate that closure of the second ring following or

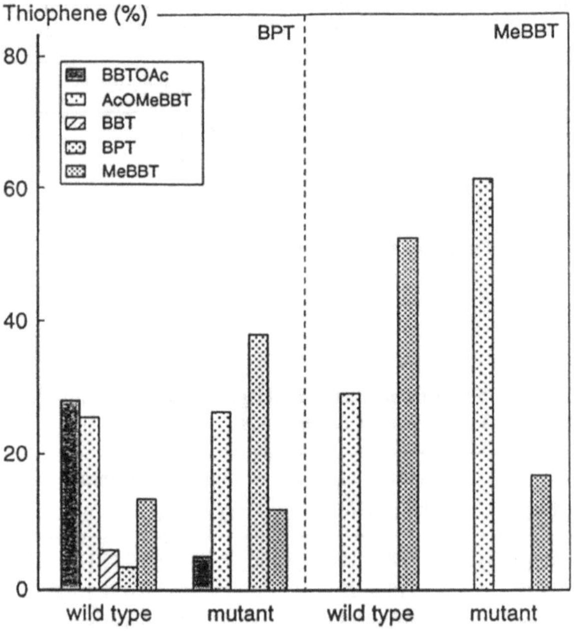

Fig. 4. Conversion of BPT (left) and MeBBT (right) to other thiophenes by wild-type and mutant roots of *Tagetes erecta*. The tissues were incubated with ^{35}S-labeled substrates and the thiophenes were analyzed by HPLC and liquid scintillation counting. Data are expressed as percentages of the total radioactivity recovered in thiophenes.

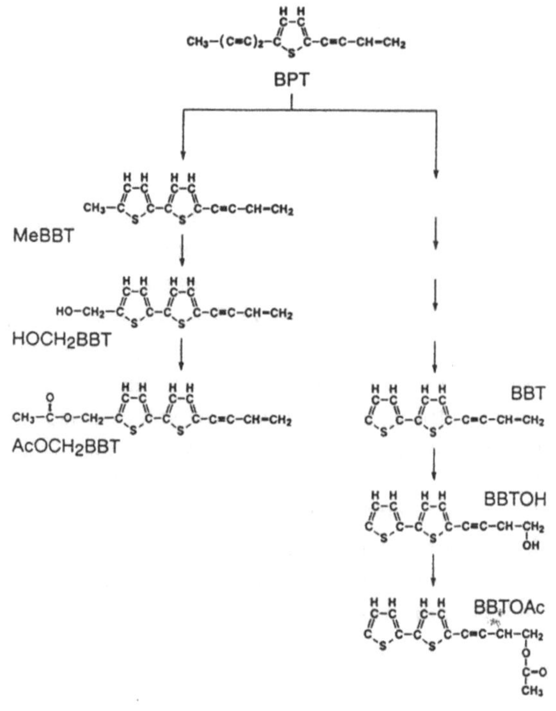

Fig. 5. Metabolic pathways in *Tagetes* leading from tridecapentaynene to the bithienyls.

Table 3. Conversion of BBT to other bithienyls.

Time of labeling (h)	Radioactivity (%)		
	BBT	BBTOH	BBTOAc
3	89 ± 2.6	2.2 ± 1.5	5.1 ± 1.5
6	85 ± 0	2.2 ± 0	9.6 ± 0
24	60 ± 2.2	2.6 ± 0	29.4 ± 1.1

Young plants were placed with the roots removed in half-strength Hoagland solution containing 0.2 MBq [^{35}S]BBT. Plants were harvested at intervals, the thiophenes were extracted from the hypocotyls, and separated by thin-layer chromatography. The radioactive spots were visualized by autoradiography, scraped off, and mixed with scintillation fluid. The radioactivity was measured and expressed as a percentage of the total count recovered in thiophenes.

coinciding with demethylation is the preferred alternative. Possibly, the presence of the end-standing methyl group makes ring closure at the acetylenic groups directly adjacent to it less efficient. If true, this would explain why the ring in BPT (Fig. 5) is formed at the third and fourth acetylenic group of tridecapentaynene.

The level in the biosynthetic scheme at which the substitution leading to the attachment of the acetoxy-group of BBTOAc takes place, is not completely certain yet. Two lines of evidence argue for BBT as the precursor of BBTOAc. When [^{35}S]BBT was fed to *Tagetes patula* hypocotyls, BBT was found to be converted to BBTOAc (Table 3). In another experiment, roots were pulsed for short culture periods with carrier-free [^{35}S] sulfate and the specific activity of three individual radiolabeled bithienyls was determined afterwards (Fig. 6). The radioactivity first appeared in BBT, then in BBTOH and finally in BBTOAc. The specific activities at 15, 30, and 60 min decreased in the same order. This result would not have been expected if sulfur had been incorporated in hydroxylated or acetylated intermediates. Substitution of BBT thus appears to be the natural process.

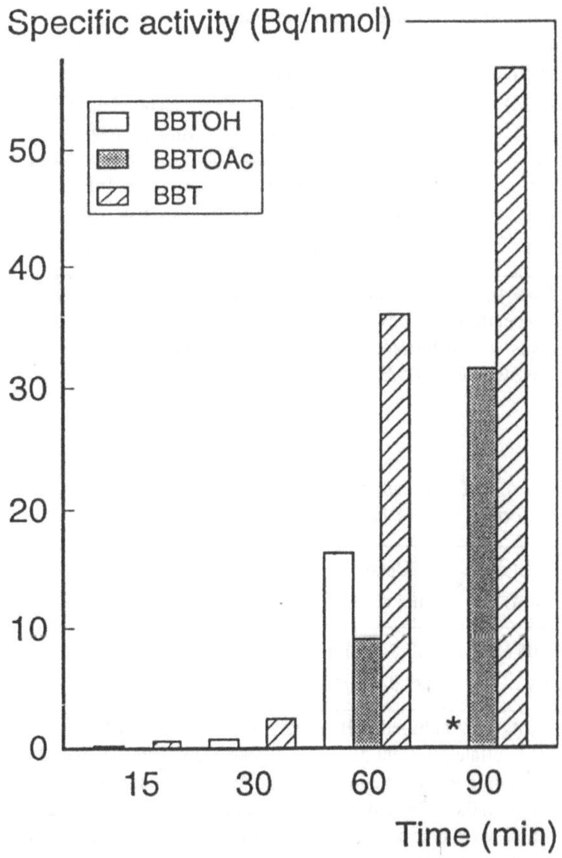

Fig. 6. Synthesis of individual thiophenes in transformed roots of *Tagetes patula.* Root tips were incubated in sulfate-free medium with carrier-free [^{35}S]sulfate and extracted after various periods. The radioactivities and the chemical concentrations of three individual bithienyls, BBT, BBTOH, and BBTOAc were measured after separation by HPLC. Data are expressed as kBq per nmol. The specific activity of BBTOH after 90 min could not be calculated because the chemical concentration was below detection.

Regulation of thiophene biosynthesis

The reduced size of the total sulfur pool in the leaves as compared to the other parts of the shoot (Table 2) opens the possibility that the thiophene-synthesizing machinery in the leaves is not saturated due to low substrate availability. If true, thiophene synthesis would proceed less efficiently in the leaves than in the hypocotyl. The lower percentage of thiophenic sulfur in the leaves than in the hypocotyl (Table 2) could, in that case, reflect the operation of an overflow mechanism and thus be indicative of a form of metabolic control.

The question of how much sulfur is available for thiophene synthesis when the sulfur supply is limited, was further studied in isolated roots cultured at various sulfate concentrations (Table 4). The concentration in the standard medium is 2 mM. Lowering this level to 0.1 mM had no appreciable effect on root growth and morphology on a solid medium. In contrast, thiophene accumulation was strongly repressed when the sulfate supply was limited. The same results were obtained from roots cultured in liquid medium even when the sulfate concentration was further reduced to 0.05 mM (data not shown). The capacity to synthe-

Table 4. Effect of sulfate on growth and thiophene accumulation in a transformed root line of *Tagetes patula*.

[Sulfate] (mM)	Length of main root (cm)	Lateral roots formed	Thiophene accumulation (μmol g FW^{-1})
0.1	6.5 ± 2	32 ± 11	0.6 ± 0.1
0.5	6.3 ± 1	34 ± 10	1.1 ± 0.2
2.0	6.7 ± 1	24 ± 8	2.3 ± 0.4

Root tips were incubated on solid medium at various sulfate concentrations. Growth, branching, and thiophene content of the roots were determined after 14 days. All data are means of six measurements ±S.E.

Table 5. Effect of sulfate on sulfur incorporation in thiophenes by transformed roots of *Tagetes patula*.

[Sulfate] (mM)	Uptake (nmol $g^{-1}h^{-1}$)	Incorporation (nmol $g^{-1}h^{-1}$)
0.05	417.5 ± 41.4	4.42 ± 0.36
2.0	387.0 ± 37.6	19.7 ± 2.0

Roots were cultured in B5 medium either at 0.05 mM or at 2.0 mM sulfate for eight days. The tissues were then transferred to medium containing 30 mM sulfate and [^{35}S]sulfate as marker. After 2h the radioactivity in the thiophenes and in the tissue as a whole were determined. The data were used to calculate sulfur uptake and incorporation.

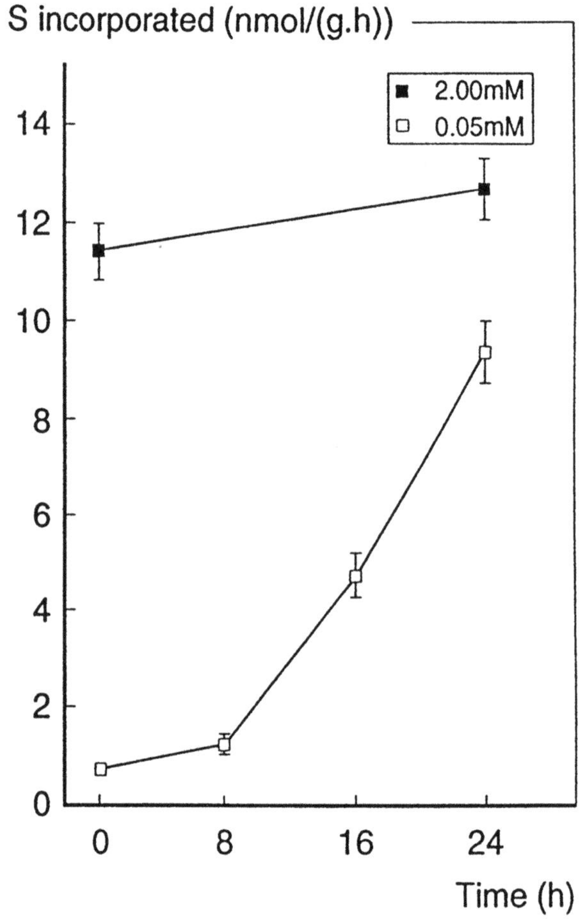

Fig. 7. Recovery of the capacity to synthesize thiophenes in roots of *Tagetes patula* starved for sulfate. Roots were precultured for nine days at 0.05 mM (open squares) or 2.0 mM (solid squares) sulfate. The tissue was then incubated in fresh medium with 2 mM sulfate. The capacity to incorporate sulfur in thiophenes was determined during 24 h at eight-hour intervals. Data are means of four measurements ±S.E.

size thiophenes under sulfur limitation was tested in roots at 0.05 mM sulfate for eight days. The tissue was then exposed to 30 mM sulfate in a medium containing [^{35}S]sulfate as a marker. The rate of synthesis was calculated from the radioactivity incorporated in thiophenes (Table 5) . In the sulfate-starved roots thiophene synthesis amounted to only 20% of the value obtained from roots grown in standard medium, an effect which is not due to reduced uptake of sulfate in the tissue. It may be concluded that the limited sulfate availability causes a shift in the partitioning of sulfur between primary and secondary metabolism. The sulfur requirements for growth and development are met at the expense of the flow of sulfur precursors used for thiophene synthesis.

The question of how this shift in favor of the primary metabolism is regulated, was addressed by culturing roots for eight days at 0.05 mM sulfate and then transferring them to 2 mM. The capacity to incorporate sulfur in thiophenes was tested at intervals during 24 h after the medium change. The hypothesis behind the experiment was that the rate of thiophene synthesis would rapidly increase if the availability of sulfurous precursors would be the rate-limiting step. However, the recovery to the level of synthesis in roots continuously cultured at 2 mM sulfate proved to be a slow process that had not been completed yet after 24 h (Fig. 7).This result supported the alternative explanation that sulfate limitation negatively influences the enzymatic machinery operative in thiophene biosynthesis. Further evidence for a molecular regulation was obtained when roots cultured at 0.05 mM sulfate were exposed to 2 mM in the presence of the transcription inhibitor cordycepin. The inhibitor completely blocked the rise in thiophene synthesis observed in uninhibited roots (Table 6). Moreover, cordycepin caused a decrease in sulfur incorporation in roots that had not been starved

Table 6. Effect of sulfate on the capacity of *Tagetes* roots to synthesize thiophenes.

Time (h)	[Sulfate] during preculture (mM)	Cordycepin added	S incorporation (nmol $g^{-1}h^{-1}$)
0	0.05	–	1.68 ± 0.35
	2.0	–	8.84 ± 0.88
24	0.05	no	5.26 ± 0.58
		yes	1.92 ± 0.25
24	2.0	no	10.13 ± 0.29
		yes	5.26 ± 0.58

Roots were precultured for 9 days in medium containing either 2 mM or 0.05 mM sulfate. The tissue was then incubated for another 24 h in fresh medium with 2 mM sulfate. Cordycepin was added to half of the cultures at 5 μg ml^{-1}. The incorporation of [^{35}S]sulfur in the thiophenes was measured at the time of the medium shift and at the end of the experiment. Data are means of 4 measurements ±S.E.

for sulfate. These results suggest that continuous transcription is required for maintaining the rate of thiophene biosynthesis and that the cell can down-regulate the secondary pathway by reducing the synthesis of the enzymes involved.

The question arises at which point in the pathway the factor sulfate affects the rate of thiophene biosynthesis. The formation of the two thiophene rings is the most likely target for this type of regulation. Experiments to attack this problem are underway.

Conclusions

Identification of intermediates and feeding experiments with presumed precursors have provided new data on the synthesis of thiophenes from polyacetylenes in *Tagetes*. The first step in the biosynthesis of the bithienyls involves addition of reduced sulfur to a C_{13}-polyacetylene at a fixed position in the chain. The monothiophene BPT formed by this reaction is converted to BBT in a sequence of reactions wherein demethylation precedes or coincides with closure of the second ring. MeBBT thus is not a precursor of BBT and BBTOAc.

The synthesis of thiophenes from polyacetylenes proved to be a suitable model system to study sulfur partitioning between primary and secondary metabolism. The flow of sulfur into thiophenes is strongly reduced when little sulfate is available to the tissue. Although this would point to operation of an 'overflow' mechanism, the regulation involves enzyme synthesis and breakdown and thus is largely at the molecular level.

References

Bohlmnn F & Hinz U (1965) Uber biogenetische Unwandlungen des Tridecen-pentains. Chem. Ber. 98:876–882

Bohlmann F, Kleine K-M & Arndt C (1964) Uber naturlich vorkommende Thiophenace-tylenverbindungen. Chem. Ber. 97:2125–2134

Bohlmann F, Arndt C, Kleine K-M & Bornowski H (1965) Die Acetylenverbindungen der Gattung *Echinops* L. Chem Ber. 98: 155–163

Bohlmann F, Wotschokowski M, Hinz U & Lucas W (1966) Uber die Biogenese einiger Thiophenverbindungen. Chem. Ber. 99: 984–989

Bohlmann F, Burkhardt T & Zdero C (1973) Naturally Occurring Acetylenes. Academic Press, London

Croes AF, Van den Berg AJR, Bosveld M, Breteler H & Wullems GJ (1989) Thiophene accumulation in relation to morphology in roots of *Tagetes patula*. Effects of auxin and transformation by *Agrobacterium*. Planta 179: 43–50

Downum KR & Towers GHN (1983) Analysis of thiophenes in the *Tageteae* (Asteraceae) by HPLC. J. Nat. Prod. 46:98–103

Galneder E & Zenk MH (1990) Enzymology of alkaloid production in plant cell cultures. In: Nijkamp HJJ, Van der Plas LHW & Van Aartrijk (Eds) Progress in Plant Cellular and Molecular Biology (pp 754–762). Kluwer Academic Publishers, Dordrecht

Jente R, Olatunji GA & Bosold F (1981) Formation of natural thiophene derivatives from acetylenes by *Tagetes patula*. Phytochemistry 20: 2169–2175

Luckner M (1990) Secondary Metabolism in Microorganisms, Plants and Animals. Fischer Verlag and Springer Verlag, Jena and Berlin

Sütfeld R (1982) Distribution of thiophene derivatives in different organs of *Tagetes patula* seedlings grown under various conditions. Planta 156: 536–540

Plant Cell, Tissue and Organ Culture **38:** 167–169, 1994.

Regulatory mechanisms of biosynthesis of betacyanin and anthocyanin in relation to cell division activity in suspension cultures

Masaaki Sakuta[1], Hiroshi Hirano[2], Koichi Kakegawa[3], Jun Suda[4], Masanori Hirose[5], Richard W. Joy IV[4], Munetaka Sugiyama[4] & Atsushi Komamine[6*]
[1]*Department of Biology, Faculty of Science, Ochanomizu University, Otsuka, Tokyo, Japan;* [2]*Sumitomo Chemical Co., Ltd., Takarazuka Research Center, Takatsuki, Osaka, Japan;* [3] *Forestry and Forest Product Research Institute, Tsukuba, Japan;* [4]*Biological Institute, Faculty of Science, Tohoku University, Sendai, Japan;* [5]*Research Laboratory for Development, Mitsubishi Oil Company, Kawasaki, Japan;* [6]*Department of Chemical and Biological Sciences, Faculty of Science, Japan Women's University, Mejiro, Tokyo, Japan (* requests for offprints)*

Key words: Anthocyanin, betacyanin, cell division, *Phytolacca*, suspension culture, *Vitis*

Abstract

Regulatory mechanisms of betacyanin biosynthesis in suspension cultures of *Phytolacca americana* and anthocyanin in *Vitis* sp. were investigated in relation to cell division activity.

Betacyanin biosynthesis in *Phytolacca* cells clearly shows a positive correlation with cell division, as the peak of betacyanin accumulation was observed at the log phase of batch cultures. Incorporation of radioactivity from labelled tyrosine into betacyanin also showed a peak at early log phase. Aphidicolin, an inhibitor of DNA synthesis, and propyzamide, an antimicrotubule drug, reduced betacyanin accumulation and inhibited the incorporation of radioactivity from labelled tyrosine into betacyanin at concentrations which were inhibitory to cell division. Both inhibitors reduced the incorporation of radioactivity from labelled tyrosine to 3,4-dihydroxyphenylalanine (DOPA), but the incorporation of labelled DOPA into betacyanin was not affected. These results suggest that the conversion of tyrosine to DOPA is coupled with cell division activity.

In contrast, the anthocyanin accumulation in *Vitis* cells showed a negative correlation with cell division. Accumulation occurred at the stationary phase in batch cultures when cell division ceased. Aphidicolin or reduced phosphate concentration induced a substantial increase in anthocyanin accumulation as well as the inhibition of cell division. Chalcone synthase (CHS) activity increased at the time of anthocyanin accumulation. Northern blotting analysis indicated that changes in CHS mRNA levels corresponded to similar changes in enzymatic activity. The pool size of endogenous phenylalanine was low during active cell division, but increased before anthocyanin began to accumulate and concomitantly with increasing levels of CHS mRNA. Exogenous supply of phenylalanine at the time of low endogenous levels induced the elevation of CHS mRNA and anthocyanin accumulation. These results indicate that the elevation of endogenous phenylalanine levels, when cell division ceases, may cause the increase in CHS mRNA levels, resulting in increased CHS activity and subsequently in anthocyanin accumulation in *Vitis* suspension cultures.

Abbreviations: CHS – chalcone synthase, CHFI – chalcone flavanone isomerase, DOPA – 3,4-dihydroxyphenylalanine, PAL – phenylalanine ammonia lyase

Introduction

Most secondary metabolites such as anthocyanins are accumulated during the stationary phase in batch suspension cultures. On the other hand, some secondary metabolites such as betacyanins are accumulated at the log phase (Sakuta et al. 1986). These facts suggest that biosynthesis of the former is correlated negatively with cell division, while that of the latter is correlated positively with cell division. In this paper, we investigated the regulatory mechanisms of biosynthesis of metabolites showing these two different accumulation patterns in batch suspension cultures. Biosynthesis of anthocyanin in suspension cultures of *Vitis* sp. was investigated as an example of the former type of accumulation pattern and that of betacyanin in *Phytolacca americana* suspension cultures as an example of the latter one.

Results and discussion

Regulatory mechanisms of anthocyanin biosynthesis in Vitis sp. suspension cultures

Suspension cultures of *Vitis* used in these experiments were cultured at 27 °C in the dark. Cells reached the stationary phase 7 days after transfer and the accumulation of anthocyanins occurred 8 days after transfer, that is, after cessation of cell division. When cell division was inhibited by aphidicolin, an inhibitor of DNA synthesis, or by reduction of the phosphate concentration in media, a rapid accumulation of anthocyanins occurred coinciding with the cessation of cell division (Hirose et al. 1990). Changes in activities of three enzymes involved in the biosynthesis of anthocyanins were investigated. The activities of PAL and CHS increased to high levels immediately after transfer to fresh medium, but decreased thereafter and remained at very low levels during the logarithmic phase. When cell division ceased, activities of these two enzymes increased and high levels of activities remained during the accumulation of anthocyanins.

The activity of CHFI did not show the increase by the transfer effect and remained at a low level in the logarithmic phase, but increased when anthocyanins accumulated.

Changes in mRNAs of PAL and CHS were investigated by northern analysis. Two hybridization signals, (PAL1 and PAL2) were detected for cDNA of PAL. Both mRNAs encoding PAL were induced in the first day after transfer, but the amounts of them decreased immediately thereafter and remained at very low levels. When cell division ceased, the amounts of them increased again. The CHS mRNA showed a similar pattern of change in its amount as that of PAL. Since changes in amounts of PAL and CHS mRNAs were coincident with those of activities of PAL and CHS, it is suggested that the activities of these enzymes may be controlled at the transcriptional level.

The size of the endogenous pool of phenylalanine was small when active cell division occurred, but it increased rapidly just before increased accumulation of anthocyanin (8 days after transfer). Noé et al. (1980) demonstrated that the pool of endogenous phenylalanine, which was enlarged as a result of treatment with L-AOPP, an inhibitor of PAL, induced PAL activity. It is possible, therefore, that in *Vitis* cultures an elevated level of endogenous phenylalanine may induce PAL and CHS activities. In order to examine this possibility, phenylalanine was added exogenously to the media 4 days after transfer when cell division was occurring actively and anthocyanins were not accumulated. Phenylalanine added at 1 and 5 mM induced more rapid increases of accumulation of anthocyanin than controls in which diluted water was added, but the effect of 5 mM phenylalanine was more pronounced. Exogenous phenylalanine added 4 days after transfer also induced increases in the activities of PAL, CHS and CHFI. Northern analysis revealed that when phenylalanine (5 mM) was added 4 days after transfer, mRNAs of PAL1 and 2 and CHS could be detected after 5 days (1 day after the addition) and the amounts of the mRNAs reached their maxima after 6 days, whereas in the control culture mRNAs of PAL and CHS were detected only after 8 days and reached their maxima after 10 days of culture, though a transient expression of the mRNAs was observed just after transfer. We investigated the induction of expression of CHS mRNA by phenylalanine in more detail by northern analysis and found that CHS mRNA could be detected shortly after addition of phenylalanine, even after 1 hr. The amino acids leucine and tyrosine had no effect on the induction of accumulation of anthocyanins.

From the results obtained, a possible mechanism for the regulation of anthocyanin biosynthesis during batch suspension cultures of *Vitis* cells is proposed. When cells divide actively the endogenous level of phenylalanine is low, because phenylalanine is utilized for biosynthesis of protein. When the cell division ceases, the level of endogenous phenylalanine increases and acts as a trigger for the initiation of transcription of PAL and CHS mRNA, resulting in de novo synthesis of PAL and CHS proteins and subsequently in the increase of the activities of these enzymes, which leads to the accumulation of anthocyanins. The increased level of endogenous phenylalanine also may supply more precursors for the biosynthesis of anthocyanins. Thus the accumulation of anthocyanins shows a negative correlation with cell division in batch suspension cultures of *Vitis*.

Regulatory mechanisms of betacyanin biosynthesis in Phytolacca americana suspension cultures

As we reported previously (Sakuta et al. 1986), betacyanins accumulate in the logarithmic phase. In contrast to most other secondary metabolites, such as anthocyanins, the maximum betacyanin content per cell was observed 5 days after transfer of *Phytolacca americana* suspension cultures. We investigated whether not only accumulation but also biosynthesis of betacyanins occurred during active cell division using labelled tyrosine. Tracer experiments revealed clearly that the peak of incorporation of radioactivity from tyrosine to betacyanin was observed 4 days after transfer, at the early log phase. When aphidicolin, an inhibitor of DNA synthesis, was added to suppress the cell division, the accumulation of betacyanin as well as its biosynthetic activity were completely suppressed. When the cell division was inhibited by propyzamide, an antimicrotubule drug, the biosynthesis of betacyanins from tyrosine was also inhibited. These results indicate that the biosynthesis of betacyanins is associated not only with DNA synthesis, but also with mitosis.

Betacyanin is believed to be synthesized from tyrosine via DOPA. We examined the incorporation of radioactivity from labelled tyrosine or DOPA to betacyanin in the presence or absence of aphidicolin or propyzamide in order to elucidate which step of biosynthesis of betacyanin is coupled with cell division. The results obtained revealed that the incorporation of radioactivity from tyrosine to betacyanins was inhibited remarkably by both inhibitors, while that from DOPA was not. It is, therefore, concluded that the step from DOPA to betacyanins is not coupled with cell division, but that from tyrosine to DOPA is coupled.

Conclusions

The regulatory mechanisms of biosynthesis of anthocyanins, which shows a negative correlation with cell division, and of betacyanins, which shows a positive correlation, were investigated in relation to cell division using *Vitis* sp. and *Phytolacca americana* suspension cultures, respectively. Concerning the regulatory mechanisms of anthocyanin biosynthesis, it is concluded that cessation of cell division causes the increase of pool size of endogenous phenylalanine which may trigger the transcription of mRNAs of enzymes involved in anthocyanin biosynthesis such as PAL and CHS, resulting in increased activities of those enzymes and in increased accumulation of anthocyanins. Thus, the accumulation of anthocyanin shows a negative correlation with cell division.

Concerning the regulatory mechanisms of biosynthesis of betacyanins, it was believed that the biosynthesis of betacyanins is not only associated with DNA synthesis but also with mitosis. It was concluded that the step of hydroxylation of tyrosine to DOPA is coupled with cell division, but the steps from DOPA to betacyanins are not. However, it still remains to be solved what event(s) of cell division is coupled or associated with hydroxylation of tyrosine to DOPA.

References

Hirose M, Yamakawa T, Kodama T & Komamine A (1990) Accumulation of betacyanin in *Phytolacca americana* cells and of anthocyanin in *Vitis* sp. cells in relation to cell division in suspension cultures. Plant Cell Physiol. 31: 267–271

Noé W, Langevartels C & Seitz V (1980) Anthocyanin accumulation and PAL activity in a suspension culture of *Daucus carota* L. Planta 149: 283–287

Sakuta M, Takagi T & Komamine A (1986) Growth related accumulation of betacyanin in suspension cultures of *Phytolacca americana*. J. Plant Physiol. 125: 337–343

Plant Cell, Tissue and Organ Culture **38:** 171–179, 1994.

The biosynthesis of rosmarinic acid in suspension cultures of *Coleus blumei*

Maike Petersen, Elisabeth Häusler, Juliane Meinhard, Barbara Karwatzki & Claudia Gertlowski
Institut für Entwicklungs- und Molekularbiologie der Pflanzen, Heinrich-Heine-Universität Düsseldorf, Universitätsstr. 1, D-40225 Düsseldorf 1, FRG

Key words: *Coleus blumei* (suspension cultures), hydroxycinnamic acid esters, phenylalanine, rosmarinic acid, secondary metabolism, tyrosine

Abstract

Suspension cultures of *Coleus blumei* accumulate very high amounts of rosmarinic acid, an ester of caffeic acid and 3,4-dihydroxyphenyllactate, in medium with elevated sucrose concentrations. Since the synthesis of this high level of rosmarinic acid occurs in only five days of the culture period, the activities of the enzymes involved in the biosynthesis are very high. Therefore all the enzymes necessary for the formation of rosmarinic acid from the precursors phenylalanine and tyrosine could be isolated from cell cultures of *Coleus blumei*: phenylalanine ammonia-lyase, cinnamic acid 4-hydroxylase, hydroxycinnamoyl:CoA ligase, tyrosine aminotransferase, hydroxyphenylpyruvate reductase, rosmarinic acid synthase and two microsomal 3- and 3′-hydroxylases. The main characteristics of these enzymes of the proposed biosynthetic pathway of rosmarinic acid will be described.

Abbreviations: DHPL, 3,4-dihydroxyphenyllactate; DHPP, 3,4-dihydroxyphenylpyruvate; pHPL, 4-hydroxyphenyllactate, pHPP, 4-hydroxyphenylpyruvate; RA, rosmarinic acid

Introduction

Rosmarinic acid (RA) is one of the most commonly occurring caffeic acid esters in the plant kingdom. It is wide spread in the families Lamiaceae, mainly the subfamily Saturejoideae (Litvinenko et al. 1975; Mölgaard & Ravn 1988), and Boraginaceae, but can also be detected in other families, in ferns (Harborne 1966; Häusler et al. 1992) and in hornworts (Takeda et al. 1990; Zinsmeister et al. 1991; Petersen, unpublished). The molecular structure of rosmarinic acid as an ester of caffeic acid and 3,4-dihydroxyphenyllactic acid (Fig. 1) was clarified by Scarpati & Oriente (1958, 1960) using extracts of *Rosmarinus officinalis*.

The production of high amounts of RA by cell cultures of *Coleus blumei* was first described by Razzaque & Ellis (1977) and Zenk et al. (1977). These suspension cultures, together with cell cultures of *Anchusa officinalis*, were the main object for investigations on the biosynthesis of this phenolic compound. The precursors from primary metabolism for the biosynthesis of RA have been identified by Ellis & Towers (1970) by feeding radioactive amino acids to plants of *Mentha*. Phenylalanine was mainly incorporated into the caffeic acid part of the molecule, whereas tyrosine and DOPA gave rise to the 3,4-dihydroxyphenyllactic acid moiety. Essentially the same results were obtained with suspension cultures of *Coleus blumei* (Razzaque & Ellis 1977). Despite the knowledge of the biosynthetic precursors of RA, the first new enzyme of RA biosynthesis, tyrosine aminotransferase, was only described in 1987 by De-Eknamkul & Ellis (1987a, b). From this year on, the enzymes of RA biosynthesis successively became elucidated and a full biosynthetic pathway of RA in cell cultures of *Coleus blumei* has been proposed recently (Petersen et al. 1993).

Rosmarinic acid production in suspension cultures of *Coleus blumei*

The amount of RA accumulated by suspension cultures of *Coleus blumei* is highly dependent on the amount of carbohydrates added to the culture media (Zenk et

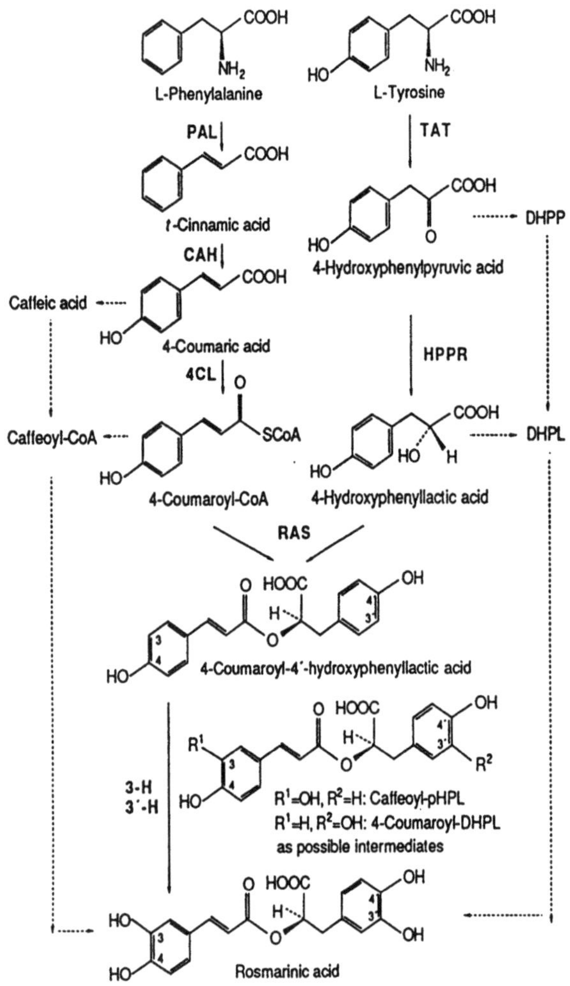

Fig. 1. Biosynthetic pathway for rosmarinic acid as deduced from the enzyme activities isolated from suspension cultures of *Coleus blumei*. PAL = phenylalanine ammonia-lyase; CAH = cinnamic acid 4-hydroxylase; 4CL = hydroxycinnamoyl:CoA ligase; TAT = tyrosine aminotransferase; HPPR = hydroxyphenylpyruvate reductase; RAS = rosmarinic acid synthase; 3H, 3′-H = 3- and 3′-hydroxylases.

al. 1977; Ulbrich et al. 1985; Petersen & Alfermann 1988; Gertlowski & Petersen 1993). In our routinely used cultivation system, cells are maintained in a modified B5-medium (Gamborg et al. 1968; Petersen & Alfermann 1988) with 2% sucrose (CB2) where they only accumulate 2–3% of their dry weight as RA. Suspension cultures transferred to the same medium with 4% sucrose (CB4) produce up to 19% of their dry weight as RA at the end of the growth phase (Fig. 2). With higher sucrose concentrations even more RA can be obtained as long as the cells tolerate the osmotic pressure of the medium (Gertlowski & Petersen 1993). Although sucrose is cleaved into glucose and fructose,

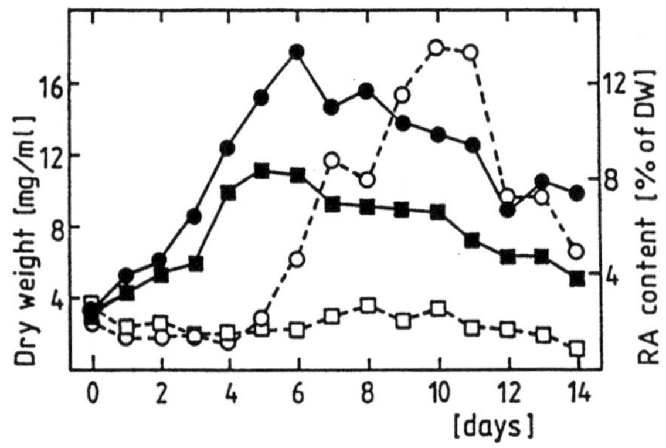

Fig. 2. Growth (depicted as dry weight accumulation: ■, ●) and rosmarinic acid accumulation (□, ○) of suspension cultures of *Coleus blumei* in media with 2% (■, □) and 4% sucrose (●, ○).

other sugars than sucrose support only lower RA accumulation.

The onset of net RA accumulation seems to be dependent on the limitation of an essential nutrient other than the carbon source. In the case of our *Coleus blumei* suspension cultures, the depletion of phosphate from the medium on day 4 to 5 of the culture period coincides with the start of RA synthesis and accumulation (Gertlowski & Petersen 1993). Reduction of the medium phosphate leads to a higher RA accumulation. Higher phosphate concentrations repress RA production in contrast to the results obtained for RA-producing cell cultures of *Anchusa officinalis* (De-Eknamkul & Ellis 1988). However, RA synthesis starts at about the same time as in media with lower phosphate levels showing that other nutrients may influence the onset of RA production as well.

Suspension cultures of *Coleus blumei* with their 'inducible' high RA accumulation have proven to be an excellent system for the investigation of the biosynthetic pathway of RA. A very high amount of this secondary compound is synthesized and accumulated in only about 5 days (from day 5 to 10) of the culture period resulting in high activities of the biosynthetic enzymes during this time.

The biosynthetic pathway of rosmarinic acid

A biosynthetic pathway for RA in suspension cultures of *Coleus blumei* derived from the enzyme activities isolated from these cells has been proposed recently (Petersen et al. 1993) and is depicted in Fig. 1. The enzyme activities involved in this pathway and their main characteristics (as far as already determined) are

Table 1. Some characteristics of the enzymes of rosmarinic acid biosynthesis in suspension cultures of *Coleus blumei*.

Enzyme	Localization	pH-optimum	Substrates	Other characteristics
Phenylalanine ammonia-lyase	soluble	7.8	phenylalanine K_m 0.16 μM	
Cinnamic acid 4-hydroxylase	membrane	7.5	cinnamic acid K_m 35 μM NADPH K_m 40 μM	cyt. P-450; inhibited by cyt. c; dependent on O_2
Hydroxycinnamic acid:CoA ligase	soluble	7.5-8.0	4-coumarate K_m 2.2 μM (caffeate) K_m 5.0 μM ATP K_m 73 (137) μM coenzyme A K_m 3.0 (6.5) μM	app. molecular mass 37 ± 2 kDa
Tyrosine aminotransferase	soluble		tyrosine 2-oxoglutarate	dependent on pyridoxal-phosphate
Hydroxyphenylpyruvate reductase	soluble	6.5-7.0	pHPP K_m 10 μM DHPP K_m 130 μM NADPH K_m 95 μM NADH K_m 190 μM	app. molecular mass 50-55 kDa; inhibited by RA and 4-coumaroyl-CoA
Rosmarinic acid synthase	soluble	7.0-7.5	4-coumaroyl-CoA K_m 20 μM caffeoyl-CoA K_m 33 μM ±-pHPL K_m 0.17 mM *R*(+)-DHPL K_m 0.37 mM	inhibited by * CoA (noncompetitively) * *S*(-)-DHPL/* RA
reverse RAS reaction			RA K_m 15 μM CoA K_m 310 μM	App. molecular mass 77 KDa
3-Hydroxylase	membrane	7.5	4-coumaroyl-3′,4′-DHPL 4-coumaroyl-4′-pHPL NADPH K_m 30 μM	inhibited by cyt. c; dependent on O_2
3′-Hydroxylase	membrane	7.5	caffeoyl-4′-pHPL 4-coumaroyl-4′-pHPL NADPH K_m 30 μM	inhibited by cyt. c; dependent on O_2

summarized in Table 1 and described in more detail in the following paragraphs.

Phenylalanine-derived pathway

Phenylalanine as the precursor for the caffeic acid moiety of RA (Ellis & Towers 1970; Razzaque & Ellis 1977) as well as the correlation of the activity of phenylalanine ammonia-lyase (PAL) with RA synthesis (Razzaque & Ellis 1977) suggested the participation of the general phenylpropanoid metabolism in the biosynthesis of RA. The respective enzymes, PAL, cinnamic acid 4-hydroxylase (CAH) and hydroxycinnamic acid:CoA ligase (4CL), have been shown to be correlated in their activities to RA synthesis and accumulation in cell cultures of *Coleus blumei* (Karwatzki et al. 1989).

Phenylalanine ammonia-lyase

Phenylalanine ammonia-lyase (PAL) catalyzes the oxidative desamination of phenylalanine and forms t-cinnamic acid and ammonia. The enzyme from suspension cultures of *Coleus blumei* is a soluble enzyme and precipitates between 40 and 55% saturation of ammonium sulphate. The affinity of the enzyme towards its substrate phenylalanine is very high with an apparent K_m-value of about 16 μM. RA up to 1 mM does not inhibit PAL activity. Highest activity in cultures in medium with 4% sucrose is found on day 6 of the culture period, 4 days earlier than the maximum RA accumulation (cf. Fig. 2).

Cinnamic acid 4-hydroxylase

Cinnamic acid 4-hydroxylase (CAH) introduces the para-hydroxyl group into the aromatic ring of t-cinnamic acid by a cytochrome P-450-dependent monooxygenation reaction. The enzyme activity essentially depends on molecular oxygen and NADPH.

CAH isolated from cell cultures of *Coleus blumei* in CB4-medium exerts its maximal activity around day 6 of the culture period as PAL. The microsomal enzyme has a pH-optimum of pH 7.5. CAH is not inhibited by phenolase inhibitors (diethyldithiocarbaminate, salicylhydroxamic acid) and potassium cyanide, but as low as 10 μM cytochrome c totally blocks the enzyme activity. This is a typical characteristic of cytochrome P-450-dependent enzymes, where cytochrome c acts as a competitive electron acceptor of NADPH:cytochrome P-450 reductase. FAD, a component of cytochrome P-450 reductase, stimulates the activity of CAH by more than 20%. NADPH is the only electron donor accepted by CAH; NADH is not used by CAH and no synergistic effect of NADH and NADPH can be observed for the enzyme from *Coleus blumei* in contrast to CAH from some other plants (e.g., Benveniste et al. 1977, 1982). The reaction was saturated at NADPH concentrations of 0.5 mM and a K_m-value of around 40 μM was determined for this substrate. The saturation concentration of CAH for cinnamic acid is at 1 mM with a K_m-value at 35 μM.

Activation of cinnamic acids
For the formation of cinnamic acid esters usually an activated cinnamic acid is essential. The activation energy can be achieved by different activation modes: the cinnamic acid can be activated as coenzyme A-thioester (e.g., flavonoid biosynthesis (Kreuzaler & Hahlbrock 1972), chlorogenic acid biosynthesis (Stöckigt & Zenk 1974)), as cinnamic acid glucose ester (e.g., biosyntheses of sinapoylmalate (Tkotz & Strack 1980) or chlorogenic acid (Kojima & Villegas 1984)) or as another cinnamic acid ester, such as chlorogenic acid (e.g., biosyntheses of dicaffeoyl quinic acid (Villegas et al. 1987) or caffeoyl glucaric acid (Strack et al.1987)). For RA biosynthesis cinnamic acid CoA-esters proved to be the only accepted activation form (Petersen 1991).

Hydroxycinnamoyl:CoA ligase
Hydroxycinnamoyl:CoA-ligase (4CL) catalyzes the activation of cinnamic acids with coenzyme A in an ATP-dependent reaction. Like PAL and CAH the enzyme from cell cultures of *Coleus blumei* in CB4-medium is most active at day 6 of the culture period. The soluble enzyme precipitates between 30 and 55% saturation of ammonium sulphate and has a pH-optimum of 7.5–8.0. The enzyme was partially purified by different purification methods, however, isoenzymes could not be resolved by these methods. The apparent molecular mass of 4CL from suspension cultures of *Coleus blumei* was at 37 ± 2 kDa. The kinetic characteristics of 4CL were determined using partially purified enzyme preparations (Karwatzki 1992). Several cinnamic acids (cinnamic, 4-coumaric, caffeic and ferulic acid) were accepted as substrates by the enzyme from *Coleus blumei*. 4-Coumarate was the best substrate with a K_m-value of 2.2 μM, the K_m value for caffeic acid was determined at 5.0 μM. Slightly different K_m-values could be observed for ATP and coenzyme A using either 4-coumarate or caffeate as substrates: the K_m-values for ATP with 4-coumarate and caffeate as substrates were at 73 μM and 137 μM, respectively and for coenzyme A at 3.0 μM and 6.5 μM, respectively (Karwatzki 1992). However, up to now different isoenzymes of hydroxycinnamoyl:CoA transferase could not be resolved.

With 4-coumaroyl-CoA as the direct precursor for the ester formation in the biosynthetic pathway of RA the final intermediate of the phenylalanine-derived pathway is reached.

Tyrosine-derived pathway

Feeding experiments with radioactive amino acids resulted in the identification of tyrosine and DOPA as precursors for the 3,4-dihydroxyphenyllactic acid moiety of RA (Ellis & Towers 1970; Razzaque & Ellis 1977). In 1987, tyrosine aminotransferase was identified as the first enzyme of the tyrosine-derived pathway of RA biosynthesis, whereas the involvement of tyrosine 3-hydroxylase and tyrosine oxidase were found to be unlikely (Ellis et al.1979; De-Eknamkul & Ellis 1987a). Therefore DOPA did no longer seem to be a direct precursor for RA.

Tyrosine aminotransferase
Tyrosine aminotransferase (TAT) catalyzes the transamination of tyrosine with help of 2-oxoglutarate giving rise to 4-hydroxyphenylpyruvate and glutamate. This pyridoxal phosphate-containing enzyme has been purified and characterized from RA synthesizing cell cultures of *Anchusa officinalis* (De-Eknamkul & Ellis 1987a, b; Mizukami & Ellis 1991). The activity of TAT is induced in cell cultures of *Coleus blumei* (from CB4-medium) together with the enzymes of the general phenylpropanoid metabolism in a coordinated way showing maximal activity at day 6 of the culture period. Since this enzyme is well characterized from cell

cultures of *Anchusa officinalis* (De-Eknamkul & Ellis 1987a, b; Mizukami & Ellis 1991) a detailed characterization of the enzyme from *Coleus blumei* has not yet been undertaken.

Hydroxyphenylpyruvate reductase

Hydroxyphenylpyruvate reductase (HPPR) reduces 4-hydroxyphenylpyruvate (pHPP) to 4-hydroxyphenyllactate (pHPL) with help of NAD(P)H (Petersen & Alfermann 1988; Häusler et al. 1991; Meinhard et al. 1992). This enzyme was first isolated and characterized from cell cultures of *Coleus blumei* where it shows its maximum activity in cell extracts from CB4-medium at day 6 of the culture period.

In a number of cases the reduction of aromatic 2-oxoacids has been described as secondary function of malate or lactate dehydrogenase (Friedrich et al. 1987, 1988). However, the enzyme from *Coleus blumei* could be proven to be a specific hydroxyphenylpyruvate reductase (Meinhard, unpublished results). Only a NADP-dependent, but no NAD-dependent oxidation of malate was detectable in enzyme extracts from *Coleus blumei*, whereas the reduction of hydroxyphenylpyruvate by HPPR uses both, NADH and NADPH. A slight and rapidly stagnating oxidation of lactate could be attributed to the activity of HPPR. Furthermore, oxamate, a specific inhibitor for lactate dehydrogenase, did not influence the activity of HPPR.

HPPR from cell cultures of *Coleus blumei* is a soluble enzyme which was partially purified by fractionated ammonium sulphate precipitation (40–60% saturation), hydroxylapatite adsorption chromatography and affinity chromatography on 2',5'-ADP-Sepharose 4B (Meinhard et al. 1992). The apparent molecular mass of this enzyme is 50–55 kDa. HPPR has a pH-optimum of pH 6.5–7.0. Besides pHPP also other hydroxyphenylpyruvates are accepted as substrates as long as they provide a free 4-hydroxyl group in the aromatic ring.

Therefore phenylpyruvate was not reduced. The K_m values for pHPP and 3,4-dihydroxyphenylpyruvate (DHPP) were determined at 10 μM and 130 μM respectively, suggesting that pHPP might be the natural substrate for HPPR (Häusler et al. 1991). Both NADH and NADPH were accepted as reduction equivalents by HPPR, but the higher affinity was determined for NADPH with a K_m value of 95 μM in contrast to 190 μM for NADH. The reaction velocities, however, were higher for NADH. Probably the enzyme can accept both electron donors in its natural environment. The activity of HPPR is inhibited by 4-coumaroyl-CoA and RA, whereas other cinnamic acids or cinnamic acid CoA-esters had no influence on HPPR activity. This inhibition by the end product of the biosynthetic pathway and the final intermediate of the phenylalanine-derived pathway could indicate a possible regulatory function of HPPR. Other inhibitors of HPPR were pyruvate and oxaloacetate at concentrations of 4 mM and 5 mM respectively. HPPR also catalyzes the reverse reaction, the oxidation of hydroxyphenyllactates with help of NAD(P).

The reaction product of HPPR, 4-hydroxyphenyllactate, is the immediate precursor for the formation of the hydroxycinnamoyl ester.

Ester formation: Rosmarinic acid synthase

The enzyme catalyzing the formation of the ester from the two precursors synthesized from phenylalanine and tyrosine is rosmarinic acid synthase (RAS) which transfers the hydroxycinnamoyl moiety from hydroxycinnamoyl-CoA to the aliphatic hydroxyl group of hydroxyphenyllactate. This enzyme was first isolated and described from cell cultures of *Coleus blumei* (Petersen & Alfermann 1988). The natural substrates of this enzyme presumably are 4-coumaroyl-CoA and 4-hydroxyphenyllactate resulting in the first ester 4-coumaroyl-4'-hydroxyphenyllactate (pC-pHPL). Therefore the systematic name of rosmarinic acid synthase (RAS) would be 4-coumaroyl-CoA:4-hydroxyphenyllactate 4-coumaroyl-transferase (Petersen 1991). Maximum activity of this enzyme during a culture period of suspension cultures of *Coleus blumei* in CB4-medium is obtained at day 7–8, somewhat later than the activities of the enzymes providing the substrates for the RAS reaction. RAS accepts a number of substrates giving rise to a variety of esters some of which have not been described to occur in nature up to now. The structures of the esters formed from 4-coumaroyl- and caffeoyl-CoA and 4-hydroxy- and 3,4-dihydroxyphenyllactate have been elucidated by ion spray mass spectrometry and tandem mass spectrometry (Petersen & Metzger 1993). Cinnamoyl-, 4-coumaroyl- and caffeoyl-CoA are accepted by RAS (the yield of the reaction products with sinapoyl- and feruloyl-CoA did not allow a further analysis). The highest affinity was determined for 4-coumaroyl-CoA with a K_m-value around 20 μM, the enzyme showed a strong substrate inhibition by this substrate. Caffeoyl-CoA was accepted with a K_m value of 33 μM and the enzyme was saturated at a concentration of 100 μM. Competition experiments with the simultaneous application of varying concentrations of 4-coumaroyl- and caffeoyl-CoA also indicated that 4-coumaroyl-CoA is the favoured substrate of RAS. Essential for the other substrate of RAS is a free 4-hydroxyl group at the aromatic ring and the R(+)-configuration of the hydroxyphenyllactate. Therefore 4-hydroxy-, 3,4-dihydroxy- and 3-methoxy-4-hydroxyphenyllactate were accepted, whereas phenyllactate and 3,4-dimethoxyphenyllactate did not serve as substrates. RAS was saturated at 1 mM of 4-hydroxy- and 3,4-dihydroxyphenyllactate (pHPL, DHPL) and K_m-values of 0.17 mM and 0.37 mM respectively, were determined. For these experiments, however, only the racemic mixture of pHPL could be used, whereas pure R(+)-DHPL was available. Since S(-)-DHPL strongly inhibited the RAS reaction, the true K_m-value for pHPL (R(+)) might even be lower. The catalytic acitivity of RAS is inhibited by pHPP and DHPP and the S(-)-isomer of DHPL, but not by cinnamic acids. Coenzyme A, however, is a strong inhibitor of the ester formation. The mechanism of this inhibition is non-competitive with inhibition constants of K_i=K_{ii}=1.7 mM. RAS catalyzes the reverse reaction, the splitting of rosmarinic acid in presence of coenzyme A into caffeoyl-CoA and 3,4-dihydroxyphenyllactate as well. This reaction needs a concentration of 50 μM of RA for saturation (K_m 15 μM), but the saturation concentration for coenzyme A is rather high at 1 mM (K_m 310 μM) indicating that this reaction might not be important *in vivo*.

3- and 3'-Hydroxylations of the aromatic rings

For a considerable time it was not clear whether RA is directly formed from caffeoyl-CoA and 3,4-dihydroxyphenyllactate or whether the introduction of the 3-hydroxyl groups of the aromatic rings occurs at a later stage of the biosynthetic pathway. Specific enzymes catalyzing the 3-hydroxylation of 4-coumaric acid or 4-coumaroyl-CoA have been described in a few cases (Kamsteeg et al. 1981; Boniwell & Butt 1986; Kneusel et al. 1989; Kojima & Takeuchi 1989), but no such enzymes are known from plants for hydroxyphenylpyruvic or –lactic acids besides phenolases, which are ubiquitous and commonly not regarded as specific (Butt 1985). From suspension cultures of *Coleus blumei* specific hydroxylases introducing the 3- and 3'-hydroxyl groups into rosmarinic acid-like esters (4-coumaroyl-4'-hydroxyphenyllactic acid, 4-coumaroyl-3',4'-dihydroxyphenyllactic acid, caffeoyl-4'-hydroxyphenyllactic acid) could be detected in microsomal preparations (Petersen et al. 1993). Monophenolic substrates like 4-coumarate, 4-coumaroyl-CoA, 4-hydroxyphenylpyruvate or 4-hydroxyphenyllactate were not accepted by these hydroxylases. Because of this strong specificity for RA-like esters it was suggested that in RA biosynthesis the introduction of the 3- and 3'-hydroxyl groups are the last two reactions of the biosynthetic pathway. These hydroxylases have a number of common characteristics with cinnamic acid 4-hydroxylase and therefore might also be cytochrome P-450-dependent monooxygenases. They require molecular oxygen and NADPH for a successful hydroxylation reaction and are strongly inhibited by cytochrome c (10 μM for a 100% inhibition), a competitive inhibitor of NADPH:cytochrome P-450 reductase. On the other hand, phenolase inhibitors (diethyldithiocarbaminate, salicylhydroxamic acid) or potassium cyanide do not negatively influence the hydroxylations. NADH can only sustain around 10% of the hydroxylation activities obtained with NADPH as electron donor. FAD, which is a component of NADPH:cytochrome P-450 reductase, stimulates the enzyme activities by 20%. Both, 3- and 3'-hydroxylase activities, were saturated with 0.5 mM NADPH and showed an apparent K_m-value of 30 μM for this substrate. Up to now it remains unclear whether those two hydroxylation reactions are catalyzed by the same enzyme or two independent enzymes. Inhibition studies with cytochrome P-450-inhibitors, however, indicate the participation of two differently sensitive enzyme systems (Petersen, unpublished results). The participation of cytochrome P-450 in the 3- and 3'-hydroxylations of RA-like esters still remains to be proven.

With the final introduction of the 3- and 3'-hydroxyl groups of the ester the end product of the biosynthetic pathway, rosmarinic acid, is formed.

Discussion and conclusions

With help of the high-producing cell suspension cultures of *Coleus blumei* and *Anchusa officinalis* the biosynthetic pathway of rosmarinic acid has been elucidated (De-Eknamkul & Ellis 1987a, b; Petersen & Alfermann 1988; Petersen et al.1993). All enzyme activities necessary for the formation of RA from the precursors phenylalanine and tyrosine (Ellis & Towers 1970) could be isolated from cell cultures of *Coleus blumei* grown in a medium with 4% sucrose which strongly enhances RA production. Since the formation of up to 20% of the cell dry weight as RA is accomplished in only about 5 days of the culture period, the activities of the involved enzymes must be very high. This is in contrast to whole plants where often lower secondary product concentrations are synthesized over a long period. This shows again the usefulness of plant cell cultures for biosynthetic studies of secondary pathways as described by Zenk (1991).

In RA biosynthesis, phenylalanine is transformed into 4-coumaroyl-CoA by the enzymes of the general phenylpropanoid pathway, phenylalanine ammonia-lyase (PAL), cinnamic acid 4-hydroxylase (CAH) and hydroxycinnamoyl:CoA ligase (4CL) resulting in 4-coumaryol-CoA as the activated precursor for the ester formation. 4-Hydroxyphenyllactate is formed from tyrosine by the activities of tyrosine aminotransferase (TAT) and hydroxyphenylpyruvate reductase (HPPR). These two enzymes have been newly detected and described as enzymes of secondary metabolism (De-Eknamkul & Ellis 1987a; Petersen & Alfermann 1988). Their substrate specificities and the induction of their activities in correlation to RA biosynthesis can be taken as evidences that TAT and HPPR are specific enzymes for this secondary metabolic pathway although the same or similar enzymes are also involved in primary pathways. The formation of the presumably first ester, 4-coumaroyl-4'-hydroxyphenyllactate, is catalyzed by rosmarinic acid synthase (RAS) under the release of coenzyme A (Petersen & Alfermann 1988; Petersen 1991). This enzyme as well as 4CL and HPPR are not specific with regard to the hydroxylation pattern of the aromatic rings of the respective substrates. Therefore the question was whether the 3-hydroxyl groups of both aromatic rings are introduced prior to or after the formation of the ester. Specific 3-hydroxylases for 4-coumaric acid or 4-coumaroyl-CoA have been described only for a few cases and the respective enzymes, with exception of the ubiquitous phenolases, do not seem to occur frequently. The identification of hydroxylase activities in microsomal preparations from cell cultures of *Coleus blumei* which are able to introduce the 3- and 3'-hydroxyl groups specifically into the aromatic rings of RA-like esters sustains the assumption that the 3- and 3'-hydroxylation reactions are the final biosynthetic steps in RA biosynthesis. This corresponds to findings in the biosynthetic pathways of flavonoids and chlorogenic acid where the introduction of the hydroxylation patterns of the aromatic rings occurs after the completion of the C-skeleton as well (Forkmann et al. 1980; Stotz & Forkmann 1982; Hagmann et al. 1983; Heller & Kühnl 1985; Larson & Bussard 1986; Kühnl et al. 1987). In a strict sense, the above described biosynthetic pathway of RA is only proven for cell cultures of *Coleus blumei*. The validity of this scheme has to be shown for other systems and whole plants as well. Some of the respective enzymes have already been found to be active in plants of *Coleus blumei* (Szabo, unpublished results). An interesting task will be to demonstrate whether the same biosynthetic pathway of RA is used in evolutionary distant plants as ferns and hornworts.

The identification of all the enzymes necessary for RA biosynthesis now enables us to investigate the regulation of this biosynthetic pathway. Under conditions which allow a high production of RA (high sugar content of the culture medium) the activities of the biosynthetic enzymes are coordinately stimulated (Karwatzki et al. 1989). The enzymes catalyzing the formation of the precursors for the ester all have their maximum activities on day 6 of the culture period, whereas the ester forming enzyme (RAS) is most active on day 7–8 and maximal RA accumulation is observed on day 10 of the culture period. In media with low sugar concentrations the activities of all the enzymes as well as RA accumulation remain low. This system is suitable for investigations on the regulation of this secondary pathway since only a limited number of enzymes is involved, but the intermediates are formed in coordinately regulated parallel pathways. For this purpose a number of enzymes of RA biosynthesis have already been purified and first experiments for the cloning of genes coding for RA biosynthetic enzymes have already been performed. Further investigations will hopefully show us whether RA biosynthesis is regulated by similar mechanisms as described for example for anthocyanin formation (e.g., Lloyd et al. 1992).

Acknowledgement

We are thankful for the financial support of the Deutsche Forschungsgemeinschaft.

References

Benveniste I, Salaün JP & Durst F (1977) Wounding-induced cinnamic acid hydroxylase in Jerusalem artichoke tuber. Phytochemistry 16: 69–73

Benveniste I, Salaün JP, Simon A, Reichhardt D & Durst F (1982) Cytochrome P-450-dependent ω-hydroxylation of lauric acid by microsomes of pea seedlings. Plant Physiol. 70: 122–126

Boniwell JM & Butt VS (1986) Flavine nucleotide dependent 3-hydroxylation of 4-hydroxyphenylpropanoid carboxylic acids by particulate preparations from potato tubers. Z. Naturforsch. 41c: 56–60

Butt VS (1985) Oxygenation and oxidation in the metabolism of aromatic compounds. In: Van Sumere CF & Lea PJ (Eds) The Biochemistry of Plant Phenolics (pp 349–365). Clarendon Press, Oxford

De-Eknamkul W & Ellis BE (1987a) Tyrosine aminotransferase: the entrypoint enzyme of the tyrosine-derived pathway in rosmarinic acid biosynthesis. Phytochemistry 26: 1941–1946

De-Eknamkul W & Ellis BE (1987b) Purification and characterization of tyrosine aminotransferase activities from *Anchusa officinalis* cell cultures. Arch. Biochem. Biophys. 257: 430–438

De-Eknamkul W & Ellis BE (1988) Rosmarinic acid: production in plant cell cultures. In: Bajaj YPS (Ed) Biotechnology in Agriculture and Forestry, Vol 4, Medicinal and Aromatic Plants I (pp 310–329). Springer-Verlag, Berlin, Heidelberg

Ellis BE, Remmen S & Goeree G (1979) Interactions between parallel pathways during biosynthesis of rosmarinic acid in cell suspension cultures of *Coleus blumei*. Planta 147: 163–167

Ellis BE & Towers GHN (1970) Biogenesis of rosmarinic acid in *Mentha*. Biochem. J. 118: 291–297

Forkmann G, Heller W & Grisebach H (1980) Anthocyanin biosynthesis in flowers of *Matthiola incana*. Flavanone 3- and flavonoid 3'-hydroxylases. Z. Naturforsch. 35c: 691–695

Friedrich CA, Morizot DC, Siciliano MJ & Ferell RE (1987) The reduction of aromatic alpha-keto acids by cytoplasmatic malate dehydrogenase and lactate dehydrogenase. Biochem. Genet. 25: 657–669

Friedrich CA, Ferell RE, Siciliano MJ & Kitto GB (1988) Biochemical and genetic identity of alpha-keto reductase and cytoplasmatic malate dehydrogenase from human erythrocytes. Ann. Hum. Gen. 52: 25–37

Gamborg OL, Miller RA & Ojima K (1968) Nutrient requirements of suspension cultures of soybean root cells. Exp. Cell Res. 50: 151–158

Gertlowski C & Petersen M (1993) Influence of the carbon source on growth and rosmarinic acid production in suspension cultures of *Coleus blumei*. Plant Cell Tiss. Org. Cult., 34: 183–190

Hagmann ML, Heller W & Grisebach H (1983) Induction and characterization of a microsomal flavonoid 3'-hydroxylase from parsley cell cultures. Eur. J. Biochem 134: 547–554

Harborne JB (1966) Caffeic acid ester distribution in higher plants. Z. Naturforsch. 21 b: 604–605

Häusler E, Petersen M & Alfermann AW (1991) Hydroxyphenylpyruvate reductase from cell suspension cultures of *Coleus blumei* Benth. Z. Naturforsch. 46c: 371–376

Häusler E, Petersen M & Alfermann AW (1992) Rosmarinsäure in *Blechnum*-Spezies. In: Haschke HP & Schnarrenberger C (Eds) Botanikertagung 1992 Berlin (p 507). Akademie Verlag, Berlin

Heller W & Kühnl T (1985) Elicitor induction of a microsomal 5-O-(4-coumaroyl)shikimate 3'-hydroxylase in parsley cell suspension cultures. Arch. Biochem. Biophys. 241: 453–460

Kamsteeg J, van Brederode J, Verschuren PM & van Nigtevecht G (1981) Identification, properties and genetic control of *p*-coumaroyl-coenzyme A 3-hydroxylase isolated from petals of *Silene dioica*. Z. Pflanzenphysiol. 102: 435–442

Karwatzki B (1992) Biochemische und immunchemische Untersuchungen zur Hydroxyzimtsäure:CoA Ligase aus rosmarinsäurebildenden Zellkulturen von *Coleus blumei* Benth. Doctoral thesis, University of Düsseldorf

Karwatzki B, Petersen M & Alfermann AW (1989) Transient activity of enzymes involved in the biosynthesis of rosmarinic acid in cell suspension cultures of *Coleus blumei*. Planta Med. 55: 663–664

Kneusel RE, Matern U & Nicolay K (1989) Formation of *trans*-caffeoyl-CoA from *trans*-4-coumaroyl-CoA by Zn^{2+}-dependent enzymes in cultured plant cells and its activation by an elicitor-induced pH-shift. Arch. Biochem. Biophys. 269: 455–462

Kojima M & Takeuchi W (1989) Detection and characterization of *p*-coumaric acid hydroxylase in mung bean, *Vigna mungo*, seedlings. J. Biochem. 105: 265–270

Kojima M & Villegas RJA (1984) Detection of the enzyme in sweet potato which catalyzes trans-esterification between 1-O-*p*-coumaroyl-d-glucose and d-quinic acid. Agric. Biol. Chem. 48: 2397–2399

Kreuzaler F & Hahlbrock K (1972) Enzymatic synthesis of aromatic compounds in higher plants: formation of naringenin (5,7,4'-trihydroxyflavanone) from *p*-coumaroyl-coenzyme A and malonyl-coenzyme A. FEBS Letters 28: 69–72

Kühnl T, Koch U, Heller W & Wellmann E (1987) Chlorogenic acid biosynthesis: Characterization of a light-induced microsomal 5-O-(4-coumaroyl)-D-quinate/shikimate 3'-hydroxylase from carrot (*Daucus carota* L.) cell suspension cultures. Arch. Biochem. Biophys. 258: 226–232

Larson RL & Bussard JB (1986) Microsomal flavonoid 3'-hydroxylase from maize seedlings. Plant Physiol. 80: 483–486

Litvinenko VI, Popova TP, Simonjan AV, Zoz IG & Sokolov VS (1975) "Gerbstoffe" und Oxyzimtsäureabkömmlinge in Labiaten. Planta Med. 27: 372–380

Lloyd AM, Walbot V & Davis RW (1992) *Arabidopsis* and *Nicotiana* anthocyanin production activated by maize regulators *R* and *C1*. Science 258: 1773–1775

Meinhard J, Petersen M & Alfermann AW (1992) Purification of hydroxyphenylpyruvate reductase from cell cultures of *Coleus blumei*. Planta Med. 58: A598–A599

Mizukami H & Ellis BE (1991) Rosmarinic acid formation and differential expression of tyrosine aminotransferase isoforms in *Anchusa officinalis* cell suspension cultures. Plant Cell Rep. 10: 321–324

Mölgaard P & Ravn H (1988) Evolutionary aspects of caffeoyl ester distribution in dicotyledons. Phytochemistry 27: 2411–2421

Petersen M (1991) Characterization of rosmarinic acid synthase from cell cultures of *Coleus blumei*. Phytochemistry 30: 2877–2881

Petersen M & Alfermann AW (1988) Two new enzymes of rosmarinic acid biosynthesis from cell cultures of *Coleus blumei*: Hydroxyphenylpyruvate reductase and rosmarinic acid synthase. Z. Naturforsch. 43c: 501–504

Petersen M, Häusler E, Karwatzki B & Meinhard J (1993) Proposed biosynthetic pathway for rosmarinic acid in cell cultures of *Coleus blumei* Benth. Planta 189: 10–14

Petersen M & Metzger JW (1993) Identification of the reaction products of rosmarinic acid synthase from cell cultures of *Coleus blumei* by ion spray mass spectrometry and tandem mass spectrometry. Phytochem. Anal., 4: 131–134

Razzaque A & Ellis BE (1977) Rosmarinic acid production in *Coleus* cell cultures. Planta 137: 287–291

Scarpati ML & Oriente G (1958) Isolamento e costituzione dell'acido rosmarinico (dal *Rosmarinus off.*). Ric. Sci. 28: 2329–2333

Scarpati ML & Oriente G (1960) Costituzione stereochimica dell'acido β(3,4-diossifenil)α-lattico dal *Rosmarinus off.* Ric. Sci. 30: 255–259

Stöckigt J & Zenk MH (1974) Enzymatic synthesis of chlorogenic acid from caffeoyl coenzyme A and quinic acid. FEBS Letters 42: 131–134

Stotz G & Forkmann G (1982) Hydroxylation of the B-ring of flavonoids in the 3'- and 5'-position with enzyme extracts from flowers of *Verbena hybrida*. Z. Naturforsch. 37c: 19–23

Strack D, Gross W, Wray V & Grotjahn L (1987) Enzymic synthesis of caffeoylglucaric acid from chlorogenic acid and glucaric acid by a protein preparation from tomato cotyledons. Plant Physiol. 83: 475–478

Takeda R, Hasegawa J & Sinozaki K (1990) Phenolic compounds from Anthocerotae. In: Zinsmeister HD & Mues R (Eds) Bryophytes. Their chemistry and chemical taxonomy (pp 201–207). Oxford Science Publications, Oxford

Tkotz N & Strack D (1980) Enzymatic synthesis of sinapoyl-l-malate from 1-sinapoylglucose and l-malate by a protein preparation from *Raphanus sativus* cotyledons. Z. Naturforsch. 35c: 835–837

Ulbrich B, Wiesner W & Arens H (1985) Large-scale production of rosmarinic acid from plant cell cultures of *Coleus blumei* Benth. In: Neumann KH, Barz W & Reinhard E (Eds) Primary and Secondary Metabolism of Plant Cell Cultures (pp 293–303). Springer-Verlag, Berlin, Heidelberg

Villegas RJA, Shimokawa T, Okuyama H & Kojima M (1987) Purification and characterization of chlorogenic acid:chlorogenate caffeoyl transferase in sweet potato roots. Phytochemistry 26: 1577–1581

Zenk MH (1991) Chasing the enzymes of secondary metabolism: Plant cell cultures as a pot of gold. Phytochemistry 30: 3861–3863

Zenk MH, El-Shagi H & Ulbrich B (1977) Production of rosmarinic acid by cell-suspension cultures of *Coleus blumei*. Naturwissenschaften 64: 585–586

Zinsmeister HD, Becker H & Eicher T (1991) Moose, eine Quelle biologisch aktiver Naturstoffe? Angew. Chemie 103: 134–151

Plant Cell, Tissue and Organ Culture **38**: 181–188, 1994.

The biosynthetic pathway of the S-alk(en)yl-L-cysteine sulphoxides (flavour precursors) in species of *Allium*

S. J. Edwards[1,2], G. Britton[1] & H. A. Collin[2]
[1]*Department of Biochemistry, University of Liverpool, P.O. Box 147, Liverpool L69 3BX, UK;* [2]*Department of Genetics and Microbiology, University of Liverpool, P.O. Box 147, Liverpool L69 3BX, UK*

Key words: *Allium cepa*, biosynthesis, compartmentation, γ-glutamyl peptides, onion, protoplasts, S-alkenyl-L-cysteine sulphoxide

Abstract

Pulse labelling experiments with $^{35}SO_4^{2-}$ fed for 24h to intact plants (shooted onion sets) of *Allium cepa* (onion) showed that > 70% of the label appeared in the S-alkenyl-L-cysteine sulphoxides within 18h, reached a maximum at 48h and thereafter decreased. The amount of label detected in the γ-glutamyl peptide fractions was below 20% of the total label at any time. It is concluded that in intact plants (at the growth stage used) the γ-glutamyl peptides are not the immediate precursors of the S-alkenyl-L-cysteine sulphoxides. The major S-alkenyl-L-cysteine sulphoxide in onion was found to be compartmentalized mainly within the endoplasmatic reticulum.

Abbreviations: AllCysSO – (+)-S-2-propenyl-L-cysteine sulphoxide; MeCysSO – (+)-S-methyl-L-cysteine sulphoxide; PrenCysSO – *trans*-(+)-S-1-propenyl-L-cysteine sulphoxide; ProCysSO – (+)-S-propyl-L-cysteine sulphoxide

Introduction

Allium species contain a high proportion (1–5% dry weight) of non-protein sulphur amino-acids (Lancaster & Shaw 1989). One class of these secondary metabolites, the S-alk(en)yl-L-cysteine sulphoxides give rise to the characteristic flavours associated with the *Allium* species. Four sulphoxides have been found to occur naturally in the *Allium* species: namely (+)-S-methyl-, (+)-S-propyl-, *trans*-(+)-S-1-propenyl-, and (+)-S-2-propenyl-L-cysteine sulphoxides (respectively Me-, Pro-, Pren- and AllCysSO). All *Allium* species are reported to contain MeCysSO (Lancaster & Shaw 1989), while ProCysSO predominates in chives, PrenCysSO in onions and AllCysSO in garlic (Edwards et al. 1994).

Flavour in *Allium* is only released when the tissue is cut or broken. Work on garlic (Cavellito & Bailey 1944; Stoll & Seebeck 1951) and on onion (Virtanen & Spåre 1961; Schwimmer 1968) established that when the plant tissue was cut the S-alk(en)yl cysteine sulphoxides were hydrolysed by the enzyme alliinase. Lancaster & Collin (1981) provided evidence that the S-alk(en)yl cysteine sulphoxides were compartmentalised within the cytoplasm and the hydrolytic enzyme alliinase located within the vacuole. The initial products of the reaction between alk(en)yl cysteine sulphoxides and alliinase are ammonia, pyruvate and alk(en)yl sulphenic acids; the last of which breaks down non-enzymatically to form a variety of disulphides, which have a much milder flavour. The quantitative and qualitative difference in the alk(en)yl cysteine sulphoxide content results in the different flavours and odours of the many *Allium* species. In onions it is particularly the presence or absence of PrenCysSO which is responsible for the lachrymatory effect of onions, and AllCysSO which produces the characteristic taste of garlic. Early work established that PrenCysSO when hydrolysed by alliinase is converted into propenyl sulphenic acid, which is chemically unstable and undergoes rearrangement to form the lachrymator compound propanal sulphoxide (Virtanen 1965).

Allium species also contain other non-volatile sulphur compounds, the γ-glutamyl peptides, which are not hydrolysed by alliinase and do not therefore contribute to flavour production. For example in the onion bulb the majority of the PrenCysSO is bound as γ-glutamyl PrenCysSO (Lancaster & Shaw 1991), which effectively removes a large proportion of the PrenCysSO from contributing to lachrymator production (Whitaker 1976). In *Allium* species there are as many as 18 sulphur-containing γ-glutamyl peptides which have been identified by Virtanen and Suzuki and their respective co-workers (reported in Lancaster & Shaw 1989). The function of the γ-glutamyl peptides in the metabolism of the plant remains unclear. Originally they were generally considered to function as reserves of nitrogen and sulphur, and have since been implicated in the transport of amino-acids in cells (Kasai & Larson 1980). In 1989 however, Lancaster and Shaw proposed that some of the γ-glutamyl peptides (e.g. γ-glutamyl PrenCysSO) may be intermediates in the biosynthetic pathway of alk(en)yl cysteine sulphoxides.

A number of pathways for the biosynthesis of the alk(en)yl cysteine sulphoxides have been proposed. Early biosynthetic studies also showed that the uptake of labelled sulphur compounds resulted in many labelled γ-glutamyl peptides as well as alk(en)yl cysteine sulphoxides. These early studies gave rise to an important question concerning the biosynthetic relationship between γ-glutamyl peptides and the alk(en)yl cysteine sulphoxides. Lancaster & Shaw (1989) used a pulse chase experiment ($^{35}SO_4$) to investigate the sequence of appearance of label in sulphur compounds in order to determine the biosynthetic relationship between flavour precursors and the γ-glutamyl peptides. Their work, performed on excised leaves, indicated that within the first 24h after the leaves were exposed to ^{35}S the label appeared predominantly in the fractions containing γ-glutamyl peptides. After 24h, as the amount of label declined in the γ-glutamyl peptides, the amount of label in the alk(en)yl cysteine sulphoxides fractions began to increase, thus indicating in this single experiment that the ^{35}S label was incorporated into the γ-glutamyl peptides *before* the alk(en)yl cysteine sulphoxides. This scheme was confirmed by Parry & Lii (1991) who showed that addition of γ-glutamyl cysteine to methacrylic acid gave rise to γ-glutamyl-S-2-carboxypropylcysteine which in onion undergoes sequential decarboxylation to γ-glutamyl-S-1-propenyl cysteine, oxidation to γ-glutamyl-S-1-propenylcysteine sulphoxide and finally cleavage by γ-glutamyl transpeptidase to PrenCysSO.

More recently however, Ohsumi et al. (1993) have shown that in differentiating tissue cultures of garlic AllCysSO is formed by the oxidation of AllCys in agreement with the pathways proposed by Suzuki et al. (1961, 1962), Sugii et al. (1963), Granroth (1970) and Turnbull et al. (1980). Thus there are two main pathways proposed so far for the synthesis of PrenCysSO. The first involves no γ-glutamyl precursors (as reported by Suzuki et al. 1961, 1962 and Sugii et al. 1963) and Granroth (1970) while the second pathway is based on the presence of γ-glutamyl alkenyl cysteine compounds as the precursors (Lancaster & Shaw 1989). It is of course possible that both biosynthetic pathways are functional.

Much of the previous work that has been performed on the biosynthetic route has been done on excised plant organs and cell tissue culture systems. To our knowledge no one has yet made a detailed study of the dynamics of the complex interaction between the two chemical groups in the intact plant. Excised leaves are effectively senescent tissue and undifferentiated tissue cultures are meristematic tissue so that the results obtained from such material may be of limited relevance to the healthy intact plant. The work presented in this paper was obtained by pulse labelling ($^{35}SO_4$) on intact growing plants in an attempt to resolve the relative importance of γ-glutamyl peptides in the biosynthesis of the sulphoxides. In addition we present data on the compartmentation of the flavour precursors and their intermediates in the cell.

Materials and methods

Plant material

Onion sets of variety Stuttgarter Riesen were obtained from Bridgemere Garden World, Cheshire, U.K. and stored at 4 °C in the dark until used. Sets were encouraged to shoot by supporting them above water for 1 week as described in Edwards et al. 1994. Water was replaced with Hoaglands and Arnon solution for 1 week then fresh culture solution was added which contained $^{35}SO_4$ (in the form $Na_2{}^{35}SO_4$, obtained from Amersham International PLC, Buckinghamshire, U.K.). When shoots were labelled prior to fractionation the plants were exposed to the same radioactive source for 7 days before being excised and the shoot material was used for preparation of protoplasts. For the pulse labelling experiments, onion sets were exposed to the label for a total of 24h, then transferred to fresh nutrient solution. Two plants were harvested at 18h, 24h, 28h, 67h and 73h after the initial exposure to radioactive solution then tissue was taken from the shoots, roots, outer brown scale, middle fleshy scale and the inner scale immediately adjacent to the shoot and extracted as described.

Extraction of S-alk(en)yl cysteine sulphoxides and γ-glutamyl peptides

S-alk(en)yl cysteine sulphoxides and γ-glutamyl peptides in the plant material were extracted in methanol:chloroform:water (12:5:3) containing 10 mM hydroxylamine (Edwards et al. 1994). The mixture was partitioned between chloroform and water and the methanol:water layer kept, dried under vacuum at 40 °C and resuspended in 100 μl distilled water. In the protoplast fractionation studies, individual gradients were extracted as above then the concentrated sample was applied to an Amberlite 120(H+) ion exchange column. The purified samples resuspended in 1 ml 0.03 M HCL, filtered through a 0.45μ millipore filter and sonicated for 15 min prior to HPLC analysis (Edwards et al. 1994). In the pulse labelling experiments the plant material was extracted in the same way but the sulphoxides and γ-glutamyl peptides were separated by Dowex 1 chromatography as described by Lancaster & Shaw (1989), sulphoxides were eluted in 0.1 M acetic acid, and the γ-glutamyl sulphoxides eluted in 1 and 2 M acetic acid. All eluates were again dried down under N_2 and resuspended in 500 μl distilled water, filtered and sonicated.

Counting of radioactivity

100 μl of the sample obtained at each point of the isolation procedure into 5 ml of scintillation fluid (Optiphase 'Hisafe' 3, LKB scintillation products), and secondly injecting 15 μl for separation of individual components by HPLC analysis and subsequent scintillation counting of the individual fractions. Samples were counted through a C14 window on an LKB 1219 Rackbeta liquid scintillation counter. Confirmation of radioactivity in the alk(en)yl cysteine sulphoxides was also obtained by TLC of 100 μl of sample on 20 × 20 cm pre-coated silica gel 60 F_{254}, 0.2 mm thick. Plates were run in solvent I, butan-2-one:pyridine:water:acetic acid (70:15:15:2) and then in the same direction in solvent II, propan-1-ol:water:n-propyl acetate:acetic acid:pyridine (120:60:20:4:1). Plates were subsequently scanned for radioactivity by means of Raytest RITA (Rapid Intelligent Radio TLC Analyser), part of the plate (5 × 20 cm) was removed and sprayed with 0.5% ninhydrin solution (in absolute ethanol). The colour was developed at 128 °C for 10 min and the ninhydrin positive spots were located.

Isolation of protoplasts

Green leaf shoot material (approx. 9 g) was cut in half to expose the internal tissue of leaves and placed face downwards into protoplast osmotic buffer (3 mM MES buffer, 0.6 M sorbitol and 3 mM $CaCl_2$ at pH 5.5) containing 2% (w/w) cellulase and 0.5% (w/w) pectinase, and left to incubate for 24h. Crude protoplasts were filtered through muslin to remove plant debris, then the filtrate of protoplast suspension was spun down at low speed for 10 min. The supernatant was resuspended in 5 ml of fresh protoplast osmoticum, then layered on top of 10 ml 16% dextran T40 and spun down for 30 min at 1 rpm. Protoplasts were collected from the Dextran/buffer interface and resuspended in fresh protoplast osmoticum. Protoplasts were then spun down for 10 min at 1 rpm and the top supernatant removed and replaced with fresh protoplast osmoticum. This procedure was repeated 3 more times to ensure a clean preparation of protoplasts. After the final spin the supernatant was removed, the protoplasts were transferred to a chilled homogeniser and 5 ml of lytic buffer (0.45 M sorbitol, 25 mM tris-HCl, 1 mM DTT, 1 mM EDTA, pH 7.0) was added plus 100 μl 0.1 M hydroxylamine and a final 3 ml of lytic buffer. Protoplasts were then homogenised by 10 strokes then left to settle before being layered onto sucrose density gradients.

Density gradients were centrifuged for 3h at 16,000 rpm at 4 °C and individual fractions (1.5 ml) were collected by means of a peristaltic pump. Individual fractions were then assayed for rotenone-insensitive NADH reductase, a marker for the endoplasmic reticulum (Collin et al. 1989), γ-glutamyl transpeptidase (Naftalin et al. 1969), chlorophyll as a marker for the chloroplasts (Wintermans & De Nots 1965), protein (Musker 1988), sucrose (refractometer), alkenyl cysteine sulphoxides and radioactivity (as above). Density gradient buffer was made up as follows 1 mM EDTA, 1 mM DTT, 0.1 mM $MgCl_2$, 50 mM tris-HCl pH 8.0. Gradients 60–15% (w/w) sucrose in 5% increments were prepared. Gradients were then left for 2h in a horizontal position at room temperature before being left overnight in a vertical position at 4 °C.

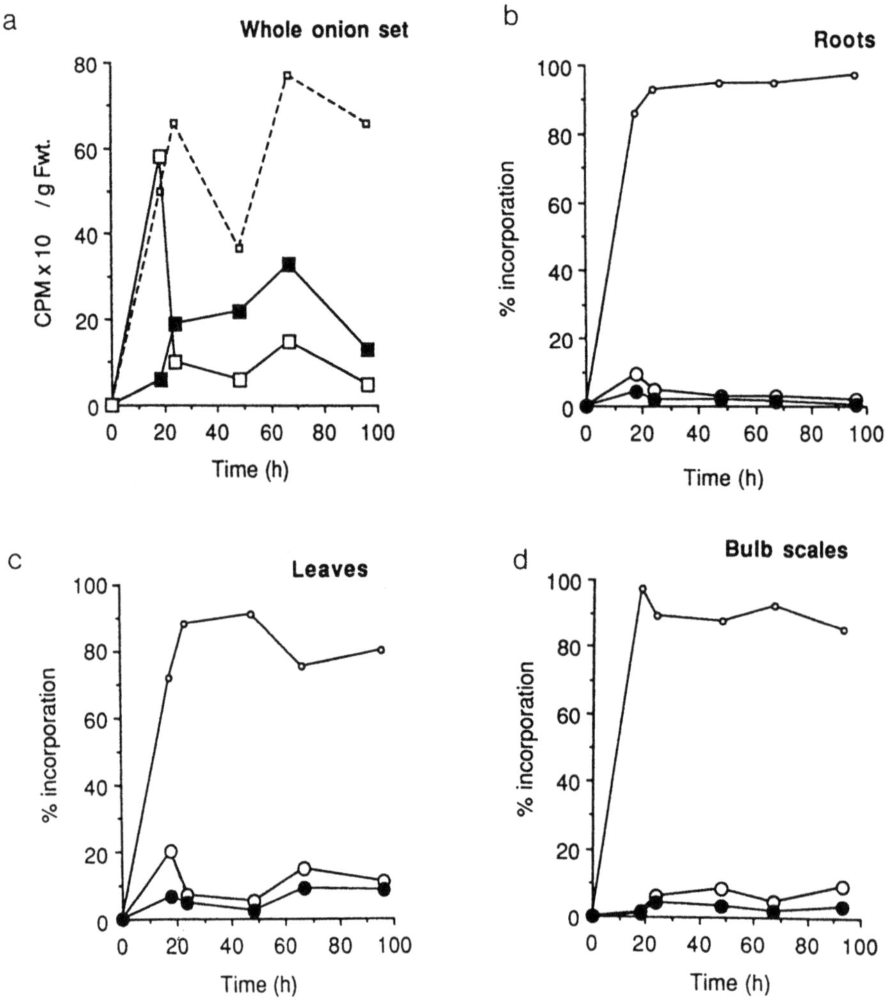

Fig. 1. Changes, with time, in incorporation of radioactivity into sulphur compounds in intact onion sets and individual plant organs after a 24h pulse of $^{35}SO_4$. Compounds were fractionated by Dowex 1 chromatography: 0.1 M acetic acid fraction (○) contains alk(en)yl cysteine sulphoxides and amino acids; 1.0 M (◯) and 2.0 M acetic acid fractions (●) contain γ-glutamyl peptides of increasing acidity. a) Total amount of ^{35}S (cpm g^{-1} FW) in individual plant organs, root (□), leaves (■) and bulb scales (☐); b) Percentage of ^{35}S in individual fractions extracted from root tissue; c) Percentage of ^{35}S in individual fractions extracted from leaf tissue; d) Percentage of ^{35}S in individual fractions extracted from bulb scale tissue.

Results and discussion

Pulse labelling of sulphur compounds in intact onion sets

High levels of radiolabel were found in the root and leaf tissue (10^4–10^5 cpm g^{-1} FW), whereas relatively little label was found to have been incorporated into bulb scale tissue, except that at 18h the level of activity in the bulbs exceeded that in either the leaves or roots (Fig. 1a). These high levels of radioactivity in the shoot may indicate that there was very little transfer of biosynthetic compounds from the bulb to the shoot, confirming the view that the biosynthetic pathway in the leaves was fully functional and not relying on preformed precursors from the bulb tissue. The highest percentage incorporation of radiolabel (in all plant organs) was always seen in the fractions containing the alk(en)yl cysteine sulphoxides (Fig. 1b-d); generally this was over 70% and frequently accounted for over 90% of the total activity. The level of radioactivity detected in this fraction increased during the labelling period (0–24h) and continued to be high until 48h after the start of the experiment (= 24h after the pulse was removed), but thereafter there was a slight decrease in the level of radioactivity. Radioactivity in the fraction containing γ-glutamyl peptides (1 and 2 M) increased slightly during the first 18h of the pulse labelling period, but

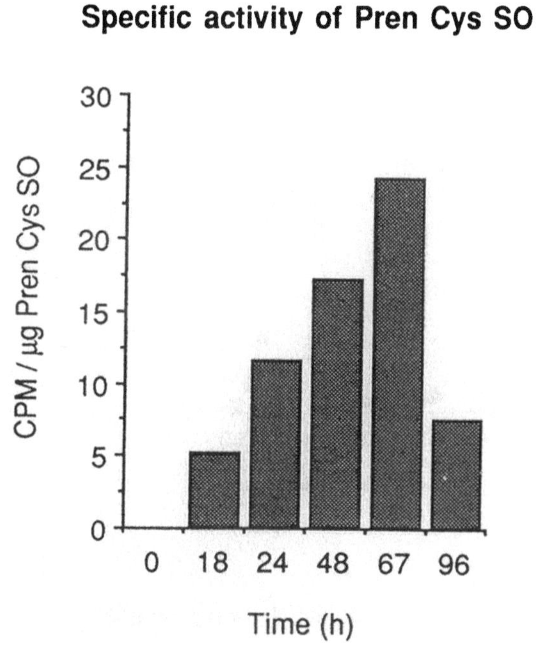

Fig. 2. Change in the amount of ^{35}S (cpm μg^{-1}) in propenyl cysteine sulphoxide isolated from intact onion leaves exposed to a 24h pulse of $^{35}SO_4$.

much less than that observed in the fractions containing sulphoxides, and thereafter declined (Fig. 1b-d). The percentage of label found in the γ-glutamyl peptide fractions (1 and 2 M acetic acid fractions) was generally less than 20%. There was little to indicate that the label had been in the γ-glutamyl peptides prior to entering the fractions containing S-alk(en)yl cysteine sulphoxides, although a shorter pulse labelling period and more intensive sampling during the pulse-labelling period may indicate otherwise. The presence of ^{35}S in the S-alkenyl cysteine sulphoxides was confirmed by TLC coupled to radioscanning. When the specific activity of PrenCysSO in leaves was measured it was found that within 18h it was already labelled and that the maximum specific activity occurred at 67h (43h after the label had been removed) after which it declined (Fig. 2). At 18 h, radioactivity in the PrenCysSO accounted for 20% of the total radioactivity recovered in the sulphoxide fraction, at 48h it accounted for 60% and at 93h it accounted for 38%.

Although specific activities for all the individual compounds in the extract are at present not available it is possible to consider radioactivity associated with each broad group of compounds as that in the alk(en)yl cysteine sulphoxides and the γ-glutamyl peptides. The results shown here obtained from intact plants differed markedly from those presented from work with excised leaves. Thus Lancaster & Shaw (1989) found that the ^{35}S label was rapidly taken up and incorporated into the fraction containing γ-glutamyl peptides and remained in the γ-glutamyl peptides for 24 h. During this period the amount of label in the alk(en)yl cysteine sulphoxides was very low but after the first 24h the percentage of label in the γ-glutamyl peptides had declined and the amount in the alk(en)yl cysteine sulphoxides began to increase. It is difficult to explain the difference between our results and those of Lancaster & Shaw (1989). One suggestion is that the different results obtained from intact leaves and detached leaves may be due to the fact that the developmental stage and physiological condition of the experimental material influences activity of the separate biosynthetic routes. In the intact plant when leaves began to senesce during bulb filling there was reported to be a decline in total alk(en)yl cysteine sulphoxide content and an increase in γ-glutamyl peptides. Conversely when the leaves were actively growing prior to bulb formation, the concentration of alk(en)yl cysteine sulphoxides was high relative to the amount of γ-glutamyl peptides present (Lancaster et al. 1984; Lancaster & Shaw 1991). The reason for changeover in the pattern of accumulation between the two chemical groups (i.e. free alkenyl cysteine sulphoxides and γ-glutamyl peptide) is not known, nor is it known whether synthesis or transformation of one to the other is responsible for their fluctuating levels. Lancaster proposed that during the early stages of shoot formation γ-glutamyl peptides were metabolised to form the free alk(en)yl cysteine sulphoxide and in support of this she was able to detect an increase in the levels of γ-glutamyl transpeptidase at this time (Lancaster et al. 1991).

The labelling of the γ-glutamyl peptides as observed by ourselves, Granroth (1970) and Suzuki et al. (1961, 1962) is not unexpected. When $^{35}SO_4$ is fed to plants, over 90% of SO_4 taken up by plants is assimilated into cysteine and over 35% of this is immediately channelled into reactions of the γ-glutamyl cycle via glutamyl cysteine and glutathione (Giovanelli et al. 1980). Further work is required to understand the full significance of the relationship between the γ-glutamyl peptides and the alk(en)yl cysteine sulphoxides.

To our knowledge no one has yet done a detailed study on the dynamics of the complex interaction between the two chemical groups, or on whether one group could be precursors to the other depending on the developmental and physiological state of the plant. Information that is available has been obtained with individual bulbs destructively harvested at different

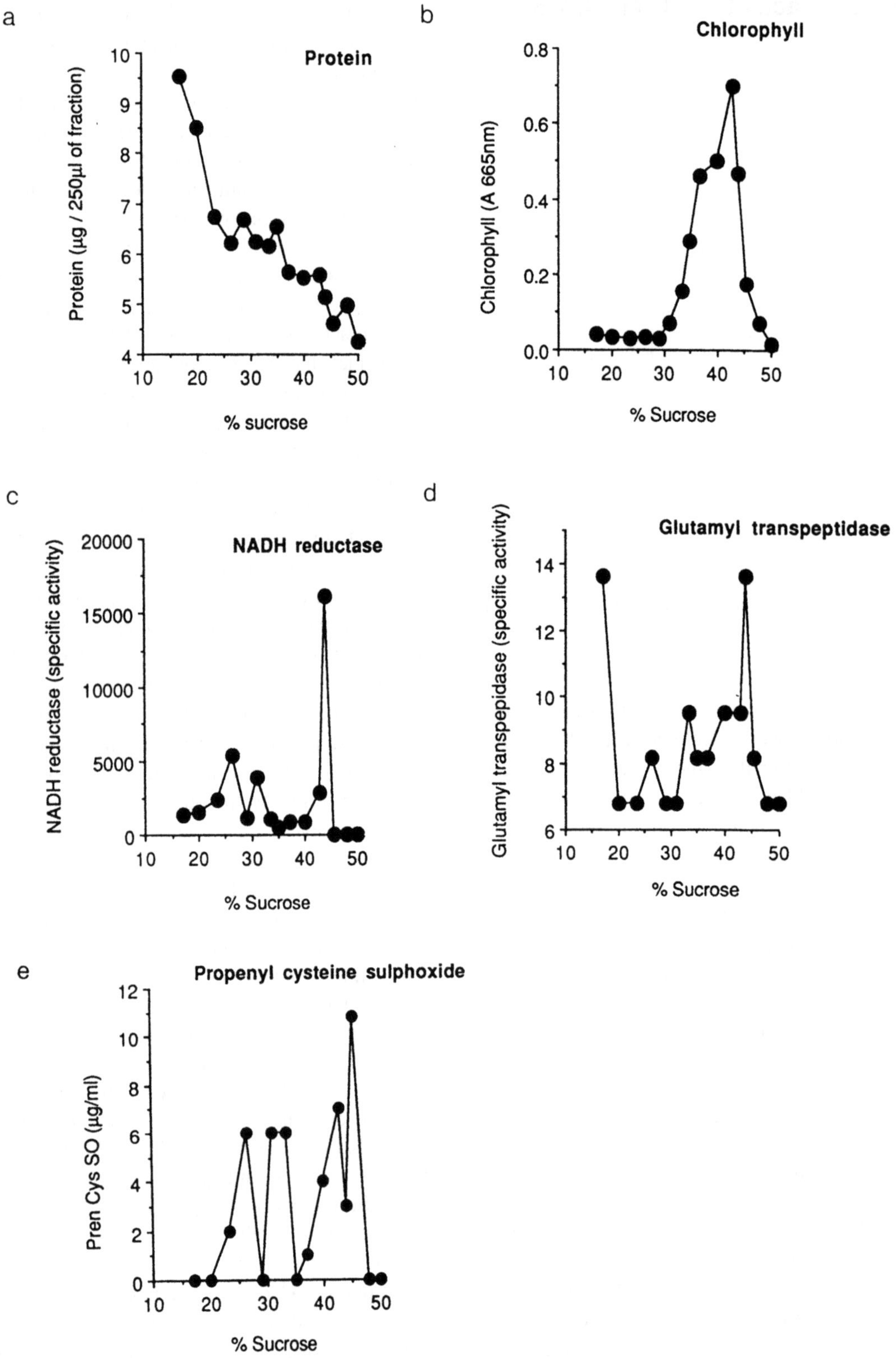

Fig. 3. Characterisation of fractions on sucrose density gradients after isopycnic centrifugation. a) protein, b) chlorophyll, c) rotenone-insensitive NADH reductase, d) γ-glutamyl transpeptidase and e) concentration of propenyl cysteine sulphoxide.

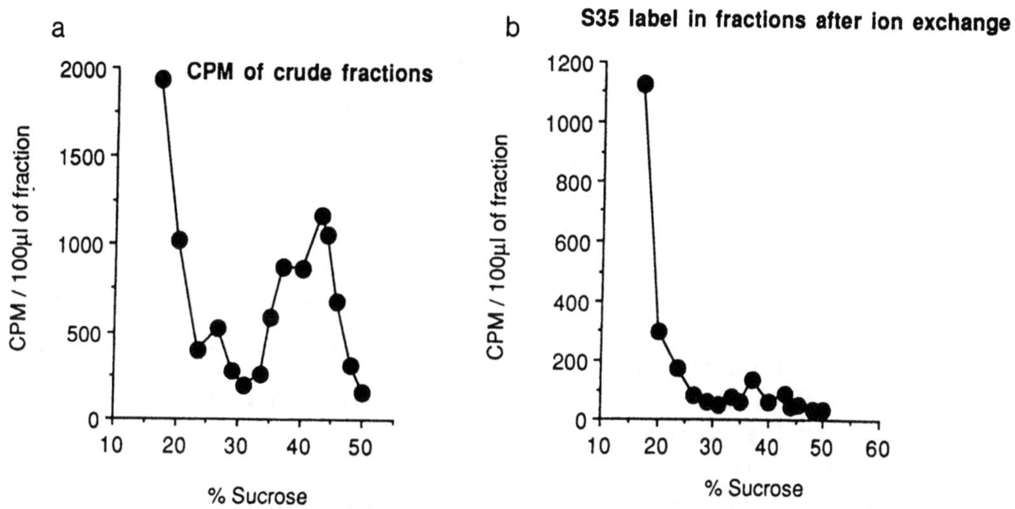

Fig. 4. Incorporation of radioactivity (^{35}S) in fractions on a sucrose density gradient after isopycnic centrifugation a) crude fraction before ion exchange and b) individual fractions after ion exchange.

times in the growing cycle and on excised leaves which may represent senescing tissue. Results obtained from such material may only provide results that describe only part of the story.

Intracellular localisation of PrenCysSO

The purpose of the second part of the work was to identify the intracellular location of PrenCysSO and intermediates in its formation. Analysis of cell fractions of leaf protoplasts showed that intact chloroplasts (measured as chlorophyll) were at a maximum on the gradient at 43% sucrose with the band extending from 30% sucrose to 45% sucrose (Fig. 3b). Endoplasmic reticulum membranes (measured as rotenone-insensitive NADH reductase) were located as two small peaks at 26% and 31% and a large peak at 44% sucrose (Fig. 3c). The peak at 26% sucrose coincided with the position of smooth ER, and the second at 31% overlapped with the beginning of the chloroplast band and the third at 44% sucrose with the chloroplast band. Each overlap of chloroplast and ER fractions had also been previously noted by Collin et al. (1989). Previously Lancaster et al. (1989) had located onion chloroplasts at 42% w/w while Collin et al. (1989) had identified onion chloroplasts at 35%, ER at 27% and 38% w/w sucrose. The distribution of PrenCysSO in the fractions showed peaks at 25, 32 and 45% w/w sucrose which coincided with the position of the ER and chloroplasts (Fig. 3e).

There was no peak which coincided with the presence of small vesicles detected at 55% w/w sucrose by Collin et al. (1989). The continuous sucrose gradient used in the present work did not extend beyond 50% sucrose. These present results showed further differences since they suggest a greater accumulation of PrenCysSO in the chloroplasts than shown by Collin et al. (1989). The enzyme, γ-glutamyl transpeptidase, which reportedly (Lancaster et al. 1991) controls the release of PrenCysSO from its bound form, γ-glutamyl PrenCysSO, was also found to be associated with the ER and the chloroplasts as well as the soluble fraction (Fig. 3d). In contrast, Lancaster et al. (1989) located γ- glutamyl transpeptidase predominantly in the soluble fraction and only a small fraction with the ER and none in the chloroplasts. Later workers suggested that γ-glutamyl transpeptidase is membrane bound (Schneider et al. 1992). The difference in the results may reflect the different physiological condition of the plant material used for the protoplast isolation. The protoplasts used in the present study were prepared on a similar time scale to that of Collin et al. (1989) except that the latter used seedling leaves as a source of material, Lancaster et al. (1989) prepared protoplasts from leaves that had been excised for 24h before protoplast isolation. The age and stage of development of leaves used for protoplast isolation might well have an effect on the activity of the cell and its components. Electron microscope examination of the protoplasts isolated directly from leaves of intact plants indicated that such protoplasts were already under stress since the protoplasts showed multiple vesicle formation and thylakoid disintegration. Care must be taken to standardise the starting material before direct comparisons can be made.

Radioactivity of the cell fractions showed the highest level in the soluble and chloroplast fractions (Fig. 4). The entry of radioactivity into specific components within the ER and chloroplasts needs to be resolved before conclusions can be drawn over sites of synthesis of PrenCysSO. Nevertheless the PrenCysSO, and the other alk(en)yl cysteine sulphoxides, by their association with membranes and organelles (vesicles, ER and chloroplasts) indicated compartmentation of sites of synthesis within the cytoplasm. The earlier work (Collin et al. 1989) and present data suggest that PrenCysSO synthesis may occur in stages in intact tissue with the first stage in the chloroplast from where intermediates are exported to the ER, possibly the smooth ER, then its products transferred to ER derived cytoplasmic vesicles for long term storage. These stages will be confirmed by the analysis of radioactivities of specific compounds in the different fractions.

References

Cavallito CJ & Bailey JH (1944) Allicin the active principle of *Allium sativum*. I. Isolation, physical properties and antibacterial action. J. Am. Chem. Soc. 66: 1950–1951

Collin HA, Musker D & Britton G (1989) Compartmentation of flavour precursor synthesis in the onion. In: Kurz WGW (Ed) Primary and secondary metabolism of plant cell cultures II (pp 125–132) Springer-Verlag, Berlin Heidelberg

Edwards SJ, Musker D, Collin HA & Britton G (1994) The analysis of S-alk(en)yl-L-cysteine sulphoxides (flavour precursors) from species of *Allium* by HPLC. Phytochem. Anal. 5: 4–9

Giovanelli J, Harvey Mudd S & Datko AH (1980) Sulfur amino acids in plants. In: Miflin BJ (Ed) The biochemistry of plants: a comprehensive treatise, Vol 5 (pp 453–505). Academic Press, New York

Granroth B (1970) Biosynthesis and decomposition of cysteine derivatives in onion and other *Allium* species. Ann. Acad. Scient. Fenn. Ser. A. 2, 154: 1

Kasai T & Larsen PO (1980) Chemistry and biochemistry of γ-glutamyl derivatives from plants including mushrooms (Basidiomycetes). In: Herz W, Grisebach H & Kirby GW (Eds) Progress in the chemistry of organic natural products, Vol 39. Springer New York

Lancaster JE & Collin HA (1981) Presence of alliinase in isolated vacuoles and of alkyl cysteine sulphoxides in the cytoplasm of bulbs of onion (*Allium cepa*). Plant Sci. Lett. 22: 169–176

Lancaster JE, McCallion BJ & Shaw ML (1984) The levels of S-alk(en)yl-L-cysteine sulphoxides during the growth of the onion (*Allium cepa* L.). J. Sci. Food Agric. 35: 415–420

Lancaster JE, Reynolds PHS, Shaw M, Domisse EM & Munro J (1989) Intracellular localisation of the biosynthetic pathway to flavour precursors in onion. Phytochemistry 28: 461–464

Lancaster JE & Shaw ML (1989) γ-Glutamyl peptides in the biosynthesis of s-alk(en)yl-L-cysteine sulphoxides (flavour precursors) in *Allium*. Phytochemistry 28: 455–460

Lancaster JE & Shaw ML (1991) Metabolism of γ-glutamyl peptides during development, storage and sprouting of onion bulbs. Phytochemistry 30: 2857–2859

Musker D (1988) Secondary Product Biosynthesis in Plant Cell Cultures Ph.D. Thesis, University of Liverpool

Naftalin L, Sexton M, Whitaker JE & Tracey D (1969) A routine procedure for estimating serum γ-glutamyl transpeptidase activity. Clin. Chim. Acta 26: 293

Ohsumi C, Hayashi T & Sano K (1993) Formation of alliin in the culture tissues of *Allium sativum*. Oxidation of S-allyl-L-cysteine. Phytochemistry 33: 107–111

Parry RJ & Lii F-L (1991) Investigations of the biosynthesis of trans-(+)-S-1-propenyl-L-cysteine sulfoxide. Elucidation of the stereochemistry of the oxidative decarboxylation process. J. Am. Chem. Soc. 113: 4704–4706

Schneider A, Martini N & Rennenberg H (1992) Reduced glutathione (GSH) transport into cultured tobacco cells. Plant Physiol. Biochem. 30: 29–38

Schwimmer S (1968) Enzymic conversion of *trans*-(+)-S-1-propenyl-L-cysteine sulphoxide to the bitter odor bearing component of onion. Phytochemistry 7: 401–404

Stoll A & Seebeck E (1951) Chemical investigations on alliin the specific principle of garlic. Adv. Enzymol. 11: 377–400

Sugii M, Nagasawa S & Suzuki (1963) Biosynthesis of S-Methyl-L-cysteine and S-Methyl-L-cysteine sulfoxide from methionine in Garlic. Chem. Pharm. Bull. 11: 135–136

Suzuki T, Sugi, M & Kakimoto T (1961) New γ-glutamyl peptides in garlic. Chem. Pharm. Bull. 9: 77–78

Suzuki T, Sugii M & Kakimoto T (1962) Metabolic incorporation of L-valine (^{14}C) into S-(2-carboxypropyl) glutathione and S-(2-carboxypropyl) cysteine in garlic. Chem. Pharm. Bull. 10: 328–331

Turnbull A, Galpin IJ & Collin HA (1980) Comparison of the onion plant (*Allium cepa*) and onion tissue culture. III. Feeding of ^{14}C labelled precursors of the flavour precursor compounds. New Phytol. 85: 483–487

Virtanen AI (1965) Studies on organic sulphur compounds and other labile substances in plants. Phytochemistry 4: 207–228

Virtanen AI & Spare C-G (1961) Isolation of the precursor of the lachrimatory factor in onion (*Allium cepa*). Suomen Kemistilehti 34: 72–74

Whitaker RJ (1976) Development of flavour, odor and pungency in onion and garlic. Adv. Food. Res. 22: 73–133

Wintermans JFGM & DeNots A (1965) Spectrophotometric characteristics of chlorophylls a and b and their pheophytins in ethanol. Biochim. Biophys. Acta. 109: 448–453

Plant Cell, Tissue and Organ Culture **38**: 189–198, 1994.

Elicitor induced secondary metabolism in *Ruta graveolens* L.

Role of chorismate utilizing enzymes

J. Bohlmann & U. Eilert
Institut für Pharmazeutische Biologie, Mendelssohnstr. 1 D-38106 Braunschweig, Germany

Key words: Anthranilate synthase, cell culture, chorismate mutase, elicitor induction, *Ruta graveolens*, shikimic acid pathway

Abstract

In vitro cultures of *Ruta graveolens* L. respond with rapid accumulation of acridone epoxides, furoquinolines and furanocoumarins, when challenged with autoclaved homogenate of the yeast *Rhodotorula rubra*. A transient increase of several enzymes of the respective biosynthetic pathways was measured but we still look for the key regulatory enzymes. We investigated whether the branch point enzymes of the shikimic acid pathway anthranilate synthase (AS) and chorismate mutase (CM) possibly play such a role. The two enzymes compete for chorismate. AS forms anthranilate, the precursor amino acid of acridone and furoquinoline alkaloids. CM channels chorismate into phenylalanine, tyrosine and phenylpropanoid biosynthesis. Elicitation resulted in a transient increase of the activity of both enzymes. Relative induction rates were 2-4 fold for AS and about 1.5 fold for CM. Constitutive CM activity, however, is about 1000 fold higher than AS activity. As in other plants 2 isoforms of CM are expected to be present in *R. graveolens*. A differential determination of the activity of the isoforms via the tryptophan activation rate proved to be ambiguous. Some evidence for the specific induction of a plastidic form of CM was obtained by inhibition of translation. The time courses of CM induction show CM not to be a key enzyme in elicitor induction of furanocoumarin accumulation. In comparison to other enzyme activities induction of anthranilate synthase activity corresponds closest to inducible acridone epoxide accumulation indicating a key role in its regulation. Induction of AS and CM was inhibited by actinomycin D and chloramphenicol while cycloheximid inhibited AS induction only.

Abbreviations: ACT – actinomycin D, AS – anthranilate synthase, CAP – chloramphenicol, CHX – cycloheximid, 4-CL – 4-coumarate CoA ligase, CM – chorismate mutase, DTT – dithiothreitol, NMT – S-adenosyl-L-methionine:anthranilic acid N-methyltransferase, PAL – phenylalanine ammonia lyase, XOMT – S-adenosylmethionine: xanthotoxol-O-methyltransferase

Introduction

Of the numerous secondary metabolites present in *Ruta graveolens* L. three different classes are derived from aromatic amino acids and thus the shikimate pathway. These are the acridone and the furoquinoline-type alkaloids which are both formed from anthranilic acid, and the coumarins and furanocoumarins, which are synthesized via phenylalanine. (Fig. 1)

These compounds are not only accumulated constitutively in the plant and its cell cultures but are also inducible by elicitation with fungal elicitors (Eilert 1989). Inducibility of product accumulation varied with the degree of differentiation of the cell culture system. The more differentiated a culture was the less was its response to elicitation. Hydroponically grown plants lacked inducible furanocoumarin and furoquinoline accumulation. Constitutive accumulation, however, was highest here.

Currently we investigate, which enzymes of the biosynthetic route are affected by elicitation to find those, which have key regulatory functions.

Concerning the furanocoumarin formation an induction was already shown for PAL, 4-CL,

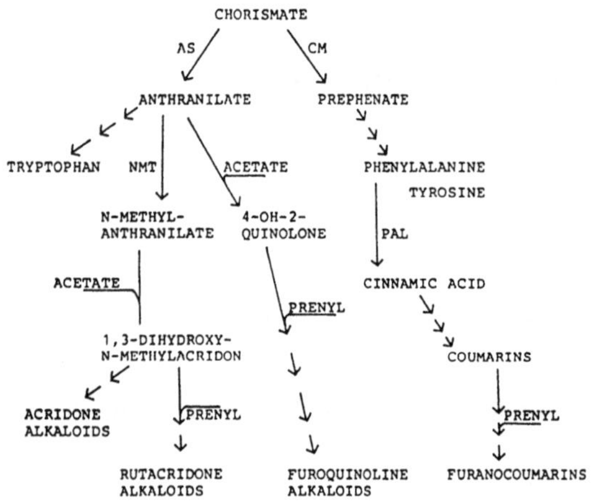

Fig. 1. Schematic biosynthetic pathways of elicitor inducible, chorismate derived secondary metabolites in *Ruta graveolens*.

(Eilert & Bohlmann 1989) and XOMT (Gibraltarskaya & Eilert 1992). The inducibility as well as the time courses match with the results obtained for elicitor induced furanocoumarin accumulation in the Apiaceae eg. parsley or *Ammi majus* (Hahlbrock & Scheel 1989).

Inducibility of enzymes of rutacridone epoxide biosynthesis has been shown for S-adenosyl-L-methionine: anthranilic acid N-methyltransferase (Eilert & Wolters 1989). The time courses of activity indicate that NMT is not a key regulatory enzyme. N-methylanthranilic acid- CoA-ligase as well as acridone synthase are reported to be inducible (Baumert et al. 1991), the published data are rather preliminary.

As pointed out, all these compounds are derived from the shikimic acid pathway, which branches at chorismate. Chorismate mutase (CM) channels chorismate into phenylalanine, tyrosine and phenylpropanoid biosynthesis. Anthranilate synthase (AS) competes with CM for the same substrate, channelling chorismate into the biosynthesis of tryptophan as well as acridone and furoquinoline alkaloids. Chorismate thus could be a key intermediate with AS and CM having key regulatory functions in channelling the flux of chorismate into two different elicitor inducible secondary metabolite branches. Therefore we investigated the effects of elicitation on the activities of these two enzymes.

Materials and methods

Cultures and culture conditions

All cell or organ cultures were grown in 250 ml Erlenmeyer flasks containing 75 ml nutrient medium and were subcultured every 4 weeks. They were kept on a gyratory shaker (120 rpm) at a room temperature of 25 °C and a 16 h/day photoperiod for light grown cultures or continuous darkness.

In vitro *systems of Ruta graveolens*

Different suspension-cultures (Rf, R-MS, R-DK, RN), shoot teratoma (RAT) and hydroponically grown photomixotrophic plants (ROH) were taken from our culture collection and cultered as described in Eilert & Wolters (1989).

Cell cultures of other Rutaceae

Suspension cultures of *Ruta chalepensis* and *Ruta macrophylla* were provided by Prof. Dr. I. Kuzovkina (Moscow) and kept on MS medium with 1 mg l^{-1} kinetin and 1 mg l^{-1} 2,4-D.

Suspension cultures of *Dictamnus albus*, *Choisya ternata* and *Skimmia japonica* were initiated in 1986 from leaf explants of garden grown plants. *D. albus* and *C. ternata* are grown on B5-medium (Gamborg et al. 1968), *S. japonica* on MS-medium (Murashige & Skoog 1962) with 0.25 mg l^{-1} kinetin and 0.25 mg l^{-1} 2,4-D and 0.5 mg l^{-1} NAA.

Suspension cultures from other families

The shoot teratoma culture of *Solanum dulcamara* was established in 1985 and is grown on hormone-free B5 medium (Eilert et al. 1987).

Suspension cultures of *Beta vulgaris* and *Chenopodium rubrum* were taken from the tissue culture collection of the Institut für Pharmazeutische Biologie, TU Braunschweig.

Elicitor preparation and elicitor experiments

The preparation of *Rhodotorula* elicitor homogenate was performed as described previously by Eilert et al. (1984). Time point of elicitation and elicitor amounts were varied and are given in the respective experiment.

Inhibitors were added 30 to 60 min prior to elicitation under sterile conditions. Control cells treated with inhibitors only were used to assess the effect of inhibitors themselves. Stock solutions of CHX, CAP and ACT (Sigma Chemical Co.) were prepared each in a concentration of 10 mg ml^{-1} in 96% ethanol. Final concentrations after addition to cell suspensions were 10 μg CHX, 5–200 μg CAP or 5 μg ACT per ml culture.

Cells were harvested at intervals given in the figures, imediately frozen in liquid nitrogen and stored at −80 °C until extraction. The experiments were at least performed in duplicates and to eliminate deviations among separate flasks all samples for a time course series were taken from one flask. All experiments were at least repeated once.

Protein extraction

Samples of 2 g frozen cells were thawed in 2 ml of the respective cold (4 °C) extraction buffer and ground in a mortar. The homogenate was centrifuged for 2 min. at 13.000 rpm (Eppendorf 5415 C centrifuge). One ml of the supernatant was desalted (PD 10 column, Pharmacia) and 2 ml eluate were collected. This protein extract was used for assays.

The extraction buffer for AS contained 0.2 M TRIS-HCl at pH 7.5 with 0.2 mM DTT, 0.2 mM Na_2EDTA, 60% glycerol, 8.0 mM $MgCl_2.H_20$ and 40 mM L-glutamine (Gaska 1981). Desalting was performed with assay buffer.

The standard extraction buffer for CM was a 50 mM potassium phosphate buffer pH 7.8 with 1 mM DTT and 30% glycerol (according to Hertel et al. 1988). This buffer was also used for desalting and in the assays. Alternatively a 0.1 M TRIS-HCL buffer pH 7.8 as used by McCue (1988) was tried.

Protein determination

Protein determination was performed according to Bradford (1976). Protein contents were calculated using reference concentrations of a mixture of bovine serum albumine and rabbit gamma globulin in the ratio of 7:3.

Determination of anthranilate synthase

The determination of anthranilate synthase was performed according to Poulsen et al. (1991). This method allows measurement of product formation as well as substrate consumption. The assay volume of 0.5 ml contained 0.5 μmol chorismate, 12.5 μmol L-glutamine, 1.5 μmol $MgCl_2$ and 125 μl protein extract. The assay buffer itself was TRIS-HCl 0.1 M, pH 7.5 supplemented with 50 μM DTT, 50 μM Na_2EDTA, 2 mM $MgCl_2.6H_20$, 10 mM L-glutamine and 5% glycerol (Gaska 1981). The reaction was started by addition of chorismate and stopped by addition of 25 μl 5 M phosphoric acid after an incubation period of 30 min at 32 °C. Blanks were made by adding phosphoric acid before the incubation. After centrifugation the supernatant was analysed by HPLC for chorismate and anthranilate. The HPLC system consisted of a model L-6200 HPLC pump (Merck-Hitachi, Darmstadt), a Rheodyne injector with a 20 μl loop, a model L-4200 UV-VIS detector (Merck-Hitachi, Darmstadt) equipped with an 11.3 μl flow cell operated at 280 nm. Fluorescence detection was performed with a model F-3000 fluorescence-spectrophotometer (Kontron, München) equipped with a 15 μl flow cell. The excitation wavelength was 340 nm and the emission wavelength 400 nm. Analyses were operated at room temperature on a 250 mm × 4 mm I.D. Nucleosil 5 C18 AB column (Macherey und Nagel, Düren) with a particle size of 8 μm at a flow rate of 1 ml min^{-1}. The eluent was the same as described in Poulsen et al. (1991). Standards used for quantitation contained 0.01 μg to 0.15 μg anthranilate per ml.

Determination of chorismate mutase (CM)

The activity of chorismate mutase was determined spectrophotometrically according to Koch et al. (1970). The assay volume of 0.4 ml contained 19.5 μmol assay buffer, 0.6 μmol chorismate and 0.1 ml protein extract. The reaction was started by addition of chorismate and stopped by addition of 0.4 ml 1 M HCl after an incubation period of 30 min. After 10 min. an aliquot of 0.6 ml was basified by addition of 2.4 ml 1.25 M NaOH. The OD at 320 nm was determined directly afterwards, 1 M NaOH served as a blank. Under acidic conditions within 10 minutes the product prephenate is completely converted to phenylpyruvate. The amount of prephenate formed was calculated using $E_{320} = 17.500\ M^{-1} \times cm^{-1}$. Chorismate contains about 7% prephenate and 1% phenylpyruvate as specified by the manufacturer (Sigma; St. Louis). Calculation of the background of prephenate of non-enzymic origin was performed according to Kuroki & Conn (1988). In these assays protein was added after addition of HCl.

Routinely the potassium phosphate buffer used for protein extraction was also employed for assaying CM activity. Alternatively 0.1 M TRIS-HCL buffer pH 7.8 according to the method by McCue (1988) was tried but yielded activities of 75% only.

Substrate concentrations of 0.1 to 0.8 μmoles/assay volume (0.25-2 mM) were used to check the effect of substrate concentration on CM activity. The effect of aromatic amino acids on the *in vitro* activity of CM was tested with L-tryptophan in a final concentration range from 0.5 to 5 mM. L-Phenylalanine and L-tyrosine were added as a mixture at 1 mM final concentration each.

Phenylalanine ammonia lyase (PAL) determination was performed using a spectrophotometric assay according to Zucker (1965).

Chemicals

Bradford reagent, BSA and RGG were purchased from BioRad, München; TRIS, DTT and glycerol from Biomol, Hamburg. Actinomycin D, cycloheximid and chloramphenicol as well as the amino acids were obtained from Fluka, Neu-Ulm. Chorismate was obtained from Sigma, St. Louis as barium salt, 80 % purity. All other chemicals used were of analytical grade and obtained from Merck, Darmstadt.

Results

Adaptation of CM assay conditions

In most plants studied two isoenzymes of CM were found (Bentley 1990, Poulsen & Verpoorte 1991). The activity only of isoform 1 (CM-1) is strongly affected by aromatic amino acids (Goers & Jensen 1984 a, b). L-Phenylalanine and L-tyrosine function as inhibitors, L-tryptophan activates CM-1. According to McCue (1988) determination of tryptophan activation rate offers a tool to differentiate between CM-1 and CM-2. The activation factor is calculated as the quotient of CM activity in the presence of tryptophan and CM activity in the absence of tryptophan. We engaged in optimization of the assay conditions in order to obtain maximum specific activities, which also might allow to distinguish between possible isoforms of CM.

Fig. 2 shows the effect of 0.5, 1.0 and 2.0 mM tryptophan in the assay at varying chorismate concentrations. Substrate saturation is reached at a chorismate concentration of 1.5 mM. The activation factor decreases with increasing chorismate concentration from 2 to 1.27. Linearization of the graphs for substrate-activity correlation according to Hanes (Bisswanger 1979) (Fig. 3) yielded a $K_{m(chorismate)}$ of 0.78 mM in the absence of tryptophan versus a $K_{m(chorismate)}$ of 0.26 mM in the presence of tryptophan. These data show, that the substrate affinity of the enzyme is increased in the presence of tryptophan.

Optimum conditions for the assay were selected with 1.5 mM chorismate and 2.0 mM tryptophan.

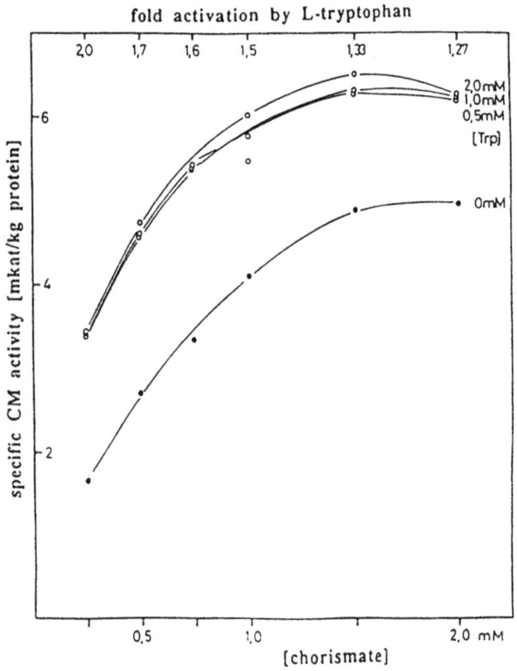

Fig. 2. *In vitro* activity of CM with different concentrations of chorismate and L-tryptophan.

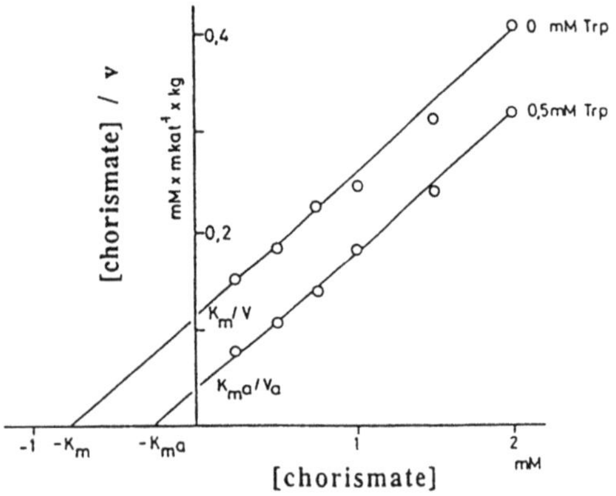

Fig. 3. Hanes plot of CM *in vitro* activity with different concentrations of chorismate and L-tryptophan.

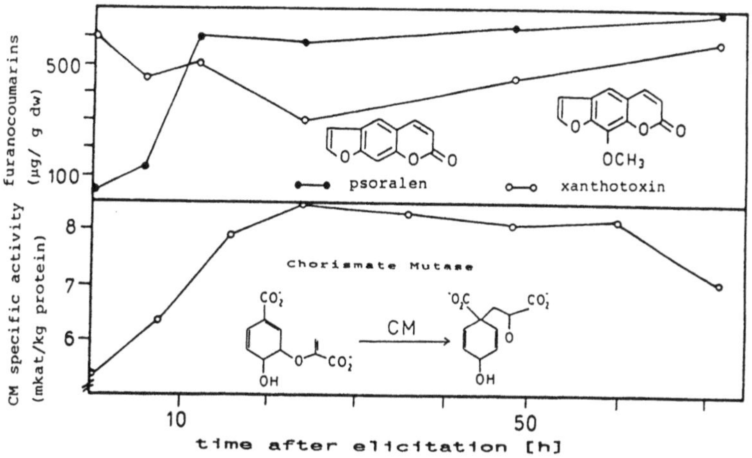

Fig. 4. Time courses of chorismate mutase specific activity and furanocoumarin accumulation in a suspension culture of *Ruta graveolens* in response to elicitation. Elicitor concentration: 1% autoclaved *Rhodotorula rubra* culture homogenate; elicitation was performed 14 days after transfer to fresh medium.

Elicitor induction of CM specific activity

In time course experiments the effect of elicitation on CM specific activity was studied. All our different *in vitro* systems of *R. graveolens* were tested. All systems responded with a rapid increase of specific activity. The induction factors, below 2, are much lower than those of other enzymes like PAL, 4-CL (Eilert & Bohlmann 1989), NMT (Eilert & Wolters 1989) or even the competing enzyme AS. Fig. 4 illustrates the typical time course of CM activity in comparison to furanocoumarin accumulation. Important to notice is, that the absolute specific activities of CM are about 1000 fold higher than those of the other enzymes mentioned.

Elicitor induction of AS specific activity

Increase of AS activity was affected by the elicitor concentration. Elicitor concentrations of 0.5 and 1.0% resulted in maximum induction of AS activities, while 0.1% proved to be suboptimum. The maximum rate of increase, only about 3.5 fold, was 10 fold lower than with other inducible enzymes like PAL, 4-CL or NMT. (Eilert & Wolters 1989; Eilert & Bohlmann 1989).

The typical time course of induction and the strong effect of differentiation on inducibility of AS activity is illustrated in Fig. 5. More differentiated cultures like the embryogenic line RN-d, shoot teratoma (RAT), and the photomixotrophic sterile hydroponic plants (ROH) show a weaker response as dedifferentiated suspension cultures (Rf, R-MS and RN). The degree of differentiation, however, was not reflected in the constitutive levels of AS in growing cultures. Specific activities in all cultures tested were always in the range of 1–2 µkat/kg protein.

Cell cultures of *R. macrophylla* and *R. chalepensis* responded similar to *R. graveolens* cultures. In cell cultures of *Choisya ternata*, *Skimmia japonica* and *Dictamnus albus*, all belonging to the Rutaceae, changes in AS specific activity were not induced. NMT-activity, however, increased as shown for *C. ternata* (Fig. 5). Suspension cultures from *Beta vulgaris*, *Chenopodium rubrum* (Chenopodiaceae) and shoot teratoma from *Solanum dulcamara* (Solanaceae) were monitored for AS induction. None of the cultures responded with an induction. Induction of PAL activity, a common response to elicitation, was observed in all cultures tested.

Inhibitor studies

The basis of elicitor stimulated enzyme induction is generally an increase of *de novo* synthesis (Eilert 1987). At the time the molecular tools to study mRNA-induction of AS were not available to us. Instead we conducted experiments with inhibitors of transcription and translation. We are aware, that this is only an indirect method with numerous factors to be considered.

ACT was used as an inhibitor of transcription. CHX served as inhibitor of cytoplasmic translation. CAP selectively inhibits 70S ribosomes of plastids and mitochondria. Plastids possess their own complete set of enzymes of aromatic amino acid biosynthesis (Bentley

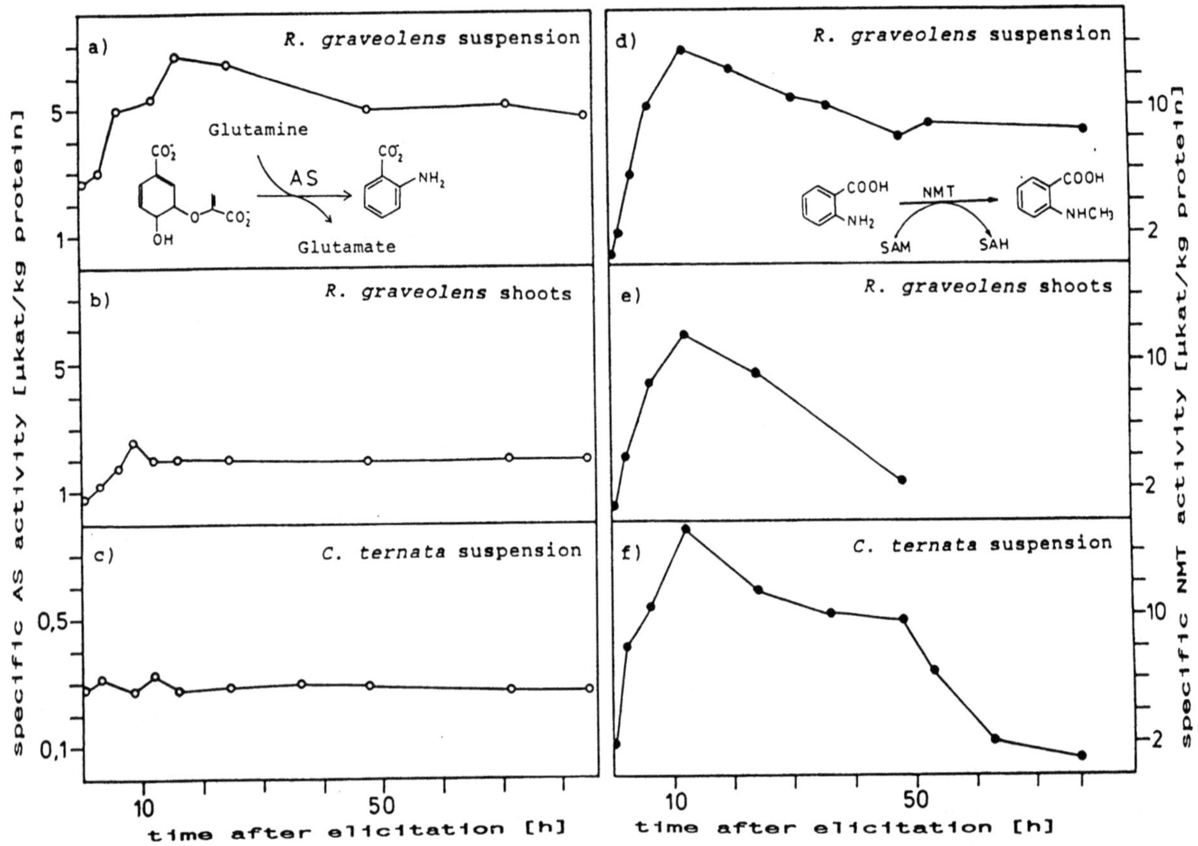

Fig. 5. Time courses of AS and NMT specific activity in a suspension culture (a, d), in shoots of hydroponically grown plants (b, e) of *Ruta graveolens* and in a suspension culture of *Choisya ternata* (c, f). Elicitor concentration: 1% autoclaved *Rhodotorula rubra* culture homogenate; elicitation was performed 14 days after transfer to fresh medium.

1990), but a dual pathway might exist in the cytosol according to Jensen et al. (1989), Hrazdina and Jensen (1990). Plastidically localized isoforms of CM have been reported from a number of plants (Kuroki & Conn 1989). The use of CHX and CAP was expected to give information on the type of ribosomes involved in the induction of AS and CM. PAL determination was included for comparison.

Fig. 6 presents the results. Elicitor induction of AS is completely inhibited in the presence of ACT, CHX and CAP. Induction of this enzyme thus involves transcription and translation steps. The experiments indicate that 70S as well as 80S ribosomes participate in AS induction. CM induction also is inhibited by ACT indicating induction of transcription. In the presence of CHX, however, CM specific activity showed a long lasting increase. Control experiments with CHX only proved this to be an effect of CHX itself. After addition of CHX CM specific activity increased but the total extractable protein content of the cells declined from 0.4 mg/ml to 0.2 mg/ml extract within 24 h. The increase of CM activity therefore is at least partially based on the decline in protein content. CAP, however, inhibited CM induction completely. On this basis we conclude that elicitor-induction of CM specific activity involves increased transcription and translation by 70S ribosomes.

Discussion

Elicitor induction results in accumulation of secondary metabolites derived from different branches of the shikimate pathway. At this branch point anthranilate synthase and chorismate mutase compete for chorismate. Time courses of the two enzyme activities were monitored in elicitor treated *in vitro* systems to gain information whether these enzymes play a key role in the regulation of elicitor induced metabolite accumulation in *R. graveolens*. Both enzymes proved to be inducible.

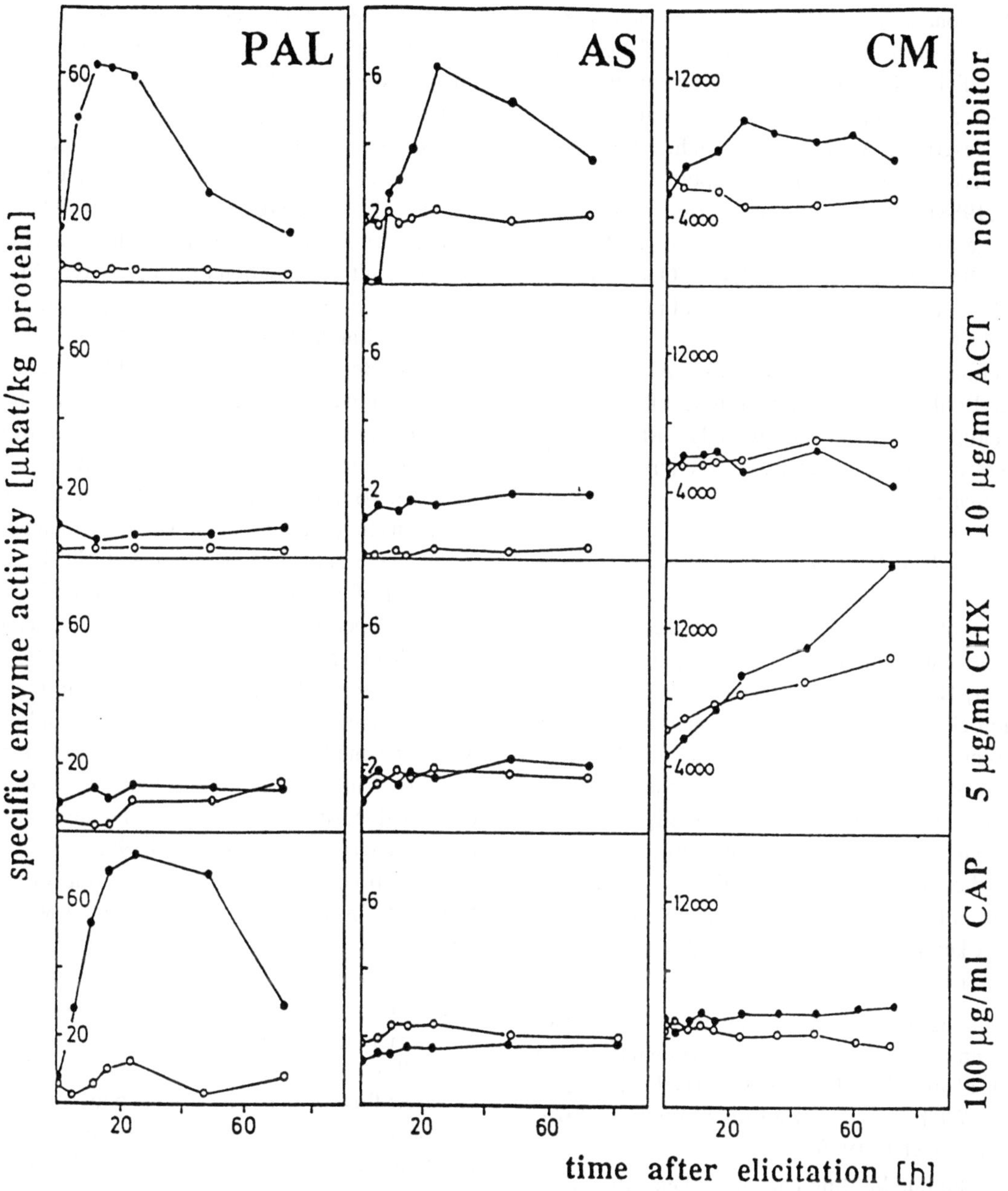

Fig. 6. Effect of actinomycin D (ACT), cycloheximid (CHX) or chloramphenicol (CAP) on the elicitor induction of AS, CM and PAL. Elicitor concentration: 1% autoclaved *Rhodotorula rubra* culture homogenate; elicitation was performed 14 days after transfer to fresh medium.

Regulatory role of CM induction

Based on the determination of the tryptophan activation rate McCue (1988) attempted to study the effect of elicitation on the activity of the two isoforms in suspension cultures of parsley. As shown for *Ruta*, too, tryptophan activation rate increased with decreasing chorismate concentration. Under low substrate concentration changes of the ratio will become most obvious. McCue (1988) chose for his study assay conditions of maximum tryptophan activation, which was at 0.3 mM chorismate. $K_{m(chorismate)}$ values were 0.09 mM for CM-1 and 1.2 mM for CM-2. Consequently McCue (1988) assayed CM-2 activity under substrate limiting conditions. The reliability of the method has to be questioned as Poulsen & Verpoorte (1991) do. Therefore we did not attempt to differentiate between isoforms. We chose assay conditions yielding maximum specific activity. In contrast to McCue (1988) we worked under saturating conditions regarding chorismate and tryptophan concentration. Our report, published in preliminary form at the 39th annual meeting of the Society for Medicinal Plant Research (Bohlmann & Eilert 1991a) is the first time, that elicitor induction of CM has been described. A wound stress induced increase of CM-activity in higher plants was reported by Kuroki & Conn (1988) for potato tubers. A differential induction of one of the isoforms was not shown (Kuroki & Conn 1988, Kuroki & Conn 1989). Our inhibitor studies, however, point to an induction of plastidically translated CM.

Induction curves of CM activity, like PAL induction (Eilert & Bohlmann 1989), do not reflect the product accumulation behaviour. Concludingly CM like PAL is not a key enzyme in the regulation of elicitor induced furanocumarin induction.

Regulatory role of AS-induction

Regulation of elicitor induced product accumulation by enzymes catalyzing late steps in the biosynthesis is reported by Hahlbrock & Scheel (1989) and Bollmann & Hahlbrock (1990) for phenylpropanoid metabolism in different systems and may hold true for *R. graveolens* as well. If acridone epoxide biosynthesis is regulated in the same way remains to be investigated. Most of these late enzymes are not even characterized yet. On the other hand control of the flow of chorismate into anthranilate would provide a suitable regulation system. A key regulatory function of AS was demonstrated for two different pathways:

1. AS is the rate limiting enzyme in tryptophan biosynthesis and regulated by tryptophan in the mode of an end product inhibition (Delmer & Mills 1968, Widholm 1974, Carlson & Widholm 1978, Brotherton et al. 1986).
2. AS is the rate limiting enzyme in benzodiazepine formation by *Penicillium cyclopium*. This mold synthesizes these alkaloids from phenylalanine and anthranilic acid (Roos & Schmauder 1989).

Does AS play a key role in the regulation of elicitor induced acridone epoxide and/or furoquinoline accumulation? Kuzovkina et al. (1987) found a correlation between constitutive AS activity and constitutive rutacridone accumulation in a number of callus strains of *R. graveolens*. Hertel et al. (1988) reproduced this work using the same cultures. In our systems we did not see such a correlation. The relative induction rates of AS activity and acridone epoxide content, however, are correlated.

NMT, the first enzyme specific for acridone alkaloid formation (Baumert et al. 1983) is a good candidate, too, for a pathway regulatory enzyme. NMT induction, in contrast, was quite similar in all *Ruta* systems tested (Fig. 5) and no close correlation to the levels of acridone epoxide accumulation became apparent (Eilert & Wolters 1989, Eilert 1989). In cell cultures from other genera of the Rutaceae elicitation did not result an induction of AS activity. NMT activity was induced but a corresponding induced accumulation of anthranilic acid derived products was not detected (Bohlmann & Eilert 1991b). Based on this comparison we suggest a key regulatory role for AS.

When presented first in preliminary form on the 38th annual meeting of the Society of Medicinal Plant Research (Bohlmann & Eilert 1990), it was the first report on elicitor induction of AS. Since then two more reports were published. Moreno et al. (1991) claim an induction of AS among other enzymes after elicitation of suspension cultures of *Catharanthus roseus*. A regulatory role is not discussed. Recently Niyogi & Fink (1992) published a study showing that one of two AS genes in *Arabidopsis thaliana* is inducible by wounding and bacterial stress. The respective enzyme possibly is involved in the biosynthesis of tryptophan derived glucosinolates which accumulate after wounding or bacterial stress (Tsuji et al. 1992).

Objections against a regulatory role of AS may be raised considering the comparatively low induction rate. Tanahashi & Zenk (1990) use this argument to rule out a regulatory role of tyrosine decarboxylase in elicitor induced benzylisoquinoline alkaloid accumulation. Here it has to be pointed out that *in vitro* activities will only give limited insight into the efficiency of substrate flow *in vivo*. The storage pool of anthranilate is probably kept very low as a consequence of the high cytotoxicity of anthranilate (Roos & Schmauder 1989; Köster et al. 1978).

Our further research is directed to gain a better understanding of the regulatory role of AS in *R. graveolens*. Purification of AS and cDNA cloning of the genes are in progress. This will allow studies of the regulation and compartmentation in detail.

Acknowledgements

We thank Mrs. C. Rindok-Barrenscheen for her excellent assistence with the maintenance of plant cell cultures. Financial support for the project by the DFG is gratefully acknowledged.

References

Baumert A, Hieke M & Gröger D (1983) N-methylation of anthranilic acid to N-methylanthranilic acid by cell-free extracts from *Ruta graveolens* tissue cultures. Planta Med. 48: 258–262

Baumert A, Maier W, Schumann B & Gröger D (1991) Increased accumulation of acridone alkaloids by cell suspension cultures of *Ruta graveolens* in response to elictors. J. Plant Physiol. 139: 224–228

Bentley R (1990) The shikimate pathway – a metabolic tree with many branches. Crit. Rev. Biochem. Mol. Biol. 25: 307–384

Bisswanger H (1979) Theorie und Methoden der Enzymkinetik. Verlag Chemie, Weinheim

Bohlmann J & Eilert U (1990) Elicitor-induction of glutamate-dehydrogenase and anthranilate synthase in *in vitro* systems of *Ruta graveolens* L., Rutaceae. Planta Med. 56: 608–609

Bohlmann J & Eilert U (1991 a) Elicitor induction of chorismate mutase in different *in vitro* systems of *Ruta graveolens*. Planta Med. 57A: 111–112

Bohlmann J & Eilert U (1991 b) S-Adenosyl-L-methionine:anthranilic acid-N-methyltransferase and elicitation response in different Rutaceae. Planta Med. 57A: 94–95

Bollmann J & Hahlbrock K (1990) Timing of changes in protein synthesis pattern in elicitor-treated cell suspension cultures of parsley (*Petroselinum crispum*). Z. Naturforsch. 45C: 1011–1020

Bradford MM (1976) A rapid and sensitive method for the quantification of microgram quantitites of protein using the principle of protein-dye binding. Anal. Biochem. 72: 248–254

Brotherton JE, Hauptmann R M & Widholm J M (1986) Anthranilate synthase forms in plants and cultured cells of *Nicotiana tabacum* L.. Planta. 168: 214–221

Carlson JM & Widholm JM (1978) Separation of anthranilate synthase from 5-methyltryptophan-susceptible and -resistant cultured Solanum tuberosum cells. Physiol. Plant. 44: 251–255

Delmer DP & Mills SE (1968) Tryptophan biosynthesis in cell cultures of *Nicotiana tabacum* Plant. Physiol. 43: 81–87

Eilert U (1987) Elicitation: Methodology and aspects of application. In: Vasil. I.K., Constabel, F. (eds.) Cell Culture and Somatic Cell Genetics of Plants. Vol. 4: Cell Culture in Phytochemistry (pp 153–196). Academic Press Inc. Harcourt Brace Jovanovich Publishers, San Diego

Eilert U (1989) Elicitor induction of secondary metabolism in dedifferentiated and differentiated *in vitro* systems of *Ruta graveolens*. In: Kurz WGW (Ed) Primary and secondary metabolism of plant cell cultures II (pp. 219–228). Springer-Verlag, Berlin.

Eilert U & Bohlmann J (1989) Effect of elicitors on enzymes of the phenylpropanoid pathway in different *in vitro* systems of *Ruta graveolens*. Planta Med. 55: 685

Eilert U & Wolters B (1989) Elicitor induction of S-adenosyl-L-methionine: anthranilic acid N-methyltransferase activity in cell suspension and organ cultures of *Ruta graveolens* L.. Plant Cell Tiss. Organ Cult. 18: 1–18

Eilert U, Ehmke A & Wolters B (1984) Elicitor-induced accumulation of acridone alkaloid epoxides in *Ruta graveolens* suspension cultures. Planta Med. 6: 459–532

Eilert U, De Luca V, Kurz WGW & Constable F (1987) Alkaloid formation by habituated and tumorous cell suspension cultures of *Catharanthus roseus*. Plant Cell Rep. 6: 271–274

Gamborg OL, Miller R A & Ojima K (1968): Nutrient requirements of suspension cultures of soybean root cell. Exp. Cell Res. 50: 151–158

Gaska JEK (1981) Anthranilate synthase and tryptophan pool size in cultured cells of *Solanum stenotum*. Diss. Rugers State Univ., New Brunswick

Gibraltarskaya E & Eilert U (1992) Elicitor induction of S-adenosyl-L-methionine: xanthotoxol-O-methyltransferase in *Ruta graveolens* in vitro. Planta Med. 58: Suppl. 1, A604

Goers SK & Jensen R A (1984a) Separation and characterization of two chorismate-mutase isoenzymes from *Nicotiana silvestris*. Planta *162*, 109–116

Goers SK & Jensen RA (1984b) The differential allosteric regulation of two chorismate-mutase isoenzymes of *Nicotiana silvestris*. Planta. 162: 117–124

Hahlbrock K & Scheel D (1989) Physiology and molecular biology of phenylpropanoid metabolism. Annu. Rev. Plant Physiol. Plant Mol. Biol. 40: 347–369

Hertel SC, Schmauder HP, Hieke M & Gröger D (1988) Alkaloidbildung und Biosynthese aromatischer Aminosäuren in Zellkulturen von Ruta graveolens L. Biochem. Physiol. Pflanzen 183: 425–437

Hrazdina G & Jensen RA (1990) Multiple parallel pathways in plant aromatic metabolism. In: Structural and organizational aspects of metabolic regulation. UCLA Symp. Ser. 133: 27–41

Jensen RA, Morris P, Bonner C & Zamir LA (1989) Biochemical interface between aromatic amino acid biosynthesis and secondary metabolism. ACS Symposium Series. 399: 89–107

Koch GLC, Shaw DC & Gibson F (1970) Tyrosine biosynthesis in *Aerobacter aerogenes*. Purification and properties of chorismate mutase – prephenate dehydrogenase. Biochim. Biophys. Acta 212: 375–386

Köster J, Ohm M & Barz W (1978) Metabolism of anthranilic acid in plant cell suspension cultures. Z. Naturforsch. 33c: 368–372

Kuroki G & Conn EE (1988) Increased chorismate mutase levels as a response to wounding in *Solanum tuberosum* L. tubers. Plant Physiol. 86: 895–898

Kuroki G & Conn EE (1989) Differential activities of chorismate mutase isozymes in tubers and leaves of *Solanum tuberosum*. Plant Physiol. 89: 472–476

Kuzovkina IN, Schmauder GP & Gröger D (1987) Activity of anthranilate synthase in callus strains of *Ruta graveolens* L. with different levels of rutacridone. Fiziol. Rast. 34: 1025–1027

McCue KF (1988) Investigation of aromatic acid metabolism in suspension cultured cells of parsley (*Petroselinum crispum*). Diss. Univ. of California, Davis.

Moreno PRH, Poulsen C, van der Heijden R & Verpoorte R (1991) Activity of some enzymes of secondary metabolism after elicitation of *Catharanthus roseus* cell cultures. Planta Med. 57A: 103

Murashige T & Skoog F (1962) A revised medium for rapid growth and bioassays with tobacco tissue cultures. Physiol. Plant 15: 473–497

Niyogi KK & Fink GR (1992) Two anthranilate synthase genes in *Arabidopsis*: defense-related regulation of the tryptophan pathway. Plant Cell 4: 721–733

Poulsen C & Verpoorte R (1991) Roles of chorismate mutase, isochorismate synthase and anthranilate synthase in plants. Phytochemistry 30: 377–386

Poulsen C, Pennings JM & Verpoorte R (1991) High-performance liquid chromatographic assay of anthranilate synthase from plant cell cultures. J. Chromatogr. 547: 155–160

Roos W & Schmauder H (1989) Positive feedback effect of benzodiazepine alkaloids on enzymes of the aromatic pathway. FEMS Microbiol. Lett. 59: 27–30

Tanahashi T & Zenk MH (1990) Elicitor induction and characterization of microsomal protopine-6-hydroxylase, the central enzyme in benzophenanthridine alkaloid biosynthesis. Phytochemistry 29: 1113–1122

Tsuji J, Jackson EP, Gage DA, Hammerschmidt R & Somerville SC (1992) Phytoalexin accumulation in *Arabidopsis thaliana* during the hypersensitive reaction to *Pseudomonas syringae pv. syringae*. Plant Physiol. 98: 1304–1309

Widholm JM (1974) Control of aromatic amino acid biosynthesis in cultured plant tissues: effect of intermediates and aromatic amino acids on free levels. Physiol. Plant. 30: 13–18

Zucker M (1965) Induction of phenylalanine deaminase by light and its relation to chlorogenic acid synthesis in tuber tissues. Plant Physiol. 40: 779–784

Plant Cell, Tissue and Organ Culture **38**: 199–211, 1994.

Constitutive and elicitation induced metabolism of isoflavones and pterocarpans in chickpea (*Cicer arietinum*) cell suspension cultures

Wolfgang Barz & Ulrike Mackenbrock
Institut für Biochemie und Biotechnologie der Pflanzen, Westfälische Wilhelms-Universität, Hindenburgplatz 55, D–4400 Münster

Key words: Elicitor, induction of enzymes, isoflavone, metabolic regulation, phytoalexin, pterocarpan

Abstract

Constitutive phenolics of chickpea cell suspension cultures are the isoflavones formononetin and biochanin A, the isoflavanones homoferreirin and cicerin and the pterocarpans medicarpin and maackiain. They accumulate as vacuolar malonylglucosides. The biosynthetic pathways to isoflavones, pterocarpans and malonylglucoside conjugates together with their enzymes are explained. Elicitation of cell cultures leads to pronounced increases in the activities of biosynthetic enzymes with differential effects on the enzymes involved in conjugate metabolism. Low elicitor doses favour pterocarpan conjugate formation whereas high doses lead to pterocarpan aglycone accumulation accompanied by vacuolar efflux of formononetin and pterocarpan malonylglucosides. Elicitor-induced changes in enzyme activities and vacuolar efflux of conjugates are prevented by application of 10^{-3}M concentrations of cinnamic acid. Cinnamate is alternatively metabolized to a glucose ester, a S-glutathionyl conjugate and to cell wall bounds forms; these reactions are intensified by elicitation. Isoflavone and pterocarpan biosynthesis and conjugate metabolism as regulated by elicitation and cinnamate is depicted in a metabolic grid to explain the complex regulatory pattern of phenolic accumulation in chickpea cell cultures.

Abbreviations: AOPP – L-α-aminooxy-β-phenylpropionic acid, BGM – biochanin A 7–0-glucoside–6″–0-malonate, FGM – formononetin 7–0-glucoside–6″–0-malonate, HPLC – high performance liquid chromatography, MaGM – maackiain 3–0-glucoside–6′–0-malonate, MeGM – medicarpin 3–0-glucoside–6′–0-malonate

Introduction

Higher plants when challenged by pathogenic microorganisms express a series of active defense mechanisms. They all aim at the inhibition of microbial growth, isolation of the pathogens in lesions and finally death of the invading microbe by the accumulation of antibiotic compounds (Lamb et al. 1989; Barz et al. 1990a; Dixon & Lamb 1990a). Detailed analyses of the chickpea (*Cicer arietinum* L.) – *Ascochyta rabiei* (teleomorph: *Mycosphaerella rabiei* Kovachevski) interaction have demonstrated that the hyersensitive response, expression of pathogenesis-related proteins together with chitinases and β–1,3-glucanases, formation of the pterocarpan phytoalexins medicarpin and maackiain together with increased polyphenol deposition are essential elements of the plant defence responses (Höhl et al. 1990; Daniel & Barz 1990; Vogelsang & Barz 1993; Barz & Welle 1992). Furthermore, constitutively produced isoflavones ('preinfectional inhibitors') are of interest for chickpea resistance due to the localisation of these fungi-toxic compounds in the outer tissue layers of the plant (Barz & Hösel 1978). Investigations on the biosynthesis of constitutive isoflavones and de-novo synthesized antimicrobial pterocarpan phytoalexins clearly demonstrated a tight metabolic linkage between these two classes of phenolic constituents (Barz & Welle 1992). The analyses revealed an interesting regulatory pattern in operation in chickpea cells between constitutive and de-novo synthesized phenolic compounds, i.e. metabolic activation of vacuolar phenolic conjugates for rapid infection-induced accumulation of phytoalexins.

Elucidation of isoflavone and pterocarpan biosynthetic pathways, characterization of enzymes and mechanisms of gene activation as well as determination of adherent regulatory pattern have been investigated in chickpea cell suspension cultures. Infection-induced changes of cellular metabolism can very efficiently be simulated in such cultures by the application of fungal polysaccharide elicitors (Barz et al. 1990a; Dixon & Lamb 1990a).

Materials and methods

Cell cultures

Chickpea (cultivar ILC 3279) cell suspension cultures (40 ml medium in 250 ml Erlenmeyer flasks) were grown as previously described (Keßmann & Barz 1987; Mackenbrock et al. 1992). Elicitation and AOPP/cinnamic acid inhibition experiments were performed with cells 3 days after transfer into new medium using published procedures (Mackenbrock et al. 1993).

Elicitor

The preparation of yeast elicitor has been described (Gunia et al. 1991).

Reference compounds

The isoflavone, pterocarpan and phenylpropanoid compounds used in the experiments described in this paper were from the institute's collection.

Quantitation of phenolics

The extraction of phenolics from cultured cells and the preparation of fractions for chromatography have been described (Keßmann & Barz 1987; Mackenbrock & Barz 1991).

Chromatographic analyses

Isoflavones and pterocarpans were analyzed by HPLC using previously described methods (Gunia et al. 1991; Keßmann & Barz 1987).

TLC analysis of cinnamic acid conjugates were performed as described by Edwards et al. (1990) and Edwards & Dixon (1991).

Enzyme assays

Cinnamic acid glucosyltransferase and glutathione S-cinnamoyltransferase were measured as described (Edwards et al. 1990; Edwards & Dixon 1991). Assays for all other enzymes mentioned in this paper have previously been described (Daniel et al. 1990; Gunia et al. 1991; Mackenbrock et al. 1992, 1993).

Protein concentrations were determined by the Bradford method with bovine serum albumin (Cohn fraction V, Sigma, Munich) as reference.

Results and discussion

Constitutive accumulation of phenolics in cell cultures

Heterotrophic chickpea cell suspension cultures established from various cultivars have turned out to be a rich source of phenolic constituents (Keßmann & Barz 1987; Weidemann et al. 1991; Barz & Welle 1992). In principle, the phenolics belong to the following classes, a) 5-hydroxyisoflavones (i.e. biochanin A), b) 5-deoxyisoflavones (formononetin), c) 2′-methoxy–5-hydroxyisoflavanones (homoferreirin, cicerin), and d) pterocarpans (medicarpin, maackiain) (structures Fig. 1).

These compounds acccumulate under normal culture conditions (Keßmann & Barz 1987) in a mainly growth-linked pattern (Barz et al. 1990b) with the 5-hydroxyisoflavonoids being the major components. All phenolics predominantly occur as 0-glucoside–6″–0-malonate conjugates (Fig. 1) which are exclusively stored in vacuoles (Mackenbrock et al. 1992). Such polar, hydrophilic conjugates of isoflavones and various other plant phenolics are well known widely occuring constituents (Barz et al. 1985). Isoflavone/isoflavanone/pterocarpan conjugate accumulation in these cultures is highly responsive to auxin regulation because cultivation under auxin-free conditions may lead to a ca. 20-fold increase in phenolic material. The bulk of these additional compounds is again being represented by 5-hydroxy-isoflavone/-isoflavanones (Vogelsang 1993).

The biosynthesis of isoflavones has extensively been studied using chickpea cultured cells so that the essential enzymes of the general phenylpropanoid pathway and chalcone synthase, chalcone isomerase as well as isoflavone synthase together with the 4′–0-methylation step have well been characterized (Barz et al. 1990a; Barz & Welle 1992). In comparison to biochanin A the pathway to formononetin (Fig. 2, upper part) involves an additional independent enzyme, chalcone reductase (CHR) which coacts with chalcone synthase (CHS) and

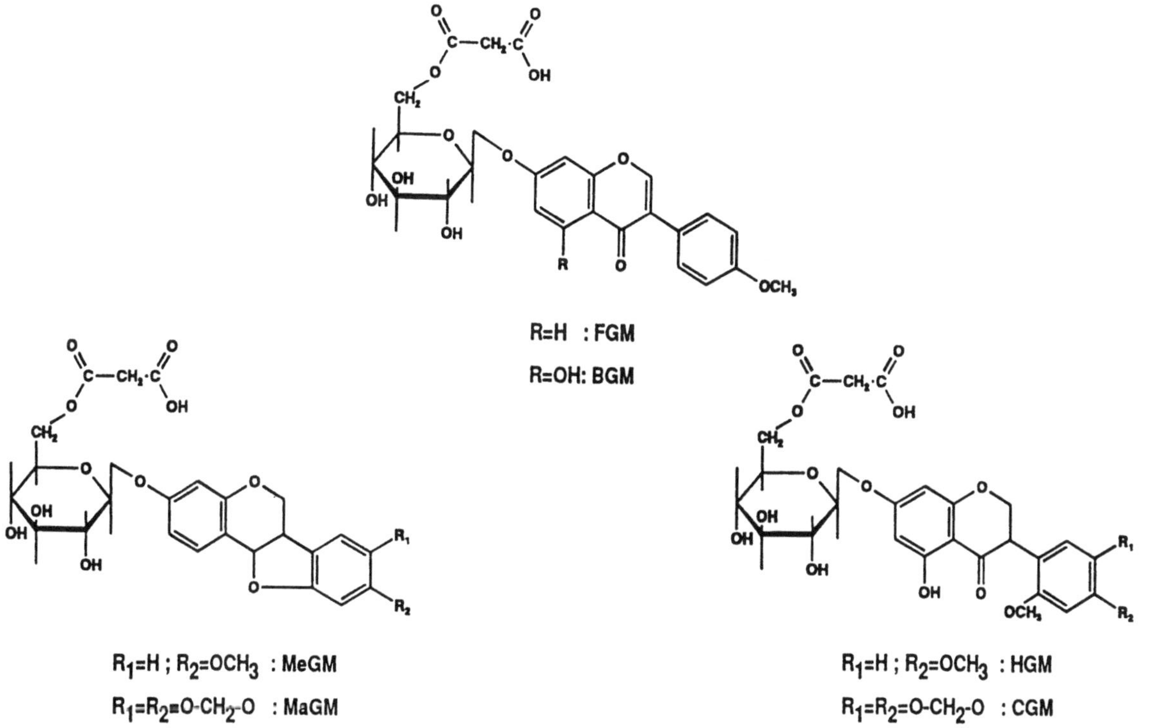

Fig. 1. Structures of malonylglucosides of the isoflavones formononetin (FGM) and biochanin A (BGM), the isoflavanones homoferreirin (HGM) and cicerin (CGM) and the pterocarpans medicarpin (MeGM) and maackiain (MaGM) constitutively formed in chickpea cell suspension cultures.

NADPH as cofactor in the formation of the intermediate 2′–4′–4-trihydroxychalcone isoliquiritigenin. This enzyme may be regarded as an important regulatory step for the channelling of substrates into the two competing pathways leading to either 5-deoxyisoflavones (daidzein, formononetin) and pterocarpans (medicarpin/maackiain) on one hand or 5-hydroxyisoflavones (biochanin A) and -isoflavanones (homoferreirin, cicerin) on the other. In chickpea CHS and CHR occur in multiple isoforms which are induced to a different extent upon elicitation of the cell cultures (Bless 1992).

Under normal growth conditions chickpea cell suspension cultures accumulate the (6aR : 11aR) pterocarpans, medicarpin and maackiain (Fig. 2), in form of the aglycones to only a very low extent if at all. However, considerable quantities of the 3–0-glucoside–6′–0-malonate conjugates (Fig. 1) are regularly formed (Weidemann et al. 1991). This has previously been interpreted as an indication for a state of partial induction of these cultures due to the specific conditions of culture growth. However, medicarpin and maackiain malonylglucosides have also been detected as normal constituents of older chickpea roots and furthermore these conjugates always co-occur with the aglycones when these are expressed as a phytoalexin response. The biosynthetic pathway leading from the formononetin intermediate to the pterocarpans (Fig. 2, lower part) has been characterized using elicited chickpea cells (see below). In other pterocarpan producing plants (glyceollin/soybean; pisatin/pea; medicarpin/alfalfa) identical sequences for such phytoalexins have been detected which also involve enzyme systems with a high degree of homology (Barz & Welle 1992; Dixon et al. 1992).

Homoferreirin and cicerin (Fig. 1) together with their malonylglucosides are synthesized from biochanin A in sequences which are highly analogous to the formation of medicarpin and maackiain from formononetin (Fig. 2). The reactions are hydroxylation of biochanin A in positions 2′ and 3′, respectively, closing of the methylenedioxyring, reduction of the intermediate isoflavones to isoflavanones with the terminal step of 2′–0-methylation. From a structural point of view this 0-methylation reaction may well be compared with the formation of a 2′-methoxychalcone (Dixon et al. 1992) in that very similar hydroxyketo substrates are being involved. The addition of the 2′–0-methylgroup

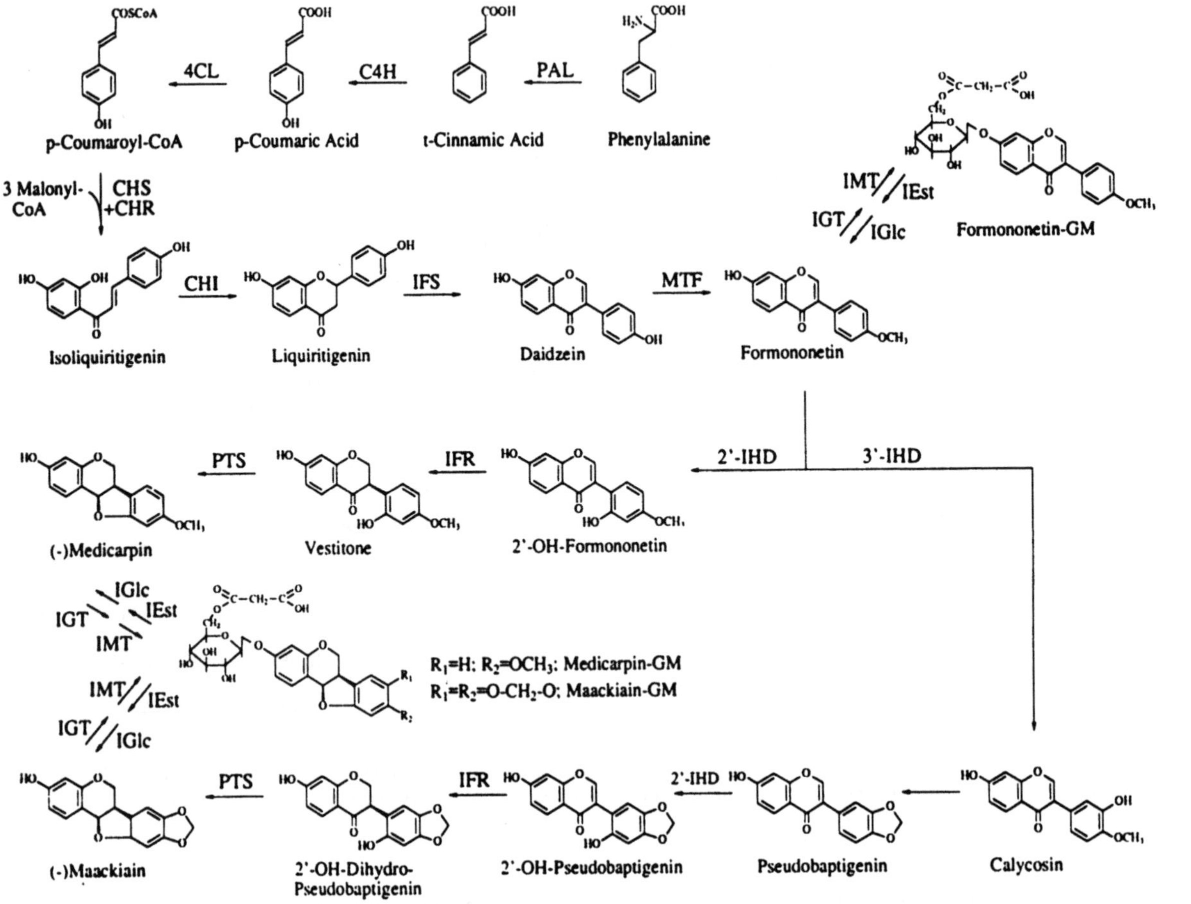

Fig. 2. Biosynthetic pathway to the isoflavone formononetin (upper half) and the pterocarpan phytoalexines medicarpin and maackiain as well as their malonylglucosides in chickpea. The enzymes are: PAL, phenylalanine ammonia lyase; C4H, cinnamic acid 4-hydroxylase; 4CL, p-coumaric acid CoA-ligase; CHS, chalcone synthase; CHR, chalcone reductase; CHI, chalcone isomerase; IFS, isoflavone synthase; MTF, isoflavone methyltransferase; IGT, isoflavone 7–0-glucosyltransferase; IMT, isoflavone glucoside malonyltransferase; IEST, isoflavone malonylglucoside malonylesterase; IGLC, isoflavone glucoside glucosidase; 2′-IHD, isoflavone 2′-hydroxylase; 3′-IHD, isoflavone 3′-hydroxylase; IFR, 2′-hydroxyisoflavone oxidoreductase; PTS, pterocarpan synthase.

in homoferreirin and cicerin formally prevents formation of pterocarpan structures; a pterocarpan synthase specific for 5-hydroxyisoflavanones and relevant products, i.e. 1-hydroxypterocarpans have so far not been detected in the chickpea system.

Great care has been taken in the characterization of the enzymes involved in the formation and the hydrolysis of the isoflavone and pterocarpan malonylglucosides (Fig. 2). These are isoflavone 7–0-glucosyltransferase (IGT), isoflavone glucoside malonyl transferase (IMT), a highly substrate-specific isoflavone malonylglucoside malonylesterase (IEST) and isoflavone glucoside glucosidase (IGLC) (Barz et al. 1990a, 1990b; Barz & Welle 1992). These enzymes were first characterized in connection with the isoflavone substrates but were later on shown to act with the pterocarpan structures also. For IEST an association with the tonoplast membrane has recently been demonstrated whereas the other three proteins appear to be soluble, cytosolic proteins (Mackenbrock et al. 1992). These four enzymes appear to act in a metabolic cycle because, as shown for the formononetin moiety, vacuolar influx and efflux of the isoflavone occur simultaneously (Barz et al. 1990b). This turnover system based on these enzymes obviously participates in regulating the pool size of the vacuolar malonylglucoside of formononetin. Therefore, a strict metabolic and spatial separation of these anabolic and catabolic enzymes must exist.

Elicitor-induced accumulation of pterocarpan phytoalexins

Chickpea cell suspension cultures treated with *A. rabiei* or yeast polysaccharide elicitor readily accumulate medicarpin and maackiain which are mainly secreted from the cells into the growth medium (Keßmann & Barz 1987; Gunia et al. 1991; Weidemann et al. 1991). Medicarpin and maackiain occur in the cell cultures in a ratio of ca. 5 : 1 and for the sake of clarity the sum of both pterocarpans will only be given in this paper. Phytoalexin formation is regarded as de-novo synthesis from early precursors of primary metabolism (Dixon & Lamb 1990; Mackenbrock & Barz 1991). Synthesis of the aglycones was found to be restricted to a period of ca. 6–20 h after elicitation followed by very rapid disappearance of the compounds from the medium. This decline in pterocarpan levels has been determined as peroxidative destruction catalyzed by extracellular peroxidases (Barz et al. 1990b). Furthermore, pterocarpan aglycone accumulation has been shown to be positively correlated with the amount of elicitor applied (Gunia et al. 1991). In contrast to this short burst of elicitor-induced aglycone formation a simultaneous and much longer lasting accumulation of medicarpin and maackiain malonylglucosides has also been found to occur (Weidemann et al. 1991). In fact, low doses of polysaccharide elicitor almost exclusively induced accumulation of the vacuolar phytoalexin conjugates (Fig. 3) and this process continued for a considerable period of time. Elicitor dose studies on the ratio of accumulating phytoalexin aglycones to pterocarpan malonylglucosides (Fig.4) revealed that low or moderate doses (5–10 mg/flask) favoured accumulation of the conjugates whereas high doses (30–80 mg) steadily led to increasing amounts of the aglycones. Figure 4 also depicts various investigations which showed that the formononetin moiety of the vacuolar-localized FGM (comp. Fig. 2) can be consumed and funnelled into pterocarpan phytoalexin biosynthesis if sufficiently high elicitor doses are being applied (Mackenbrock & Barz 1991). Furthermore, this vacuolar efflux appears to be highly substrate-specific because the pool sizes of the vacuolar BGM were not affected by elicitation (Fig. 4). The contribution of formononetin derived from FGM to phytoalexin biosynthesis became even more prominent when the introductory enzyme of pterocarpan formation, PAL, was inhibited by L-α-aminooxy-β-phenylpropionic acid (L-AOPP) during elicitation. Since under these conditions no phenylpropanoid precursors were available for de-novo pterocarpan formation the elicitor-induced accumulation of phytoalexins was now quantitatively covered by formononetin derived from FGM (Mackenbrock & Barz 1991).

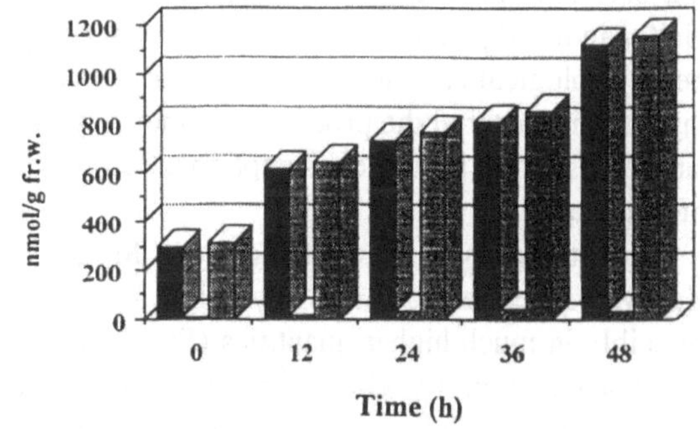

Fig. 3. Time course of elicitor-induced (10 mg per flask) accumulation of pterocarpan phytoalexin aglycones and pterocarpan conjugates (the sum of medicarpin and maackiain) in chickpea cell suspension cultures. The bars represent:

■ phytoalexin conjugates,
▨ phytoalexin aglycones, and
▩ total amount of pterocarpans

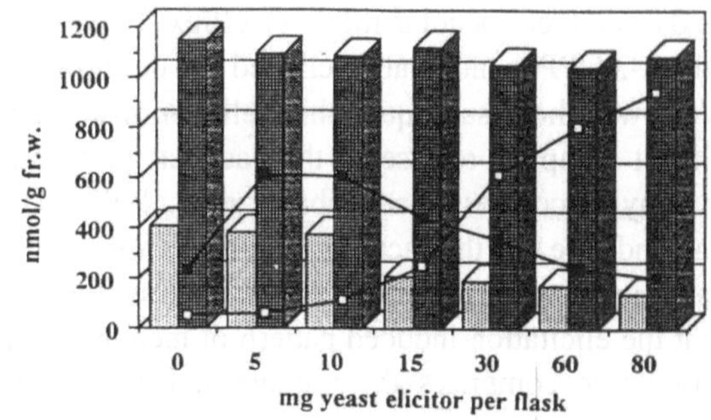

Fig. 4. Effect of elicitor dose on the accumulation of pterocarpan aglycones, pterocarpan conjugates and the malonylglucosides of formononetin and biochanin A in chickpea cell cultures 12 h after elicitation. The bars represent:

▒ formononetin malonylglucoside,
■ biochanin A malonylglucoside

The curves indicate:

–□– phytoalexin aglycones,
–■– phytoalexin conjugates.

The consumption of constitutive isoflavone conjugates for the induced formation of phytoalexins has also been observed in other systems than chickpea. In *Phytophthora megasperma* infected seedlings of soybean accumulation of the glyceollin phytoalexins is accompanied by a substantial turnover of a daidzein 7-0-glucoside-6″-0-malonate (Graham et al. 1990). The isoflavone aglycone is then thought to be used for glyceollin formation. Such data obtained for soybean and chickpea provide evidence for a tight metabolic link between constitutive isoflavone metabolism and elicitor-induced pterocarpan formation. Depending on the physiological situation and the actual demand for phytoalexin material the process of de-novo synthesis can be sustained by the consumption of constitutive intermediates.

The pterocarpan malonylglucosides MeGM and MaGM synthesized constitutively in low amounts and inducibly in much higher quantities (Fig. 3) also represent a reservoir for elicitor-induced accumulation of pterocarpan aglycones. Thus, conjugate material formed under the impact of low elicitor doses is readily consumed for aglycone formation with subsequent secretion of the phytoalexins into the growth medium when high elicitor doses are applied (Fig. 4) (Mackenbrock et al. 1993). Again, the elicitor quantity, i.e. the intensity of the elicitation effect, determines the ratio of conjugate to aglycone formation and alternatively leads to either vacuolar influx or efflux. Experiments with L-AOPP in moderately elicited cell cultures clearly showed that a subsequent high elicitor dose leads to almost complete release of the vacuolar pterocarpan malonylglucosides (Mackenbrock et al. 1993). These data indicate that the vacuolar conjugates are metabolically linked to the process of phytoalexin synthesis and that the elicitation-induced pattern of metabolic regulation also comprises the vacuole and the tonoplast transport systems.

Elicitor-induced changes of enzyme activities

Elicitor-treated chickpea cell suspension cultures have been very helpful in the elucidation of the enzymology of pterocarpan biosynthesis.

The pterocarpan-specific branch leading from formononetin to medicarpin and maackiain (Fig. 2) starts with cytochrome P_{450} monooxygenases for isoflavone 2′- and 3′-hydroxylation (2′-IHD, 3′-IHD). These inducible, microsomal enzyme activities have thoroughly been characterized in cell cultures and plants (Clemens et al. 1993) and they were recognized as distinct enzymes for the two position-specific reactions. Expression of isoflavone 2′-hydroxylase has been found to be closely related to phytoalexin accumulation and it may be regarded as an important regulatory step (Gunia et al. 1991; Clemens et al. 1993).

Reduction of the two 2′-hydroxyisoflavones shown in Fig. 2 is catalyzed by a NADPH : isoflavone oxidoreductase (IFR) leading to chiral 2′-hydroxyisoflavanones. This enzyme has meanwhile been cDNA-cloned from chickpea and alfalfa and its elicitor-induced expression was studied at the mRNA level (Tiemann et al. 1993).

The terminal enzyme reaction for the formation of the dihydrofuran ring of pterocarpans (PTS) (Fig. 2) only accepts the 3 R-configurated isoflavanone intermediates and it is together with 2′-IHD, 3′-IHD and IFR coordinately induced upon elicitation of cell cultures or infection of plants (Barz & Welle 1992).

Elicitation experiments in chickpea cell suspension cultures revealed that the enzymes of the complete biosynthetic sequence (Fig. 2) comprising the general phenylpropanoid pathway, chalcone synthase, isoflavone synthase, including the 4′-0-methylation step and the pterocarpan specific branch from formononetin to medicarpin and maackiain are all subject to induction (Daniel et al. 1990). Although formononetin biosynthesis proceeds constitutively in the chickpea cultures a pronounced increase in the enzyme activities of these enzymes has also been measured. The adherent aspects of metabolic regulation and the involvement of isoforms at certain reactions steps (CHS, CHR, see above) have been reviewed (Barz & Welle 1992). This pattern of a coordinate induction of a long and complex biosynthetic pathway appears to be a general feature of phytoalexin formation from early precursor of primary metabolism (Hahlbrock & Scheel 1989).

In view of the elicitor dose-dependent, differential accumulation versus metabolism of formononetin and pterocarpan malonylglucosides during phytoalexin aglycone formation (Fig. 4) more detailed investigations on the elicitation of essential biosynthetic enzymes (PAL, C4H, CHS, 2′-IHD and 3′-IHD) and the enzymes involved in conjugate metabolism (IGT, IMT, IEST, IGLC) have been performed. Using a wide range of elicitor doses (10–80 mg/flask) changes in enzyme activities during the first 10–12 h in comparison to the controls were recorded. Elicitor doses of ca. 80 mg/flask had previously been shown to result in maximum expression of elicitor-induced changes in enzyme activities in the chickpea cultures (Gunia et al. 1991; Mackenbrock et al. 1993).

In case of the 5 key biosynthetic enzymes a coordinated and linear increase in enzyme activities over this range of elicitor doses could be measured. The data are shown in Table 1 in the columns 'control' and 'elicitor'. Although the absolute values of elicitor-induced levels of enzyme activities vary the results clearly showed that these enzymes are subject to strong induction. As mentioned before this process is accompanied by the formation of either pterocarpan malonylglucosides or pterocarpan aglycones (Fig. 4). These data will later be used for the discussion of the phenomenon of cinnamate-caused modulation of enzyme activities.

The analyses of the enzymes involved in the metabolism of malonylglucoside metabolism (Table 2) revealed a more complex pattern of elicitor-caused changes in enzyme activities. Low elicitor doses (10 mg/flask) appeared to increase the activities of IGT and IMT whereas IEST and IGLC were repressed in their activities. In contrast, high elicitor doses (80 mg/flask) caused strong repression of IGT and IMT below the control values with IEST and IGLC being induced to significantly higher values. These fluctuations of enzyme activities over the indicated range of elicitor doses should be evaluated in comparison to the data shown in Fig. 4. Low elicitor doses which lead to the preferential formation of pterocarpan malonylglucosides also increase the activities of IGT and IMT. Thus, the biosynthetic branch of conjugate formation and vacuolar influx appears to be intensified. On the other hand, a strong elicitation effect represented by high elicitor doses (60–80 mg/flask) results both in an increase of IEST and IGLC activities and also in increased rates of conversion of pterocarpan and formononetin malonylglucosides to pterocarpan aglycones; in essence, vacuolar efflux appears to be favoured. The differential elicitor-caused behaviour of the enzymes of conjugate metabolism (Table 2) perfectly agrees with the data (Fig. 4) obtained for the differential accumulation of the various phenolic compounds. However, it is presently unknown whether enzyme induction or repression processes are involved in the modulation of the four enzyme activities.

The regulatory system for the differential accumulation of aglycones versus conjugates obviously involves the enzymes responsible for conjugate formation and hydrolysis and also comprises those factors which control the storage and efflux capability of the vacuole. The mechanisms for uptake, storage and release of secondary products presently known favour the assumption that specific tonoplast carrier systems exist (Wink 1993). In case of the isoflavonoid malonylglucosides such carriers will then be linked with the enzymes involved in conjugate formation and hydrolysis.

Regulation of isoflavonoid metabolism by cinnamic acid

The mobilisation of the formononetin and the pterocarpan moieties of the vacuolar conjugates for pterocarpan aglycones accumulation is presently described as a metabolic grid with the vacuolar conjugates being reservoir pools for phytoalexin formation (Fig. 2). As a suitable model explaining the regulation of vacuolar efflux of malonylglucosides and the changes in enzyme acitivities (Table 1 and 2) the regulatory potential of cinnamic acid was thought to be of importance. This acid is known to participate in the regulation of enzymes of the phenylpropanoid pathway (Dixon et al. 1980; Bolwell et al. 1986, 1988), to mediate the metabolic flux through this sequence (Shields et al. 1982) and to affect gene expression by influencing the translational process of PAL and CHS (Bolwell et al. 1986, 1988; Dixon & Lamb 1990b). Therefore, the cytoplasmic concentration of cinnamic acid may also represent a regulatory component both for controlling vacuolar influx and efflux (Fig. 2) as well as for the modulation of the enzyme activities involved in the synthesis and the hydrolysis of pterocarpan and formononetin conjugates (Table 2).

Experiments to test this hypothesis are depicted in Fig. 5. Chickpea cell suspension cultures were treated with a high elicitor dose (80 mg/flask) and concomitantly with increasing amounts of cinnamic acid (10^{-4} to 10^{-3}M). Elicited control cells (Fig. 5 B) showed in comparison to non-treated cells (Fig. 5 A) a pronounced accumulation of pterocarpan aglycones and a substantial decrease of the constitutively formed pterocarpan conjugates and FGM. This elicitor effect is quantitatively reversed upon application of increasing levels of exogenous cinnamic acid (Fig. 5 C-E). Finally, 10^{-3} M cinnamic acid completely inhibited the strong elicitor effect because a situation as seen with the non treated cells (Fig. 5 A) has again been measured. Essentially as seen in the previous experiments (Fig. 4) the vacuolar pool of BGM was not affected.

Inhibition of both vacuolar efflux and hydrolysis of malonylglucoside conjugates by high cinnamic acid concentrations was further substantiated by elicitation experiments in the presence of cinnamic acid and L-AOPP for PAL inhibition. Again, no phytoalexin aglycones were formed under these conditions indicating that cinnamate appears to act at the vacuolar efflux system (data not shown).

To further elucidate the inhibitory mechanism of high cinnamic acid concentrations on conjugate metabolism possible modifications of elicitor-induced modulations of enzyme activities by cinnamate were also recorded. The data in Table 1 show that a high elicitor dose (80 mg/flask) strongly increases the activities of PAL, C4H, CHS, 2′-IHD,3′-IHD, IEST and

Table 1. Changes of enzyme activities (μkat/kg protein) in chickpea cell suspensions after application of 80 mg yeast elicitor/flask and reversion of enzyme induction/representation by simultaneous application of cinnamic acid (10^{-3}M).

Enzyme	Control	Elicitor	Elicitor + Cinnamic acid
Phenylalanine Ammonialyase (PAL)	10	28	11
Cinnamic acid 4-Hydroxylase (C4H)	7	17	8
Chalcone Synthase (CHS)	2,5	4,8	3
Isoflavone 2'-Hydroxy-lase (2'-IHD)	0,8	7,8	2,2
Isoflavone 3'-Hydroxy-lase (3'-IHD)	1,8	5,3	2,6
Isoflavone 7-0-Glucoside Malonyltransferase (IMT)	1.800	1.100	1.500
Isoflavone Malonyl-Glucoside Malonyl-Esterase (IEST)	20.000	35.000	20.000
Isoflavone Glucoside Glucosidase (IGLC)	300	480	300

Table 2. Changes of enzyme activities (μkat/kg protein) involved in Isoflavone/Pterocarpan Malonylglucoside metabolism in chickpea cell suspensions after application of different amounts of yeast elicitor. Enzymes were assayed 10 h after elicitation.

Enzyme	Elicitor dose (mg)/flask			
	0	10	30	80
Isoflavone-7-0-Glucosyl-Transferase (IGT)	30	46	21	18
Isoflavone Glucoside Malonyltransferase (IMT)	2.000	2.450	1.300	1.080
Isoflavone Malonyl-Glucoside Malonyl-esterase (IEST)	18.000	13.000	19.000	26.000
Isoflavone Glucoside Glucosidase (IGLC)	100	100	120	158

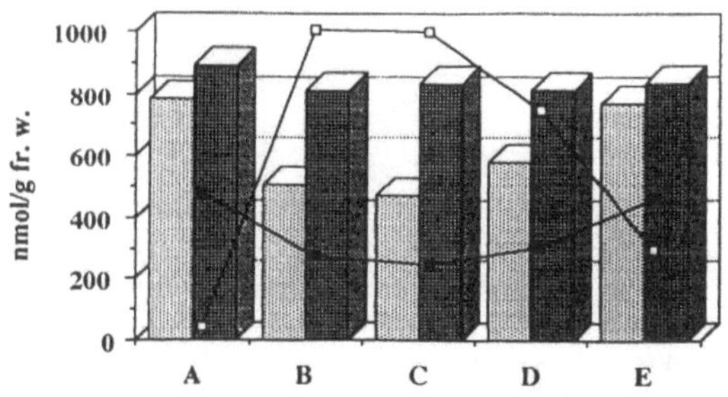

Fig. 5. Effect of cinnamic acid on the elicitor (80 mg per flask) induced metabolism of isoflavone and pterocarpan conjugates measured 12 h after treatment. Experimental conditions were: A: unelicited cells; B: elicitor-treated cells; C: elicitor + 10^{-4} M cinnamate; D: elicitor + 5 × 10^{-4} M cinnamate; E: elicitor + 10^{-3} M cinnamate. The bars represent:

▩ formononetin malonylglucoside
■ biochanin A malonylglucoside

The curves indicate:

–□– phytoalexin aglycones
–■– phytoalexin conjugates.

IGLC whereas the activities of IGT and IMT were found to be reduced. Upon simultaneous application of 10^{-3} M cinnamic acid and elicitor to such cells (Table 1, column 'elicitor + cinnamic acid') these changes of enzyme activities were all quantitatively prevented. These results indicate a pronounced regulatory potential of cinnamic acid on pterocarpan biosynthetic enzymes, proteins involved in the metabolism of malonylglucosides and on vacuolar efflux systems. Furthermore, the data again support the assumption that the various enzymes of the pterocarpan biosynthetic pathway (Fig. 2) and the vacuolar pools of FGM, MeGM and MaGM together with the relevant enzymes all belong to one metabolic grid.

The effects here described for cinnamic acid appear to be highly specific for this particular compound because similar experiments as presented in Fig. 5 and Table 1 using appropriate concentrations of p-coumaric acid failed to show any significant changes exerted by this phenylpropanoid intermediate.

Cinnamate metabolism in elicited cell cultures

Application of cinnamic acid to chickpea cell cultures in comparatively high doses (Fig. 5) caused pronounced regulatory effects on isoflavone/pterocarpan metabolism but failed to lead to any significant toxic effects or cell necrosis. Therefore, metabolic reactions for the detoxification of excess quantities of the exogenously applied acid are most likely functioning in these cells.

In agreement with studies by Edwards et al. (1990) and Edwards & Dixon (1991) on French bean cell cultures metabolism of cinnamate fed to cultured chickpea cells mainly resulted in

a) the accumulation of wall-bound, ethanol-insoluble material,
b) formation of cinnamic acid glucoside and
c) glutathionyl-S-cinnamic acid; respectively.

These results were obtained by feeding (10^{-5}–10^{-3}M) ^{14}C-cinnamic acid to the cultures (uptake ca. 85–90%/20h) with detailed TLC and HPLC analyses of the soluble conjugates formed using appropriate reference compounds. Incorporation of radioactivity into insoluble cell structures appeared to be the quantitatively most prominent portion (35–45%). Incorporation of ^{14}C-cinnamic acid into isoflavones and pterocarpans proceeded to a very low degree only whereas the two conjugates were formed at a level of 2–4%.

Subsequent enzymic investigations aimed at determination of the UDP-glucose dependent cinnamic acid glucosyltransferase (CGT) and the glutathione S-cinnamoyl transferase (GSCT) in the chickpea cultures. Furthermore, induction of these enzyme activities by elicitation and the possible regulatory effects of cinnamate were also measured. Enzyme assays using protocols by Edwards et al. (1990) and Edwards & Dixon (1991) showed that chickpea cell suspension cultures express both enzyme activities. Constitutive levels of CGT were found to be 7–8 μkat/kg protein and both elicitor and cinnamic acid were shown to exert a strong inducing effect independently from each other (Fig. 6). Maximum increase by approximately a factor 3 was measured 8 h after application of the elicitor and 12 h after treatment with cinnamic acid. Simultaneous treatment of cultures with combinations of elicitor (i.e. 80 mg/flasks and cinnamate (10^{-3} M) revealed a slight synergistic effect in the increase of CGT by a factor of 3.4. The data in Fig. 6 clearly document that high cinnamate levels (5 × 10^{-4}/10^{-3}M) may lead to an enhanced detoxification pathway for this regulatory compound. On the other hand such high cinnamate doses strongly repress elicitor-induced responses, namely, increase of biosynthetic enzyme activities (Table 1), phytoalexin accumulation (Fig. 5) as well as vacuolar efflux of formononetin and pterocarpan moieties from malonyl conjugates.

Assays for GSCT revealed a rather similar situation as seen for CGT because the constitutive levels (6 μkat/kg protein) were increased to 10–11 μkat/kg pro-

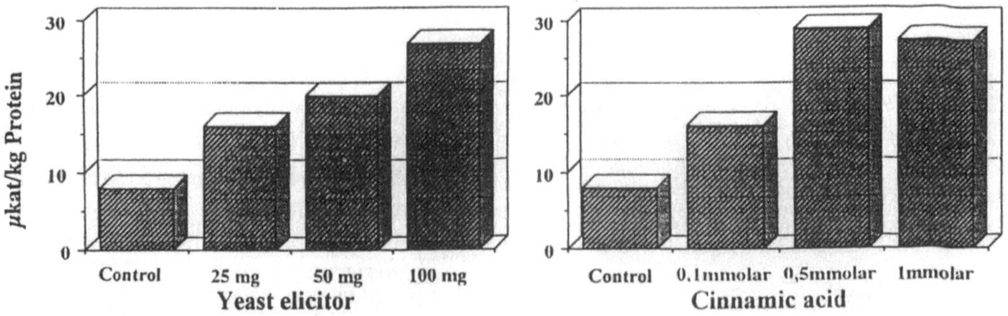

Fig. 6. Elicitor- (left) and cinnamic acid-induced (right) increase of cinnamic acid glucosyltransferase acitivity in chickpea cell suspension cultures.

tein both by yeast-elicitor (100 mg) or cinnamic acid (10^{-3}M). The maximum values were reached in ca. 12 h, but synergistic effects of the two inducers were much less pronounced. These GSCT data again reflect the metabolic function of cinnamate in that excess concentrations of this regulatory compound induce an increase in the detoxification reaction. Finally, elicitation of chickpea cell cultures (40–100 mg/flask; preincubation 8 h) increased the ability of the cells for transfer of exogenous cinnamic acid into insoluble wall-bound structures almost twofold (data not shown in detail).

In summary, chickpea cell suspension cultures were shown to posses at least three potent reaction sequences for the conversion/detoxification of cinnamic acid which are inducible both by elicitation and cinnamate itself. Elicitor-induction of glucosyltransferases (Cosio et al. 1985; Hahlbrock & Scheel 1989; Edwards et al. 1990) and glutathione S-cinnamoyl transferase (Edwards & Dixon 1991) together with induction of such enzymes by cinnamate itself (Edwards & Owen 1988) have also been shown in other plant systems. Furthermore, comparison of Fig. 5 with the data on CGT and GSCT document the existence of cinnamic acid repressible and inducible pathways in chickpea cells. Incorporation of cinnamate in wall-bound structures may on one hand lead to more resistent, less digestible wall structures and on the other hand lower the cytoplasmic concentration of this compound. This decrease in cinnamate level will in turn reduce the inhibition of phenylpropanoid and phytoalexin metabolism as caused by this acid.

Conclusions

In chickpea cell cultures the complete biosynthetic pathway from phenylalanine to the pterocarpans via the intermediate step of an isoflavone (Fig. 2), the branches leading to the vacuolar constituents FGM, MeGM and MaGM, respectively, together with the enzymes involved in malonylglucoside formation and hydrolysis as well as the reactions leading to cinnamic acid conjugates and cinnamate cell wall binding may be summarized in one metabolic grid (Fig. 7). This grid comprises a system of diverging reactions in which the enzymes are all highly responsive to elicitation and to regulation by cinnamic acid.

The intensity of elicitation by low or high doses influences the extent to which the conjugates FGM, MeGM and MaGM, respectively, are either preferentially formed (low/moderate doses) or consumed for the accumulation of pterocarpan aglycones. The elicitor-caused production of conjugates is favoured both by a simultaneous increase in the activities of the enzymes IGT and IMT and a reduction of the activities of IGLC and IEST. In contrast, high activities of the enzymes necessary for pterocarpan aglycone formation (PAL, C4H, CHS, 2′-IHD, 3′-IHD) as induced by high elicitor doses are accompanied by an intensified release of vacuolar conjugates and a reversion of the ratio of enzyme activities involved in conjugate formation versus hydrolysis. This regulatory pattern most likely also involves elicitor-caused changes in the affinity or the transport-rate of appropriate tonoplast carriers.

In addition to the previously described regulatory effects of cinnamic acid on the differential expression of PAL and CHS (Bolwell et al. 1986, 1988; Dixon & Lamb 1990b) our data now show that a similiar situation appears to exist with the cytochrome P_{450} monooxygenases C4H, 2′-IHD and 3′-IHD, respectively. Whether gene respression or translational processes are involved remains to be investigated. Especially interesting are the observations that high cytoplasmic concentrations of cinnamate (formed at times of high expression of PAL) also prevent the elicitor-caused modulations of the activities of IGT, IMT, IEST

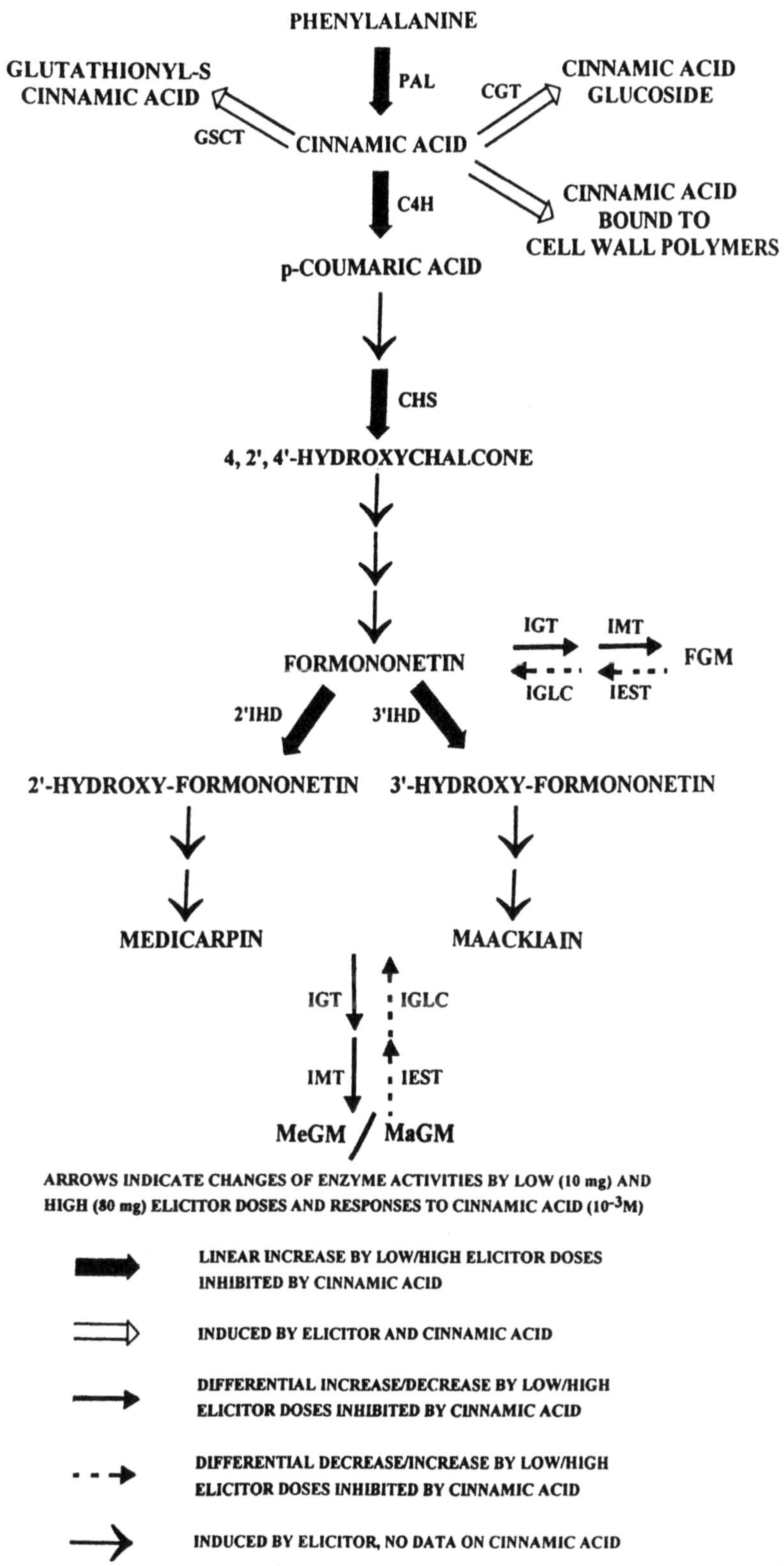

Fig. 7. Postulated metabolic grid comprising isoflavone and pterocarpan biosynthetic routes, formation of vacuolar-localized malonylglucosides of formononetin and medicarpin/maackiain as well as enzyme reactions for cinnamic acid conjugation and cell wall binding. Enzyme activities are differentially amenable by elicitation and/or cinnamic acid treatment. Enzymes are: GSCT, glutathione S-cinnamoyl transferase; CGT, cinnamic acid glucosyltransferase. For the other enzymes see legend of Fig. 2.

and IGLC (Table 2) and that the regulatory potential of cinnamic acid obviously also affects the postulated tonoplast carrier systems for the conjugates. Again, the exact mechanisms are presently unknown. Furthermore, the hypothetical metabolic grid (Fig. 7) also comprises three elicitor-amenable/cinnamate intensified diverging reactions (GSCT, CGT, wall-bound cinnamate) which allow for the conversion of high cytoplasmic concentrations of cinnamic acid to most likely non-active forms.

In essence, the investigations on medicarpin and maackiain accumulation in chickpea cell suspension cultures revealed that the biosynthetic pathway cannot be explained by a mere unidirectional, de-novo expressed synthetic route from precursors of primary metabolism but rather by a metabolic grid (Fig. 7) including constitutively formed conjugates and a regulatory pattern governed by differential changes of enzyme activities and pathway intermediates.

Acknowledgements

Financial support by Deutsche Forschungsgemeinschaft (grant Ba 280/14–1; postdoctoral fellowship to UM) and Fonds der Chemischen Industrie is gratefully acknowledged.

References

Barz W & Hösel W (1978) Metabolism and degradation of phenolic compounds in plants. Rec. Adv. Phytochem. 12: 339–369

Barz W, Köster J, Weltring KM & Strack D (1985) Recent advances in the metabolism and degradation of phenolic compounds in plants and animals. Ann. Proc. Phytochem. Soc. Europe 25: 307–347

Barz W & Welle R (1992) Biosynthesis and metabolism of isoflavones and pterocarpan phytoalexins in chickpea, soybean and phytopathogenic fungi. In: Stafford HA & Ibrahim RK (Eds) Recent Advances in Phytochemistry, Vol. 26 (pp 139–164). Plenum Press, New York

Barz W, Bless W, Börger-Papendorf G, Gunia W, Mackenbrock U, Meier D, Otto Ch & Süper E (1990a) Phytoalexins as part of induced defence reactions in plants: their elicitation, function and metabolism. In: Ciba Foundation Symposium 154, Bioactive Compounds from Plants, (pp 140–156). Wiley, Chicester

Barz W, Beimen A, Dräger B, Jaques U, Otto Ch, Süper E & Upmeier B (1990b) Turnover and storage of secondary products in cell cultures. In: Charlwood BV & Rhodes MJC (Eds) Secondary Products from Plant Tissue Culture, Vol 30 (pp 79–102). Proc. Phytochem. Soc. Europe, Oxford Science Publications

Bless W (1992) Untersuchungen zur Biosynthese der Pterocarpan-Phytoalexine in der Kichererbse (*Cicer arietinum* L.). Nachweis und Charakterisierung einer Pterocarpansynthase und einer Chalkonreduktase. Ph. D. thesis, University of Münster

Bolwell GP, Cramer CL, Lamb CJ, Schuch W & Dixon RA (1986) L-phenylalanine ammonia-lyase from *Phaseolus vulgaris*: modulation of the levels of active enzyme by trans-cinnamic acid. Planta 169: 97–197

Bolwell GP, Mavanad M, Millar DJ, Edwards KJ, Schuch W & Dixon RA (1988) Inhibition of mRNA levels and activities by trans-cinnamic acid in elicitor-induced bean cells. Phytochemistry 27: 2109–2117

Clemens S, Hinderer W, Wittkampf U & Barz W (1993) Characterization of cytochrome P_{450}-dependent isoflavone hydroxylases from chickpea. Phytochemistry 32: 653–657

Cosio EG, Weissenböck G & McClure JW (1985) Acifluorfen-induced isoflavonoids and enzymes of their biosynthesis in mature soybean leaves. Whole leaf and mesophyll responses. Plant Physiol. 78: 14–19

Daniel S & Barz W (1990) Elicitor-induced metabolic changes in cell cultures of chickpea (*Cicer arietinum* L.) cultivars resistant and susceptible to *Ascochyta rabiei*, II. Planta 182: 279–286

Daniel S, Tiemann K, Wittkampf U, Bless W, Hinderer W & Barz W (1990) Elicitor-induced metabolic changes in cell cultures of chickpea (*Cicer arietinum* L.) cultivars resistant and susceptible to *Ascochyta rabiei*, I. Planta 182: 270–278

Dixon RA & Lamb CJ (1990a) Molecular communication in interactions between plants and microbial pathogens. Ann. Rev. Plant Physiol. Plant Mol. Biol. 41: 339–367

Dixon RA & Lamb CJ (1990b) Regulation of secondary metabolism at the biochemical and genetic levels. Proc. Phytochem. Soc. Europe 30: 61–79

Dixon RA, Browne T & Ward M (1980) Modulation of L-phenylalanine ammonia lyase by pathway intermediates in cell suspension cultures of dwarf French bean (*Phaseolus vulgaris* L.). Planta 150: 279–285

Dixon RA, Choudlary AD & Dalkin K (1992) Molecular biology of stress-induced phenylpropanoid and isoflavonoid biosynthesis in alfalfa. In: Stafford HA, Ibrahim RK (Eds). Recent Advances in Phytochemistry, Vol 26 (pp 91–138) Plenum Press, New York

Edwards R & Owen WJ (1988) Regulation of glutathione S-transferases of *Zea mays* in plants and cell cultures. Planta 175: 99–106

Edwards R & Dixon RA (1991) Glutathione s-cinnamoyl transferases in plants. Phytochemistry 30: 79–84

Edwards R, Mayandad M & Dixon RA (1990) Metabolic fate of cinnamic acid in elicitor-treated cell suspension cultures of *Phaseolus vulgaris*. Phytochemistry 29: 1867–1873

Graham TL, Kim JE & Graham MY (1990) Role of constitutive isoflavone conjugates in the accumulation of glyceollin in soybean infected with *Phytophthora megasperma*. Molec. Plant – Microbe Interact. 3: 157–166

Gunia W, Hinderer W, Wittkampf U & Barz W (1991) Elicitor-induction of cytochrome P_{450} monooxygenases in cell suspension cultures of chickpea (*Cicer arietinum* L.) and their involvement in pterocarpan phytoalexin biosynthesis. Z. Naturforsch. 46c: 58–66

Hahlbrock K & Scheel D (1989) Physiology and molecular biology of phenylpropanoid metabolism. Ann. Rev. Plant Physiol. 40: 347–369

Höhl B, Pfautsch M & Barz W (1990) Histology of disease development in resistant and susceptible cultivars of chickpea (*Cicer arietinum* L.) inoculated with spores of *Ascochyta rabiei*. J. Phytopathol. 129: 31–45

Keßmann H & Barz W (1987) Accumulation of isoflavones and pterocarpan phytoalexins in cell suspension cultures of different cultivars of chickpea (*Cicer arietinum* L.). Plant Cell Rep. 6: 55–59

Lamb CJ, Lawton MA, Dron M & Dixon RA (1989) Signal and transduction mechanisms for activation of plant defences against microbial attack. Cell 56: 215–224

Mackenbrock U & Barz W (1991) Elicitor-induced formation of pterocarpan phytoalexins in chickpea (*Cicer arietinum* L.) cell suspension cultures from constitutive isoflavone conjugates upon inhibition of phenylalanine ammonia lyase. Z. Naturforsch. 46c: 43–50

Mackenbrock U, Vogelsang R & Barz W (1992) Isoflavone and pterocarpan malonylglucosides and β–1,3-glucan- and chitin-hydrolases are vacuolar constituents in chickpea (*Cicer arietinum* L.). Z. Naturforsch 47c: 815–822

Mackenbrock U, Gunia W & Barz W (1993) Accumulation and metabolism of medicarpin and maackiain malonylglucosides in elicited chickpea (*Cicer arietinum* L.) cell suspension cultures. J. Plant Physiol. 142: 385–391

Shields SE, Wingate VPM & Lamb CJ (1982) Dual control of phenylalanine ammonialyase production and removal by its product cinnamic acid. Eur. J. Biochem. 123: 389–395

Tiemann K, Filmer B, Inzé D, van Montagu M & Barz W (1993) Phytoalexin biosynthesis in chickpea (*Cicer arietinum* L.). cDNA cloning and regulation of NADPH : isoflavone oxidoreductase (IFR). In: Fritig B & Legrand M (Eds) Mechanisms of Plant Defense Responses (pp 320–323) Kluwer Academic Publishers

Vogelsang R & Barz W (1993) Purification, characterization and differential hormonal regulation of a β–1,3-glucanase and two chitinases from chickpea (*Cicer arietinum* L.). Planta 189: 60–69

Vogelsang R (1993) Biochemische und molekularbiologische Untersuchungen über Abwehrreaktionen der Kichererbse (*Cicer arietinum* L.) nach Infektion mit dem Schadpilz *Ascochyta rabiei*. Ph. D. thesis, University of Münster

Weidemann C, Tenhaken R, Höhl U & Barz W (1991) Medicarpin and maackiain 3–0-glucoside–6′–0-malonate conjugates are constitutive compounds in chickpea (*Cicer arietinum* L.) cell cultures. Plant Cell Rep. 10: 371–374

Wink M (1993) The plant vacuole: a multifunctional compartment. J. Exp. Bot. Supplement. 44: 231–246

Plant Cell, Tissue and Organ Culture **38**: 213–220, 1994.

Regulation of isoflavonoid metabolism in alfalfa

Nancy L. Paiva, Abraham Oommen, Maria J. Harrison & Richard A. Dixon
Plant Biology Division, The Samuel Roberts Noble Foundation, P.O. Box 2180, Ardmore, OK 73402, USA

Key words: Glomus versiforme, isoflavone reductase, medicarpin, *Medicago sativa*, phytoalexin, *Phoma medicaginis*

Abstract

Isoflavonoids are believed to play important roles in plant-microbe interactions. During infection of alfalfa (*Medicago sativa*) leaves with the fungal pathogen *Phoma medicaginis*, rapid increases in mRNA levels and enzyme activities of isoflavone reductase, phenylalanine ammonia-lyase, chalcone synthase and other defense genes are observed within 1 to 2 hours. The phytoalexin medicarpin and its antifungal metabolite sativan increase beginning at 4 and 8 hours, respectively, along with other isoflavonoids. In contrast, during colonization of alfalfa roots by the symbiotic mycorrhizal fungus *Glomus versiforme*, expression of the general phenylpropanoid and flavonoid genes phenylalanine ammonia-lyase and chalcone synthase increases while mRNA levels for the phytoalexin-specific isoflavone reductase decrease. The total isoflavonoid content of colonized roots increases with time and is higher than that of uninoculated roots, but the accumulation of the antifungal medicarpin is somehow suppressed.

An isoflavone reductase genomic clone has been isolated, promoter regions have been fused to the reporter gene β-glucuronidase, and the promoter-reporter fusions have been transformed into tobacco and alfalfa. Using histological staining, we have studied the developmental and stress-induced expression of this phytoalexin-specific gene in whole plants at a more detailed level than other methods allow. The isoflavone reductase promoter is functional in tobacco, a plant which does not synthesize isoflavonoids. Infection of transgenic alfalfa plants by *Phoma* causes an increase in β-glucuronidase staining, as does elicitation of transgenic alfalfa cell cultures, indicating that this promoter fusion is a good indicator of phytoalexin biosynthesis in alfalfa.

Abbreviations: CA4H – cinnamic acid 4-hydroxylase, CHI – chalcone isomerase, CHOMT – chalcone O-methyltransferase, CHS – chalcone synthase, 4CL – 4-coumarate:CoA ligase, COMT – caffeic acid O-methyltransferase, FGM – malonylated glucoside of formononetin, GUS – β-glucuronidase, IFOH – isoflavone 2′-hydroxylase, IFR – isoflavone reductase, IFS – isoflavone synthase, IOMT – isoflavone 4′-O-methyltransferase, MGM – medicarpin 3-O-glucoside-6″-O-malonate, PAL – L-phenylalanine ammonia-lyase, PTS – pterocarpan synthase, VAM – vesicular arbuscular mycorrhizal, X-gluc – 5-bromo-4-chloro-3-indolyl-β-D-glucuronide

Introduction

Isoflavonoids are thought to play important roles in the interactions of leguminous plants with microorganisms. Much of our recent work has focused on the regulation of the biosynthesis of medicarpin, the major isoflavonoid-derived phytoalexin in alfalfa (*Medicago sativa*). We are also interested in the regulation of this pathway in relation to several other pathways which share common intermediates (see Fig. 1). The products of these pathways have been implicated in such varied processes as defense, nodulation, fungal symbiosis, UV protection, flower color, allelopathy, and mechanical strength. The biosynthesis of many of these compounds occurs in plant cell cultures, either constitutively or following the addition of elicitors; however, we must return to the whole plant level if we wish to discover the biological significance of these compounds or understand their developmentally regulated accumulation.

In addition to our interest in understanding the signals that cause the plant to initiate the synthesis of defense compounds during plant-pathogen interactions, we are also interested in how a plant distin-

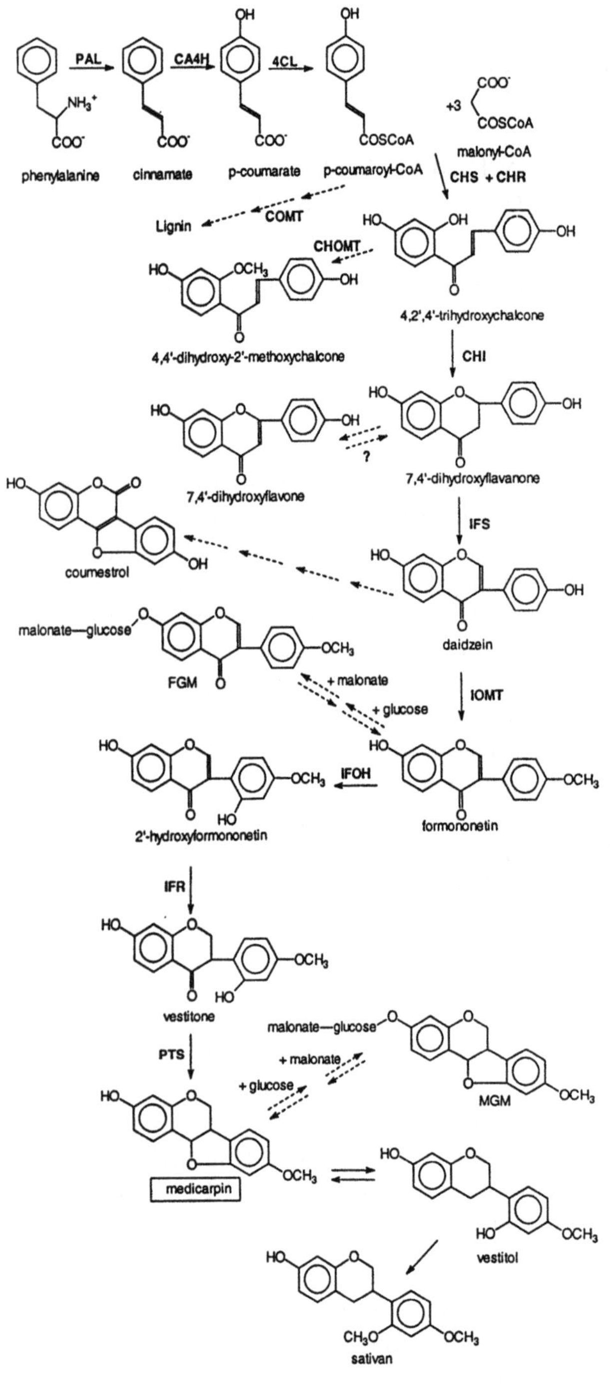

Fig. 1. Biosynthesis of the isoflavonoid phytoalexin medicarpin and related isoflavonoids in alfalfa (Dewick & Martin 1979a; Dewick & Martin 1979b). The enzymes in the medicarpin pathway are: PAL, CA4H, 4CL, CHS, CHI, IFS, IOMT, IFOH, IFR, and PTS. Two 'branch pathway' enzymes are shown: COMT, an enzyme involved in lignin biosynthesis; CHOMT, which produces a *nod*-gene-inducing chalcone. 7,4′-Dihydroxyflavone and additional isoflavonoids (coumestrol, sativan, vestitol, FGM, and MGM) which were observed in alfalfa are included on the pathway, adjacent to their corresponding precursor.

guishes between pathogens and symbionts, and how the plant's phenolic composition might be altered during such interactions. This chapter will focus on recent data on changes in isoflavonoid composition of alfalfa during attack by a fungal pathogen and during colonization by a fungal symbiont. These changes correlate with the expression of genes encoding biosynthetic enzymes for these metabolites. We also briefly describe a phytoalexin-specific promoter-reporter gene construct which we are using to investigate the cellular localization of phytoalexin synthesis.

Background: Characterization of medicarpin-related genes in alfalfa suspension cultures and whole plants

Alfalfa suspension cultures have been the subject of most of our investigations of the biochemistry and molecular biology of medicarpin biosynthesis (Fig. 1). A thorough review of this topic was recently published (Dixon et al. 1992). For many of the enzymes in the pathway (PAL, CA4H, CHS, CHI, IFR), cDNA clones have been isolated from an elicitor-induced alfalfa cell suspension library (Gowri et al. 1991b; Fahrendorf & Dixon 1993; Junghans et al. 1993; Paiva et al. 1991). Northern analysis using these cDNA probes revealed that there is a rapid increase in transcripts encoding these enzymes, usually apparent within 1 to 4 hours following the addition of elicitor. All of the activities in the proposed pathway are detectable in elicited cell cultures, at levels 2 to 30-fold higher than in unelicited cell cultures. Increases in enzyme activities are first apparent 1 to 2 hours after the first increases in their mRNA. Near-maximal extractable enzyme activity is usually reached within 10 to 15 hours, followed by either a rapid decay in activity (as in the case of PAL) or stable level of activity (as in the case of IFR).

Before elicitation, the cell cultures contain no free medicarpin, but may contain a small amount of MGM. Following elicitation, most of the MGM is rapidly hydrolyzed to form free medicarpin. Within 6 hours the level of total medicarpin begins to increase, at first mainly due to accumulation of free medicarpin. The levels of 7,4′-dihydroxyflavone and other unidentified flavonoids also increase in the cultures (N.L. Paiva & L.W. Sumner, unpublished results). Several hours after elicitation, free medicarpin levels drop, while MGM levels increase, suggesting conjugation of the medicarpin aglycone. The roots of healthy young alfalfa plants also contain no free medicarpin, but do contain significant amounts of MGM (Kessmann et al. 1990). Both roots and cell cultures also contain FGM but no detectable free formononetin (N.L. Paiva and L.W. Sumner, unpublished results). Although it is possible that this FGM could serve as a reservoir of isoflavonoid precursor for medicarpin biosynthesis, labelling data in chickpea cell cultures revealed that under normal elicitation conditions, formononetin from FGM is not significantly incorporated into medicarpin (Kessmann & Barz 1987). Recently, FGM has been shown to have *nod* gene-inducing activity for *Rhizobium meliloti*, the species of Rhizobium which nodulates alfalfa and many other legumes (Dakora et al. 1993).

Several intermediates in the pathway leading to medicarpin are also precursors of other metabolites synthesized in alfalfa cell cultures and plants. We have recently isolated alfalfa cDNA clones encoding two phenolic O-methyltransferase enzymes specific for two of these branch pathways, caffeic acid O-methyltransferase (COMT) and chalcone O-methyltransferase (CHOMT) (Gowri et al. 1991a; Maxwell et al. 1993). These catalyze the formation of ferulic acid from caffeic acid in the lignin pathway (COMT) and 4,4′-dihydroxy, 2′-methoxychalcone from 2′,4,4′-trihydroxychalcone (isoliquiritigenin 2′-O-methyltransferase, CHOMT). The 2′-methoxychalcone is the most potent of the *R. meliloti nod* gene inducers released from alfalfa roots. Current efforts are focusing on the engineering of altered lignin and *nod* gene-inducer levels using these cDNAs in transgenic plants. Significant reductions in lignin levels have been obtained by expressing a COMT antisense gene in transgenic tobacco (W. Ni et al. 1994). These cDNA clones can also be used to monitor the relative expression of genes in these interconnected pathways.

The alfalfa cell cultures have served as a good model system for the study of the phytoalexin pathway and its regulation. These cultures are a source of relatively homogeneous tissue which allows synchronous and high expression of the genes of the medicarpin pathway following elicitation, and have been the source material for isolation of all of our alfalfa cDNA clones. We have been able to verify a close correlation of metabolite accumulation, enzyme activity, and mRNA levels. While cell cultures are still the ideal source for the isolation of as yet uncharacterized phytoalexin genes, we have recently been exploring levels of gene expression and metabolite accumulation in whole plant systems.

Phoma medicaginis-Medicago sativa interaction

Phoma medicaginis, a well-known fungal pathogen of alfalfa, causes spring black stem and leaf spot disease. It can seriously decrease both the yield and quality of alfalfa hay during cool, moist periods in the spring and fall. Early studies demonstrated that medicarpin accumulated in *Phoma*-infected alfalfa leaves (Higgins 1972). We are interested in characterizing this system for a number of reasons. First, we would like to determine the response of our previously characterized defense response genes in a natural plant-pathogen interaction, and see how this compares with their responses in the cell culture system. Second, we are carrying out a number of projects aimed at manipulating the amount or structure of the phytoalexins in alfalfa and these studies require suitable pathogens with which to challenge the modified plants. Third, we have generated transgenic alfalfa plants containing the IFR promotor driving the GUS reporter gene (see discussion below). After first characterizing the temporal and spatial expression of the endogenous IFR gene in *Phoma*-infected non-transformed plants, we can then compare this with the expression of the reporter gene. If the patterns are the same, such transgenic plants may allow us to track the induction of phytoalexin biosynthesis at the cellular level in response to a range of pathogenic and symbiotic microbes.

Initially, whole alfalfa plants (cv Apollo) were sprayed with *P. medicaginis* spores suspended in a dilute Tween-20 solution. The plants were then enclosed in plastic bags to increase the humidity, and maintained in a growth chamber for 5 to 10 days. Characteristic brown to black lesions developed on the leaves, indicative of a successful infection. Younger leaves, which emerged after spraying, and leaves on control (Tween-20 sprayed) plants, did not show lesions. HPLC analysis revealed the presence of medicarpin in leaves with lesions, but not in healthy leaves. Significant amounts of sativan and coumestrol, along with traces of 4-O-methoxymedicarpin and possibly vestitol, were also present in infected leaves; these compounds were never detected in elicited alfalfa cell suspension cultures (N.L. Paiva, unpublished results). All but coumestrol are known phytoalexins biosynthetically derived from medicarpin, and we therefore predicted all the genes in the medicarpin pathway (PAL, CHS, CHI, IFR, PTS, etc; see Fig. 1) would be induced in *Phoma*-treated plants.

In order to facilitate time course studies and to increase the number of infected cells, a detached-leaf system was utilized. Younger leaves (fully expanded trifoliates) were cut from several plants, then dipped in either *Phoma*/Tween spore suspension or dilute Tween alone. Leaves were then placed in petri dishes containing damp filter paper and the dishes sealed with parafilm. At various time intervals, the leaves were collected, frozen, and assayed for phytoalexins, enzyme activities and mRNA levels.

The first traces of medicarpin were observed at 4 hours after inoculation with *Phoma* (Fig. 2A). By 8 hours both medicarpin and sativan were detectable, and the total amount of phytoalexins increased thereafter. No MGM accumulation was observed. Tween-treated leaves showed little or no accumulation of these phytoalexins during this time.

Two 'phytoalexin specific' enzymes in the medicarpin pathway, IFR and PTS, reached low but detectable levels within 2 hours in *Phoma*-treated leaves, and activities increased rapidly thereafter (Fig. 2B and C). In Tween-treated leaves, little or no activity was detected in the first 20 hours. Significant activity was observed at 4.5 days in Tween-treated leaves, perhaps due to induction by the Tween itself or other stresses.

Northern analysis with the cDNA probes for PAL, CHS, CHI, and IFR revealed increases in the corresponding mRNA levels within 1 hr after pathogen-inoculation, and higher levels for the next 20 hours. In Tween-treated leaves, the mRNA levels for these enzymes increased slightly, but were well below the levels in infected leaves.

Overall, the observed kinetics of rapid mRNA increase, followed by increased enzyme and, later, metabolite levels, are all consistent with previously characterized plant pathogen interactions and the changes observed in elicited cell cultures. Current experiments are focussing on examining the response of other defense-related genes (such as COMT and chitinase) and on confirming that the transformable alfalfa cultivar Regen SY responds in a similar fashion to the non-transformable cultivar Apollo used in these experiments. As they become available, our genetically-engineered alfalfa plants harboring reporter genes or producing modified phytoalexins will be challenged with *Phoma*, and the responses compared with those above. We will then be able to determine whether or not our phytoalexin manipulations have an effect on the invasion by the pathogen, and whether or not our reporter gene constructs are providing an accurate representation of IFR expression.

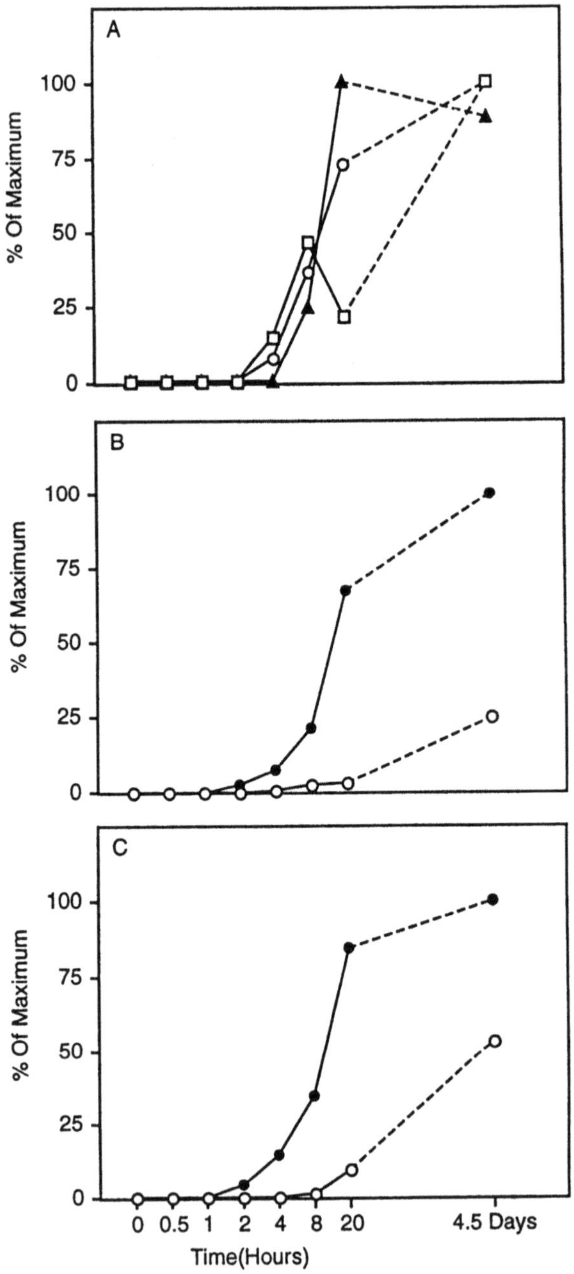

Fig. 2. Levels of phytoalexins and related enzyme activities in alfalfa leaves after inoculation with *Phoma medicaginis* spore suspension. Levels are reported as the percentage of the maximum observed in the experiment. (A) Levels of sativan (▲), medicarpin (□), or total phytoalexins (○) in *Phoma*-treated leaves; (B) Levels of isoflavonone reductase (IFR) enzyme activity in *Phoma*-treated (•) or control Tween-treated (○) leaves. (C) Levels of pterocarpan synthase (PTS) enzyme activity in *Phoma*-treated (•) or control Tween-treated (○) leaves.

Vesicular arbuscular mycorrhizal fungus-*Medicago truncatula* interaction

Vesicular arbuscular mycorrhizal (VAM) fungi are soil fungi which form beneficial symbiotic associations

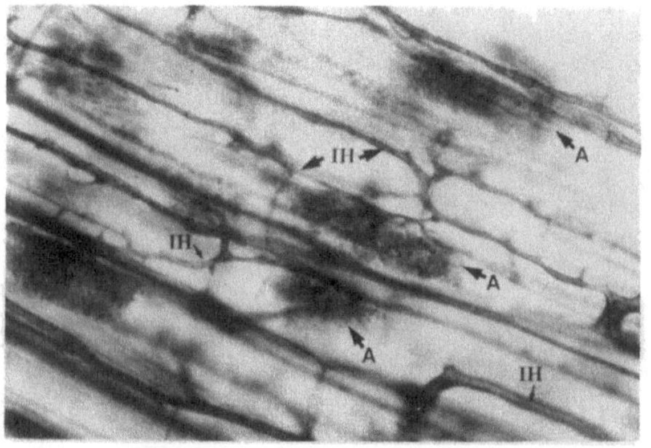

Fig. 3. Magnified view of a leek (*Allium porrum*) root heavily colonized by *Glomus versiforme*. Hyphae were stained with Chlorazol Black E. Internal hyphae (IH) general grow between the cortical cells. Arbuscules (A), structures consisting of finely branched hyphae, form within the cell walls without penetrating the host tonoplast membrane (Mauseth 1988). Arbuscules participate in nutrient exchange, while vesicles (not shown) contain lipids thought to act as energy reserves. Identical structures were observed in colonized *Medicago sativa* and *M. truncatula* roots.

with the roots of alfalfa and a number of other plants. In exchange for carbohydrate energy, mycorrhizal fungi are believed to confer many benefits on the host plant, including increased phosphorus and mineral uptake. Mycorrhizal plants also show increased drought resistance and enhanced disease resistance. Colonization by mycorrhizal fungi involves invasion of the root cortical cells by the fungal hyphae, and the formation of specialized structures within the root cortical cells called arbuscules and vesicles (See Fig. 3). These processes are also reminiscent of early events in the infection by fungal pathogens. We are interested in determining how the plant distinguishes symbiont from pathogen, what types of genes might be mycorrhizal-specific, and whether flavonoids are either affected by or play a role in the symbiosis.

Flavonoid composition and mRNA levels were determined at different times after inoculation of *Medicago truncatula* roots with *Glomus versiforme* (Harrison & Dixon 1993). *M. truncatula* is a diploid medic species which has root isoflavonoid profiles very similar to those of *M. sativa*. *G. versiforme* is a species of VAM fungus previously observed as an active colonizer of alfalfa. Because the fungus first infects, then continually spreads to colonize new sites, the time points cannot be classed as discreet phases in the developmental process. Instead they are referred to as 'initiating' (7 to 13 days) when few arbuscules have developed and 'well developed' (40 days) when many of the cortical cells contain arbuscules.

Uninoculated *M. truncatula* roots contain FGM and MGM as the predominant isoflavonoids. In heavily colonized roots these are still the major peaks in the HPLC profile; in fact, their concentration increases 3 to 4 fold on a fresh weight basis. Free medicarpin and formononetin levels are low or not detected in uninoculated roots. In the initial stages of colonization, approximately 10% of the medicarpin is present as the free aglycone, and a small amount of free formononetin is also detected. In the well developed stages of colonization, the free medicarpin and formononetin decrease to barely detectable levels, despite the increases in MGM and FGM. Both 7,4′-dihydroxyflavone and coumestrol increase early in inoculated roots and are still detectable during the well developed interaction; in contrast, these latter two compounds are not detectable in uninoculated roots.

Northern analysis indicated that PAL and CHS transcripts both increased approximately 2-fold during colonization by the mycorrhizal fungus. IFR transcripts, however, showed a slight increase, followed by a significant decrease (50%) as the interaction progressed, compared with uninoculated roots.

These responses are clearly very different from those observed in plant-fungal pathogen interactions. The slight increase in PAL and CHS mRNA levels is minor compared to that observed for these transcripts in the elicited cell culture system (e.g. 50-fold induction for PAL), but is probably sufficient to account for the increased accumulation of isoflavonoid conjugates. The decrease in IFR transcript level, as well as the lack of accumulation of free medicarpin, indicate that the symbiotic fungus has either avoided triggering the defense response system, or has actually suppressed the defense response.

Analysis of the isoflavone reductase promoter

From Southern blot analysis, IFR appears to be encoded by a single gene in alfalfa (Paiva et al. 1991), unlike some of the earlier pathway genes such as PAL and CHS (Gowri et al. 1991b; Junghans et al. 1993). We have isolated a genomic clone for IFR from our alfalfa cv Apollo genomic library. Sequence analysis of the promoter region revealed little or no similarity to previously studied PAL or CHS promoters from bean (*Phaseolus vulgaris*) (A. Oommen, N.L. Paiva, and R.A. Dixon, unpublished results). Regions of the promoter have been subcloned into pBI101 to produce binary vectors wherein the IFR promoter controls the expression of the GUS reporter gene (Jefferson et al. 1987b). The lengths of the promoter in these constructs were chosen based on the availability of restriction sites in the genomic clones and binary vector; one such construct contains over 700 bp of the IFR promoter as well as most of the 5′-untranslated region. Using *Agrobacterium*-mediated transformation, stably transformed tobacco (*Nicotiana tabacum*) and alfalfa plants have been regenerated. Tobacco was chosen due to the ease and speed of transformation in this species. We were also curious to observe how an isoflavonoid-specific promoter would behave in a plant which produces no isoflavonoids or pterocarpans. Although alfalfa requires 4 to 6 months for transformation and regeneration, we are very eager to compare expression of the GUS reporter gene with that of the endogenous IFR gene.

Tobacco plants containing the IFR-GUS construct show strong X-gluc staining in roots, in stems especially at petiole junctions, in young leaves, and in certain flower parts. When present, the staining is generally associated with specific cell layers, usually adjacent to xylem elements or near meristematic zones. The root expression was consistent with the northern analysis of IFR expression previously shown for alfalfa plants (Paiva et al. 1991), but the high expression in young leaves and stems was surprising. Thus, while this isoflavonoid-pathway promoter is indeed active in a solanaceous plant, the expression does not appear to be developmentally controlled in the same manner as in alfalfa. A cross-section of an X-gluc stained tobacco root is illustrated in Fig. 4A and B; the staining is largely restricted to the root cortex (see discussion below).

Unstressed alfalfa plants bearing the IFR-GUS reporter construct have exhibited X-gluc staining only in the roots. Expression is very high immediately behind the root tip meristem, but is also evident further back along the root. Expression is also high at the sites of lateral root initiation. This root expression is consistent with the previously described northern analysis and the constitutive accumulation of MGM throughout the root system of alfalfa. Cross sections of young roots reveal that the GUS activity is highest in the endodermis and inner cortex region (Fig. 4C and D), but not in the inner stele or epidermis. These delicate alfalfa roots would be difficult to dissect in order to confirm the exact location of MGM accumulation by phytochemical means, but the cortex-specific

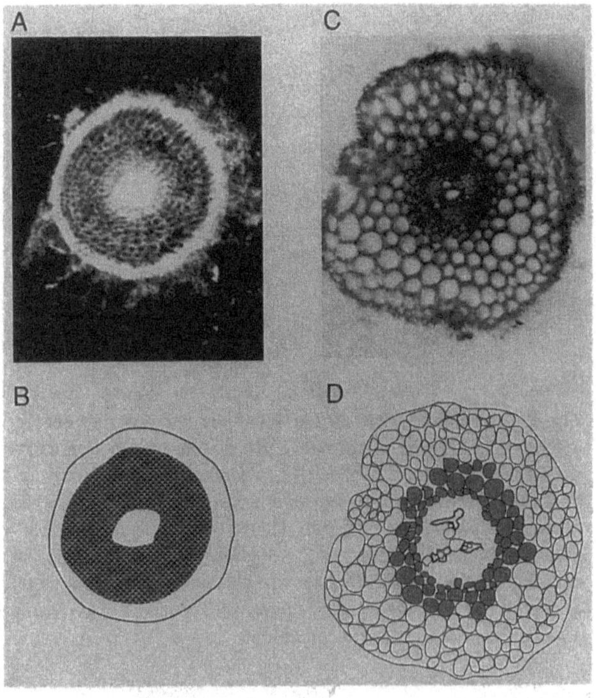

Fig. 4. Cross sections of roots of transgenic tobacco and alfalfa plants bearing the IFR promoter-GUS constructs. Thin sections were stained with X-gluc, a histochemical substrate of GUS (Jefferson 1987a; Martin et al. 1992). The presence of GUS activity is revealed by the formation of a blue (BrCl-indigo) precipitate. (A) tobacco root cross-section. (B) schematic diagram of tobacco root cross-section, with cells exhibiting blue staining shaded in grey (root cortex region). (C) alfalfa root cross-section. (D) schematic diagram of alfalfa root cross-section, with cells exhibiting GUS activity shaded in grey (inner cortex and endodermis).

expression pattern is consistent with two independent experiments. First, recent *in situ* hybridization experiments on young *Medicago truncatula* and *M. sativa* roots have detected the presence of IFR mRNA in the same cortex region (Harrison and Dixon 1994). (*M. truncatula* roots constitutively accumulate MGM just as *M. sativa* roots do). Second, histochemical analysis revealed that formononetin (a precursor of medicarpin) accumulates predominantly in the cortex of the roots of chickpea, a species which also accumulates MGM constitutively (Wiermann 1981). As formononetin is only one step away from the substrate of IFR (2′-hydroxyformononetin), we would expect to find formononetin and isoflavone reductase in the same tissues.

In preliminary experiments, we have observed strong X-gluc staining in *Phoma*-treated transgenic alfalfa leaves, while no X-gluc staining was observed in untreated or Tween-treated transgenic leaves or

Phoma-treated untransformed leaves. We also initiated callus cultures from the leaves of transgenic tobacco and alfalfa plants, and have just begun elicitation experiments with these transgenic cell lines. While no X-gluc staining has yet been observed in elicited tobacco cell cultures, a strong increase in X-gluc staining was apparent in transgenic alfalfa cells treated with yeast-elicitor. We are currently determining the kinetics of the elicitation of GUS activity, and will compare this to the activation of the endogenous IFR gene, both at the mRNA and protein levels. We have also prepared cultures and plants bearing a shorter (350 bp) promoter construct, and these, along with other deletion lines, will be similarly characterized in an attempt to determine which regions of the promoter are required for the elicitation responsiveness. In the case of an *Arabidopsis* chitinase promoter only 192 bp were required to confer both pathogen inducibility and tissue specific expression, and only 326 bp of a bean CHS promoter confers induction by elicitation (Samac & Shah 1991; Dron et al. 1988). However, in other systems, it has been found that longer promoter segments (1500 to 2000 bp) confer higher levels. If proper regulation is conferred by either promoter construct, they can not only be used to drive reporter genes and new defense gene constructs, but also be dissected more closely to provide clues as to which *cis*-elements are important in the regulation.

Conclusions

We have characterized the isoflavonoid levels and the expression of genes involved in isoflavonoid biosynthesis in both a plant-pathogen and a plant-symbiont interaction, and found striking differences between the two. In response to a fungal leaf pathogen, phytoalexins were rapidly synthesized, while in response to a fungal root symbiont, no free phytoalexins accumulated while other isoflavonoids did accumulate. This correlated well with increased expression of general phenylpropanoid genes (PAL, CHS) in both systems, and differences in the induction of a defense-specific gene (IFR). As we further characterize these interactions, we hope to gain an understanding of the differences between the regulation of these two types of processes. Also, the distinct and reproducible changes in metabolite levels during these interactions suggest roles for isoflavonoids in controlling the interactions. For example, just as the increases in the antifungal medicarpin and sativan during the *Phoma* interaction are thought to help control the spread of the pathogen, the increase in FGM and other isoflavonoids may act as a stimulus for further colonization by the mycorrhizal fungus.

Using new tools such as the IFR promoter-GUS fusions as a marker for phytoalexin biosynthesis, we will be able to identify the cells in which defense compounds are actively synthesized in response to treatment with pathogens, symbionts, abiotic inducers, and other stresses. Suspension cultures from these plants will allow us to examine the possible regulation of medicarpin biosynthesis at the level of IFR transcription by pathway intermediates and culture conditions. A detailed functional dissection of the IFR promoter will provide the first information on the *cis*-elements and *trans*-acting factors regulating a gene specifically involved in isoflavonoid synthesis.

References

Dakora FD, Joseph CM & Phillips DA (1993) Alfalfa (*Medicago sativa* L.) root exudates contain isoflavonoids in the presence of *Rhizobium meliloti*. Plant Physiol. 101: 819–824

Dewick PM & Martin M (1979a) Biosynthesis of pterocarpan and isoflavan phytoalexins in *Medicago sativa*: the biochemical interconversion of pterocarpans and 2′-hydroxyisoflavans. Phytochemistry 18: 591–596

Dewick PM & Martin M (1979b) Biosynthesis of pterocarpan, isoflavan and coumestan metabolites of *Medicago sativa*: chalcone, isoflavone and isoflavanone precursors. Phytochemistry 18: 597–602

Dixon RA, Choudhary AD, Dalkin K, Edwards R, Fahrendorf T, Gowri G, Harrison MJ, Lamb CJ, Loake GJ, Maxwell CA, Orr J & Paiva NL (1992) Molecular biology of stress induced phenylpropanoid and isoflavonoid biosynthesis in alfalfa. In: Stafford H & Ibrahim RK (Eds) Phenolic Metabolism in Plants (pp 91–137). Plenum Press, New York

Dron M, Clouse SD, Dixon RA, Lawton MA & Lamb CJ (1988) Glutathione and fungal elicitor regulation of a plant defense gene promoter in electroporated protoplasts. Proc. Natl. Acad. Sci. USA 85: 6738–6742

Fahrendorf T & Dixon RA (1993) Molecular cloning of the elicitor-inducible cinnamic acid 4-hydroxylase cytochrome P450 from alfalfa. Arch. Biochem. Biophys. 305: 509–515

Gowri G, Bugos RC, Campbell WH, Maxwell CA & Dixon RA (1991a) Molecular cloning and expression of alfalfa S-adenosyl-L-methionine: caffeic acid 3-O-methyltransferase, a key enzyme of lignin biosynthesis. Plant Physiol. 97: 7–14

Gowri G, Paiva NL & Dixon RA (1991b) Stress responses in alfalfa (*Medicago sativa* L.) 12. Sequence analysis of phenylalanine ammonia-lyase (PAL) cDNA clones and appearance of PAL transcripts in elicitor-treated cell cultures and developing plants. Plant Mol. Biol. 17: 415–429

Harrison MJ & Dixon RA (1993) Isoflavonoid accumulation and expression of defense gene transcripts during the establishment of vesicular arbuscular mycorrhizal associations in roots of *Medicago truncatula*. Mol. Plant Microbe Interact. 6: 643–654

Harrison MJ & Dixon RA (1994) Spatial patterns of expression of flavonoid/isoflavonoid pathway genes during interactions between roots of *Medicago truncatula* and the mycorrhizal fungus *Glomus versiforme*. Plant J., in press

Higgins VJ (1972) Role of the phytoalexin medicarpin in three leaf spot diseases of alfalfa. Physiol. Plant Pathol. 2: 289–300

Jefferson RA (1987) Assaying chimeric genes in plants: The GUS gene fusion system. Plant Mol. Biol. Rep. 5: 387–405

Jefferson RA, Kavanagh TA & Bevan MW (1987) GUS fusions: β-glucuronidase as a sensitive and versatile gene fusion marker in higher plants. EMBO J. 6: 3901–3907

Junghans H, Dalkin K & Dixon RA (1993) Stress responses in alfalfa (*Medicago sativa* L.). XV. Characterization and expression patterns of members of a subset of the chalcone synthase multigene family. Plant Mol. Biol. 22: 239–253

Kessmann H & Barz W (1987) Accumulation of isoflavones and pterocarpan phytoalexins in cell suspension cultures of different cultivars of chickpea (*Cicer arietinum*). Plant Cell Rep. 6: 55–59

Kessmann H, Edwards R, Geno PW & Dixon RA (1990) Stress responses in alfalfa (*Medicago sativa* L.) V. Constitutive and elicitor-induced accumulation of isoflavonoid conjugates in cell suspension cultures. Plant Physiol. 94: 227–232

Martin T, Schmidt R, Altmann T & Frommer WB (1992) Non-destructive assay systems for detection of β-glucuronidase activity in higher plants. Plant Mol. Biol. Rep. 10: 37–46

Mauseth JD (1988) Plant Anatomy, Benjamin/Cummings, Menlo Park

Maxwell CA, Harrison MJ & Dixon RA (1993) Molecular characterization and expression of alfalfa isoliquiritigenin 2′-O-methyltransferase, an enzyme specificially involved in the biosynthesis of a transcriptional activator of *Rhizobium meliloti* nodulation genes. Plant J., 4: 971–981

Ni W, Paiva NL & Dixon RA (1994) Reduced lignin in transgenic plants containing an engineered caffeic acid O-methyltransferase antisense gene. Transgenic Res., in press

Paiva NL, Edwards R, Sun Y, Hrazdina G & Dixon RA (1991) Molecular cloning and expression of alfalfa isoflavone reductase, a key enzyme of isoflavonoid phytoalexin biosynthesis. Plant Mol. Biol. 17: 653–667

Samac DA & Shah DM (1991) Developmental and pathogen-induced activation of the *Arabidopsis* acidic chitinase promoter. Plant Cell 3: 1063–1072

Wiermann R (1981) Secondary plant products and cell and tissue differentiation. In: Stumpf PK & Conn EE (Ed) The Biochemistry of Plants, Vol 7 (pp 85–116). Academic Press, New York

Plant Cell, Tissue and Organ Culture **38**: 221–225, 1994.

Regulation of phenylalanine ammonia-lyase genes in carrot suspension cultured cells

Yoshihiro Ozeki[1] & Junko Takeda[2]
[1]*Department of Biology, College of Arts and Sciences, The University of Tokyo, Komaba, Meguro-ku, Tokyo 153, Japan;* [2]*Department of Agricultural Chemistry, Faculty of Agriculture, Kyoto University, Sakyo-ku, Kyoto 606, Japan*

Key words: Anthocyanin, carrot, chalcone synthase, *Daucus carota*, 2,4-dichlorophenoxyacetic acid, phenylalanine ammonia-lyase

Abstract

Induction of anthocyanin synthesis occurs during metabolic differentiation in carrot suspension cultured cells grown in medium lacking 2,4-dichlorophenoxyacetic acid (2,4-D), and is closely correlated with embryogenesis. Anthocyanin synthesis may also be induced by light-irradiation under different culture conditions. The phenylalanine ammonia-lyase (PAL) gene (TRN-PAL), which was transiently induced by the transfer effect, was also rapidly induced after light-irradiation. However, TRN-PAL was not involved in anthocyanin synthesis. A second PAL gene, ANT-PAL, was involved in anthocyanin synthesis. ANT-PAL was induced during metabolic differentiation in medium lacking 2,4-D parallel with the induction of chalcone synthase (CHS). PAL genes in the carrot genome are expressed differentially depending on the nature of the environmental stimulus, e.g. transfer effect and light, and other parameters which also affect anthocyanin synthesis.

Abbreviations: CHS – chalcone synthase, 2,4-D – 2,4-dichlorophenoxyacetic acid, GUS – β-glucuronidase, Luc – firefly luciferase, PAL – phenylalanine ammonia-lyase, UV – ultraviolet

Introduction

Plant cell cultures are often used as model systems to investigate the regulation of secondary metabolism by environmental and developmental stimuli, in particular the mechanism of resistance to UV irradiation and elicitation (Dixon et al. 1983; Hahlbrock & Scheel 1989). This is because the cells can be grown in a stable and homogenous environment with respect to e.g. temperature, light, nutrients, etc. This reduces the complexity of the experimental system compared to intact plants and facilitates analysis of regulatory mechanisms. Secondary metabolism may be induced in cultured cells as a defense response towards environmental stress in the same way as in intact plants. Other secondary metabolic pathways, especially those which synthesize commercially useful metabolites, are expressed only in differentiated cells in intact plants and not in cultured cells because they are undifferentiated in medium containing auxin. These activities are strongly repressed in most undifferentiated cultured plant cells.

We have established a system of carrot suspension cultures in which anthocyanin synthesis can be induced under defined conditions (Ozeki & Komamine 1981). In this system, undifferentiated carrot cells, which did not synthesize anthocyanin, were subcultured in medium containing 2,4-D. When cells were transferred to a medium without 2,4-D, anthocyanin synthesis was induced 4–5 days after culturing in the dark, together with embryogenesis. Investigation of physiological properties suggested a close correlation between the induction of embryogenesis, 'morphological differentiation', and anthocyanin synthesis, 'metabolic differentiation' (Ozeki & Komamine 1981, 1986). A second system was established by Takeda (1988) using the same carrot suspension cultured cells, in which anthocyanin synthesis was induced by light-irradiation in 2,4-D free diluted medium. The regulatory mechanism of light induced anthocyanin synthesis was thought to be different from the induction of anthocyanin synthesis by morphological differentiation.

In order to elucidate the mechanism of induction of anthocyanin synthesis by developmental or environmental stimuli in carrot suspension cultures, we have investigated the regulation of PAL. PAL is a key regulatory enzyme of phenylpropanoid metabolism catalyzing the first reaction leading to phenolic compounds such as lignin, flavonoids, anthocyanin, furanocoumarin, and isoflavonoid phytoalexins (Dixon et al. 1983; Jones 1984). PAL is highly regulated during development and in response to a range of environmental factors including elicitation, wounding, and light (Kuhn et al. 1984; Edwards et al. 1985; Minami et al. 1989; Tanaka et al. 1989; Ohl et al. 1990; Estabrook & Sengupta-Gopalan 1991; Gowri et al. 1991; Yamada et al. 1992). Most investigators have focused on one PAL gene which is regulated by both developmental and environmental stimuli. For example, studies on the PAL promoter in parsley have indicated the existence of both tissue-specific and stress-specific regulation (Lois et al. 1989). Analysis of PAL cDNA sequences in carrot suspension cultured cells revealed that a specific PAL gene, TRN-PAL, was induced by a transfer effect irrespective of the presence or absence of 2,4-D. When anthocyanin synthesis was induced in differentiated cells in medium lacking 2,4-D, the second PAL gene, ANT-PAL, was induced (Ozeki et al. 1990b). This shows that different PAL genes are present on the carrot genome and specific PAL genes are induced in response to developmental or environmental stimuli.

To investigate the mechanism of PAL gene regulation by developmental and environmental stimuli in carrot suspension cultured cells, expression of PAL genes during anthocyanin synthesis was studied following differentiation and light-irradiation.

Materials and methods

Cultures

Carrot (*Daucus carota* cv. Korodagosun) suspension cultured cells were maintained as previously reported (Ozeki & Komamine 1981). Cells were subcultured in modified Lin and Staba's medium (Fujimura & Komamine 1975) containing 0.5 μM 2,4-D. For induction of anthocyanin synthesis as well as embryogenesis, cells were transferred to the medium lacking 2,4-D but containing 1 μM zeatin and 4% sucrose and then cultured in the dark (Ozeki & Komamine 1981, 1985). Induction of anthocyanin synthesis by light was carried out as previously described (Takeda 1988). Carrot cells, subcultured as above, were transferred to medium lacking 2,4-D but containing only one-tenth of the concentration of nutrients and 4% sucrose. In this diluted medium, little anthocyanin was synthesized in the dark. When the culture flask was exposed to white light 5 days after the transfer, rapid anthocyanin synthesis was initiated. In both culture systems, cells were harvested using a buchner funnel, quickly frozen in liquid nitrogen and stored at −80 °C.

mRNA extraction, Northern blotting and primer extension

Total RNA was extracted from carrot cells using a modified phenol-SDS method (Ozeki et al. 1990a). Poly(A)$^+$RNA was purified using oligo-dT cellulose. The amounts of PAL and CHS mRNA were determined densitometrically from signals on the X-ray film of Northern and dot blots of mRNA probed with carrot PAL and CHS cDNAs (Ozeki et al. 1990b). Primer extension using PAL gene specific primers was performed using standard methods (Sambrook et al. 1989). Twenty μg of poly(A)$^+$RNA was used as a template for primer extension. Oligonucleotides for PAL and CHS specific primers were synthesized as follows: TRN-PAL, 5′-GTT-TCC-TAG-CAC-AAC-ATT-CTT-GTT-CTC-ACA-3′, corresponding to the complementary position from +136 bp to +107 bp; ANT-PAL, 5′-CGT-TCT-CAT-GAT-GCC-CGT-TCG-TAT-ACG-CCA-3′, corresponding to that from +197 bp to +168 bp; CHS, 5′-ACT-CAT-TCA-CAG-TCA-CCA-TCT-CTA-GCC-AGA-3′, corresponding to that from +91 bp to +62 bp.

Analysis of expression of promoters using luciferase in carrot protoplasts

The promoters, *gTRN-PAL* corresponding to TRN-PAL cDNA, *gANT-PAL* corresponding to ANT-PAL, and *gCHS* corresponding to CHS cDNA were obtained from carrot genomic DNA library (Ozeki et al. unpublished data). The putative promoter region from −2334 bp to +144 bp of *gTRN-PAL*, that from −2448 bp to +233 bp of *gANT-PAL*, and that from −2357 bp to 96 bp of *gCHS* were translationally fused to the firefly luciferase (*Luc*) gene, terminated by a *NOS*-terminator. These constructs were subcloned into pBluescript SK+.

Protoplasts of carrot suspension cultured cells were prepared using Cellulase ONOZUKA RS and Macerozyme R-10. Plasmids harboring promoter-*Luc*

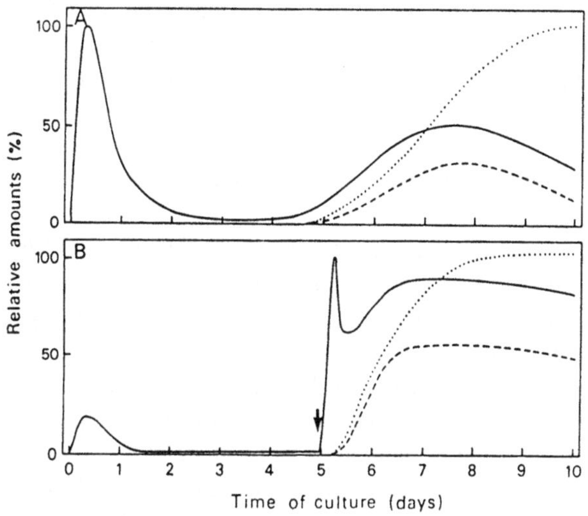

Fig. 1. Induction of mRNA for PAL (solid lines) and CHS (broken lines) of cells during metabolic differentiation in the original medium (*A*) and of cells cultured in the diluted medium by light-irradiation (*B*). Dotted lines show accumulation of anthocyanin. Arrow in (*B*) shows the time of light-irradiation. The results were obtained from three independent experiments.

constructs were introduced into protoplasts by electroporation. After the electroporation, protoplasts were cultured for 24 h in medium containing or lacking 2,4-D under light or dark conditions before harvesting by centrifugation. Protoplasts were disrupted by sonication and debris were removed by centrifugation. The supernatants were used for the measurement of Luc activity expressed from the PAL and CHS promoters using PicaGene^{TM} Luminescence Kit (Wako Pure Chem., Osaka, Japan) in a liquid scintillation counter according to the supplier's manual.

Results and discussion

Two stimuli induced anthocyanin synthesis in carrot suspension cultured cells

In the original system, anthocyanin synthesis was initiated 4–5 days after the culture in the dark, when the cells were transferred to the basal medium lacking 2,4-D but containing zeatin and cultured. Anthocyanin accumulated gradually until 5–7 days after initiation in the dark (Fig. 1). In this system, light-irradiation did not affect initiation of anthocyanin synthesis or accumulation of anthocyanin. However, when cells were transferred to one tenth diluted medium lacking 2,4-D, little anthocyanin synthesis was observed even 14 days after culture in the dark (data not shown). However, when cells were irradiated with light 5 days after transfer to the diluted medium, anthocyanin synthesis was immediately and dramatically induced. Anthocyanin synthesis started about 6 h after onset of irradiation and the rate of accumulation was very rapid compared to the original system (Fig. 1B).

In both systems, changes in PAL and CHS mRNA were measured (Fig. 1). PAL mRNA increased transiently and rapidly within 6 h after transfer into the fresh medium by transfer effect, and disappeared thereafter (Ozeki et al. 1990a) both in diluted and non-diluted fresh medium. In the original system, PAL mRNA increased again 4–5 days after transfer, when anthocyanin synthesis was initiated (Fig. 1A). PAL mRNA slowly increased and reached a maximum 2–3 days after initiation, concomitant with the changes in CHS mRNA (Fig. 1A). In comparison to the original system, the amount of PAL mRNA remained low in diluted medium without 2,4-D throughout dark incubation when anthocyanin synthesis was not initiated. When cells were irradiated with light 5 days after transfer, induction of PAL mRNA was initiated (Fig. 1B). Large scale PAL gene transcription occurred after irradiation, in the same way as observed directly after transfer. Maximum level of PAL mRNA was reached after about 6 h, afterwards the PAL mRNA level first showed a rapid decrease without reaching the low level before irradiation. Subsequently PAL mRNA level slightly increased again 24–36 h after the induction (Fig. 1B). At the same time, the CHS mRNA level reached a maximum and anthocyanin was rapidly synthesized. These results suggest that there are two types of PAL induction, first just after light-irradiation, and second concomitant with CHS induction and anthocyanin accumulation.

Analysis of expression of two PAL genes by primer extension

We investigated which PAL genes, TRN-PAL induced by transfer effect, or ANT-PAL induced during anthocyanin synthesis (Ozeki et al. 1990b), were induced during induction of anthocyanin synthesis by light-irradiation using primer extension with TRN- and ANT-PAL specific primers. Poly(A)$^+$RNA was prepared from PAL-induced cells immediately after transfer to fresh medium and from cells synthesizing anthocyanin from the original system. Primer extension was performed using these mRNAs as templates (Fig. 2). Specific primers hybridized to TRN- or ANT-PAL mRNA and gave signal strengths which corresponded

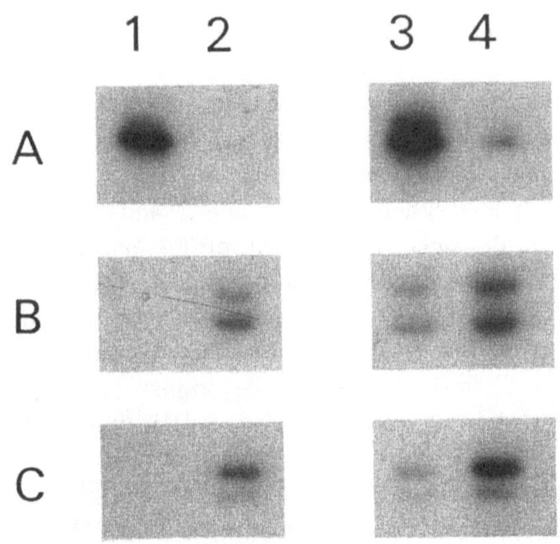

Fig. 2. Primer extension of poly(A)$^+$RNAs using TRN-PAL specific primer (*A*), ANT-PAL specific primer (*B*), and CHS specific primer (*C*). Poly(A)$^+$RNAs were prepared from: cells transferred to fresh medium containing 2,4-D and cultured 6 h (lane 1); anthocyanin synthesizing cells cultured in the original medium lacking 2,4-D for 7 days (lane 2); cells cultured in the diluted medium for 5 days, 6 h (lane 3) and 36 h (lane 4) after the onset of light-irradiation.

to the amounts of TRN- or ANT-mRNAs (Ozeki et al. 1990b). In the case of cells cultured in diluted medium without 2,4-D, poly(A)$^+$RNAs were prepared from cells 6 h and 36 h after light-irradiation. The TRN-PAL primer gave a strong signal with mRNA from cells 6 h after irradiation, when PAL mRNA was transiently and rapidly induced (Fig. 2). mRNA from cells 36 h after irradiation showed a faint signal to TRN-PAL primer (Fig. 2). Although ANT-PAL primer gave a little signal with mRNA from cells 6 h after irradiation, the signal using this primer was strong with mRNA from cells 36 h after irradiation, when anthocyanin was accumulating rapidly (Fig. 2). These results show that, in the case of the induction of anthocyanin synthesis by light-irradiation, at first the TRN-PAL gene was transiently and rapidly induced; then after transient TRN-PAL induction, the ANT-PAL gene was induced concomitant with CHS gene induction (Fig. 2C).

Many researchers have reported that elicitors or UV-irradiation can induce flavonoid synthesis, for example, phytoalexins (Hahlbrock et al. 1976; Schröder et al. 1979; Lawton et al. 1983). In all cases, stress induces PAL transcription at first, followed by the induction of CHS transcription, resulting in flavonoid synthesis. However, our results show that the TRN-PAL gene is initially induced by stress, and that it is not involved in anthocyanin synthesis. Later, the other PAL gene, ANT-PAL, was induced concomitantly with the induction of CHS gene. The ANT-PAL gene product catalyzes anthocyanin synthesis. However, the level of ANT-PAL is small, so it is not easy to detect its increment.

Regulation of PAL promoter by light and 2,4-D

In order to study the regulatory mechanism of TRN-PAL induction by transfer and light-irradiation and that of ANT-PAL induction during anthocyanin synthesis, genomic clones, *gTRN-* and *gANT-PAL*, corresponding to TRN- and ANT-PAL respectively, have been obtained. Chimeric gene fusions were constructed from the 5′-flanking regions of *gTRN-* and *gANT-PAL* gene and *Luc* or *GUS*. Because the *GUS* gene product is highly stable in plant cells, repression of promoter activity by 2,4-D and dark treatment might not be reported. *Luc* gene product is less stable than the *GUS* gene product (Millar et al. 1992) and it is particularly suitable as a reporter gene for the analysis of down-regulation by 2,4-D and dark treatment. The chimeric construct harboring the *gTRN-PAL* promoter was electroporated into protoplasts prepared from carrot cells cultured in medium containing 2,4-D in the dark. After electroporation, protoplasts were cultured in medium containing 2,4-D for 24 h under the light or dark condition. Light-irradiation activated the *gTRN-PAL* promoter 3.6-fold compared to the dark-control.

Protoplasts were prepared from anthocyanin synthesizing cells in medium lacking 2,4-D and *gANT-PAL* and *gCHS* promoter constructs were transferred by electroporation. Protoplasts were cultured with or without 2,4-D and/or light or dark, and expression of luciferase activity was determined (Table 1). *gANT-PAL* and *gCHS* promoters were the most active after light irradiation in the absence of 2,4-D. When anthocyanin synthesis was induced by light-irradiation, dark treatment effectively repressed anthocyanin accumulation (Takeda 1990). 2,4-D repressed *gANT-PAL* promoter activity to a level 53% that of the control under light condition. Dark treatment was effective in repressing *gANT-PAL* promoter activity to 30%. The putative regulatory elements affected by light and 2,4-D are currently under molecular analysis using deletion mutants of the *gANT-PAL* promoter region.

Conclusions

In carrot suspension cultured cells, two PAL genes are present: one is rapidly and transiently induced

Table 1. Effect of light and 2,4-D on luciferase expression from the *gANT-PAL* and *gCHS* promoters. Protoplasts were prepared from anthocyanin synthesizing cells 2 days after light irradiation. *gANT-PAL* and *gCHS Luc* constructs were electroporated and cultured under light (L) or in the dark (D) and with or without 2,4-D. *Luc* activity was determined 24 h after electroporation. Luciferase expression in light exposed protoplasts is presented as 100%.

	Luciferase activity (%)			
Promoter	L	L + 2,4-D	D	D + 2,4-D
gANT-PAL	100	53	30	39
gCHS	100	52	43	19

by environmental stimuli, the other is induced during anthocyanin synthesis (metabolic differentiation). The PAL gene product activities are the same in both cases. The expression of genes may depend on response to environmental or developmental stimuli. These results have implications for identification of target genes for metabolic engineering using antisense RNA technology.

References

Dixon RA, Dey PM & Lamb CJ (1983) Phytoalexins: Enzymology and molecular biology. Advances in Enzymology and Related Areas of Molecular Biology 55: 1–135

Edwards K, Cramer CL, Bolwell GP, Dixon RA, Schuch W & Lamb CJ (1985) Rapid transient induction of phenylalanine ammonia-lyase mRNA in elicitor-treated bean cells. Proc. Natl. Acad. Sci. USA 82: 6731–6735

Estabrook EM & Sengupta-Gopalan C (1991) Differential expression of phenylalanine ammonia-lyase and chalcone synthase during soybean nodule development. Plant Cell 3: 299–308

Fujimura T & Komamine A (1975) Effects of various growth regulators on the embryogenesis in a carrot suspension culture. Plant Sci. Lett. 5: 359–364

Gowri G, Paiva NL & Dixon RA (1991) Stress responses in alfalfa (*Medicago sativa* L.) 12. Sequence analysis of phenylalanine ammonia-lyase (PAL) cDNA clones and appearance of PAL transcripts in elicitor-treated cell cultures and developing plants. Plant Mol. Biol. 17: 415–429

Hahlbrock K & Scheel D (1989) Physiology and molecular biology of phenylpropanoid metabolism. Ann. Rev. Plant Physiol. Plant Mol. Biol. 40: 347–369

Hahlbrock K, Knobloch KH, Kreuzaler F, Potts JRM & Wellmann E (1976) Coordinated induction and subsequent activity changes of two groups of metabolically interrelated enzymes. Light-induced synthesis of flavonoid glycosides in cell suspension cultures of *Petroselinum hortense*. Eur. J. Biochem. 61: 199–206

Jones DH (1984) Phenylalanine ammonia-lyase: Regulation of its induction, and its role in plant development. Phytochemistry 23: 1349–1359

Kuhn DN, Chappell J, Boudet A & Hahlbrock K (1984) Induction of phenylalanine ammonia-lyase and 4-coumarate:CoA ligase mRNAs in cultured plant cells by UV light or fungal elicitor. Proc. Natl. Acad. Sci. USA 81: 1102–1106

Lawton MA, Dixon RA, Hahlbrock K & Lamb CJ (1983) Elicitor induction of mRNA activity. Rapid effects of elicitor on phenylalanine ammonia-lyase and chalcone synthase mRNA activities in bean cells. Eur. J. Biochem. 130: 131–139

Lois R, Dietrich A, Hahlbrock K & Schulz W (1989) A phenylalanine ammonia-lyase gene from parsley: structure, regulation and identification of elicitor and light responsive cis-acting elements. EMBO J. 8: 1641–1648

Millar AJ, Short SR, Chua NH & Kay SA (1992) A novel circadian phenotype based on firefly luciferase expression in transgenic plants. Plant Cell 4: 1075–1087

Minami E-i, Ozeki Y, Matsuoka M, Koizuka N & Tanaka T (1989) Structure and some characterization of the gene for phenylalanine ammonia-lyase from rice plants. Eur. J. Biochem. 185: 19–25

Ohl S, Hedrick SA, Chory J & Lamb CJ (1990) Functional properties of a phenylalanine ammonia-lyase promoter from *Arabidopsis*. Plant Cell 2: 837–848

Ozeki Y & Komamine A (1981) Induction of anthocyanin synthesis in relation to embryogenesis in a carrot suspension culture: Correlation of metabolic differentiation with morphological differentiation. Physiol. Plant 53: 570–577

Ozeki Y & Komamine A (1985) Effects of inoculum density, zeatin and sucrose on anthocyanin accumulation in a carrot suspension culture. Plant Cell Tissue Organ Cult. 5: 45–53

Ozeki Y & Komamine A (1986) Effects of growth regulators on the induction of anthocyanin synthesis in a carrot suspension culture. Plant Cell Physiol. 27: 1361–1368

Ozeki Y, Komamine A & Tanaka Y (1990a) Induction and repression of phenylalanine ammonia-lyase and chalcone synthase enzyme proteins and mRNAs in carrot suspension cultures regulated by 2,4-D. Physiol. Plantarum 78: 400–408

Ozeki Y, Matsui K, Sakuta M, Matsuoka M, Ohashi Y, Kano-Murakami Y, Yamamoto N & Tanaka Y (1990b) Differential regulation of phenylalanine ammonia-lyase genes during anthocyanin synthesis and by transfer effect in carrot cell suspension cultures. Physiol. Plantarum 80: 379–387

Sambrook J, Fritsch EF & Maniatis T (1989) Molecular Cloning, a Laboratory Manual, Second Ed. Cold Spring Harbor Laboratory Press, New York

Schröder J, Kreuzaler F, Schafer E & Hahlbrock K (1979) Concomitant induction of phenylalanine ammonia-lyase and flavanone synthase mRNAs in irradiated plant cells. J. Biol. Chem. 254: 57–65

Takeda J (1988) Light-induced synthesis of anthocyanin in carrot cells in suspension. I. The factors affecting anthocyanin production. J. Exp. Bot. 39: 1065–1077

Takeda J (1990) Light-induced synthesis of anthocyanin in carrot cells in suspension. II. Effects of light and 2,4-D on induction and reduction of enzyme activities related to anthocyanin synthesis. J. Exp. Bot. 41: 749–755

Tanaka Y, Matsuoka M, Yamamoto N, Ohashi Y, Kano-Murakami Y & Ozeki Y (1989) Structure and characterization of a cDNA clone for phenylalanine ammonia-lyase from cut-injured roots of sweet potato. Plant Physiol. 90: 1403–1407

Yamada T, Tanaka Y, Sriprasertsak P, Kato H, Hashimoto T, Kawamata S, Ichinose Y, Kato H, Shiraishi T & Oku H (1992) Phenylalanine ammonia-lyase genes from *Pisum sativum*: Structure, organ-specific expression and regulation by fungal elicitor and suppressor. Plant Cell Physiol. 33: 715–725

Plant Cell, Tissue and Organ Culture **38**: 227–234, 1994.

Accumulation of anthraquinones in *Morinda citrifolia* cell suspensions

A model system for the study of the interaction between secondary and primary metabolism

Marc J.M. Hagendoorn, Linus H.W. van der Plas* & Gert J. Segers
*Department of Plant Physiology, Agricultural University Wageningen, Arboretumlaan 4, 6703 BD Wageningen, The Netherlands (*requests for offprints)*

Key words: Anthraquinones, auxins, *Morinda citrifolia*, respiration, sugar metabolism

Abstract

Cell suspensions of *Morinda citrifolia* were cultivated in a B5-medium containing 4% sucrose as the sole carbon source and 1 mg l^{-1} naphthyl acetic acid (NAA) or 1 mg l^{-1} 2,4-dichloro-phenoxyacetic acid (2,4-D). Both auxins were able to support growth but only in the presence of NAA anthraquinone production was observed. 2,4-D inhibited the production in NAA cultures. Anthraquinone synthesis took place in the growth and the stationary phase and amounts of 0.2–0.4 mmol (about 100–200 mg) g^{-1} dry weight could be reached.

Under both growth conditions sucrose was hydrolyzed extracellularly by invertase. From the resulting monosaccharides, glucose was taken up preferentially and an appreciable uptake of fructose only took place when medium glucose was exhausted. Sugar uptake rates were similar when cells were grown in NAA and in 2,4-D medium but the intracellular sugar contents (expressed on a dry weight basis) differed considerably. The presence of sucrose, glucose and fructose was demonstrated under both growth conditions. The amounts of sucrose and glucose were much lower in the 2,4-D cells than in the NAA-cells especially during the growth phase. Fructose contents were low and comparable, while in NAA cells an unknown sugar (possibly the sugar moiety of the glycosylated anthraquinones) was observed especially at the end of the growth phase and in the stationary phase. The differences in sugar concentrations were even larger due to the lower water contents of the NAA cells.

Respiration of 2,4-D cells was much higher than that of NAA cells during the growth phase. A sharp increase in sugar contents (mainly sucrose) occurred in the 2,4-D cells at the end of the growth phase and corresponded with the fall in respiratory activity.

A possible correlation between the lack of production of anthraquinones in 2,4-D cells and a less efficient growth metabolism in these cells is discussed.

Abbreviations: AQ – anthraquinones, 2,4-D – 2,4-dichloro-phenoxy-acetic acid, DW – dry weight, FW – fresh weight, NAA –, naphthyl acetic acid, pCPO – p-chloro-phenoxy-acetic acid

Introduction

Secondary metabolism is dependent on primary metabolism in a number of ways: indirectly, because the latter provides the building blocks necessary for growth and the production of cell material as well as for the production of the various enzymes of the secondary metabolic pathways; and directly, because precursors and energy necessary for the production of secondary metabolites are derived from the primary metabolism.

As growth and secondary metabolite production both compete for the same pools of substrates, these processes often are separated in time: production starts after completion of the growth phase. Examples for such a behavior can be found in plants (pigment production in flowers) as well as in in vitro systems (for instance rosmarinic acid production in *Coleus*-cells, Ulbrich et al. 1985). Direct study of the competition between primary and secondary metabolism is only possible in a limited number of *in vitro* systems because systems with a quantitatively significant production are rare. This condition is met by suspension cultures of *Morinda citrifolia* cells: this system, already described by Zenk et al. (1975) is able to produce large amounts of AQ (more than 10–20 % on a DW basis). At least 11 different AQ are described for cell cultures of *Morinda citrifolia* (a.o. rubiadin, moridone and lucidin, Leistner 1973, 1975, Inoue et al. 1981). Most of the AQ of *Morinda citrifolia* cells (more than 90%) are glycosylated especially as -O-glucosylxylosyl (Inoue et al. 1981) and subsequently stored in the vacuole (Zenk et al. 1975).

Production of AQ and growth both take place at the same time, making these cells very suitable for the study of the interaction between primary and secondary metabolism (Leistner 1985). Another advantage of this system is the ease by which AQ-production by *Morinda citrifolia* cells can be manipulated using different kinds of auxin (i.e 2,4-D or NAA).

In this paper we describe a number of differences between the primary metabolism of producing and non-producing *Morinda citrifolia* cells. Special attention will be paid to sugar metabolism and respiration.

Materials and methods

Tissue culture

The *Morinda citrifolia* L. (Rubiaceae) cultures (Zenk et al. 1975) were maintained in 250 ml Erlenmeyer flasks, containing 60 ml of B5-culture medium (Gamborg et al. 1968) supplemented with sucrose (40 g l^{-1}), kinetin (0.2 mg l^{-1}) and auxin (1mg l^{-1} NAA or 1mg l^{-1} 2,4-D). The cells were subcultured every 14 days (10 ml culture with 50 ml fresh medium). At the start of the experiments 25 g l^{-1} FW was used as inoculum.

Fresh weight/dry weight

FW was determined after harvesting the cells on a Büchner funnel applied with a paper filter. For determination of DW the cells were dried at 60 °C for 24 hours.

Anthraquinone determination

For AQ determination 0.02–0.2 g FW was boiled in 5 ml 80% aqueous ethanol for 10 min. After centrifugation (1500 *g*, 5 min) the supernatant was collected and the pellet was again boiled in 5 ml 80% aqueous ethanol. The supernatants were pooled and made up to 10 ml. In this way more than 99% of the AQ could be extracted from the cells. The absorption at 434 nm was determined and the AQ content was estimated using a millimolar extinction coefficient of 5.5 (Zenk et al. 1975, Koblitz 1988). For further calculations a mean molecular weight value of 534 for the glycosylated AQ was used (based on a sugar moiety of glucose and xylose, Inoue et al. 1981).

Determination of carbohydrates

For determination of the cellular carbohydrate content, circa 10 mg DW was boiled for 15 minutes in 75% aqueous methanol. As an internal standard raffinose was added. Subsequently the methanol was evaporated under vacuum at 20 °C The dried pellet was taken up in water and the suspension was centrifuged in an Eppendorf centrifuge (5 min, 14 000 rpm). The supernatant was directly injected in a Dionex HPLC system (Dionex corporation, Sunnyvale, USA), equipped with a Carbopac PA 100 column and a pulse amperometric detection system. Samples were eluted isocratically with 0.1 N NaOH. Carbohydrates were identified by comparing retention times with sugar standards.

Sugars in the medium were determined in a comparable way by injecting directly medium samples after appropriate dilution. Medium sucrose, glucose and fructose were also determined enzymatically with test kit 716260, purchased from Boehringer Mannheim Gmbh (Germany).

In vivo respiration

In vivo respiration was determined according to Van der Plas et al. (1988). To determine the capacities and activities of the different respiratory pathways, 0.5 mM potassium cyanide and 5 mM benzohydroxamate were used.

Results

Growth and secondary metabolism of Morinda citrifolia *cell suspensions*

When cells of *Morinda citrifolia* were cultivated heterotrophically on a B5-medium containing 4% sucrose as the sole carbon-source, growth occurred both in the presence of NAA and 2,4-D (1 mg l^{-1}), although the DW increase with NAA was larger than with 2,4-D (Fig.1).

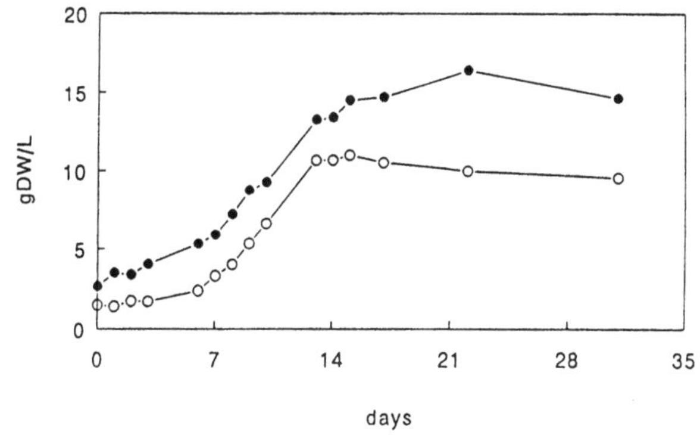

Fig. 1. Time course of growth of *Morinda citrifolia* cells grown in the presence of 1 mg l^{-1} NAA (closed circles) and 1 mg l^{-1} 2,4-D (open circles). Growth is expressed as DW. Measurements from two separate experimental series are depicted.

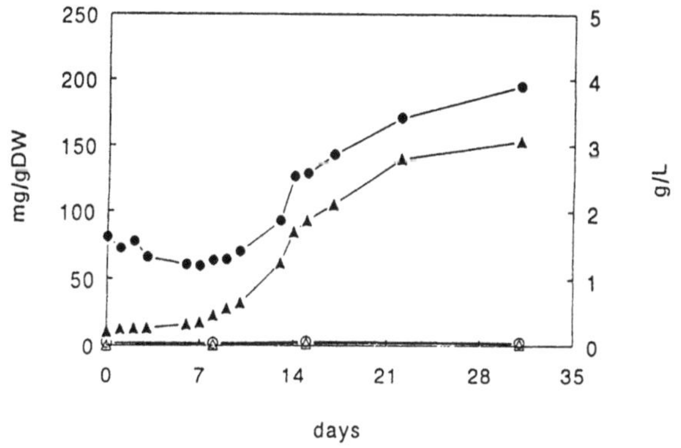

Fig. 2. Time course of antraquinone production of *Morinda citrifolia* cells grown in the presence of 1 mg l^{-1} NAA (closed symbols) and 1 mg l^{-1} 2,4-D (open symbols). Anthraquinone content is expressed as mg g^{-1} DW (circles) and as g l^{-1} cell culture (triangles). Triplicate measurements from two separate experimental series are depicted.

The production of AQ was dependent on the type of auxin used: production was completely absent in 2,4-D medium while during growth with NAA a continuous synthesis of AQ was observed, leading to AQ amounts of 0.2 to 0.4 mmol per gram of DW (10 – 20 % of the DW, Fig. 2). During the log phase production in NAA cells kept up with growth, leading to a more or less stable concentration in the cells; when growth stopped, the production continued leading to a further increase in the concentration of AQ (Fig. 2). Transfer of 2,4-D cells to NAA medium did not lead immediately to AQ-production: significant production only started after the second subculturing in NAA medium and high concentrations of AQ were only found after the third subculturing. The gradual increase in AQ-concentration after several subculturings might be harmful for the cells: we repeatedly observed that the

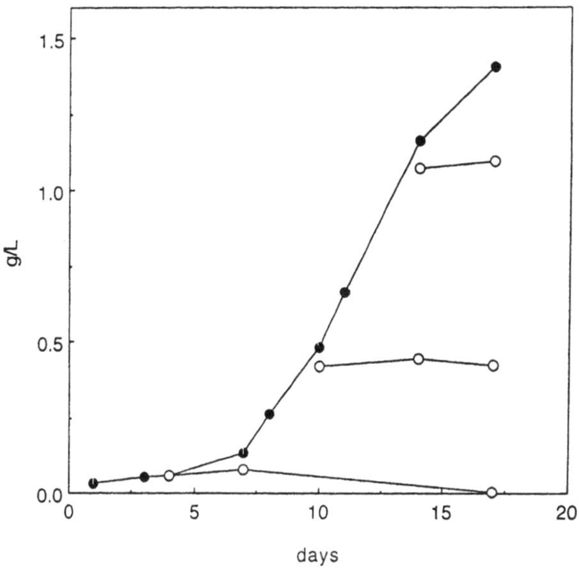

Fig. 3. Time course of antraquinone production of *Morinda citrifolia* cells grown in the presence of 1 mg l^{-1} NAA (closed symbols) and after addition of 1 mg l^{-1} 2,4-D (open symbols) at 4, 10 and 14 days after inoculation. Anthraquinone content is expressed as g l^{-1} cell culture. Triplicate measurements are depicted.

viability of the cells decreased with increasing AQ-concentrations: growth did not longer occur when inoculation was carried out with cells containing more than 0.3 mmol AQ per gram DW. Extra addition of 2,4-D to cells growing in NAA containing media had no significant effect on growth (expressed as DW increase) of the cells, but the effect on AQ-production was very dramatic: production stopped within one day and the amount of AQ in the culture stayed at the level found at the day of 2,4-D addition (Fig. 3).

Consequences of the production of AQ for the sugar metabolism and respiration of Morinda citrifolia *cells*

A quantitatively important synthesis of secondary metabolites as observed for these *Morinda citrifolia* cells is only possible if there is a large and continuous flow of substrates from the primary metabolism to the secondary metabolic pathway. As the cells were grown heterotrophically on sucrose as the sole carbon-source, attention was paid to differences in the uptake and degradation of sugars as well as to differences in the various sugarpools inside the cells.

It has already been described for many types of cell suspensions that sucrose is hydrolyzed extracellularly into glucose and fructose before being taken up by the cells. This phenomenon also was observed for our

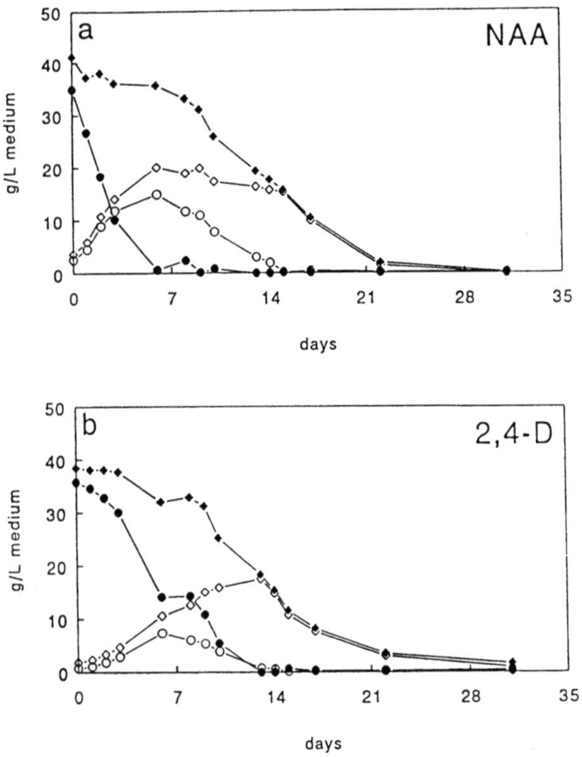

Fig. 4. Time course of medium sugar degradation of *Morinda citrifolia* cells grown in the presence of 1 mg l^{-1} NAA (a) and of 1 mg l^{-1} 2,4-D (b). Data are shown for sucrose (closed circles), glucose (open circles), fructose (open diamonds) and total sugars (summation of sucrose, glucose and fructose, closed diamonds), all expressed as g l^{-1} medium. Duplicate measurements of two independent experimental series are depicted.

Morinda citrifolia cells: the concentration of sucrose rapidly fell, especially in NAA cells (Fig. 4). The resulting glucose and fructose replaced the sucrose in the medium; both were taken up by the cells but in 2,4-D as well as in NAA medium the cells showed a strong preference for the uptake of glucose. Finally both monosaccharides were completely taken up by the cells and after about 3 weeks no sugar could be detected in the medium. Although the growth curves of *Morinda citrifolia* cells in both media were not identical (Fig. 1), there were no differences in the rate of total sugar uptake (Fig. 4).

However, the sugar contents of the cells showed considerable differences (Fig. 5). Sucrose, glucose and fructose all were present under both growth conditions, but sucrose was present in the largest amount (forming more than 10 % of the biomass of the NAA cells in the first three weeks of growth, Fig. 5a). The amount of glucose was in NAA cells generally twice as high as in 2,4-D cells (Fig. 5b), a situation also observed for sucrose in the first two weeks. Fructose amounts were lowest (less than 5% on a DW basis) and not significantly different for both cell types (Fig. 5c).

At the end of the growth phase (about 14 days after inoculation, Fig. 1) a considerable amount of sugar was still present in the medium (more than 10 g l^{-1}, Fig. 4). The uptake of sugar continued after the cessation of growth leading to a considerable increase of the size of the endogenous sugar pools in the 2,4-D cells. This was first visible in the glucose pool and manifested itself especially in the increase of the size of the sucrose pool (from about 50 to about 200 mg per gram of DW, Fig. 5a). In both growth conditions fructose did not show any changes at all at the end of the growth phase (Fig. 5c).

The changes in glucose and sucrose in the NAA cells at the same time were less obvious. In NAA cells sucrose, a typical storage sugar, dominated. Striking were also the relatively large amounts of glucose in these cells: in 2,4-D cells glucose and fructose occurred in almost equal amounts during most of the time while in NAA cells much more glucose was found than fructose. Next to these three sugars a fourth type of sugar was observed, almost exclusively in NAA cells. The concentration of this component increased continuously during the growth and the stationary phase corresponding to the course of the production of AQ in both phases (Fig. 5d). At the end of the stationary phase (about 4 weeks after inoculation) soluble sugars were virtually absent in the 2,4-D cells, while in the NAA-cells still relatively large amounts of glucose and this compound "X" were found.

When respiration is measured in cells of *Morinda citrifolia* growing in NAA or 2,4-D medium considerable differences were observed: respiratory activities of 2,4-D-cells were much higher than those of NAA cells in the period of DW increase (Fig. 6). The fall in respiration at the end of the growth phase in these cells corresponded with the increase in their sugar content (Fig. 5a & b).

An indication for differences in the metabolism of 2,4-D and NAA cells could also be found in the osmotic behavior of the cells. The FW/DW ratio for *Morinda citrifolia* cells growing in NAA and 2,4-D medium was very different. Fig. 7 shows that this ratio varied from 15 to 20 in 2,4-D cells, the lowest values being found in the log phase. In NAA cells much lower values were observed: throughout the growth cycle FW/DW-ratios were about 10–12. Addition of 2,4-D to NAA cells also led to changes in the FW/DW-ratio: a gradual increase in the water content was observed resulting in

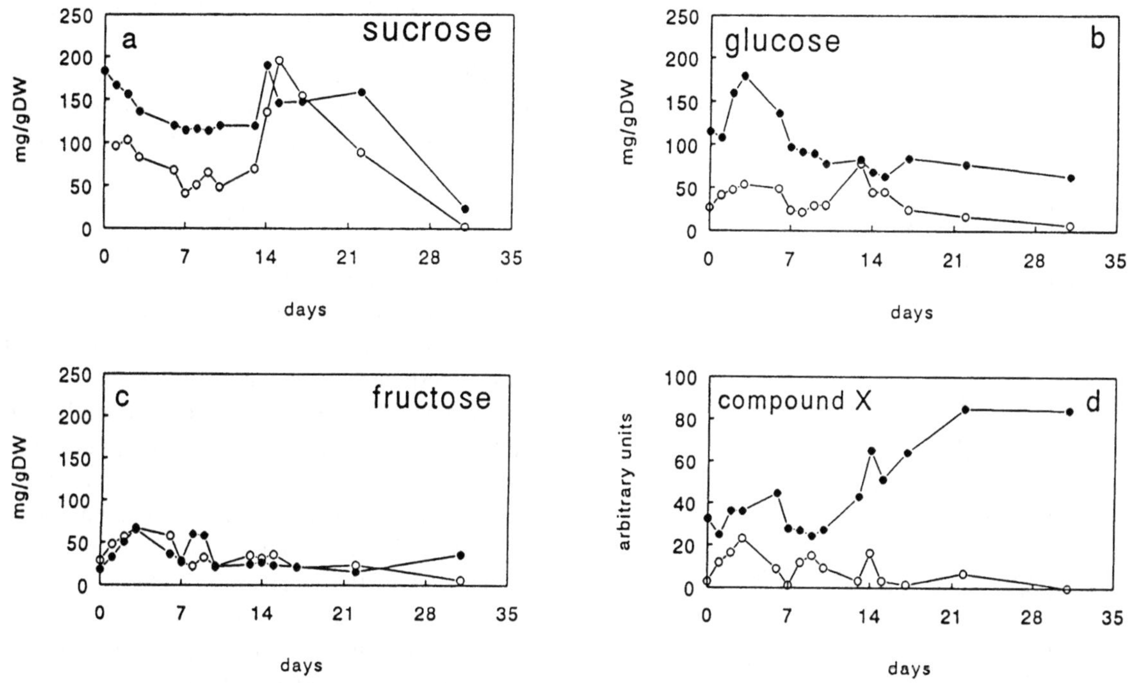

Fig. 5. Time course of changes in sugar contents of *Morinda citrifolia* cells grown in the presence of 1 mg l^{-1} NAA (closed circles) and of 1 mg l^{-1} 2,4-D (open circles). Data are shown for sucrose (a), glucose (b), fructose (c) and the unknown compound X (d, possibly the glucose-xylose-containing sugar moiety of the glycosylated AQ, see text). The data on sucrose, glucose and fructose are expressed as mg g^{-1}DW, the data on compound "X" are expressed in arbitrary units. Duplicate measurements of two independent experimental series are depicted.

the FW/DW-values of about 20 that are characteristic for 2,4-D cells. These values were reached in a few days, also when the concentration of AQ in the cells was already high at the moment of 2,4-D addition.

Discussion

The ability of a wide range of auxins to induce AQ-production in cells of *Morinda citrifolia* was already tested by Zenk et al. (1984). Independent of their effects on growth, considerable differences in the effects of auxins on AQ-synthesis can be observed. AQ-synthesis, for instance, did not occur in the presence of pCPO, while a considerable synthesis was observed when the related compound p-chlorophenylacetic acid was used. The above described results for the effects of 2,4-D and NAA fit very well with the results described by Zenk et al. (1975): apparently this property is very stable throughout the years. Also the produced quantities are comparable with the values reported in literature for this cell line (see Koblitz 1988 for a review).

Apparently NAA is able to switch on the expression of genes responsible for the various enzymes necessary for the production of AQ. The absence of production with 2,4-D might be either a lack of induction or a repression/inhibition of these enzymes. To be able to distinguish between these two possibilities we added 2,4-D to producing NAA cells in various stages of growth (Fig. 3). This addition led to an almost immediate cessation of the AQ-production; as the amounts of AQ in the culture did not decrease there was apparently no appreciable degradation too. Other auxins that were not able to induce production (like for instance pCPO) also showed the same behavior, i.e. an inhibition of the production of AQ in NAA cells (results not shown). The rapid effect of the addition of 2,4-D on production without any appreciable effect on growth indicates that 2,4-D indeed inhibits the synthesis of AQ and not only lacks the capacity to induce this pathway. The effect of 2,4-D depended on the concentration: even a ratio of 1 : 10 for 2,4-D and NAA still resulted in almost complete inhibition of the AQ-production (results not shown).

One of the possible explanations for the lack of production in the presence of 2,4-D is a lack of suffi-

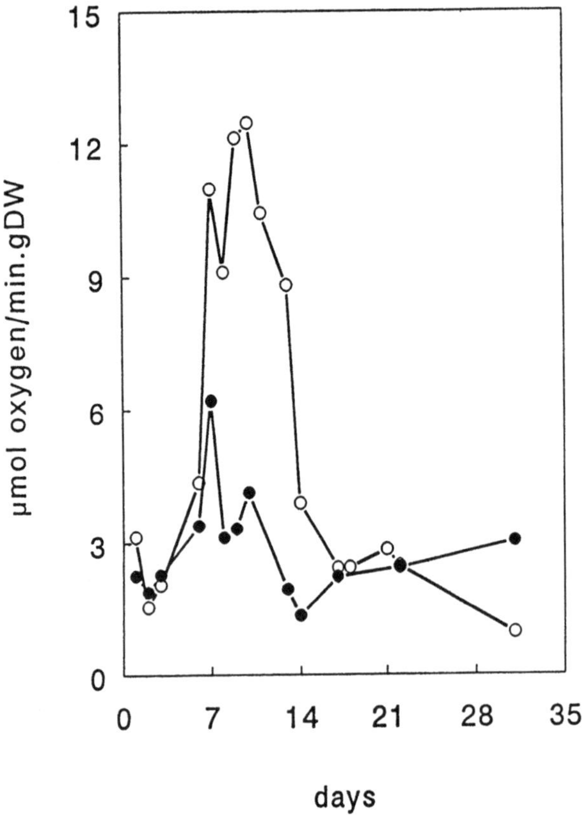

Fig. 6. Time course of total respiration of *Morinda citrifolia* cells grown in the presence of 1 mg l^{-1} NAA (closed circles) and of 1 mg l^{-1} 2,4-D (open circles). Data are expressed as μmol oxygen uptake g^{-1} DW min^{-1}. Each point represents the means of 8 measurements derived from two independent experimental series.

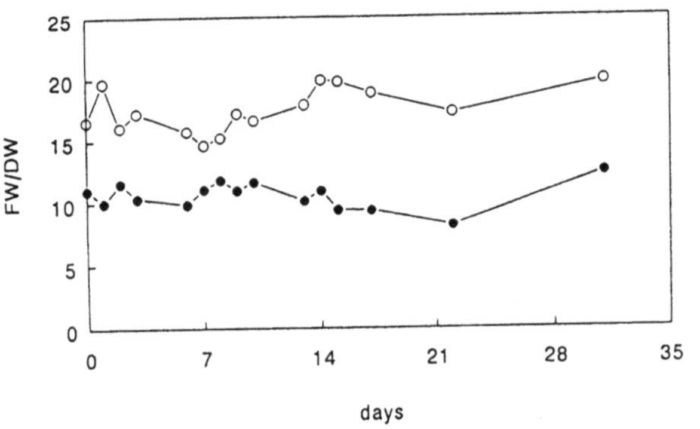

Fig. 7. Time course of the FW/DW-ratio of *Morinda citrifolia* cells grown in the presence of 1 mg l^{-1} NAA (closed circles) and of 1 mg l^{-1} 2,4-D (open circles). Measurements of two separate experimental series are depicted.

cient carbon-skeletons: a shortage of substrates for this secondary metabolite pathway might prevent an 'overflow ' in the direction of the synthesis of AQ. Such a lack of substrates might manifest itself in changes in the sugar pools and in the sugar metabolism.

The overall rate of disappearance of sugars from the medium was almost identical both for producing and non-producing cells. Both in NAA and in 2,4-D medium sucrose is hydrolyzed extracellularly; the resulting monosaccharides are taken up, with a strong preference for glucose. For many types of cell suspensions such an extracellular hydrolysis of sucrose already has been described (Fowler & Stepan-Sarkissan 1985, Callebaut et al. 1987, Dijkema et al. 1988). The enzyme invertase is responsible for this hydrolysis; it is excreted into the apoplast/cell wall and consequently is found in the growth medium of cell suspension cultures. The faster rate of sucrose hydrolysis observed for NAA cells might be caused by the larger inoculum and consequently the larger amount of invertase already present at the start of the incubation. Preferential uptake of glucose, after hydrolysis of the medium sucrose, also is a common phenomenon for plant cell cultures and has been described for a series of species, ranging from carrot and cucumber to sugar cane (Thom et al. 1981, Callebaut et al. 1987, Dijkema et al. 1988). Due to the high medium sugar concentrations (40 g l^{-1}), the carbon source is not the limiting nutrient in cell cultures of *Morinda citrifolia* : at the end of the growth phase the medium still contained 10–15 g l^{-1} sugar (almost exclusively fructose).

Although at first only glucose and later almost exclusively fructose was taken up from the medium, sucrose, glucose and fructose were all present intracellularly under both growth conditions. Apparently sucrose is resynthesized inside the cells, this storage sugar generally being present in the largest amount. However, large differences in sugar contents were observed: especially the amounts of glucose and sucrose were much smaller in the 2,4-D cells. The flow of sugar to the sugar degradation pathways may be so rapid in 2,4-D cells during the growth phase that these cells have no possibility to build up a sugar storage pool of sufficient size as long as growth continues: most of the imported sugar is immediately used. This corresponds with the situation described for bean cells by Botha et al. (1992). In NAA cells the removal of sugar from these pools to the various cellular processes in which this sugar is used is less rapid: in these cells a sugar pool of considerable size is present all the time and no large shifts are observed at the end of the growth phase. The differences between the two growth conditions become even stronger when the sugar concentrations on a fresh weight basis are taken into account: the FW/DW ratio of 2,4-D cells is almost twice as high as that of NAA cells, resulting in sugar concentrations in the vacuoles of NAA cells that are considerably higher than already could be concluded from the differences in sugar content on a DW basis.

The relatively low water content of the NAA cells in combination with the relatively high concentrations of sugars and metabolites, indicate that osmotic potential and turgor might differ under these two growth conditions. Preliminary results indeed indicate a cellular osmotic potential (at the end of the growth phase) that is twice as high in NAA cells as in 2,4-D cells and a turgor pressure for these cells that is three times as high. The reason for these differences is not clear: the higher turgor value might be caused by thicker and/or more rigid cell walls in these cells. The lower osmotic potential of the 2,4-D cells might also be caused by an increased leakage of ions and other components from the cells into the medium. From calculations carried out for e.g. *Petunia* cells it is clear that the (re)uptake of ions from the medium is an important component of the maintenance respiration of the cells (De Gucht & Van der Plas 1992; see also Van der Werf et al. 1988). An increased leakage of ions and the subsequent need for active reuptake might explain (part of) the high respiration values observed for these 2,4-D cells.

There is no direct relation between the low FW/DW-ratio of NAA-cells and their AQ-content: an increased water uptake and an adaptation of the FW/DW-ratio was also observed when 2,4-D was added to NAA grown cells with already a high AQ-content. Striking is also the relatively large amount of glucose in the NAA cells in the stationary phase when most of the storage reserves (i.e. sucrose) are already exhausted. A second sugar that was found especially in the stationary phase was a compound "X" that was virtually absent in the 2,4-D cells. This "X" most probably was a disaccharide containing xylose, presumably xylose-glucose, the sugar moiety that was described by Inoue et al. (1981) for the four most important glycosylated AQ of *Morinda citrifolia*. The presence of this compound "X" and glucose in the stationary phase possibly is connected with the synthesis of AQ-glycosides that is still conspicuous in these cells (see Fig. 2).

The general impression from these sugar data is, that the need for sugar as a source for energy and building blocks is relatively lower in the NAA cells especially in the growth phase, leading to the relatively high amounts of glucose and sucrose in these cells and the lack of effect of the cessation of growth on the amount of sugars in these cells. This result is somewhat unexpected as the growth (DW increase) of these NAA cells is larger and secondary metabolite production is only found in these cells and not in the 2,4-D cells.

A possible cause for the differences between 2,4-D and NAA cells might be either a low ATP-production on a sugar basis (due to a more inefficient sugar degradation and/or energy production in respiration) or a high need for ATP for growth and/or maintenance (due to a 'less efficient' metabolism). The contribution of the alternative pathway mediated respiration (which is energetically inefficient) to the high oxygen uptake of log phase 2,4-D cells was relatively small (results not shown). As a consequence the cytochrome pathway mediated respiration was also much higher for cells grown in 2,4-D medium when compared with cytochrome path activity of NAA cells. There was no clear difference in residual oxygen uptake (which is supposed to be extra-mitochondrial and to yield no ATP) between the two growth conditions (results not shown). Apparently, this indicates that the cause of the lower DW production and smaller sugar pools in the 2,4-D cells is not a more inefficient sugar degradation and ATP-production but a less efficient growth metabolism: these cells seem to have a higher need for sugars (energy/ATP) for growth and maintenance, for instance for an active (re)uptake of ions and other metabolites, leaked from the cells, see above. One should also bear in mind that part of the cell components of the NAA cells is relatively "cheap": after three weeks more than 30% of the DW still consist of sugars and even after 4 weeks this is about 15–20%, while these figures in 2,4-D cultures are about 15 and less than 5% respectively (Fig. 5). Also about half of the weight of the glycosylated AQ is made up of "cheap" sugars.

Conclusions

From our experiments it can be concluded that there are clearly two separate effects of auxins on the cells of *Morinda citrifolia*: an effect on growth which is the same for "NAA-like" and "2,4-D like" auxins and an effect on production of AQ that is different: production is only observed in cells growing in the presence of NAA and comparable auxins while 2,4-D-like auxins inhibit this production.

The lack of production of AQ in the 2,4-D cells might be connected with the less efficient growth metabolism: during the growth phase the sugar contents are low and the respiration is high. The need for energy might be relatively low in NAA cells because of the large amounts of "cheap" sugars and relatively high in 2,4-D cells due to possible leakage of ions and other metabolites. As a consequence, 2,4-D cells might have not enough substrate left for the production of secondary metabolites. Whether such a decrease in production is regulated by a simple overflow mechanism or whether a more specific and subtle regulatory mechanism for the adjustment of primary and secondary metabolism is involved, cannot be concluded from these experiments and will be subject for further research.

Acknowledgements

The cell cultures of *Morinda citrifolia* were a generous gift of the Division of Pharmacognosy (Center for Bio-Pharmaceutical Sciences, Leiden; prof. dr. R.Verpoorte & Dr. R. van der Heijden).

References

Botha FC, O'Kennedy MM & Du Plessis S (1992) Activity of key enzymes involved in carbohydrate metabolism in *Phaseolus vulgaris* cell suspension cultures. Plant Cell Physiol. 33: 477–483

Callebaut A, Motte JC & De Cat W (1987) Substrate utilization by embryogenic and non-embryogenic cell suspension cultures of *Cucumis sativus* L. J. Plant Physiol. 127: 271–280

De Gucht LPE & Van der Plas LHW (1992) Growth kinetics of glucose limited *Petunia hybrida* cell suspensions in continuous culture. Determination of experimental values for the costs of growth and maintenance. In: Lambers H & Van der Plas LHW (Eds) Molecular, Biochemical and Physiological Aspects of Plant Respiration (pp 619–624). SPB Academic Publishing The Hague

Dijkema C, De Vries SC, Booij H, Schaafsma TJ & Van Kammen A (1988) Substrate utilization by suspension cultures and somatic embryos of *Daucus carota* L. measured by ^{13}C NMR. Plant Physiol. 88: 1332–1337

Fowler MW & Stepan-Sarkissan G (1985) Carbohydrate source, biomass productivity and natural yield in cell suspension cultures. In: Neumann K-H, Barz W & Reinhard E (Eds) Primary and Secondary Metabolism of Plant Cell Cultures (pp 66–73). Springer-Verlag, Berlin Heidelberg

Gamborg OL, Miller RA & Ojima V (1968) Nutrient requirements of suspension cultures of soybean root cells. Exp. Cell. Res. 50: 151–158

Inoue K, Nayeshiro H, Inouye H & Zenk M (1981) Anthraquinones in cell suspension cultures of *Morinda citrifolia*. Phytochemistry 20: 1693–1700

Koblitz H (1988) Anthraquinones. In: Constabel F & Vasil IK (Eds) Cell Culture and Somatic Cell Genetics of Plants, Vol 5 (pp 113–139). Academic Press, Orlando

Leistner E (1973) Biosynthesis of moridone and alizarin in intact plants and cell suspension cultures of Morinda citrifolia. Phytochemistry 12: 1669–1674

Leistner E (1975) Isolierung, Identifizierung und Biosynthese von Anthrachinonen in Zellsuspensionskulturen von *Morinda citrifolia*. Planta Med., Suppl. pp 214–224

Leistner E (1985) Biosynthesis of chorismate derived quinones in plant cell cultures. In: Neumann K-H, Barz W & Reinhard E (Eds) Primary and Secondary Metabolism of Plant Cell Cultures (pp 215–224). Springer-Verlag, Berlin Heidelberg

Thom M, Maretzki A, Komor E & Sakai WS (1981) Nutrient uptake and accumulation by sugarcane cell cultures in relation to the growth cycle. Plant Cell Tissue Organ Cult. 1: 3–14

Ulbrich B, Wiesner W & Arens H (1985) Large-scale production of rosmarinic acid from plant cell cultures of *Coleus blumei* Benth. In: Neumann K-H, Barz W & Reinhard E (Eds) Primary and Secondary Metabolism of Plant Cell Cultures (pp 293–303). Springer-Verlag, Berlin Heidelberg

Van der Plas LHW, De Gucht LPE, Bakels RHA & Otto B (1988) Growth and respiratory characteristics of batch and continuous cell suspension cultures derived from fertile and male sterile *Petunia hybrida*. J. Plant Physiol. 130: 449–460

Van der Werf A, Kooijman A, Welschen R & Lambers H (1988) Respiratory energy costs for the maintenance of biomass, for growth and for ion uptake in roots of *Carex diandra* and *Carex acutiformis*. Physiol. Plant. 72: 483–491

Zenk MH, El Shagi H & Schulte U (1975) Anthraquinone production by cell suspension cultures of *Morinda citrifolia*. Planta Med., Suppl. pp 79–101

Zenk MH, Schulte U & El Shagi H (1984) Regulation of anthraquinone formation by phenoxyacetic acids in *Morinda citrifolia* cell cultures. Naturwissenschaften 71: 266

Plant Cell, Tissue and Organ Culture **38:** 235–240, 1994.

Calystegines as a new group of tropane alkaloids in Solanaceae

Birgit Dräger, Christoph Funck, Almut Höhler, Gabriele Mrachatz, Adolf Nahrstedt, Andreas Portsteffen, Angela Schaal & Rainer Schmidt
Institut für Pharmazeutische Biologie und Phytochemie, Hittorfstr. 56, D–48149 Münster, Germany

Key words: Calystegine, tropane alkaloid, tropinone reductase

Abstract

Calystegines are a new group of polyhydroxy alkaloids with a nortropane skeleton. They were detected in *Atropa belladonna* root cultures by chromatographic methods (TLC, GC) and identified by NMR and mass spectroscopy. Their occurrence was examined in several species of the Solanaceae. The biosynthesis of these compounds is suggested to proceed *via* the tropane alkaloid pathway, the first metabolite being pseudotropine. A pseudotropine-forming tropinone reductase was isolated and characterized from *Atropa belladonna* root cultures. Further evidence is given for the significance of tropinone and pseudotropine in calystegine formation by feeding experiments that increased calystegine formation. ^{15}N-tropinone was shown to be incorporated into calystegines.

Abbreviations: GC — gas chromatography, TBON — 8-thiabicyclo[3.2.1]octan-3-one, TLC — thin-layer chromatography

Introduction

Calystegines are a new group of alkaloids that has recently been identified from *Calystegia sepium* (Goldmann et al. 1990). The structures that were elucidated are given in Fig. 1. All calystegines have a nortropane skeleton and several hydroxy groups in various positions. The trihydroxylated calystegines are summarized as calystegine A-group, tetrahydroxy derivatives form the calystegine B-group. The hydroxy groups make the compounds hydrophilic; in usual alkaloid extraction procedures that contain a lipophilic extraction step from an aqueous alkaline medium they remain in the aqueous phase. That may explain why they have not been found earlier.

Calystegines were initially detected as silver-positive spots on paper electrophoresis that was done as test for opines in *Agrobacterium*-transformed root cultures of *Calystegia sepium.* Systematic screening of a number of intact plant organs revealed calystegines also in *Atropa belladonna* roots (Tepfer et al. 1988). Further investigations led to the assumption that calystegines have a role as nutritional mediators, i.e. they are excreted into the soil and attract bacteria that are of advantage in the rhizosphere of the plant (Tepfer et al. 1988).

Interest also arose in the aminoketal nortropane structure of calystegines and several synthetic approaches were made (Boyer et al. 1992, Duclos et al. 1992). Calystegine A_3 was synthesized as racemic mixture and enantiomerically pure (+) and (−)-calystegines B_2 were successfully prepared. The equilibrium between the open and closed ring structure was shown to be dependent on the position and the stereochemistry of the hydroxy groups, but still it is not possible to determine systematically which structures isomerize and which do not.

The calystegine structures suggest that they are formed by the tropane alkaloid pathway as depicted in Fig. 2. All calystegines found in *Calystegia sepium* have an equatorial hydroxy group at carbon 3. That is of particular interest because an enzyme activity has been found in several Solanaceae (Dräger et al. 1988, Portsteffen et al. 1992, Hashimoto et al. 1992) that reduces tropinone to form pseudotropine exclusively. Pseudotropine though does not accumulate in the root cultures nor does it isomerize to tropine, thus indicating further metabolism. Pseudotropine could be the substrate for calystegine formation as shown in Fig. 2.

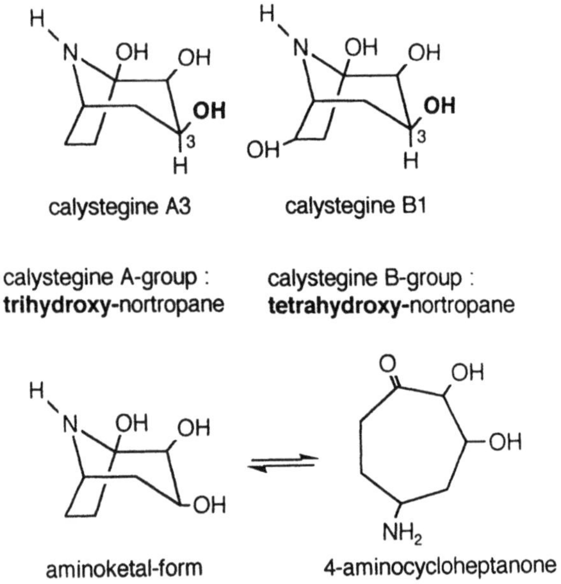

Fig. 1. Calystegine structures.

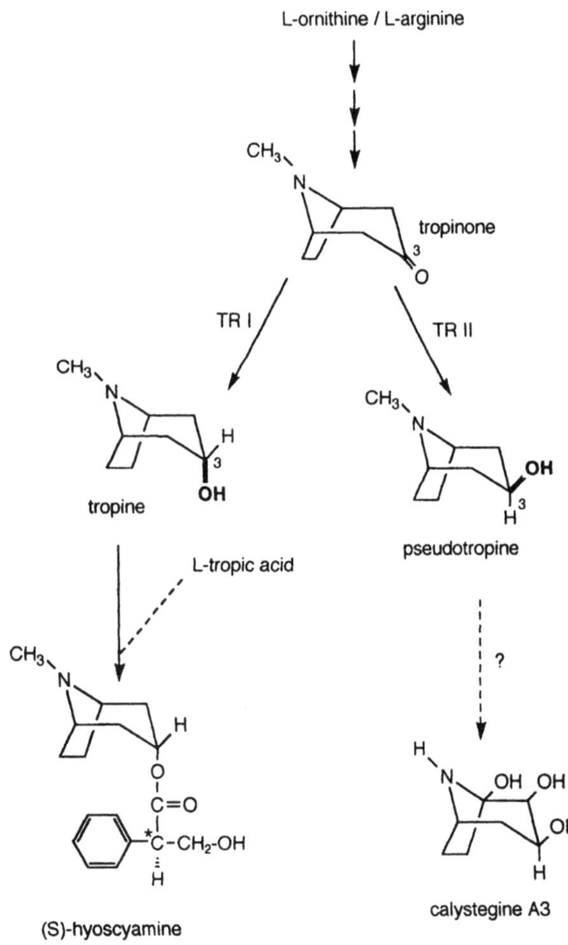

Fig. 2. Some steps of the tropane alkaloid pathway.

Results and discussion

Identification and quantification of calystegines

We tested transformed root cultures of *Atropa belladonna* by high voltage paper electrophoresis as used by Goldmann et al. (1990) and found compounds with similar behavior as the calystegines. With calystegine A_3 and a mixture of calystegine B_1 and B_2 as reference compounds (kindly provided by Dr. Goldmann) we established TLC systems on silica gel for calystegines and a GC system for the trimethylsilyl ethers. The compounds from *Atropa belladonna* root cultures showed the same behavior as the original calystegines in all chromatographic systems. To assess the identity we isolated the calystegine fraction from 500 g cultured *A. belladonna*-roots and purified calystegines by ion exchange and partition chromatography. We separated calystegine A from calystegine B. The calystegine A-fraction contained 85% calystegine A_3, the calystegine B-fraction contained a mixture of tetrahydroxy-isomers. The mass spectrum with chemical ionization (MS-CI, ammonia as reactant gas) showed one single mass of 160 (M + 1) for the calystegine A-fraction. The calystegine B-fraction showed a base peak at 176 (M + 1) and a minor signal (ca. 14%) with the mass 158 (M + 1$-H_2O$). ^{1}H-NMR and ^{13}C-NMR analysis of fraction A gave the same signals as published by Goldmann et al. (1990). In contrast to what was found with *Calystegia sepium* calystegine A_3 is the major compound here and calystegines B occur in smaller quantities.

Derivatisation is necessary for GC of calystegines. Trimethylsilyl ethers are formed with hexamethyldisilazane and trichlorosilane at room temperature. Under these mild conditions only the hydroxy groups react, the secondary amine group remains free. Stronger silylating agents also affect the bridgehead nitrogen which then probably partially cleaves off, the result is always a mixture of derivatives. Detection was done by flame ionization (FID) and nitrogen-sensitive thermionic detection (PND). For quantitative analysis the tissue extracts were purified by cation exchange on Dowex 50-WX8 resin before GC assay to provide cleaner samples with better separation. Quantification was done with octadecane as inner standard for FID and desoxynojirimycin for PND.

We determined the accumulation in the cultured roots over a 16-day-period (Fig. 3) and found a pattern frequently seen with secondary products, highest contents were measured in the late growth phase. The

Table 1. Accumulation of calystegine A_3 in cultured roots and intact plant tissues of *Solanaceae*.

Plant	Organ	Cal. A_3 content $\mu g\ g^{-1}$ FW
Atropa belladonna	cultured roots	55–60
	intact roots	20–30
	leaves	40–60
	fruits with seeds	6
Hyoscyamus niger	cultured roots	100–120
	intact roots	22
	leaves	2–4
Hyoscyamus albus	cultured roots	30–160
Hyoscyamus muticus	cultured roots	112
Hyoscyamus pusillus	cultured roots	43
Hyoscyamus aureus	cultured roots	79
Calystegia sepium	cultured roots	59
(Convolvulaceae)	leaves	10
Datura stramonium	cultured roots	no calystegines detected
Lycopersicum esculentum	leaves	no calystegines detected
Solanum melongena	intact roots	no calystegines detected
Solanum nigrum	roots, fruits, leaf	no calystegines detected

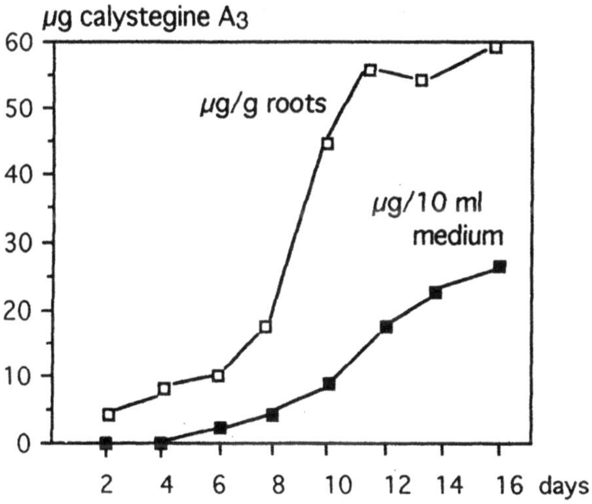

Fig. 3. Accumulation of calystegine A_3 in cultured roots and culture medium of *Atropa belladonna*.

culture medium was examined because Tepfer et al. (1988) postulated a role of the calystegines in the rhizosphere which would require excretion. Calystegines were found in the culture medium, but only in small amounts (Fig. 3). We do not yet know whether they are excreted or washed out from lysed root cells.

Several root cultures and intact plant organs of *Solanaceae* were examined for calystegines by TLC and GC (Table 1). Calystegines were found in the genera *Atropa* and *Hyoscyamus*, apart from their presence in *Calystegia*, *Convolvulaceae*, systematically closely related to *Solanaceae* (order '*Solanales*', Cronquist 1988). A large number of *Solanaceae* still waits to be examined. Apparently the calystegines occur not only in the root but also in the leaves. The values are given for calystegine A_3 only, but from comparison with authentic calystegine B we found that in *Hyoscyamus albus* the content of calystegines of the B-group is even higher.

Evidence for calystegine formation via the tropane alkaloid pathway

In the tropane alkaloid pathway tropinone reduction is a branching point (Fig. 2) and pseudotropine could be the first specific precursor for calystegines. We isolated and characterized the pseudotropine forming tropinone reductase from *Atropa belladonna* root cultures (Dräger & Schaal 1994). The purification is outlined in Table 2. Molecular weight analysis by gel filtration showed 75000 ± 3000 Da for the intact enzyme. With chromatofocussing the enzyme showed an iso-

Table 2. Purification steps of pseudotropine forming tropinone reductase from *Atropa belladonna*.

	Protein	Activity	Specific activity	Activity enrichment
crude protein	99.7 mg	18.6 nkat	186.7 μkat kg^{-1}	1
40%–75% $(NH_4)_2SO_4$-satn.	45.0 mg	15.1 nkat	335.4 μkat kg^{-1}	1.8
after Butyl-HIC	2.3 mg	4.7 nkat	2039.4 μkat kg^{-1}	10.9
after TMAE-anion exchange	232.0 μg	4.1 nkat	17472.0 μkat kg^{-1}	93.6

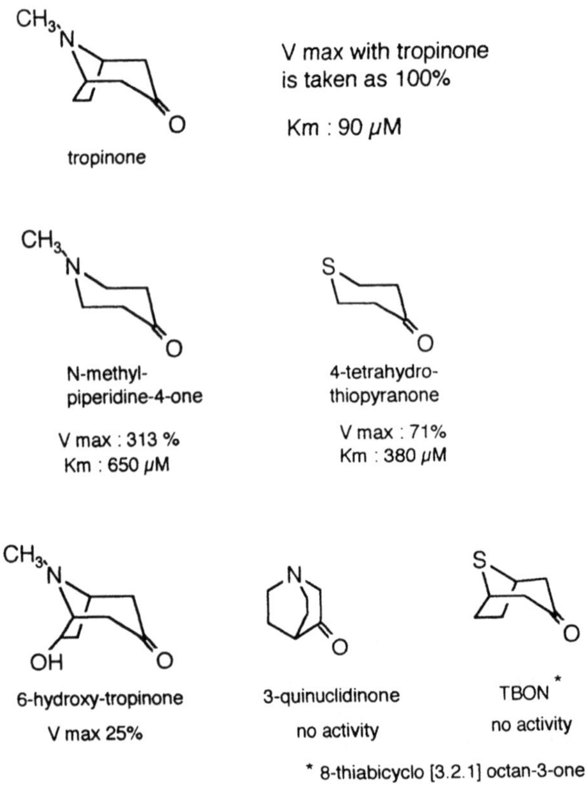

Fig. 4. Substrate analogues for pseudotropine forming tropinone reductase from *Atropa belladonna*.

electric point at pH 5.2. The enzyme is stable for 5–6 months when frozen at −20 °C, provided the buffer contains 20%–30% glycerol and SH-group protecting agents. With gel filtration as well as with chromatofocussing the activity strongly decays indicating that at the IEP as well as with the omission of glycerol which is necessary for gel filtration, the purified enzyme becomes unstable. The pH-optimum of the reduction is at pH 6.0. Tropinone reductase is quite specific for tropinone. We tested a number of similar compounds (Fig. 4). N-methylpiperidin-4-one is the only substrate that we found to be reduced faster than tropinone, but it shows a higher K_m-value indicating lower affinity to the active site. None of the compounds with structures similar to tropinone was inhibitory to tropinone reduction. In contrast, TBON (8-thiabicyclo[3.2.1]octan-3-one), the sulphur analogue of tropinone, is a potent inhibitor of tropine forming tropinone reductase *in vivo* and *in vitro* (Parr et al. 1991, Dräger et al. 1992). NADPH is the only possible cosubstrate, with NADH the enzyme is inactive. The Km-value for NADPH is 20.8 μM.

Comparing this enzyme from *Atropa belladonna* root cultures with other pseudotropine forming tropinone reductases we find similar properties as the enzyme described from *Hyoscyamus niger* (Dräger et al. 1988, Hashimoto et al. 1992), a plant that also accumulates calystegines. But it should be mentioned that pseudotropine forming tropinone reductase has also been found in *Datura stramonium* root cultures (Portsteffen et al. 1992), where no calystegines could be detected.

Goldmann et al. (1990) showed already that putrescine, an early metabolite of the tropane alkaloid biosynthesis, is incorporated into calystegines indicating that in fact the tropane alkaloid pathway is involved. To approach to the putative branching point between calystegine formation and hyoscyamine formation we applied TBON, an inhibitor of the tropine forming tropinone reductase, to 4-day-old root cultures of *Atropa belladonna*. Thus more tropinone should be available for the formation of pseudotropine and eventually calystegines. The results (Fig. 5) show the expected decrease in hyoscyamine content and also an increase in calystegine accumulation. The experiment was corroborated by feeding pseudotropine alone. Addition of 1 mM or 2 mM pseudotropine to 7-day-old cultures resulted in 160–180 μg calystegine A_3 g^{-1} FW after 14 days of growth (control: 53 μg g^{-1} FW). These findings strongly suggest that there is

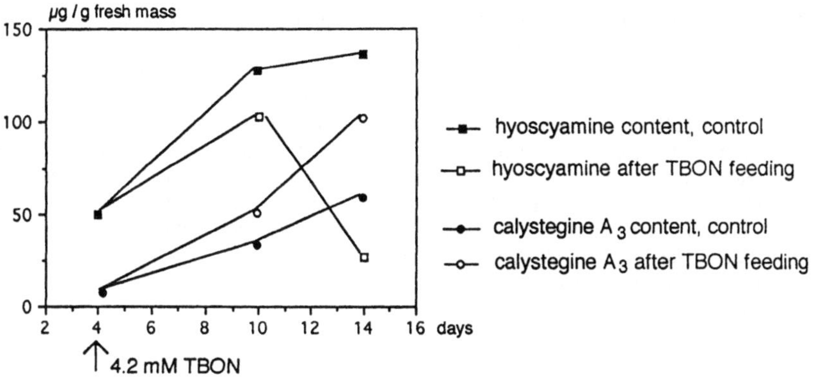

Fig. 5. Effect of TBON on accumulation of hyoscyamine and calystegine A_3 in root cultures of *Atropa belladonna*.

a biosynthetic connection between pseudotropine and calystegines.

The classical approach for the elucidation of biosynthetic steps in secondary metabolism is the feeding of labelled precursors. We applied ^{15}N-tropinone which was synthesized by Henry D. Boswell, Glasgow, and was a kind gift from Dr. Richard Robins, Norwich. To achieve high incorporation of the ^{15}N-tropinone into the calystegines we blocked tropine formation by simultaneous feeding of TBON. 7-day-old cultures were fed and after 7 days the calystegines were worked up by ion exchange chromatography and partition chromatography. With mass spectroscopy with chemical ionization the calystegine A-fraction showed only one set of signals at 160/161 indicating that there were only calystegine A or isomers thereof in the fraction. Signal 161 (M + 1 + 1, deriving from ^{15}N and natural abundance of ^{13}C) was enriched compared with signal 161 in calystegine A-fractions from untreated cultures (M + 1 + 1, deriving only from natural abundance of ^{13}C). Thus it can be concluded that tropinone is incorporated into calystegines.

Future prospects

Chemotaxonomical and ecological questions concerning the calystegines wait to be answered. A systematic screening of all tropane alkaloid containing plants and all *Solanaceae* and related families will enable statements concerning the co-occurrence of calystegines with other tropane alkaloids. Tepfer et al. (1988) have already described one role of the calystegines as nutritional mediators in the rhizosphere. The occurrence of the calystegines in leaves and fruits prompts to think of further significance. Tropane alkaloids are known to be transported from the root to the leaves. Transport and storage also need to be investigated for the calystegines. The suggested biosynthesis *via* tropinone and pseudotropine needs further proofs. The next steps will be the search for the enzymes that demethylate and hydroxylate pseudotropine to form calystegines.

Acknowledgements

We thank R. Hahn, Institute for Pharmaceutical Biology, Münster, for helpful hints with the MLCCC system and Dr. J. Lauterwein, Institute of Organic Chemistry, Münster, for recording the NMR spectra and help with the interpretation. Further we wish to thank Dr. A. Goldmann, INRA, Versailles, for samples of calystegine A_3 and calystegine B, and Dr. R. J. Robins, IFR, Norwich, for TBON and ^{15}N-tropinone. B.D. acknowledges financial support of the Deutsche Forschungsgemeinschaft.

References

Boyer FD, Ducrot PH, Henryon V, Soulié & Lallemand JY (1992) Studies on the synthesis of the 1-hydroxynortropane system; part II. Synthesis of calystegine A_3 and physoperuvine. Synlett 357–359

Cronquist A (1988) The evolution and classification of flowering plants. The new Botanical Garden, Bronx NY, pp 420

Dräger B, Hashimoto T & Yamada Y (1988) Purification and characterization of pseudotropine forming tropinone reductase from *Hyoscyamus niger* root cultures. Agric. Biol. Chem. 52: 2663–2667

Dräger B, Portsteffen A, Schaal A, McCabe PH, Peerless ACJ & Robins RJ (1992) Levels of tropinone reductase activities influence the spectrum of tropane esters found in transformed root cultures of *Datura stramonium*. L. Planta 188: 581–586

Dräger B & Schaal A (1994) Tropinone reduction in *Atropa belladonna* root cultures. Phytochemistry 35: 1441–1447

Duclos O, Mondange M, Duréault A & Depezay JC (1992) Polyhydroxylated nortropanes starting from D-glucose: Synthesis of homochiral (+) and (−)-calystegines B_2. Tetrahedron Lett. 33: 8061–8064

Goldmann A, Milat ML, Ducrot PH, Lallamand JY, Maille M, Lepingle A, Charpin I & Tepfer D (1990) Tropane derivatives from *Calystegia sepium*. Phytochemistry 29: 2125–2127

Hashimoto T, Nakajima K, Ongena G & Yamada Y (1992) Two tropinone reductases with distinct stereospecificities from cultured roots of *Hyoscyamus niger*. Plant Physiol. 100: 836–845

Parr AJ, Walton NJ, Bensalem S, McCabe P & Routledge W (1991) 8-Thiabicyclo[3.2.1]octan-3-one as a biochemical tool in the study of tropane alkaloid biosynthesis. Phytochemistry 30: 2607–2609

Portsteffen A, Dräger B & Nahrstedt A. (1992) Two tropinone reducing enzymes from *Datura stramonium* transformed root cultures. Phytochemistry 31: 1135–1138

Tepfer D, Goldmann A, Pamboukdjian N, Maille M, Lepingle A, Chevalier D, Dénarié J & Rosenberg C (1988) A plasmid of *Rhizobium meliloti* 41 encodes catabolism of two compounds from root exudate of *Calystegia sepium*. J. Bacteriol. 170: 1153–1161

Plant Cell, Tissue and Organ Culture **38:** 241–247, 1994.

Esterification reactions in the biosynthesis of tropane alkaloids in transformed root cultures

Richard J Robins[1], Peter Bachmann[3], Abigael C J Peerless[1] & Sylvie Rabot[2]
[1] *Institute of Food Research, Norwich Laboratory, Norwich Research Park, Colney, Norwich NR4 7UA, UK*
[2] *UEPSD, Bâtiment 440, INRA, Centre de Recherches de Jouy, F-78352 Jouy-en-Josas Cedex, France*
[3] *P. Bachmann, Institut für Pharmazeutische Biologie, Mendelssohnstrasse 1, D-38106 Braunschweig, Germany.*

*Key words:*Acryl-coenzyme A transferase, biosynthesis, coenzyme A thioester, *Datura*, pseudotropine, transformed root cultures, tropane alkaloids, tropine

Abstract

Transformed root cultures of *Datura stramonium* and of related species contain both aliphatic and aromatic tropane esters. It has been shown that these esters are produced by the action of several acyl transferases that transfer the acidic moiety to tropan-3α-ol (tropine) or tropan-3β-ol (pseudotropine) from various acyl-coenzyme A thioesters. The presence of these enzymes has been examined in a range of tropane-alkaloid-producing and non-tropane-alkaloid-producing species. Activities that esterify tropine appear to be confined to species that accumulate tropane alkaloids, whereas a number of species that do not accumulate tropane alkaloids possess some ability to esterify pseudotropine.

The present state of knowledge of these enzymes is reviewed. One of these activities, tigloyl-Coenzyme A:pseudotropine acyl transferase, has been purified to near homogeneity and the properties of this enzyme are summarized.

Abbreviations: CoA – coenzyme A, gc – gas chromatography, ms – mass spectrometry

Introduction

Datura stramonium and several other solanaceous genera produce a wide range of alkaloids that are esters of tropine (tropan-3α-ol) or pseudotropine (tropan-3β-ol) (Lounasmaa 1988). These isomeric tropan-3-ols are synthesised by the stereospecific reduction of tropinone by two separate enzymes (Portsteffen et al. 1992; Hashimoto et al. 1992). Tropinone is formed from putrescine, the first step of the alkaloid pathway (Fig. 1) being the *N*-methylation of this diamine by putrescine *N*-methyltransferase (Walton et al. 1994). The major tropane alkaloids of pharmaceutical importance (Evans 1990) are hyoscyamine and scopolamine (Fig. 2), in both of which tropine is esterified with a tropic acid moiety derived from phenylalanine. The levels of accumulation of the different alkaloids in root cultures vary considerably. Some species accumulate as end products compounds that in other species are further metabolized (Parr et al. 1990; Parr 1992). In most species the spectrum of alkaloids is dominated by tropine-derived esters but in a few cases pseudotropine-derived esters can be dominant (Beresford & Woolley 1974).

Datura root cultures of a number of species produce hyoscyamine as the major alkaloid (Parr et al. 1990). Among the minor bases present, 3α-acetoxyltropane (acetyltropine) and 3α-tigloxyltropane (tiglyltropine) are usually found. If tropine is fed to cultures of *D. stramonium*, acetyltropine accumulates to a high level (Robins et al. 1991a). When pseudotropine is fed, the level of pseudotropine increases very substantially and some esters of pseudotropine, which are hardly detectable in control tissue, accumulate (Dräger et al. 1992). Feeding tropinone, the direct precursor of tropine and pseudotropine (Fig. 1) leads to some increase in both groups of products but the tropine-derived group predominates.

Recently, it has been shown that the enzyme activities responsible for forming esters of tropine and pseudotropine are acyl-CoA-dependent acyl transferases (Robins et al. 1991b; Rabot & Robins 1992). This paper reviews the present state of work in this area.

Esterifying activities examined

Work has concentrated on six activities:

Acyl transferase activity	Product (Fig. 2)
acetyl-CoA:tropine	acetyltropine
acetyl-CoA:pseudotropine	acetylpseudotropine
tigloyl-CoA:tropine	tiglyltropine
tigloyl-CoA:pseudotropine	tiglylpseudotropine
phenylacetoyl-CoA:tropine	phenylacetoyltropine
phenylacetoyl-CoA:pseudotropine	phenylacetoylpseudotropine

These activities have been examined in the 20–80% (w/v) saturated ammonium sulphate fraction of a number of species. Activity was assayed by incubating a suitable amount of desalted extract in buffer at pH 9.0 (200 Mm glycine) with tropine (1.4 mM) or pseudotropine (1.4 mM) and the appropriate CoA thioester (1.3 Mm). The formation of product was determined by gc and gc/ms as described in Dräger et al. (1992).

Distribution in plant species

The levels of the six acyl transferase activities of interest were tested in 10-day-old tissue from a range of species (Table 1). The cultures were all grown and sub-cultured essentially as described for *D. stramonium* in Robins et al. (1991a). In almost all cases, the ability to form acetyl-, tiglyl- and phenylacetoyl-esters of pseudotropine was found to be higher than the comparable activities with tropine as acyl acceptor. The ratio of the different activities within a species varied considerably. Also, the extractable activity from different species showed a wide range of values.

The levels of the appropriate esters of tropine and pseudotropine in tissue of the same age are indicated in Table 2. As can be seen, there is a moderate correlation between the ability of species to accumulate the aliphatic esters and the levels of the acetyltropine- and tiglyltropine-forming activities. Thus, *D. stramonium, D. wrightii, D. innoxia* and the *Brugmansia* hybrid all possess the acetyl-CoA and tigloyl-CoA transferases that esterify tropine and accumulate acetyltropine and tiglyltropine. Similarly, *N. physaloides, S. tuberosum, N. tabacum* and *V. officinalis* have neither the esterifying activities nor these products. *H. muticus* and *A. belladonna* show inconsistent behaviour in apparently having no activity yet accumulating the alkaloidal esters. The esterification of tropine with phenylacetoyl-CoA shows a comparable correlation with the accumulation of hyoscyamine and other aromatic esters (data not shown). The level of activity of this enzyme is very low in all species, however, even though hyoscyamine is the major product accumulated.

The ability to form pseudotropine esters shows little correlation with the accumulation of these products. Many of the cultures examined accumulate no appreciable quantities of pseudotropine esters. Even when fed pseudotropine or high levels of tropinone, esters of pseudotropine accumulate to a very much smaller extent than might be predicted on the basis of the level of acyl transferase activities that can be determined

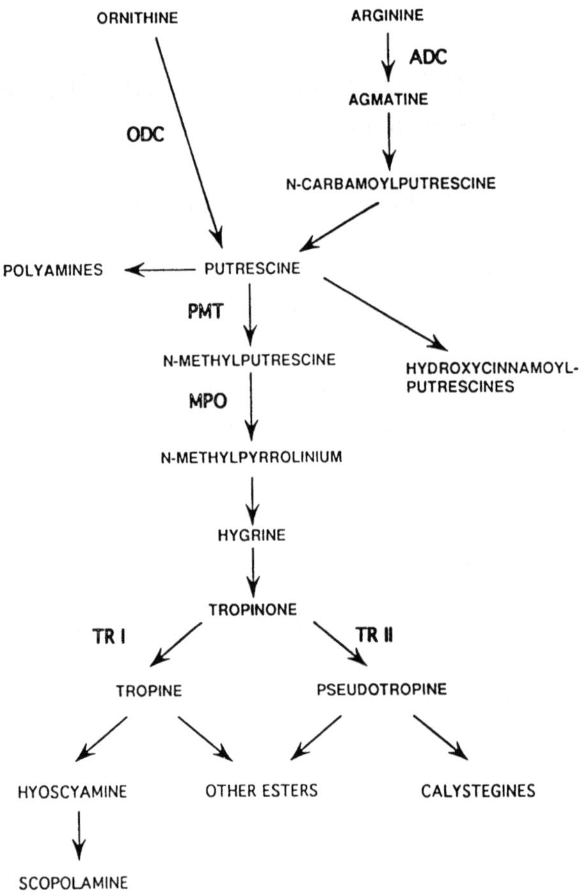

Fig. 1. Schematic pathway showing the route by which tropane alkaloids are biosynthesized.The indicated enzyme activities are: ODC, ornithine decarboxylase; ADC arginine decarboxylase; PMT, putrescine *N*-methyltransferase; MPO, *N*-methylputrescine oxidase; TR I, tropine-forming tropinone reductase; TR II, pseudotropine-forming tropinone reductase.

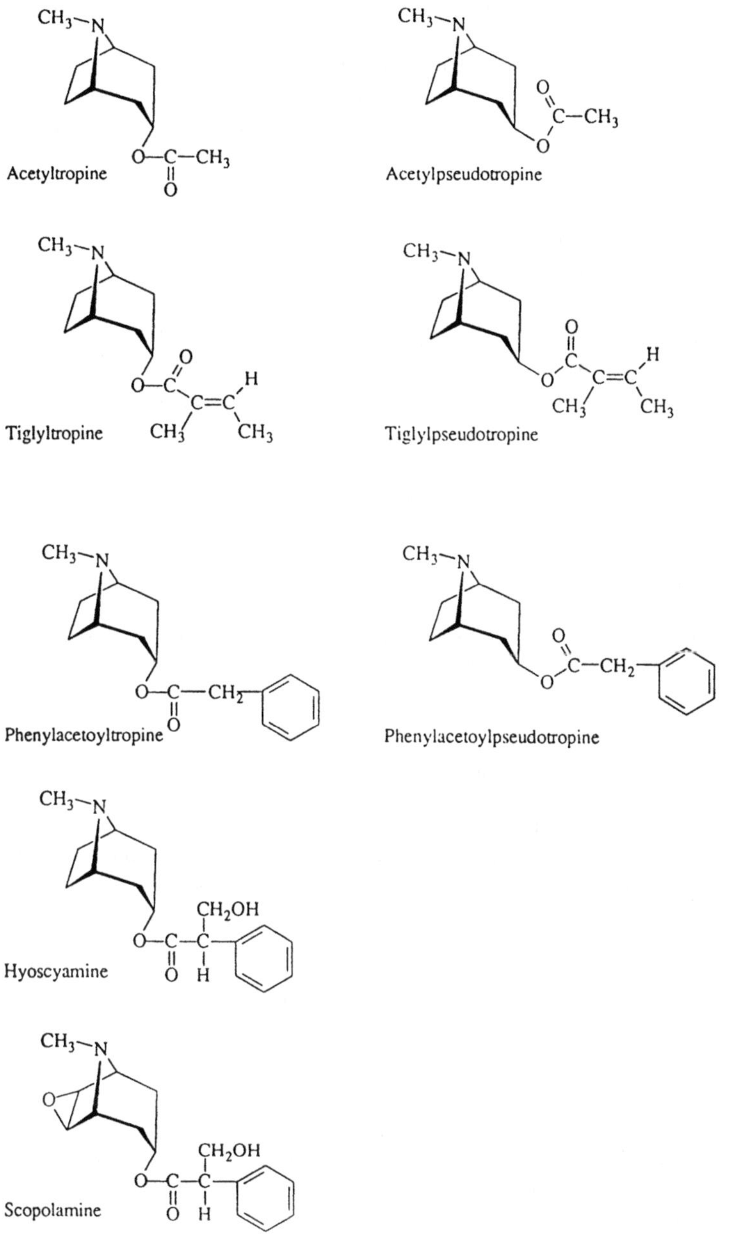

Fig. 2. Structures of the alkaloids made by the various ester-forming activities described and of hyoscyamine and scopolamine.

in vitro. These activities with pseudotropine are also found in a number of species that do not form tropane alkaloids at all, bringing into question their significance for tropane metabolism in general. Nevertheless, as illustrated below, the tigloyl-CoA:pseudotropine acyl transferase activity shows properties compatible with its being specific to this group of products.

Separation of the acyl transferase activities

Fig. 3 shows schematically how the acyl transferase activities in a crude extract of *D. stramonium* roots can be separated into a number of fractions. The activities that acylate tropine or pseudotropine with acetyl-CoA and tigloyl-CoA thioesters can be completely separated by ammonium sulphate fractional precipitation followed by anion-exchange chromatography (Robins et al. 1991b). The phenylacetoyl-CoA:pseudotropine

Table 1. Levels of tropine- and pseudotropine-utilizing acyl-coenzyme A acyl transferase activities in various species.

Species[a]	Acyl transferase activity[b] (pkat mg protein^{-1}) forming: Acetyl-tropine	Acetyl-pseudotropine	Tiglyi-tropine	Tiglyl-pseudotropine	Phenylacetoyl-tropine	Phenylacetoyl-pseudotropine
Atropa belladonna	0.0	1.1	0.0	68.3	0.00	5.4
Brugmansia candida						
X *B. aurea* DB5	4.0	14.3	1.4	70.1	0.22	0.2
Datura innoxia	29.9	82.1	7.4	242.1	0.13	55.2
Datura quercifolia D101	0.0	0.0	0.5	8.9	0.08	2.8
Datura stramonium D15/5	7.1	32.9	4.0	273.3	0.08	37.4
Datura wrightii D111/3	56.1	77.1	12.6	318.6	1.86	83.0
Hyoscyamus muticus H61/21	0.0	21.6	0.0	100.3	0.00	2.0
Nicandra physaloides NP2/2	0.0	0.0	0.0	87.6	0.00	0.0
Nicotiana tabacum SC58	0.0	0.0	0.0	17.0	0.00	0.0
Solanum tuberosum	0.0	0.0	0.0	2.6	0.00	0.0
Valeriana officinalis	0.0	0.0	0.0	3.7	0.00	0.0

[a] Root cultures of these species were grown in the same conditions as described for *D. stramonium* in Robins et al. (1991a) and were harvested at 10 days from sub-culture

[b] The activity was determined (see text) using appropriate volumes of a 20 to 80% saturated ammonium sulphate cut, following de-salting on a PD-10 column pre-equillibrated with 50 mM potassium phosphate buffer (pH 7.0) containing EDTA (5 mM), DTT (3 mM) and sucrose (100 mM).

Table 2. Levels of tropine alkaloids in ten-day-old cultures of various species.

Species[a]	Alkaloid[b] (nmol g FM^{-1}) Tri	Pstri	AcTri	AcPstri	TigTri	TigPstri	PhacTri	PhacPstri
Atropa belladonna	325	98	0	271	208	0	0	0
Brugmansia candida								
X *B. aurea* DB5	56	13	105	3	122	44	<1	2
Datura innoxia D44/103	95	6	157	4	20	7	0	0
Datura quercifolia D101	116	8	186	13	31	13	0	0
Datura stramonium D15/5	118	3	97	0	49	0	15	0
Datura wrightii D111/3	44	3	512	11	98	53	<1	2
Hyoscyamus muticus H61/21	2	91	<1	76	17	<1	0	0
Nicotiana tabacum SC58[c]	0	0	0	0	0	0	0	0
Nicandra physaloides NP2/2	<1	0	<1	2	16	0	0	0
Solanum tuberosum[c]	0	0	0	0	0	0	0	0
Valeriana officinalis[c]	0	0	0	0	0	0	0	

[a] Root cultures of these species were grown in the same conditions as described for *D. stramonium* in Robins et al. (1991a) and were harvested at 10 days from sub-culture

[b] The alkaloid level was determined as described in Robins et al. (1991a). Abbreviations used are: Tri=tropine; Pstri=pseudotropine; AcTri=acetyltropine; AcPstri=acetylpseudotropine; TigTri=tiglyltropine; TigPstri=tiglypseudotropine; PhacTri=phenylacetoyltropine; PhacPstri=phenylacetoylpseudotropine.

[c] Some minor peaks in gc confirmed not to be these alkaloids by gc/ms.

acyl transferase activity is largely separated from the tigloyl-CoA:pseudotropine acyl transferase activity by hydrophobic interaction chromatography. All the ability to esterify tropine with phenylacetoyl-CoA is lost during any of the concentration and purification procedures so far attempted (Bachmann P & Robins R J, unpublished).

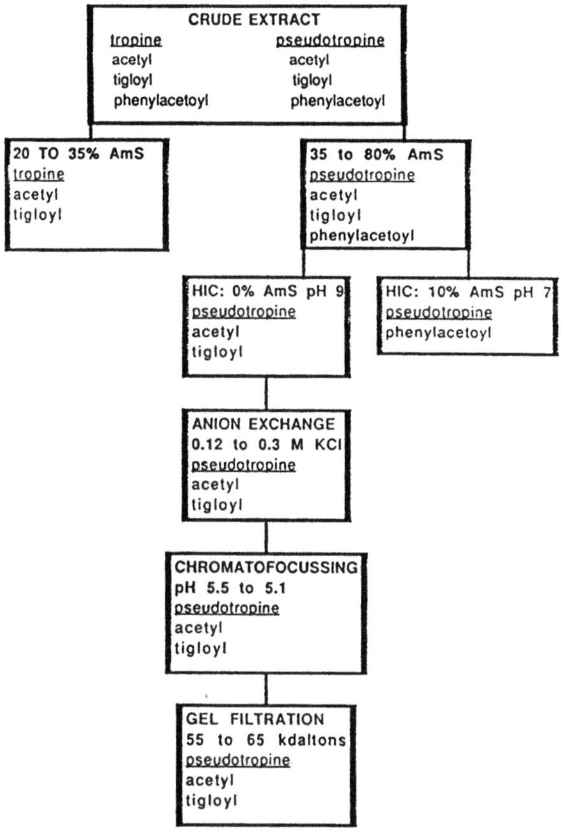

Fig. 3. Separation scheme illustrating the steps involved in fractionating the various acyl-CoA acyl transferases.

Purification and properties of the tigloyl-CoA:pseudotropine acyl transferase activity from *D. stramonium*

Of the four activities cited above, only one, tigloyl-CoA:pseudotropine acyl transferase, has been purified to near homogeneity using root tissue grown in 15 l fermenters essentially as described by Hilton and Rhodes (1990). This enzyme can form both acetylpseudotropine and tiglylpseudotropine from pseudotropine. The activities forming these two products have been co- purified about 300-fold (Fig. 3) by ammonium sulphate precipitation, hydrophobic- interaction, anion-exchange, chromatofocusing and gel-filtration chromatography (Rabot & Robins 1992; Rabot S & Robins R J unpublished). The properties of this enzyme are summarized in Table 3. Since the two activities purify in parallel and show competitive kinetics for tigloyl-CoA and acetyl-CoA, they are presumed to be properties of the same enzyme.

The enzyme shows quite different specificity characteristics for its two co-substrates. It can transfer an acyl group to pseudotropine from a range of saturated and unsaturated CoA thioesters, including tigloyl- (100%), *iso*-butyroyl- (10%), β-methylcrotonoyl- (10%), *n*-propionoyl- (5%), acetyl- (4%) and *iso*-valeroyl-CoA (2%). No activity was found with acetoacetyl-CoA or crotonoyl-CoA. In contrast, the enzyme is very specific for the acyl-group acceptor. Of a wide range of compounds used, the enzyme only catalyzed the tigloylation of pseudotropine (100%), and 4-hydroxy-1-methyl-piperidine (14%). No activity was found with tropine, indicating that a β-hydroxyl group is required, or with norpseudotropine or 4-hydroxypiperidine, suggesting that the presence of a *N*-methyl group is necessary (Rabot S & Robins RJ unpublished). Thus, it appears that the enzyme is specific for the alkaloid acceptor but can transfer an acyl moiety to pseudotropine from a range of acyl-CoA thioester donors. Similar properties have been described for quinolizidine-alkaloid-tigloylating enzymes identified in crude extracts of *Lupinus* species (Wink & Hartmann 1982; Saito et al. 1993).

No attempt has yet been made to assess the specificity of the tigloyl-CoA:pseudotropine acyl transferase from other species. It may be that the low levels of transferase activity found in the genera examined that do not accumulate tropane alkaloids, such as *S. tuberosum* and *N. tabacum*, represent a minor activity of a non-specific enzyme.

Activity with pseudotropine and phenylacetoyl-CoA is clearly a distinct enzyme as the two activities can be largely separated by hydrophobic-interaction chromatography. Furthermore, the purified tigloyl-CoA:pseudotropine acyl transferase is inactive with phenylacetoyl-CoA.

Conclusions

Several activities are present in root extracts of *Datura* and related genera. So far, only one of these, the tigloyl-CoA:pseudotropine acyl transferase, has been characterized in any detail.Whether the acetyl-CoA:tropine acyl transferase and the phenylacetoyl-CoA:pseudotropine acyl transferase will show similar characteristics - accepting a narrow range of acyl group acceptors and a broad range of acyl donors - has yet to be established. It seems reasonable to suggest, however, that the wide spectrum of minor bases observed in the alkaloidal extracts of these roots is simply due to the low selectivity of the enzymes for acyl-CoA thioesters. Hence, the relative availability of different

Table 3. Properties of the tigloyl-CoA:pseudotropine acyl transferase isolated from *D. stramonium* root cultures.

Purification		c. 300-fold
Ratio tigloyl:acetyl transferase	-crude	19.78
	-100-fold purified	23.81
Apparent K_m values	-pseudotropine	0.34 mM
	-acetyl-CoA	0.36 mM
	-tigloyl-CoA	1.31 mM
pH for maximal activity		9–10
Substrate specificity	-alcohol acceptor	high
	-acyl donor	low
Kinetics	-tigloyl-CoA versus acetyl-CoA	competitive
	-acetyl-CoA versus tigloyl-CoA	competitive

CoA thioesters may be an important factor in determining the extent to which each product accumulates. This suggestion is indirectly supported by the data obtained from cultures to which tropinone or tropine has been fed, in which acetyltropine, derived from the most freely-available thioester, acetyl-CoA, accumulates to a high level (Robins et al. 1991a).

In contrast, the role *in vivo* of the pseudotropine-acylating activities is not clear. Only very low levels of pseudotropine esters accumulate in any of the species examined, yet extracts of several of these species show a high potential for forming these products from pseudotropine *in vitro*. In *D. stramonium*, the high extractable activity of tropinone reductase I relative to tropinone reductase II (Portsteffen et al. 1992) correlates well with the observation that only a small amount of pseudotropine accumulates (Dräger et al. 1992). This is not the only reason that pseudotropine esters do not accumulate, however, since, when pseudotropine is fed to cultures (Dräger et al. 1992) the acetyl and tigloyl esters do not accumulate to nearly as high concentrations as that reached by acetyltropine when tropine is fed (Robins et al. 1991a).

Phenylacetoyltropine is a natural minor component of the alkaloid spectrum of several tropane-alkaloid-producing species (Table 2). It has been demonstrated that this alkaloid can be made *in vitro* from phenylacetoyl-CoA and tropine by extracts of *D. stramonium* roots. We originally argued that this esterification might represent a minor action of a putative enzyme making hyoscyamine from tropoyl-CoA and tropine. The phenylacetoyl-CoA:tropine acyl transferase activity is, however, present at only very low levels and has proved intransigent to purification or concentration. Furthermore, we and other workers have failed to identify any tropoyl-CoA:tropine acyl transferase activity (for a review see Robins & Walton 1994). Recently, evidence that littorine is formed prior to hyoscyamine has been presented (Robins et al. 1994a, b). This would suggest that a phenyllactoyl-CoA:tropine acyl transferase should be present. No such activity has yet been detected, however (Robinson T & Robins R J unpublished). Clearly the formation of hyoscyamine from tropine requires considerably more investigation.

Acknowledgements

We wish to thank the following: Dr P McCabe (Glasgow, UK) for supplies of substrates and substrate analogues; Professor T Robinson (Amherst, USA) for the synthesis of acyl-CoA thioesters; Chris Waspe and Dr P Wilson (IFR) for the fermenter-grown cultures; John Eagles (IFR) for the gc/ms analyses; our colleagues Drs M J C Rhodes, A J Parr and N J Walton for much advice and discussion; Dr B Dräger (Münster, Germany) for discussions and critically reading the manuscript. Dr S Rabot gratefully acknowledges a 6-month INRA/AFRC Fellowship at IFR(N).

References

Beresford, P. J. & Woolley, J. G. (1974) Biosynthesis of tigloidine in *Physalis peruviana*. Phytochemistry 13: 2143–2144

Dräger, B., Portsteffen, A., Schaal, A., McCabe, P. H., Peerless, A. C. J. & Robins, R. J. (1992) Levels of tropinone-reductase activities influence the spectrum of tropane esters found in transformed root cultures of *Datura stramonium* L. Planta 188: 581–586

Evans, W. C. (1990) Datura, a commercial source of hyoscine. Pharm. J. 244: 651–653

Hashimoto, T., Nakajima, K., Ongena, G. & Yamada, Y. (1992) Two tropinone reductases with distinct stereospecificities from cultured roots of *Hyoscyamus niger*. Plant Physiol. 100: 836–845

Hilton, M. G. & Rhodes, M. J. C. (1990) Growth and hyoscyamine production of 'hairy root' cultures of *Datura stramonium* in a modified stirred tank reactor. Appl. Microbiol. Biotechnol. 33: 132–138

Lounasmaa, M. (1988) The tropane alkaloids. In: Brossi, A. (Ed) The Alkaloids, vol. 33 (pp 1–81). Academic Press, Orlando

Nickon, A. & Fieser, L. F. (1952) Configuration of tropine and pseudotropine. J. Amer. Chem. Soc. 74: 5565–5570

Parr, A. J. (1992) Alternative metabolic fates of hygrine in transformed root cultures of *Nicandra physaloides*. Plant Cell Rep. 11: 270–273

Parr, A. J., Payne, J., Eagles, J., Chapman, B., Robins, R. J. & Rhodes, M. J. C. (1990) Variation in tropane alkaloid accumulation within the Solanaceae and strategies for its exploitation. Phytochemistry 29: 2545–2550

Portsteffen, A., Dräger, B. & Nahrstedt, A. (1992) Two tropinone reducing enzymes from *Datura stramonium* transformed root cultures. Phytochemistry 31: 1135–1138

Rabot, S. & Robins, R. J. (1992) Pseudotropine:tigloyl-CoA acyl transferase: an ester-forming enzyme from *Datura stramonium* transformed root cultures. In: (van Beek, T. A., ed.), Phytochemistry and Agriculture (p. 67). Abstracts of Phytochemical Society/Royal Netherlands Chemical Society meeting, Wageningen, April 1992

Robins, R. J. & Walton, N. J. (1994) The Biosynthesis of the Tropane Alkaloids. In: Cordell, G. A. (Ed.) The Alkaloids, vol. 44 (pp 115–187). Academic Press, Orlando

Robins, R. J., Parr, A. J., Bent, E. G. & Rhodes, M. J. C. (1991a) Studies on the biosynthesis of tropane alkaloids in *Datura stramonium* L. transformed root cultures 1. The kinetics of alkaloid production and the influence of feeding intermediate metabolites. Planta 183: 185–195

Robins, R. J., Bachmann, P., Robinson, T., Rhodes, M. J. C. & Yamada, Y. (1991b) The formation of 3α- and 3β-acetoxytropanes by *Datura stramonium* transformed root cultures involves two independent acetyl-CoA-dependent acyl transferases. FEBS Letters 292: 293–297

Robins, R. J., Woolley, J. G., Ansarin, M., Eagles, J. & Goodfellow, B. J. (1994a) Phenyllactic acid but not tropic acid is an intermediate in the biosynthesis of tropane alkaloids in *Datura* and *Brugmansia* transformed root cultures. Planta, 194: 86–94

Robins, R. J., Bachmann, P. & Woolley, J. G. (1994b) Biosynthesis of hyoscyamine involves an intramolecular rearrangement of littorine. J. Chem. Soc. Perkin Trans. I: 615–619

Saito, K., Suzuki, H., Takamatsu, S. & Murakoshi, I. (1993) Acyltransferases for lupin alkaloids in *Lupinus hirsutus*. Phytochemistry 32: 87–91

Walton, N. J., Peerless, A. C. J., Robins, R. J., Rhodes, M. J. C., Boswell, H. D. & Robins, D. J. (1994) Purification and properties of putrescine *N*-methyltransferase from transformed roots of *Datura stramonium*. Planta, 193: 9–15

Wink, M. & Hartmann, T. (1982) Enzymatic synthesis of quinolizidine alkaloid esters: a tigloyl-CoA: 13-hydroxylupanine *O*-tigloyltransferase from *Lupinus albus* L. Planta 156: 560–565

Plant Cell, Tissue and Organ Culture **38:** 249–256, 1994.

Characterization of *Coptis japonica* cells with different alkaloid productivities

Fumihiko Sato, Norimatsu Takeshihta, Hiroyuki Fujiwara, Yasuyuki Katagiri, Liping Huan & Yasuyuki Yamada
Department of Agricultural Chemistry, Kyoto University, Kyoto 606–01, Japan

Key words: Berberine, biosynthetic activity, *Coptis japonica*, cultured cells, high-alkaloid productivity, Ranunculaceae

Abstract

Several cell lines of *Coptis japonica* with different alkaloid productivities were characterized to obtain information on how a high metabolite production is established. High and low metabolite producing cells, except those from one cell line, showed similar growth kinetics and a similar pattern of nutrient uptake. Amino acid contents, especially that of tyrosine, differed between cell lines, but no correlation was found between the amino acid or tyrosine levels and alkaloid production. Since the addition of tyrosine did not increase the production of berberine, this primary substrate is apparently not the limiting factor for high production in cultured *Coptis* cells. The addition of berberine to the medium revealed that low-producing cells also have the ability to store alkaloid, and that low productivity is not due to decomposition of alkaloids which have been produced. The direct measurement of the biosynthesis of berberine using ^{14}C-tyrosine clearly showed that high-producing cells had a higher biosynthetic activity of berberine from tyrosine than low-producing cells. The measurement of enzyme activities in berberine biosynthesis indicated that the early steps of berberine biosynthesis are important in the increased production of berberine.

Introduction

Selecting high metabolite producing cells is a practical means for establishing a high metabolite producing system (Zenk et al. 1977; Yamamoto et al. 1982; Sato & Yamada 1984). However, it is unknown how a high metabolite productivity can be established in selected cells. In the present study the characteristics of several cell lines with different alkaloid productivities were analyzed to obtain information on establishing high metabolite production.

Materials and methods

Cultured cells

Cultured *Coptis* (*Coptis japonica* Makino var. Dissecta (Yatabe) Nakai) cells with different berberine productivities were maintained in Linsmaier & Skoog (1967) liquid medium (containing 10 μM 1-naphthalene acetic acid and 0.01 μM 6-benzyladenine on a rotary shaker (ca. 100 rpm) in the dark at 25 ± 1 °C as described previously (Yamada & Sato 1981).

Growth measurement

In the time course experiments, about 0.5 g of cultured *Coptis* cells were inoculated in 25 ml of media and cultured for the period indicated. Cells were harvested, freeze-dried and weighed.

Chemical analysis of medium, sugar and amino acids

Ammonium was measured by the indophenol method (Weatherburn 1967). The amount of nitrate was determined as ammonium after the reduction of nitrate with Devarda's Metal in a Conway micro-diffusion unit (Bremner 1965; Conway & Byrne 1933). Phosphate was determined by the molybdenum blue method

(Murphy & Riley 1962). Sugar concentration in the medium was determined by the 0.2% anthrone/sulfuric acid method (Morris 1948). Amino acid (Yemm & Cocking 1955) and tyrosine contents (Ceriotti & Spandrio 1957) were determined spectrophotometrically with ninhydrin reagent and 1-nitroso-2-naphthol, respectively. Free phosphate, neutral sugars and amino acids in cells were extracted with hot 90% EtOH. Sugars and amino acids in cell extracts were first fractionated with ion-exchange column chromatography, and the concentrations of sugars and amino acids were then determined.

Measurement of alkaloid contents and addition of berberine or tyrosine

Alkaloids were extracted from cells with 90% MeOH until the cells lost their yellow color. The alkaloid levels were determined by HPLC as described previously (Sato & Yamada 1984).

Berberine (1.2 mg or 22 mg berberine chloride) was added to the culture (25 ml) after two weeks of growth. After one additional week of culture, cells were harvested, washed and extracted. Cultured tobacco (*Nicotiana tabacum* cv. Samsun NN) cells were used as a reference. Tyrosine was added in a similar manner.

Tracer experiment with ^{14}C-tyrosine

U-^{14}C-Tyrosine (NEN) was added to the culture medium at a concentration of 3.7 MBq ml^{-1} (0.1 μM) after 2 weeks of culture. After 4 additional days of culture, cells were harvested, washed and extracted. Extracts was analyzed by HPLC. Each alkaloid fraction was collected manually, pooled and counted with a liquid scintillation counter.

Determination of biosynthetic activity

Cells were homogenized in liquid N_2, and enzymes were extracted with 200 mM Tris-HCl (pH 7.5) containing 20 mM mercaptoethanol, 10 mM sodium ascorbate and 5% (w/v) polyvinylpolypyrrolidone, after the temperature of the homogenized cells reached ca 0 °C. The homogenates were centrifuged at 25000 g for 30 min, enzymes were desalted through a PD-10 column (Pharmacia) equilibrated with 50 mM Tris-HCl (pH 7.5) containing 10 mM mercaptoethanol.

Six enzyme activities involved in berberine biosynthesis were measured in the following reaction mixtures (50 μl). Some enzymes were renamed due to the recent results obtained in incorporation experiments (Muller & Zenk 1992). The reaction mixture for norcoclaurine-6-O-methyltransferase (= norlaudanosoline-6-O-methyltransferase; Rueffer et al. 1983; 6-OMT), 3'-hydroxy-N-methyl-(S)-coclaurine-4'-O-methyltransferase (6-O- methylnorlaudanosoline-4'-O-methyltransferase; Frenzel & Zenk 1990b; 4'-OMT) and coclaurine-N-methyltransferase (= norreticuline-N-methyltransferase; Frenzel & Zenk 1990a; NMT) consisted of 100mM Tricine-NaOH (pH 7.0), 2 mM SAM, 100 mM sodium ascorbate, 1 mM substrate and 20 μl of enzyme preparation. Substrates for 6-OMT, 4'-OMT and NMT were (R,S)-6-O-methylnorlaudanosoline, (R,S)-6-O-methylnorlaudanosoline and (R,S)-norreticuline, respectively. The reaction mixture for berberine bridge enzyme (Steffens et al. 1984; BBE), scoulerine-9-O-methyltransferase (Muemmler et al. 1985, Sato et al. 1993; SMT) and tetrahydroberberine oxidase (Yamada & Okada 1985; THBO) consisted of 300 mM Tris-HCl (pH 8.5) and 1 mM substrate. The reaction mixture for SMT also contained 100 mM sodium ascorbate and 2.0 mM SAM. Substrates for BBE, SMT and THBO were (R,S)-reticuline, (R,S)-scoulerine and tetrahydroberberine, respectively. The assay mixture was incubated at 30 °C for 30 min., and the reaction was terminated by the addition of 50 μl of MeOH. After the proteins were precipitated at 15000*g* for 5 min., the amount of alkaloid produced was determined by the following HPLC system; mobile phase, CH3CN-HOAc-H2O (18:1:82 for 6-OMT, 4'-OMT and NMT, 32:1:68 for BBE, SMT and THBO) and 5 mM sodium 1-octanesulfonate; Column, Nova-Pak C18; flow rate 1.0 ml min^{-1}; detection absorbance 280 nm.

Results and discussion

Growth kinetics and alkaloid production

By repeated selection, several cell lines were established which showed different berberine productivities (Fig. 1). Cell line 156-1 accumulated the highest amount of berberine, while ABA was a moderate-producer. Non-selected cells and 81T-105T showed low productivities. The differences in production were maintained throughout the culture. When cell growth in the different cell lines was compared, the cells were divided into two groups; normal growing cells (156-1, ABA, non-selected) and slow growing cells (81T-

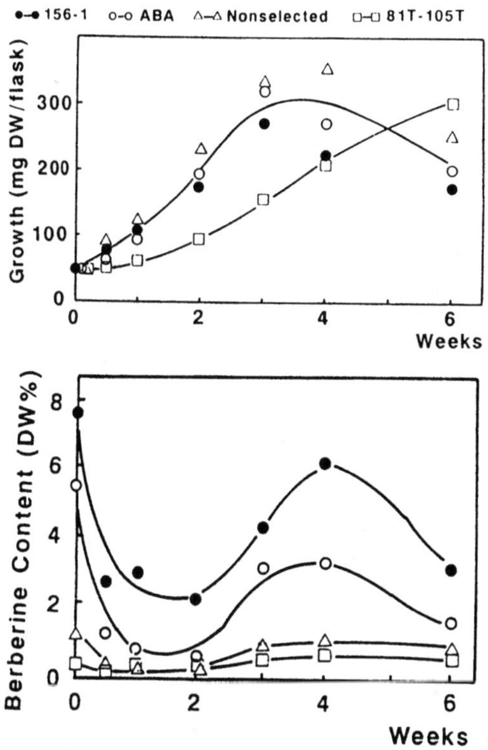

Fig. 1. Growth kinetics and alkaloid production in different cell lines

105T). The growth rate did not correlate with berberine production.

Differences in growth rate reflected the uptake of sugar from the medium and the endogenous phosphate contents (Fig. 2). As compared to normal growing cells, slow growing cells showed slow sugar uptake. The different cell lines showed differences in their free sugar concentrations but no correlation was found with the growth kinetics or alkaloid production. All cell lines absorbed phosphate within 1 week after culture, but in slower growing cells a higher inorganic phosphate contents were found. Again, no correlation was found with the alkaloid production.

Slow growing cells showed a relatively slow uptake of inorganic nitrogen (Fig. 3). On the other hand, the amino acid and tyrosine concentrations were relatively higher in slow growing cells than in the other cells. The cell lines showed different levels, and different changes in the levels of amino acids during culture.

As mentioned above, the uptake of sugar or inorganic nitrogen and the concentration of free phosphate all reflected the growth rate. However, none of the parameters we measured correlated with alkaloid production.

Addition of tyrosine

One of the limiting steps in the production of secondary metabolites should be the competition for the substrate between secondary metabolism and primary metabolism. Therefore, we added tyrosine to the medium to increase the flow from primary metabolites to secondary metabolism. When tyrosine, a primary substrate for berberine biosynthesis, was added to high- and low-producing cells, at a final concentration of 0.02% or 0.1% (w/v), a small increase (less than 10%) in alkaloid production was observed. This result indicated that the primary metabolite pool does not limit productivity, and/or that the compartmentation of enzymes (Galneder et al. 1988) and the inhibitory effects of added compounds on cell viability may reduce the use of the added compounds in secondary metabolism.

Ability of cultured cells to store berberine

Alkaloid-producing cells have varying capacities to take up their metabolites. Cultured tobacco cells, which do not contain berberine alkaloid, accumulated berberine at a concentration of 0.64 mg/total cells, when a low concentration of berberine (1.2 mg as chloride/25 ml culture) was added (Table 1). However, almost all of the tobacco cells died when a higher concentration of berberine (22 mg as chloride/25 ml culture) was added.

To examine whether a high production of secondary metabolites corresponded to a high storage capacity of those metabolites, berberine was added to cultures of either high-producing 156-1 cells or low-producing non-selected cells (Table 1). Both high and low berberine producing cultured *Coptis* cells accumulated the added berberine, even when it was added at a high concentration. Although high-producing cells accumulated more berberine than low-producing cells, and although these cell lines showed different final intracellular berberine concentrations, the differences in the absorption and accumulation of berberine were not as large as the difference in the original productivity. This indicates that low-producing cells and high-producing cells have similar capacities for storing berberine. We also concluded that berberine was not readily decomposed and would be the end-product in *Coptis* cells: the difference between the amount of berberine taken up from culture medium (21.9 mg/high-producing cells vs. 14.5 mg/low-producing cells) was similar to the increase in the amount of intracellular berberine (17.9

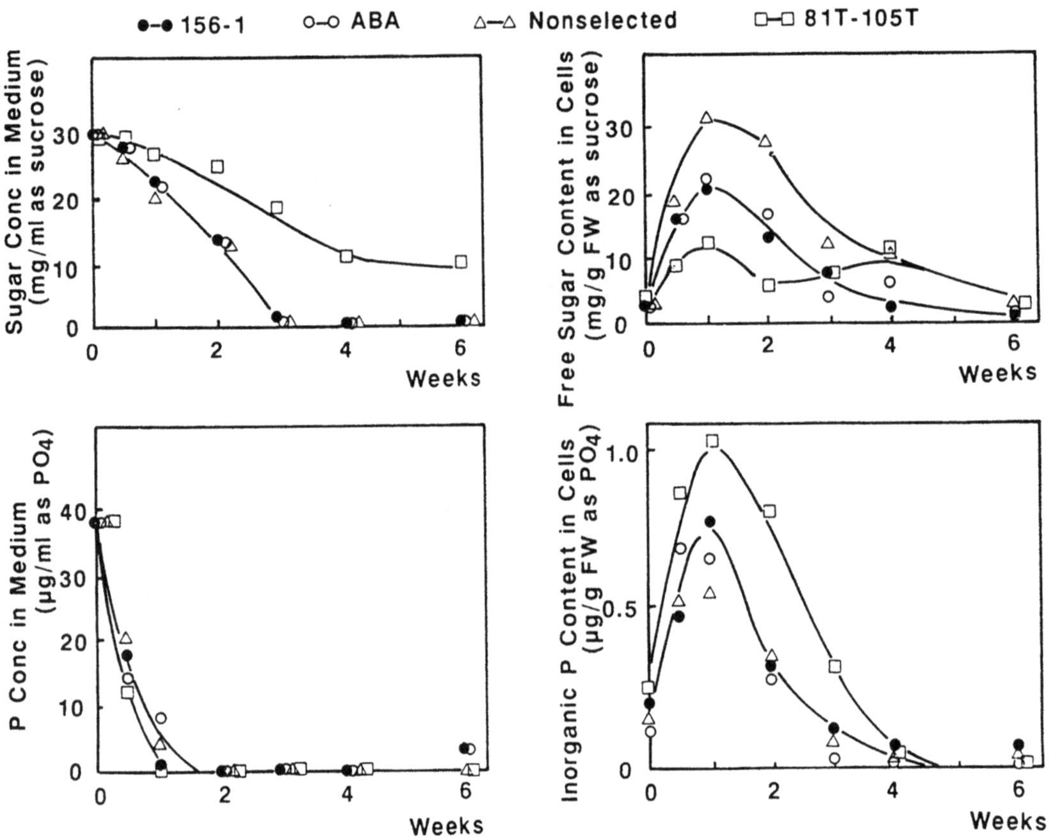

Fig. 2. Uptake of sucrose and phosphate in cultured *Coptis* cells with different berberine productivities

Table 1. Amounts of alkaloid in cells after adding berberine to medium

	Alkaloid amounts in cells (mg/flask)		
	Tobacco Berberine	Non-selected Coptis Berberine: Columbamine: Palmatine	High producing 156–1 Coptis Berberine: Columbamine: Coptisine
Control	N.D.	1.7 (0.2) : 2.2: 0.35	10.3 (0.1): 2.9 : 9.5
Berberine +1.2mg/flask	0.64 (0.55)	2.9 (0.6): 1.7 : 0.35	12.7 (0.1): 2.8 : 9.7
Berberine +22mg/flask	1.5 (20.5)	10.7 (7.5): 0.87: 0.1	28.2 (0.1): 1.8 : 9.1

N.D. : not detected. Alkaloid amounts are indicated as chloride salts. Numbers in parentheses indicate the amounts of alkaloid in the medium

Alkaloid contents in inoculum were as follows: in nonselected cells, the amounts of berberine, columbamine and palmatine were 1.3, 1.0, and 0.46 mg flask^{-1}. In 156-1 cells, the amounts of berberine, columbamine and coptisine were 4.1, 0.8, 4.1 mg flask^{-1}.

Fresh weights of 156-1, nonselected *Coptis*, and tobacco cells were 2.28 g, 1.77 g and 0.45 g/25 ml of culture, respectively, when berberine was added to the medium. After one additional week of culture, fresh weights of each cells were 3.05 g, 2.60 g and 2.82 g, without the addition of berberine; 2.80 g, 2.62 g and 0.66 g with the addition of 1.2 mg berberine; 2.88 g, 2,69 g and 0.52 g with the addition of 22 mg berberine, respectively.

mg/high-producing cells vs. 9.0 mg/low-producing cells).

For reasons not yet understood, low-producing cells could not take up all of the berberine available in medium. Recently, Sato et al. (1992) suggested that a high concentration of malic acid contributed to the formation of a highly water-soluble salt of berberine and that the production of berberine in cultured cells

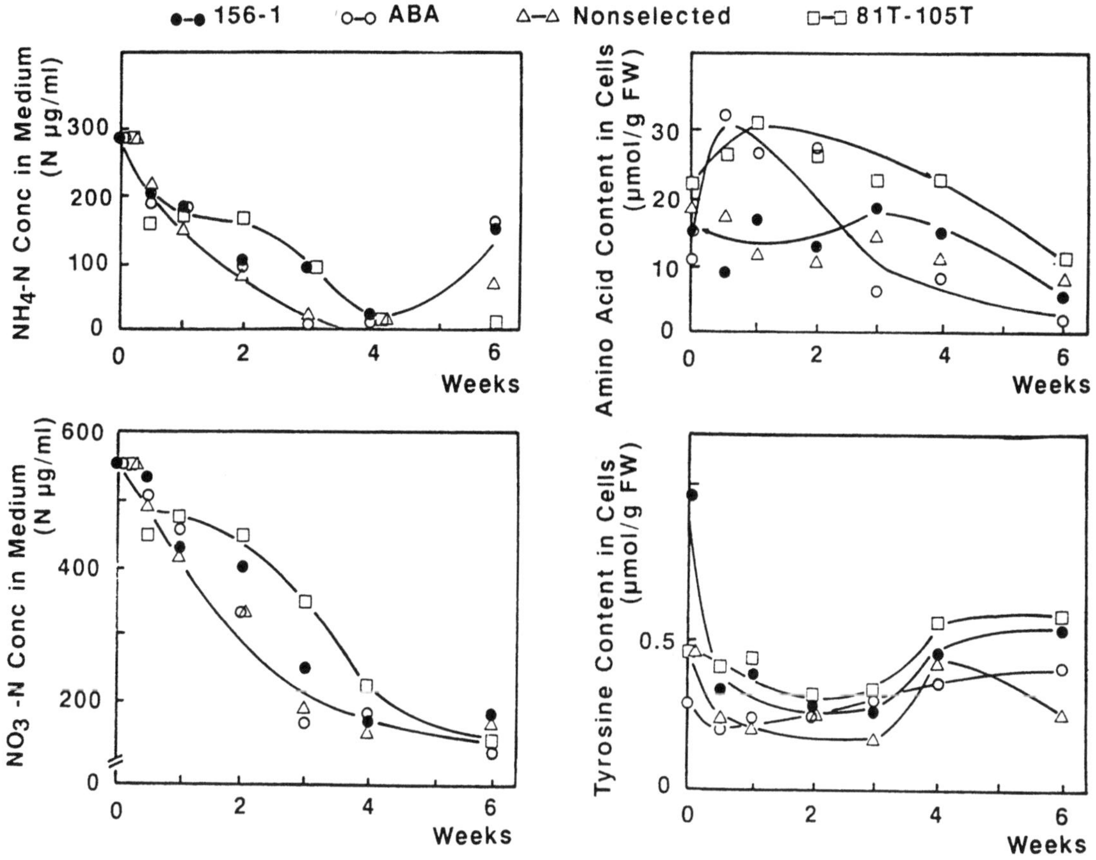

Fig. 3. Uptake of inorganic nitrogen and changes in amino acid concentrations in cultured *Coptis* cells with different alkaloid productivities

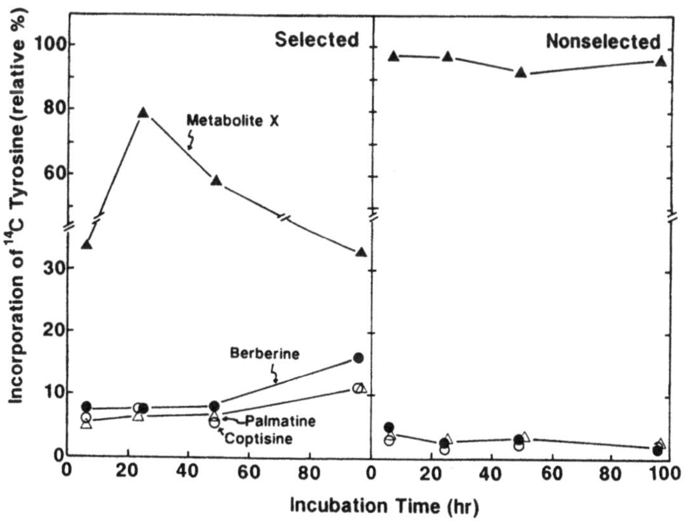

Fig. 4. Incorporation of ^{14}C-tyrosine into metabolites in high-berberine producing 156-1 cells and low-producing nonselected cells.

was accompanied by a steady increase of malic acid during the entire culture period. Organic acid concentrations should be determined in different cell lines in future.

During an additional week of culture, when a high concentration of berberine was added to the medium, the production of other alkaloids, i.e., columbamine and palmatine, was inhibited, while that of coptisine

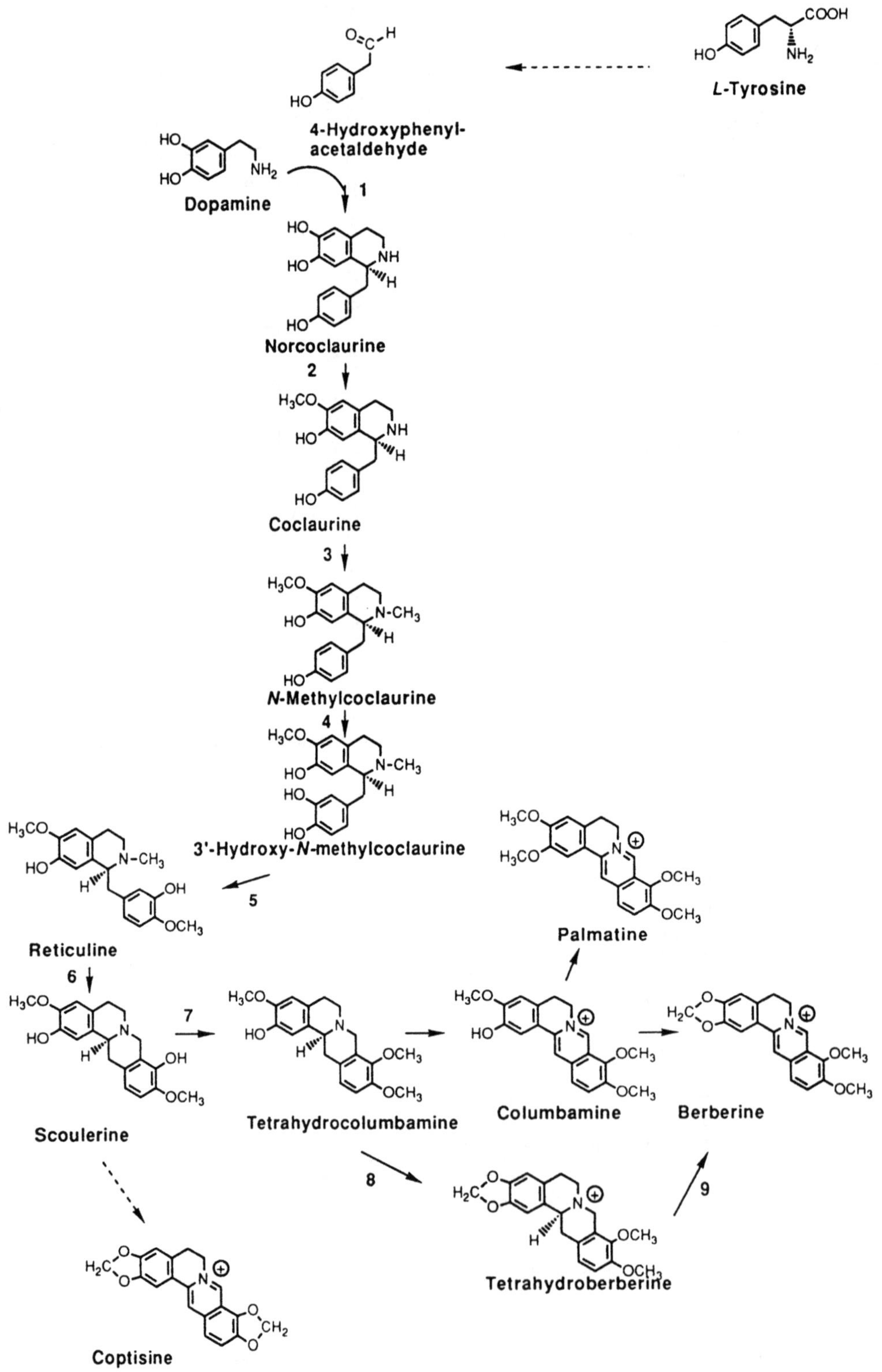

Fig. 5. The biosynthetic pathway to berberine. Numbers refer to the names of the enzymes given in Table 2.

Table 2. Biosynthetic enzyme activities in high (156-1) and low (nonselected) berberine producing *Coptis* cells.

Enzyme	Activity (nmol min^{-1} g^{-1} FW)[1]	
	156-1	nonselected
1. Norcoclaurine synthase	–	–
2. Norcoclaurine-6-O-methyltransferase (6-OMT)[2]	5.2	3.1
3. Coclaurine-N-methyltransferase (NMT)	–	–
4. Phenolase	–	–
5. 3′-Hydroxy-N-methylcoclaurine-4′-O-methyl-transferase (4′-OMT)[2]	0.2	nd
6. Berberine bridge enzyme (BBE)	1.5	0.5
7. Scoulerine-9-O-methyltransferase (SMT)	16.6	12.9
8. Methylenedioxyring forming enzyme	–	–
9. Tetrahydroberberine oxidase (THBO)	8.3	25.4

Enzymes were prepared from cultured cells at the middle of the log phase.
[1] nd = not detected; – = not measured.
[2] Substrates for norcoclaurine-6-O-methyltransferase and 3′-hydroxy-N-methylcoclaurine-4′-O-methyltransferase were (R,S)-norlaudanosoline and (R,S)-6-O-methylnorlaudanosoline.

was unaffected. This result supports a previous finding that berberine inhibits *O*-methylation catalyzed by SMT (Sato et al. 1993), which reduces the biosynthesis of columbamine, palmatine and berberine.

Tracer experiment

High-alkaloid producing cells took up twice as much ^{14}C-tyrosine as low producing cells (data not shown). Furthermore, high-producing cells showed clearly higher biosynthetic activity from tyrosine than low-producing cells (Fig. 4).

Measurement of biosynthetic enzymes in berberine biosynthesis

The activities of enzymes in a biosynthetic pathway usually differ and the enzymes with lowest activity may make significant contributions to the control of the rate of biosynthesis. Therefore, if the rate-limiting enzymes can be identified, productivity can be enhanced by isolating the cDNA of the rate-limiting enzyme(s) (Dittrich & Kutchan 1991) and transforming plant cells to either over-express the rate-limiting enzyme using the isolated cDNA and a strong promoter (Yun et al. 1992), or to improve an enzyme property using protein engineering techniques (Stark et al. 1992).

Comparison of biosynthetic enzyme activities in the high alkaloid producing cells and low producing cells showed that the early steps of berberine biosynthesis are rate-limiting in *Coptis* cells (Fig. 5 and Table 2): a. the enzyme activities in the early steps of berberine biosynthesis (from norcoclaurine to scoulerine) were lower than those in the later steps, and b. the differences between the activities of the high- and low-producing cells were greater during the early steps than during the later steps. Measurement of three enzymes during cell growth confirmed this conclusion (data not shown).

Conclusion

The characteristics of *Coptis* cells with different alkaloid productivities were determined. Neither the growth characteristics, nor the amino acid concentrations, nor the capacity of the cells to store the alkaloid reflected alkaloid production. Biosynthetic activities, especially enzyme activities during the early steps of berberine biosynthesis, are believed to be rate-limiting.

Acknowledgements

This research was supported in part by a Grant-in-Aid for Scientific Research on Priority Areas (04252102) from the Ministry of Education, Science and Culture, Japan.

References

Bremner JM (1965) Inorganic forms of nitrogen. In: Black CA et al. (Eds.) Methods of Soil Analysis (pp. 1179–1237) Amer. Soc. Agron. Inc., Publ.

Ceriotti G & Spandrio L (1957) Colorimetric determination of tyrosine. Biochem. J. 66: 607–610

Conway EJ & Byrne A (1933) An absorption apparatus for the microdetermination of certain volatile substances I. The microdetermination of ammonia. Biochem. J. 27: 419–429

Dittrich H & Kutchan TM (1991) Molecular cloning, expression and induction of berberine bridge enzyme, an enzyme essential to the formation of benzophenanthridine alkaloids in the response of plants to pathogenic attack. Proc. Natl. Acad. Sci. USA. 88: 9969–9973

Frenzel T & Zenk MH (1990a) Purification and characterization of three isoforms of S-adenosyl-L-methionine:(R,S)-tetrahydrobenzylisoquinoline-N-methyltransferase from *Berberis koetineana* cell cultures. Phytochemistry 29: 3491–3497

Frenzel T & Zenk MH (1990b) S-Adenosyl-L-methionine: 3′-hydroxy-N-methyl-(S)-coclaurine-4′-O-methyltransferase, a regio- and stereoselective enzyme of the (S)-reticuline pathway. Phytochemistry 29: 3505–3511

Galneder E, Rueffer M, Wanner G, Tabata M & Zenk MH (1988) Alternative final steps in berberine biosynthesis in *Coptis japonica* cell cultures. Plant Cell Rep. 7:1–4

Linsmaier EM & Skoog F (1967) Organic growth factor requirements of tobacco tissue cultures. Physiol. Plant. 18: 100–127

Morris DL (1948) Quantitative determination of carbohydrates with Dreywood's reagent. Science 107: 254–255

Muller MJ & Zenk MH (1992) The norcoclaurine pathway is operative in berberine biosynthesis in *Coptis japonica*. Planta Med. 58: 524–527

Muemmler S, Rueffer M, Nagakura N & Zenk MH (1985) S-adenosyl-L-methionine: (S)-scoulerine-9-O-methyltransferase, a high stereo- and regio-specific enzyme in tetrahydroprotoberberine biosynthesis. Plant Cell Rep. 4: 36–39

Murphy J & Riley JP (1962) A modified single solution method for the determination of phosphate in natural waters. Anal. Chim. Acta 27: 31–36

Rueffer M, Nagakura N & Zenk MH (1983) Partial purification and properties of S-adenosylmethionine: (R,S)-norlaudanosoline-6-O-methyltransferase from *Argemone platyceras* cell cultures. Planta Med. 49: 131–137

Sato F & Yamada Y (1984) High berberine-producing cultures of *Coptis japonica* cells. Phytochemistry 23: 281–285

Sato H, Taguchi G, Fukui H & Tabata M (1992) Role of malic acid in solubilizing excess berberine accumulating in vacuoles of *Coptis japonica*. Phytochemistry 31: 3451–3454

Sato F, Takeshita N, Fitchen JH, Fujiwara H & Yamada Y (1993) S-Adenosyl-L-methionine:scoulerine-9-O-methyltransferase from cultured *Coptis japonica* cells. Phytochemistry 32: 659–664

Stark DM, Timmerman KP, Barry GF, Preiss J & Kishore GM (1992) Regulation of the amount of starch in plant tissues by ADP glucose pyrophosphorylase. Science 258: 287–292

Steffens P, Nagakura N & Zenk MH (1984) The berberine bridge forming enzyme in tetrahydroprotoberberine biosynthesis. Tetrahedron Lett. 25: 951–952

Weatherburn MW (1967) Phenol-hypochlorite reaction for determination of ammonia. Anal. Chem. 39: 971–974

Yamada Y & Okada N (1985) Biotransformation of tetrahydroberberine to berberine by enzymes prepared from cultured *Coptis japonica* cells. Phytochemistry 24: 63–65

Yamada Y & Sato F (1981) Production of berberine in cultured cells of *Coptis japonica*. Phytochemistry 20: 545–547

Yamamoto Y, Mizuguchi R & Yamada Y (1982) Selection of a high and stable pigment-producing strain in cultured *Euphorbia millii* cells. Theor. Appl. Genet. 61: 113–116

Yemm W & Cocking EC (1955) The determination of amino acids with ninhydrin. Analyst 80: 209–213

Yun D-J, Hashimoto T & Yamada Y (1992) Metabolic engineering of medicinal plants; Transgenic *Atropa belladonna* with an improved alkaloid composition. Proc. Natl. Acad. Sci. USA. 89: 11799–11803

Zenk MH, El-Shagi H, Arens H, Stöckigt J, Weiler EW & Deus B (1977) Formation of the indole alkaloids serpentine and ajmalicine in cell suspension cultures of *Catharanthus roseus*. In Barz W et al. (Eds.) Plant Tissue Culture and Its Biotechnological Applications (pp 27–43) Springer, Heidelberg

Plant Cell, Tissue and Organ Culture **38**: 257–262, 1994.

Traits of transgenic *Atropa belladonna* doubly transformed with different *Agrobacterium rhizogenes* strains

Mondher Jaziri[1], Kayo Yoshimatsu[2], Jacques Homès[1] & Koichiro Shimomura[2]
[1]*Université Libre de Bruxelles, Laboratory of Plant Morphology, 1850 chaussée de Wavre, B-1160 Brussels, Belgium;* [2]*Tsukuba Medicinal Plant Research Station, National Institute of Health Sciences, 1 Hachimandai, Tsukuba, Ibaraki, 305 Japan*

Key words: Agrobacterium, Atropa belladonna, double transformation, phytohormone content, tropane alkaloids, viviparous leaves

Abstract

Hairy root cultures of *Atropa belladonna* L. were established by infection either with *Agrobacterium rhizogenes* ATCC 15834 or MAFF 03–01724, and transgenic plants were obtained from both hairy root cultures. Doubly transformed roots were induced by re-infection of the leaf segments of transgenic *Atropa belladonna* plants (*A. rhizogenes* 15834) with MAFF 03–01724. Shoots and viviparous leaves were regenerated from the doubly transformed roots. The genetic transformation was determined by the opine assay (agropine, mannopine and/or mikimopine) and polymerase chain reaction. Physiological changes and tropane alkaloid biosynthesis in the hairy roots (singly and doubly transformed) were investigated. The alkaloid content in the doubly transformed root strain was intermediate as compared to the root strains which were singly transformed. On the other hand endogenous IAA levels in doubly transformed roots were significantly decreased compared to both singly transformed roots.

Abbreviations: BA – benzyladenine, IAA – indoleacetic acid, NAA – naphthaleneacetic acid, PCR – polymerase chain reaction, *t*-ZR – trans-zeatin

Introduction

Infection of plant cells by *Agrobacterium rhizogenes* usually results in root formation at the sites of infection. This morphogenic event is due to the transfer of genetic information from bacteria to plant cells. The integration and expression of the transferred DNA (T-DNA) of Ri plasmids cause metabolic changes involved in auxin and cytokinin syntheses (Zambryski et al. 1989). Two regions, TL- and TR-DNA, were found to be integrated and stably conserved in the plant genome (Jouanin 1984). The TR and/or TL-DNA may be present in a different copy number(s) (Merlo et al. 1980; Thomashow et al. 1980). Until now the exact number of the T-DNA fragments as well as their orientation in the plant genome (sense or anti-sense) have not been clarified. Theoretically, a large number of T-DNA copies can be integrated into plant genome, but the possible number of the integration is still obscure. Our objective in this communication is to contribute toward the solution of the questions whether the number of the T-DNA copies in the plant genome can be increased and additional traits can be induced by re-transformation with *A. rhizogenes* and how the double transformation may affect the primary and secondary metabolism of plant cells. *Atropa belladonna* has been used for *in vitro* studies as a model plant because of its morphogenic potential. It is one of the widely used medicinal plants which contain tropane alkaloids. The morphological, physiological and the biochemical traits of doubly transformed tissue cultures of *Atropa belladonna* are discussed.

Materials and methods

Plant material

Sterile plants of *Atropa belladonna* L. were established by shoot tip culture of field-growing plants and maintained on hormone-free half-strength Murashige and Skoog (1/2 MS) (Murashige & Skoog 1962) solid medium at 25 °C under 16 h light/day, 4000 Lux.

Establishment of non-transformed root culture

Root tips (ca. 1 cm length) from non-transformed sterile *Atropa belladonna* plantlets (propagated by shoot tip culture) were transferred onto 1/2 MS liquid medium supplemented with IAA or NAA (0.5 mg l^{-1}) and cultured at 25 °C in the dark with subculturing at 4 weeks intervals.

Bacterial strains

Agrobacterium rhizogenes ATCC 15834 and MAFF 03–01724 were grown on YEB agar medium (beef extract (5 g l^{-1}), yeast extract (1 g l^{-1}), peptone (5 g l^{-1}), sucrose (5 g l^{-1}), $MgSO_4 \cdot 7H_2O$ (0.5 g l^{-1}), the pH was adjusted to 7.2). The bacteria were subcultured in YEB liquid medium at 25 °C in the dark on a rotatory shaker (100 rpm) for 15 h before use for transformation of plants.

Induction of hairy roots

Leaf segments (ca. 0.5 × 0.5 cm) of *Atropa belladonna* plants were transferred into 1/2 MS liquid medium (20 ml/100 ml Erlenmeyer flask) and co-cultured with *A. rhizogenes* ATCC 15834 or MAFF 03–01724 (200 μl bacteria suspension/flask) at 25 °C in the dark on a rotatory shaker (100 rpm). After one day, the leaf segments were placed on 1/2 MS solid medium containing 0.5 g l^{-1} Claforan (Hoechst Japan Ltd.) and incubated at 25 °C in the dark. After 3 to 4 weeks, emerging root tips were transferred to hormone-free 1/2 MS solid medium without antibiotic and maintained in the same conditions with subculturing at 3–4 weeks intervals.

Regeneration of transgenic plants

Transformed root segments (*A. rhizogenes* ATCC 15834) were cultured onto 1/2 MS solid medium supplemented with 15 g l^{-1} sucrose, NAA (0.5 μM) and BA (5 μM) at 25 °C under 16 h light/day (4000 Lux). After 4–6 weeks, adventitious shoots were regenerated from the root segments. Shoots were spontaneously regenerated from the hairy roots (*A. rhizogenes* MAFF 03–01724) cultured on hormone-free medium over 2 months. Both shoots were transferred onto hormone-free 1/2 MS solid medium supplemented with 30 g l^{-1} sucrose in culture tubes. The rooted plants were propagated as sterile shoot cultures under the same conditions.

Establishment of the doubly transformed cultures

The doubly transformed cultures were established by the co-culture method using leaf segments from the transgenic *Atropa belladonna* (*A. rhizogenes* ATCC 15834) plants and *A. rhizogenes* MAFF 03–01724. The conditions for the experiments were the same as described in section 'Induction of Hairy Roots'. The 1/2 MS medium used for the experiments was supplemented with 30 g l^{-1} sucrose and adjusted to pH 5.7 before autoclaving at 120 °C for 15 min. Solid media were solidified with 2 g l^{-1} Gelrite (Merck et Co, USA).

Detection of opines

Fresh tissue (100 to 200 mg fresh wt.) was homogenised with a plastic rod in a microtube, and the crude extract (50 μl) obtained after centrifugation (6000 g for 1 min) was subjected to high-voltage paper electrophoresis according to the method described by Petit et al. (1983). The detection of mikimopine was performed with the Pauly reagent (Isogai et al. 1990) and the detection of agropine and mannopine was visualised by alkaline silver nitrate reagent.

DNA extraction and PCR analysis

Template DNA was prepared from 10 to 20 mg fresh root material according to the method described by Edward et al. (1991). The oligonucleotide primers TL-DNA (*rol* A-1 and *rol* B-2) and TR-DNA (*Ags*-1 and *Ags*-2) for *A. rhizogenes* ATCC 15834 (Slightom et al. 1986) and T-DNA of MAFF 03–01724 were synthesized with DNA synthesizer (Applied Biosystem model 380 A). PCR was carried out in 50 μl of total volume containing 2.5 μl of DNA sample, 10 μl of each primer, 8 μl of each dNTPs, 10 mM of Tris HCl pH 8.8, 1.5 mM of $MgCl_2$, 50 mM of KCl, 0.1% of triton X-100, 0.5 units of Perfect Match DNA Polymerase Enhancer (Stratagene) and 2.5 units of *Taq* polymerase (Promega). For the positive control of TL- and TR-DNA, the relevant cosmid pLJ1 and pLJ85 (Jouanin 1984) were used as the template respectively. PCR amplification was performed under the following conditions: initial denaturation at 94 °C for 7 min followed by 30 cycles of denaturation at 94 °C for 1 min, annealing at 55 °C for 2 min, extension at 72 °C for

Table 1. Recapitulative data concerning the genetic transformation of normal, transformed and doubly transformed root clones of *Atropa belladonna*.

	Opine assay agr./man.	PCR analysis mik.	T-DNA MAFF	TL-DNA ATCC	TR-DNA ATCC
Root clones:					
NAb	−	−	−	−	−
RiAb	+	−	−	+	−
MAb	−	+	+	−	−
RiAbM1	+	+	+	+	−
RiAbM3	+	−	−	+	−

agr: agropine; man: mannopine; mik: mikimopine.
+/−: presence/absence.

3 min with a final extension at 72 °C for 10 min. The PCR reaction mixture (10 μl) was electrophoresed on 0.8% agarose gel and visualised by ethidium bromide staining at 312 nm.

Sample preparation and HPLC analysis

After 3 weeks of culture, fresh wt. and dry wt. (after lyophilization) were determined. About 50 mg of dried sample was extracted with 5 ml of $CHCl_3$-MeOH-28% NH_4OH (15:5:1) using sonication (20 min). The further sample preparation was the same as described previously by Shimomura et al. (1991) and Ishimaru & Shimomura (1989). The alkaloid extracts were dissolved in 150 μl of MeOH and 10 μl was injected into HPLC using the same system as described by Shimomura et al. (1991) and 5 tropane alkaloids were quantified.

Analysis of phytohormones

Determination of IAA, t-ZR and DHZR was performed by an enzyme immunoassay using monoclonal antibodies as described by Weiler et al. (1986). Kit's were purchased from Phytodetek™ (Idetek, Inc., USA).

Results and discussion

Establishment of non-transformed cultures

From the sterile plants of *Atropa belladonna* propagated by shoot tip culture, non-transformed root cultures (NAb) were established by transferring root tips (ca. 1 cm length) from the sterile plants on the same solid or liquid medium supplemented with IAA (0.5 mg l^{-1}). After 2 months of culture, adventitious shoots were regenerated from the non-transformed root cultures. The regenerated shoots were isolated and rooted again on hormone-free 1/2 MS solid medium. These plantlets showed the similar phenotypical characteristics as those propagated by shoot tip culture.

Transformation with A. rhizogenes

Root induced on *Atropa belladonna* leaf segments co-cultured with *A. rhizogenes* strains grew on hormone-free medium, showing typical features of hairy root syndrome, plagiotropic growth and hormone independence. The genetic transformation was demonstrated by opine assay (agropine and mannopine for strains transformed with *A. rhizogenes* ATCC 15834 (RiAb) and mikimopine for root strains transformed with *A. rhizogenes* MAFF 03–01724 (MAb) and by PCR analysis (Table 1). The PCR amplification of the root strains transformed with *A. rhizogenes* ATCC 15834 showed that only the TL region of the T-DNA was detected in the transformed root cultures. Thomashow et al. (1980) showed that the TR-DNA fragment is not present in all transformed clones, the TR-DNA fragment might be deleted or rearranged.

Regeneration of transgenic plants

Transformed shoots were regenerated sporadically from hairy root segments of the strain transformed with ATCC 15834, cultured on 1/2 MS medium containing NAA (0.5 μM) and BA (5 μM). In contrast, after long-term cultivation (2 months, without subculture) shoots were regenerated spontaneously from the hairy roots of MAFF 03–01724, cultured on hormone-free medium. Both shoots were transferred onto hormone-free 1/2 MS solid medium. The rooted transgenic plants were propagated as sterile shoot cultures under the same conditions. The transgenic plants thus obtained showed some morphological differences as compared to the non-transformed plants: thick and wrinkled leaves with an apparent high chlorophyll content and short internodes for the transgenic plants from ATCC 15834); lanceolated and smaller green-pale coloured leaves for the transgenic plants from MAFF 03–01724. These observations indicate that several developmental parameters caused by the expression of T-DNA genes of Ri plasmid have been altered. Such alterations were previously reported in transgenic plants and could result from auxin and cytokinin imbalance activity (Schmülling et al. 1988; Spena et al. 1987; Estruch et al. 1991a).

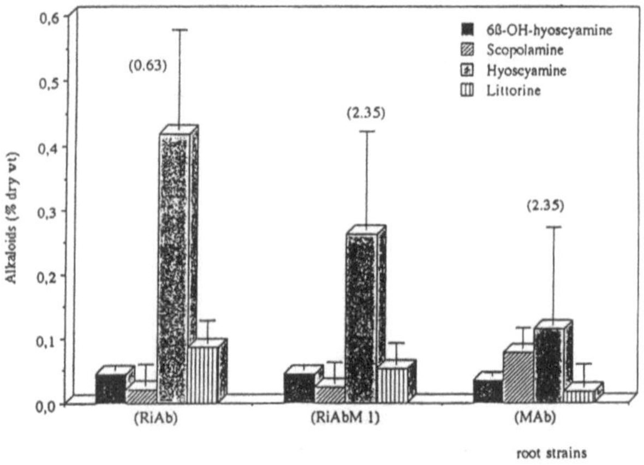

Fig. 1. Growth and tropane alkaloid content in hairy root cultures of *Atropa belladonna* transformed with *A. rhizogenes* ATCC 15834 and/or MAFF 03–01724. Hairy roots (ca. 20 mg fresh wt.) were cultured in hormone-free 1/2 MS medium (3% sucrose, 50 ml/100 ml flask) for 3 weeks. Number in parentheses show the fresh weight (g).

Establishment of doubly transformed cultures

In order to be able to confirm the integration of the plasmid DNA after the double transformation, the bacteria selected for the second infection were different from the ones used to obtain the transgenic plants. Twenty leaf segments from transgenic *Atropa belladonna* (ATCC 15834) were co-cultured with *A. rhizogenes* MAFF 03–01724. For the controls, 20 leaf segments from the same transgenic plants were cultured under the same conditions without *Agrobacterium*. The emerging hairy roots from the leaf segments (3–4 weeks after infection) were transferred separately onto hormone-free 1/2 MS solid medium containing Claforan. After subculture 4 times with antibiotic, actively growing root segments (1–2 cm length) were transferred onto the same medium without antibiotic. From the control experiment, numerous roots were regenerated (more than 50 roots/leave segment) whereas from the doubly infected leaf segments only 1–2 roots/leaf explant were regenerated. Six out of the total hairy root clones obtained after the double infection were selected for their good growth and used for further experiments. After long-term cultivation (2–3 months) adventitious shoots and viviparous leaves were regenerated from the doubly transformed cultures.

Analysis of the doubly transformed cultures

The genetic transformation of the doubly transformed cultures (RiAbM) was demonstrated by the opine assay and PCR analysis (Table 1). The opine assay showed that only one hairy root clone (RiAbM1) out of 6 clones tested was doubly transformed, it contained agropine, mannopine and mikimopine. The adventitious shoots and viviparous leaves were also found to be doubly transformed. The PCR analysis of the RiAbM1 clone showed that T-DNAs from both bacteria used were present in the sample (Table 1).

Growth and tropane alkaloids content

By HPLC the following tropane alkaloids could be simultaneously determined: 6β-OH-hyoscyamine, scopolamine, hyoscyamine, and littorine. The growth and the alkaloid content in hairy roots of *Atropa belladonna* transformed with *A. rhizogenes* ATCC 15834 and/or MAFF 03–01724 were determined after 3 weeks culture in hormone-free 1/2 MS liquid medium (Fig. 1). The highest content of alkaloids, particularly hyoscyamine, was found in the hairy root strain transformed with *A. rhizogenes* ATCC 15834. On the other hand, the growth of this strain was very slow (0.63 g dry wt/flask) as compared to the hairy roots transformed with *A. rhizogenes* MAFF 03–01724 and to the RiAbM1 strain (2.35 g dry wt/flask for both strains). The integration of the T-DNAs from both bacteria affect the growth and the alkaloid production of the hairy root cultures. The alkaloids content in the doubly transformed roots was intermediate as compared to the singly transformed roots, while it dis-

Table 2. IAA and t-ZR content in normal and transformed root cultures of *Atropa belladonna* (pmol/g fresh wt.).

Root strains	IAA	t-ZR
normal root*	23.2	2.79
root transformed with *A. rhizogenes* ATCC 15834	57.3	25.86
root transformed with *A. rhizogenes* MAFF 03–01724	77.7	14.86
doubly transformed root culture (RiAbM1)	ld	23.00

*: normal root cultured on 1/2 MS solid medium containing NAA (0.5 mg/l).
ld: value lower than the limit detection of the assay.

played the good growth as the hairy roots transformed with *A. rhizogenes* MAFF 03–01724.

Phytohormone analysis

The analysis of phytohormones in doubly transformed root strain (RiAbM1), normal root culture (NAb) and hairy root strains transformed either with *A. rhizogenes* ATCC 15834 (RiAb) or MAFF 03–01724 (MAb) revealed differences in their concentrations (Table 2). As expected, higher amounts of t-ZR (and DHZR, data not shown) were detected in transformed cultures as compared to the normal root cultures. On the other hand, no trace of free and conjugated IAA was detected in the doubly transformed root strain (RiAbM1). In comparison with the normal root culture, the IAA level was 2.5 and 3.4 times higher in hairy root strains transformed either with *A. rhizogenes* ATCC 15834 or MAFF 03–01724 respectively. Other researchers measuring IAA in transformed root cultures of carrot observed a significant increase in the concentration of IAA compared with the initial concentration of IAA in normal root cultures (Epstein et al. 1991). The low amount of IAA in the doubly transformed cultures is surprising. It remains to be seen whether the suppression of IAA synthesis is a consequence of interactions between the different, endogenous and exogenous IAA genes. As mentioned above, after long-term cultivation in hormone-free medium, the doubly transformed root strain regenerates viviparous leaves that form, after subculture, adventitious buds on their surfaces or edges. This manifestation of totipotency of differentiated leaf cells is a phenomenon occurring either as part of a normal developmental process or as a teratological event. Estruch et al. (1991b) reported that the vivipary is acquired by tobacco leaves that are somatic genetic mosaics for the expression of a cytokinin-synthesizing gene. The regenerated shoots from the viviparous leaves we obtained after the double transformation of *Atropa belladonna* exhibited loss of apical dominance and were unable to root. This observation is also reported in *Nicotiana* Sp. in which the expression of the *ipt* gene under the control of the 35S RNA promoter from cauliflower mosaic virus (CaMV) increases the cytokinin content up to 137 times in transgenic shoots (Estruch et al. 1991a). In the case of the doubly transformed *Atropa belladonna* cultures, the formation of the viviparous leaves is a direct consequence of a phytohormone imbalance (Table 2) which might be caused by transcriptional inactivation (sense or antisense) of the different IAA genes expression and/or by a change of T-DNA by methylation and expression which could be associated with the phenotypic variation observed (vivipary) and phytohormone content (especially the decrease of the IAA level). The sensitivity of the doubly transformed strain to exogenous IAA added to the culture medium is now under investigation.

Acknowledgement

The authors wish to thank Dr. L Jouanin (Laboratoire de Biologie Cellulaire, INRA, Versailles, France) for providing pLJ1 and pLJ85 and Mr. M. Matsumoto for PCR analysis.

References

Edward K, Johnstone C & Thompson C (1991) A simple and rapid method for the preparation of plant genome DNA for PCR analysis. Nucl. Acids Res. 19: 1349

Epstein E, Nissen SJ & Sutter EG (1991) Indole-3-acetic acid and indole-3-butyric acid in tissues of carrot inoculated with *Agrobacterium rhizogenes*. J. Plant Growth Regul. 10: 97–100

Estruch JJ, Chriqui D, Grossmann K, Schell J & Spena A (1991a) The plant oncogene rol C is responsible for the release of cytokinins from glycoside conjugates. EMBO J. 10: 2889–2895

Estruch JJ, Prinsen E, Van Onckelen H, Schell J & Spena A (1991b) Viviparous leaves produced by somatic activation of an active cytokinin-synthesizing gene. Science 254: 1364–1367

Ishimaru K & Shimomura K (1989) 7 beta-hydroxyhyoscyamine from *Duboisia myoporoides-D. Leichhardtii* hybrid and *Hyoscyamus albus*. Phytochemistry 28: 3507–3509

Isogai A, Fukuchi N, Hayashi M, Kamada H, Harada H & Suzuki A (1990) Mikimopine, an opine in hairy roots of tobacco induced by *Agrobacterium rhizogenes*. Phytochemistry 29: 3131–3134

Jouanin L (1984) Restriction map of agropine-type Ri plasmid and its homologies with Ti plasmids. Plasmid 12: 91–102

Merlo DJ, Nutter RC, Montaya AL, Garfinkel DJ, Drummond MH, Chilton MD, Gordon MP & Nester EW (1980) The boundaries and copy numbers of Ti plasmid T-DNA vary in crown gall tumors. Mol. Gen. Genet. 177: 637–643

Murashige T & Skoog F (1962) A revised medium for rapid growth and bioassays with tobacco tissue cultures. Physiol. Plant. 15: 473–497

Petit A, David C, Dahl GA, Ellis JG, Guyon P, Casse-Delbart F & Tempé J (1983) Further extension of the opine concept: plasmids in *Agrobacterium rhizogenes* cooperate for opine degradation. Mol. Gen. Genet. 190: 204–214

Schmülling T, Schell J & Spena A (1988) Single genes from *Agrobacterium rhizogenes* influence plant development. EMBO J. 7: 2621–2629

Shimomura K, Sauerwein M & Ishimaru K (1991) Tropane alkaloids in the adventitious and hairy root cultures of solanaceous plants. Phytochemistry 30: 2275–2279

Slightom JL, Durant-Tardif M, Jouanin L & Tepfer D (1986) Nucleotide sequence analysis of TL-DNA of *Agrobacterium rhizogenes* agropine type plasmid. Identification of open reading frames. J. Biol. Chem. 261: 108–121

Spena A, Schmülling T, Konez C & Schell JS (1987) Independent and synergic activity of rol A, B and C loci in stimulating abnormal growth in plant. EMBO J. 6: 3897–3899

Thomashow MF, Nutter R, Postle K, Chilton MD, Chilton FR, Blattner A, Powell MP, Gordon MP & Nester EW (1980) Recombination between higher plant DNA and Ti Plasmid of Agrobacterium tumefaciens. Proc. Natl. Acad. Sci. USA 77: 6448–6452

Weiler EW, Eberle J, Mertens R, Atzorn R, Feyerabend M, Jourdan PS, Annscheidt A & Wieczorek U (1986) Antisera-and monoclonal antibody-based immunoassay of plant growth hormones. In: Wang T (Ed) Immunology in Plant Sciences (pp 27–58). Cambridge University Press, London

Zambryski P, Tempé J & Schell J (1989) Transfer and function of T-DNA genes from *Agrobacterium* Ti and Ri plasmids in plants. Cell 56: 193–201

Plant Cell, Tissue and Organ Culture **38:** 263–272, 1994.

Effect of nitrogen and sucrose on the primary and secondary metabolism of transformed root cultures of *Hyoscyamus muticus*

Kirsi-Marja Oksman-Caldentey, Nina Sevón, Leena Vanhala & Raimo Hiltunen
Pharmacognosy Division, Department of Pharmacy, P.O. Box 15, FIN–00014 University of Helsinki, Finland

Key words: Fatty acids, hyoscyamine, *Hyoscyamus muticus*, nitrogen, sucrose, transformed root cultures

Abstract

The effect of carbon and nitrogen sources on two well-established hairy root clones, LBA1S and C58A, of *Hyoscyamus muticus* strain Cairo, were investigated. Both clones exhibited completely different patterns with regards to their growth rate, hyoscyamine accumulation, and fatty acid contents. Clone C58A grew faster and yielded more biomass (17.4 g l^{-1}, in 21 days), but produced less hyoscyamine. The maximum hyoscyamine content (120 mg l^{-1}) in clone LBA1S was reached in 28 days. Neither of the clones could use lactose or fructose as the sole carbon source, nor ammonium as the sole nitrogen source. The growth in the medium containing glucose was significantly reduced compared to that containing sucrose. Clone LBA1S was sensitive to the changes in sucrose concentration and an increase in ammonium in the culture medium, whereas C58A tolerated these changes better but was more sensitive to the increase in total nitrogen. Lipid synthesis was active in the exponential growth phase, and the total fatty acid content varied from 5 to 34 mg g^{-1} of dry root material. The major fatty acids were linoleic, palmitic and linolenic. There were considerable differences in the total amount of lipids and in their relative ratios when different nutrients were applied.

Abbreviations: DW — dry weight, FA — fatty acids, FFA— free fatty acids, FW — fresh weight

Introduction

The basic nutritional requirements of cultured plant cells are normally very similar to those of the whole plant. However, the nutrient media successfully used for cell cultures often have to meet specific requirements. Nitrogen and carbon are two of the elements essential for plant cell and tissue cultures. Even though photoautotrophic cultures exist which satisfy their carbon requirement exclusively through photosynthesis, most cultures lack the capacity to synthesize their own supply of carbohydrates (see review *e.g.* Hüsemann 1985). Nitrogen is the element essential for the synthesis of the macromolecules, DNA, RNA and proteins. Of all the nutrients, the form of nitrogen (oxidized or reduced, organic or inorganic) probably has the most pronounced effects on the growth and differentiation of cultured tissue (Gamborg et al. 1968).

The lipids in heterotrophic and photoautotrophic cell suspension cultures have been widely investigated (*e.g.* Radwan et al. 1979; Barz et al. 1980), but very little is known about the lipid composition in transformed root cultures (Toivonen et al. 1992). Temperature alteration as well as other environmental stresses or changes in nutrient composition may induce modifications in lipid composition (Weber & Mangold 1988). The chemical and physical properties of lipids may thus influence the activities of the enzymes involved in the biosynthesis of the secondary products. The modification of membrane lipids in plant cells could thus be used to improve the release of secondary products into the culture medium (Neidleman 1987; Weber & Mangold 1988).

The transformed roots of *Hyoscyamus muticus* L. are fast-growing in hormone-free medium, genetically stable and produce secondary metabolites characteristic of the intact plant (Oksman-Caldentey et al. 1989; Sevón et al. 1992). Thus they might offer a suitable *in vitro* system for studying primary metabolism *e.g.* lipid synthesis. In contrast to cell suspension cultures, secondary metabolite accumulation in hairy roots is not readily manipulated by alterations in medium composition (Payne et al. 1987). However, we have earlier demonstrated that by adding chitosan to the medium, the hyoscyamine content in *H. muticus* hairy roots could be increased 5-fold (Sevón et al. 1992).

Here we report the effect of carbon and a range of inorganic nitrogen sources on the growth, fatty acid composition and hyoscyamine production of two different hairy root clones of *H. muticus*.

Material and methods

Bacterial strains and cell culture material

Agrobacterium rhizogenes strain LBA 9402 (gift from Prof. U. Nyman, Copenhagen) grown on YMB medium (Sevón et al. 1992) and *A. tumefaciens* C58CI pRTGUS104 (constructed by U. Commandeur, BBA, Braunschweig) grown on solid Luria medium (5 g l^{-1} yeast extract, 10 g l^{-1} tryptone, 10 g l^{-1} NaCl, 10 g l^{-1} Difco Bacto agar, pH 7.0) were used to obtain hairy roots on *Hyoscyamus muticus* strain Cairo. The origin of the plant material, transformation method, and cultivation conditions of the root clones have been described elsewhere (Oksman-Caldentey et al. 1991; Sevón et al. 1992). Clones LBA1S and C58A were chosen for further study.

Growth of the root clones

The growth of both root clones was followed by transferring 100 mg ± 5 mg of fresh roots to 20.0 ml of liquid B50 medium (Oksman-Caldentey et al. 1991) containing 3% sucrose in 100 ml conical flasks (30 flasks). Three flasks were removed once or twice a week (see Fig. 1) and filtered. The remaining medium volume was measured. The roots were washed with water and lyophilized (dry weight). The growth was expressed as g l^{-1}. All the samples were stored at −20 °C until analysed.

Changing the carbon source

Hairy root clones were cultivated for one passage (15 days) in autoclaved B50 basal medium containing different concentrations of sucrose (10, 20, 30, 40, 50, 60 and 80 g l^{-1}) or without sucrose. Thereafter, 100 mg ± 5 mg (FW) of root tips were excised and transferred to 20.0 ml of the corresponding fresh liquid media. Similar experiments were performed using glucose, fructose and lactose individually, each at concentrations of 5, 10, 20, 30, 40, 50 and 60 g l^{-1}. After 21 days the roots were harvested as mentioned above and the culture growth (three replicate flasks) was measured. The fatty acids and hyoscyamine contents were determined quantitatively as described below.

Changing the nitrogen content and nitrogen source

In the first experiments hairy root clones were cultivated for one passage (15 days) in autoclaved B50 medium containing 3% sucrose and different concentrations of total nitrogen, given as $(NH_4)_2SO_4$ and KNO_3 (0, 1, 10, 27, 50, 60, 75 and 100 mM). The ammonium:nitrate ratio (2:25) was maintained as in the B50 basal medium. Thereafter, the roots were subcultured, grown and harvested as described for the sucrose experiments.

In a second series of experiments the NH_4^+ and NO_3^- ratio was changed to 0 + 5, 0 + 27, 2 + 25, 10 + 17, 13.5 + 13.5, 17 + 10, 25 + 2 and 27 + 0 mM. Potassium and sulphate levels were kept constant. After 15 days the roots (100 mg ± 5 mg) were transferred to the corresponding fresh media (20.0 ml). After 21 days the roots were harvested as described for the sucrose experiments.

Hyoscyamine determination

Quantitative hyoscyamine determination was performed on the dried root material and media using the radioimmunoassay described earlier (Oksman-Caldentey et al. 1987) with minor modifications (Sevón et al. 1992). The hyoscyamine contents were expressed as mg l^{-1}.

Lipid analysis

For the separation of the lipid classes twenty mg of dry root material was extracted with 3 ml of methanol:chloroform (1:1). The extract was filtered through a Sep-pak® silica cartridge (Millipore) and TLC separation of the whole lipid fraction was performed on HPTLC silica gel plates $60F_{254}$ (Merck, Darmstadt) and developed with *n*-hexane – diethyl ether – glacial acetic acid (80:20:2). The plate was dipped into cupric acetate monohydrate-phosphoric acid (Fried 1991) and detection of the lipid classes was performed at 350 nm using a Shimadzu CS 9000 densitometer.

The fatty acid composition of each lipid class was analysed according to the method of Seppänen-Laakso et al. (1990). The lipid fractions (phospho-

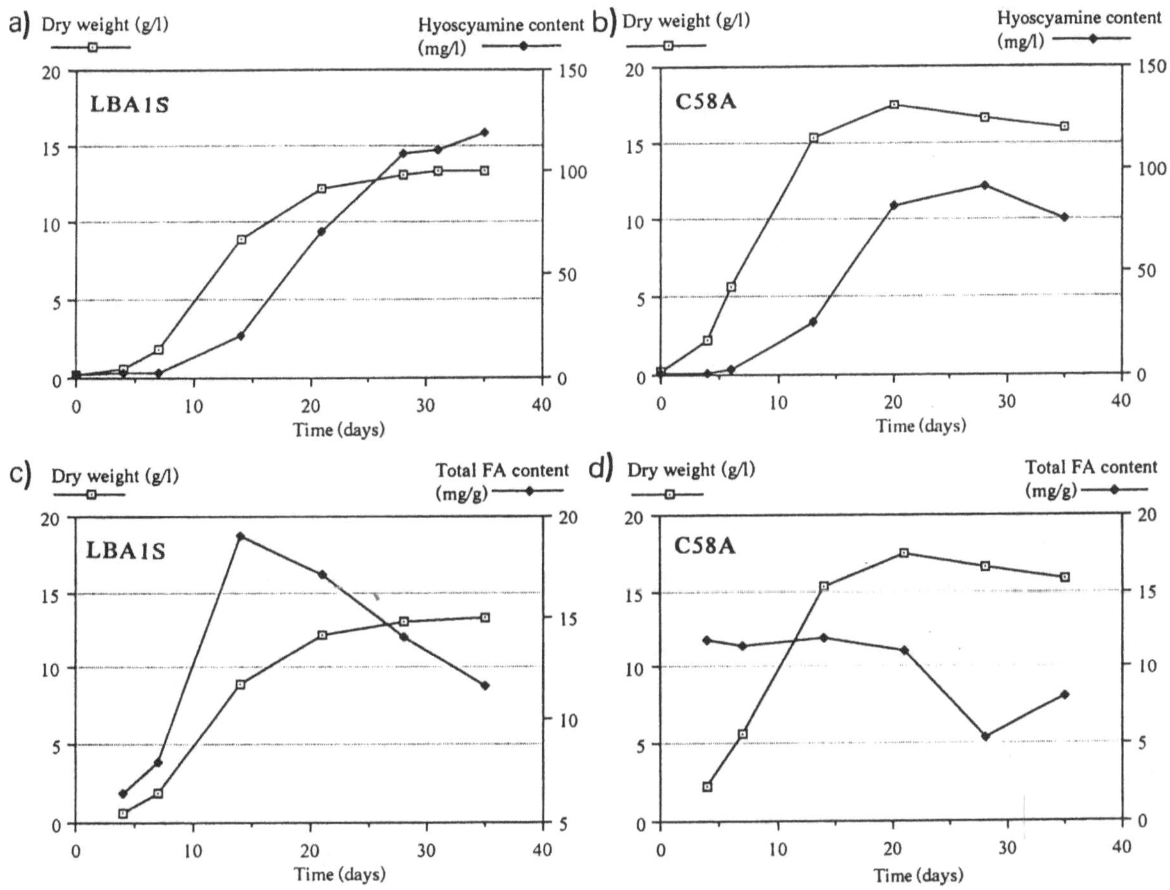

Fig. 1. Time course of growth and hyoscyamine accumulation of the hairy root clones LBA1S (a) and C58A (b) during five weeks cultivation. Panels (c) and (d) show the total fatty acid content of clones LBA1S and C58A, respectively, during the same time period.

lipids, triglycerides, free fatty acids) were scraped off the plate and extracted with methanol:chloroform (1:1) before transmethylation. Heptadecanoic acid (C17:0) (50 μg) and triarachidate (100 μg) were added to the root samples (20 mg $\pm$ 5 mg) as internal standards. Derivatization was performed by the method of Seppänen-Laakso et al. (1990) with minor modifications. Briefly, sodium methoxide in dry methanol (1.0 ml, 0.5 N) and a couple of boiling stones were added to the samples dissolved in 1.5 ml of petroleum ether, and the transesterification carried out in a water bath at 40 °C for not more than 5 minutes. The excess of sodium methoxide was neutralized with a 15% aqueous solution of sodium hydrogen sulphate (2 ml). The fatty acid methyl esters (FAME) and free fatty acids (FFA) were analysed from the upper petroleum ether layer (1 μl aliquot for GC analysis).

Transmethylated and free fatty acids were analysed simultaneously on a DANI 3865 GC using a programmed temperature vaporizer (PTV) sampling technique with split mode. The column used was NB-351 (25 m, 0.32 mm i.d., film thickness 0.20 μm) fused silica. The analysis conditions were as follows: carrier gas H_2, split ratio 50:1, oven temperature programmed from 100 °C to 235 °C at 13 °C min^{-1}, FID 250 °C, and PTV injector temperature increased from 70 °C to 250 °C in 8 seconds. The peak areas were measured with a Shimadzu C-R3A integrator. Identification of the fatty acids was based on reference substances and the retention times of co-chromatographed, known fatty acid mixtures. The average precision of the determinations was 3.0% (C.V.) ranging from 2.2 to 4.3 per cent.

Table 1. The major fatty acids (%) in different lipid classes of the transformed root cultures of *Hyoscyamus muticus*. PL = phospholipids, TG = triglycerides, FFA = free fatty acids.

Fatty acid	PL	TG	FFA
Palmitic acid	20	25	53
Stearic acid	3	4	8
Oleic acid	2	2	7
Linoleic acid	39	40	16
Linolenic acid	10	13	4
Others	26	16	12

Results and discussion

Characterization of clones LBA1S and C58A

The two well-established transformed root culture clones, LBA1S and C58A of *Hyoscyamus muticus*, represent clones with an average growth rate and high stability with regards to their hyoscyamine production. LBA1S had been cultivated in B50 basal medium for 60 and C58A for 30 transfer passages, respectively. Morphologically the roots of clone C58A were much thicker, shorter and more branched than those of LBA1S. Slight callus formation was occasionally observed in the roots of clone C58A.

The time course of growth and hyoscyamine production of both clones was followed for 35 days (Fig. 1a and b). Clone C58A grew faster and reached its growth maximum (17.4 g l^{-1}) after 21 days. The hyoscyamine content followed the growth with a delay of about one week, reaching a maximum (92 mg l^{-1}) after 28 days of incubation. Clone LBA1S grew slower and continued to grow until five weeks. It produced less biomass in the stationary phase (13.3 g l^{-1}), but contained more hyoscyamine (120 mg l^{-1}). The calculated growth and hyoscyamine accumulation rates were 0.63 g l^{-1} d^{-1} and 4.29 mg l^{-1} d^{-1} for LBA1S, and 0.83 g l^{-1} d^{-1} and 3.29 mg l^{-1} d^{-1} for C58A, respectively. Only small quantities of hyoscyamine (less than 9 % of the total hyoscyamine content) were released into the medium.

Lipid composition of the transformed root clones

Since very little is known about the fatty acids of transformed root cultures a preliminary study of the whole lipid composition was necessary. The major lipid class was phospholipids, representing 41.7% of the total lipids, followed by triglycerides (30.2%), and free fatty acids (11.3%). Phospholipids (PL) and triglycerides (TG) consisted mainly of the same fatty acids, and in almost the same ratio, linoleic acid (C18:2) being the major component (about 40%) followed by palmitic (C16:0) and linolenic acids (C18:3) (Table 1). Stearic (C18:0) and oleic acids (C18:1) were minor components. From the free fatty acids (FFA) the major component was palmitic acid followed by linoleic acid.

The GC method used in this study simplified the FA analysis of different root samples because both the bound and free fatty acids were detected simultaneously by a single GC analysis. Table 2 shows the whole fatty acid composition of 21-day-old transformed root culture clones LBA1S and C58A. The fatty acid composition pattern was found to be similar in both clones although the relative ratios and absolute amounts varied in some cases. According to the Student's t-test, the relative proportions of linoleic and behenic (C22:0) acids were very significant and significantly ($p = 0.003$; $n = 6$ and $p = 0.024$; $n = 6$, respectively) higher in LBA1S than in C58A, whereas the relative proportion of linolenic acid was highly significantly ($p = 0.0001$; $n = 6$) lower in LBA1S than in C58A. In addition to the major components, small amounts of vaccenic (C18:1), lignoseric (C24:0), and hexadecatrienoic acid (C16:3) were also detected (Table 2). Hexadecatrienoic acid is known to be rather typical in small quantities for Solanaceous plants (Matsuzaki et al. 1984). The reason for the composition of fatty acids being restricted to these few major species may be that the lipid classes in rapidly growing plant cells are predominantly polar lipids, namely phospholipids and glycolipids, which are the main membrane lipids (MacCarthy & Stumpf 1980). The fatty acid composition in our transformed roots resembled that recently found in the hairy roots of *Catharanthus roseus* (Toivonen et al. 1992). However, the relative amounts of fatty acids differed considerably, especially in the case of linoleic and oleic acids, the latter being one of the major components in *C. roseus* (over 25 %).

The total fatty acid content varied from 5.3 to 19.0 mg g^{-1} (DW) during the time course of growth of both investigated clones. The amount of free fatty acids did not exceed 2.2 mg g^{-1} (DW). Maximal fatty acid formation occurred either at the end of the exponential growth phase or throughout the active growth phase (Fig. 1 c and d). This might be explained by the increasing amount of actively dividing meristematic root cells, and thus obviously greater amounts of mem-

Table 2. Comparison of the fatty acid composition in two hairy root clones, LBA1S and C58A, of *Hyoscyamus muticus*.

Fatty acids	Clone LBA1S		Clone C58A	
	Fatty acid content (n = 6)			
	%	mean ± SD (mg g^{-1})	%	mean ± SD (mg g^{-1})
C 16 : 0	20.4	4.48 ± 0.52	21.5	5.13 ± 0.62
C 16 : 3	tr.	tr.	0.5	0.14 ± 0.01
C 18 : 0	2.7	0.61 ± 0.06	2.7	0.63 ± 0.07
C 18 : 1 n − 9	1.4	0.32 ± 0.05	1.0	0.25 ± 0.05
C 18 : 1 n − 7	tr.	tr.	0.9	0.22 ± 0.05
C 18 : 2	61.2**	13.46 ± 1.57	53.5	13.06 ± 1.19
C 18 : 3	11.3***	2.49 ± 0.26**	15.7	3.59 ± 0.74
C 22 : 0	1.5*	0.34 ± 0.08	1.1	0.26 ± 0.03
C 24 : 0	tr.	tr.	0.9	0.25 ± 0.03
$\sum$ FFA	0.8	0.16 ± 0.08	1.2	0.28 ± 0.09

*** $p < 0.001$; ** $p < 0.01$; * $p < 0.05$; tr. = traces

Table 3. Fatty acid composition during the time course of growth of the transformed root clones LBA1S (A) and C58A (B).

A	FA content (mg g^{-1})					
Fatty acids	4 days	7 days	14 days	21 days	28 days	35 days
C 16 : 0	1.46	1.66	4.02	3.41	2.87	2.48
C 18 : 0	0.16	0.10	0.47	0.44	0.46	0.47
C 18 : 1	0.19	0.11	0.24	0.21	0.19	0.16
C 18 : 2	3.74	4.46	10.31	9.37	7.51	6.47
C 18 : 3	0.89	1.11	2.02	1.53	0.96	0.60
$\sum$ FA	6.41	7.94	19.04	17.21	14.02	11.59
$\sum$ FFA	–	0.50	2.14	2.25	2.03	1.41

B	FA content (mg g^{-1})					
Fatty acids	4 days	7 days	14 days	21 days	28 days	35 days
C 16 : 0	2.70	2.65	2.23	1.95	1.18	1.79
C 18 : 0	0.40	0.26	0.24	0.27	0.23	0.36
C 18 : 1	0.36	0.33	–	–	–	–
C 18 : 2	6.63	5.63	4.76	4.44	2.76	4.64
C 18 : 3	1.79	1.09	1.00	0.84	0.35	0.54
$\sum$ FA	11.88	11.42	11.97	11.09	5.38	8.02
$\sum$ FFA	–	1.46	3.74	3.59	0.87	0.69

brane lipids. The same kind of increase in fatty acids during the exponential growth phase has been reported for the suspension cells of many other plant species, and in one case, the hairy roots of *C. roseus* (e.g. MacCarthy & Stumpf 1980; Toivonen et al. 1992). Table 3 shows the formation of individual fatty acids in the investigated clones during the time course of growth. Even though the maximum for all the fatty acids in clone LBA1S was reached after 14 days (Table 3A), the amount of stearic acid appeared to increase towards the end of the growth phase. In the metabolism of fatty acids, stearic acid is the precursor of oleic, linoleic and linolenic acids. It is obvious that the desaturation activity decreases during the time course of the stationary phase, resulting in the decrease of oleic, linoleic and linolenic acids after 21 days of incubation. The fatty acid production of clone C58A was different (Table 3B). In general, the absolute amounts of individual fatty acids were highest in the beginning of the growth, as has already been seen in Fig. 1d. However, the relative proportions of the fatty acids fluctuated, and only linolenic acid clearly decreased. After one week incubation oleic acid was no longer detected in the roots of C58A.

Effect of sucrose and other carbon sources

Neither of our clones could utilize lactose or fructose as the sole carbon source. Many suspension cultures are known to utilize sucrose and glucose equally well, fructose being less efficient (Fowler 1982). However, Uozumi et al. (1991) recently reported that fructose is an excellent carbon source for the growth of carrot hairy roots. Fructose can also enhance secondary metabolite production in some cases (Jung et al. 1992). In cell suspension cultures the sucrose in the medium is rapidly converted to glucose and fructose, and glucose is utilized first (Nikolova et al. 1991). In our experiments the growth of C58A was very poor in glucose-containing media: it grew only slightly when the glucose content was 10 g l^{-1}, and growth was completely inhibited at all other concentrations. In the case of clone LBA1S, glucose at a concentration of 40 g l^{-1} resulted in a growth level that was about 80% of the optimal growth level in the sucrose-containing medium. The production of hyoscyamine was significantly lower than that in the sucrose-containing media. A similar inability to grow with glucose as the sole carbon source has earlier been reported for hairy roots of *Daucus carota*, *Armoracia rustica* and *Datura stramonium* (Taya et al. 1989; Payne et al. 1987).

The influence of sucrose on the growth of these two clones was completely different. Clone LBA1S achieved maximum growth at a sucrose concentration of 30 g l^{-1} (Fig. 2). Clone C58A continued to grow without any callusing roots even at a sucrose concentration of 80 g l^{-1}, where the rate of growth was more than double that of clone LBA1S. The growth rate of C58A in this case can be calculated as high as 1.2 g l^{-1} per day. However, LBA1S produced about twice as much hyoscyamine as C58A, but seemed to be more sensitive to the changes in sucrose concentration than C58A. Our findings support the results found in *Psoralea* hairy roots (Nguyen et al. 1992), but are in contrast to those of Christen et al. (1992), who reported

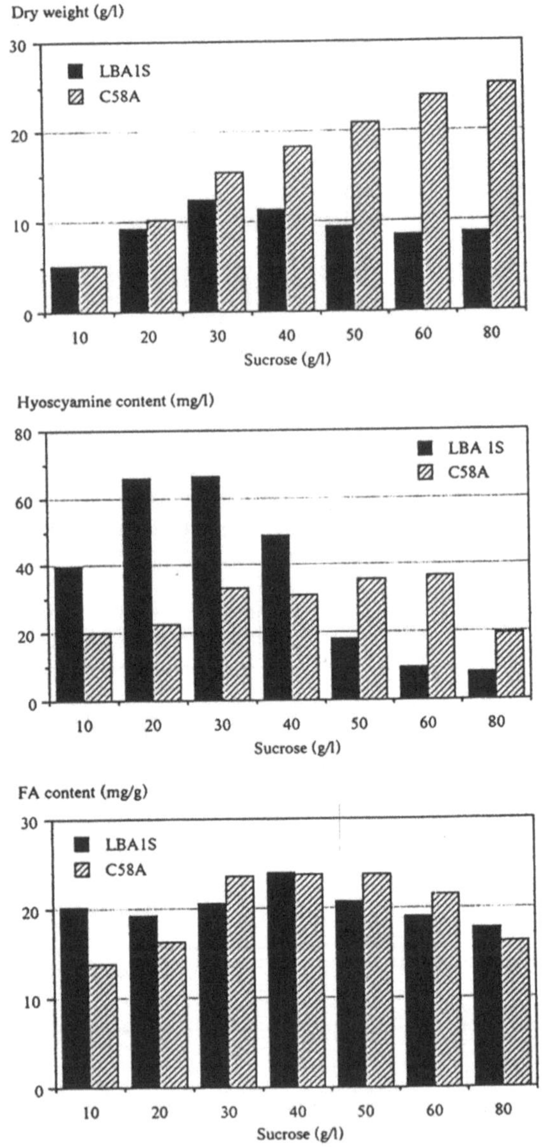

Fig. 2. Effect of different sucrose concentrations on the growth, hyoscyamine accumulation and total fatty acid content in hairy root clones LBA1S and C58A.

that different sucrose concentrations did not affect the growth of *H. albus* hairy roots.

The production of fatty acids in the root tissue was not influenced very much by changes in the sucrose concentration (Fig. 2). The production of fatty acids was still rather high, even at the concentration where the growth of clone LBA1S was not very good (at 80 g l^{-1}). The same kind of phenomenon was earlier seen in Fig. 1c, where the fatty acid content remained rather high still in the late stationary phase when growth was no longer active. When we investigated the production of individual fatty acids, all of them clearly reached a maximum at a sucrose concentration of 40 g l^{-1} in clone LBA1S, and at a sucrose concentration ranging from 30 to 50 g l^{-1} in clone C58A. Interestingly, neither clones produced any detectable amount of oleic acid when the sucrose concentration was under 30 g l^{-1}.

Effect of inorganic nitrogen

Most of the media used for plant cell cultures contain either inorganic or organic nitrogen, or a combination of both. Inorganic nitrogen is usually given as nitrate or ammonium. The total nitrogen requirements of the two hairy root clones of *H. muticus* are presented in Table 4. In the first experiments nitrogen was given as nitrate and ammonium in the same ratio as in B5 medium (Gamborg et al. 1968), which contains 2 mM ammonium and 25 mM nitrate. It can be seen that when the total nitrogen concentration was over 10 mM, the growth was optimal for both clones (Table 4). Hyoscyamine production in LBA1S was almost constant at this nitrogen level, whereas clone C58A was more sensitive to an increase in nitrogen. Even though the growth was still good at high nitrogen concentrations, hyoscyamine production rapidly decreased when the total nitrogen content was over 75 mM. However, there exist great differences between cell and tissue cultures of various plant species as regards their ability to use nitrogen. The optimal concentration for *D. stramonium* hairy roots was 20 mM (Payne et al. 1987), whereas the cell cultures of *Holarrhena antidysenterica* required 60 mM for the optimal accumulation of conessine (Panda et al. 1992).

The transformed roots of *H. muticus* could not use ammonium as the sole nitrogen source (data not shown). The ammonium-nitrate ratio had a remarkable effect on the growth and hyoscyamine production of the roots as well as on the release of hyoscyamine into the medium. Maximum growth was achieved in both clones when the concentration of ammonium was no higher than 2 mM (Fig. 3a, b). In contrast to the first experiments, clone LBA1S was more sensitive to ammonium increase, whereas clone C58A tolerated half the total nitrogen given in the form of ammonium. The effect of nitrogen as ammonium on the hyoscyamine contents followed the same pattern as for growth. When the ammonium concentration was less than 13.5 mM the release of hyoscyamine was less than 12% but above this concentration the hyoscyamine release increased gradually to 33%. Physiological studies carried out on many cell cultures indicate that ammonium is utilized before nitrate (Kurz & Constabel 1985). Ammonium is usually unsuitable as the sole nitrogen source, as was the case in our experiments. This is probably because under such conditions the pH of the medium tends to fall below 5 during culture. The drop in pH may also restrict the availability of nitrogen. Nigra et al. (1990) concluded that the ratio between ammonium and nitrate has a greater effect than total nitrogen on solasodine production in *Solanum* suspension cultures. This was also demonstrated in our experiments.

The maximal total fatty acid content was achieved when the ammonium concentration was over 13.5 mM, representing about 50% more than that attained under standard conditions (Fig. 3c, d). Even the highest ammonium concentrations (up to 27 mM) had almost no inhibiting effect on fatty acid formation in both strains. The same pattern was obtained for the individual fatty acids as for the total fatty acid contents. The production of fatty acids was favoured when half of the nitrogen was given as ammonium. The only exception was oleic acid. The amount of oleic acid was very low in general, but its production increased when growth was less active. There seems to be a negative correlation between lipid synthesis and hyoscyamine accumulation especially when most of the nitrogen is given as ammonium (Fig. 3c, d).

Conclusions

It can be concluded from the results presented in this study that the two transformed root culture clones, LBA1S and C58A, of *H. muticus*, exhibited different patterns for growth, and hyoscyamine and fatty acid production. In general, clone C58A grew better, but produced less hyoscyamine. Clone LBA1S was more sensitive to changes in the sucrose concentration and an increase in ammonium in the media, whereas C58A

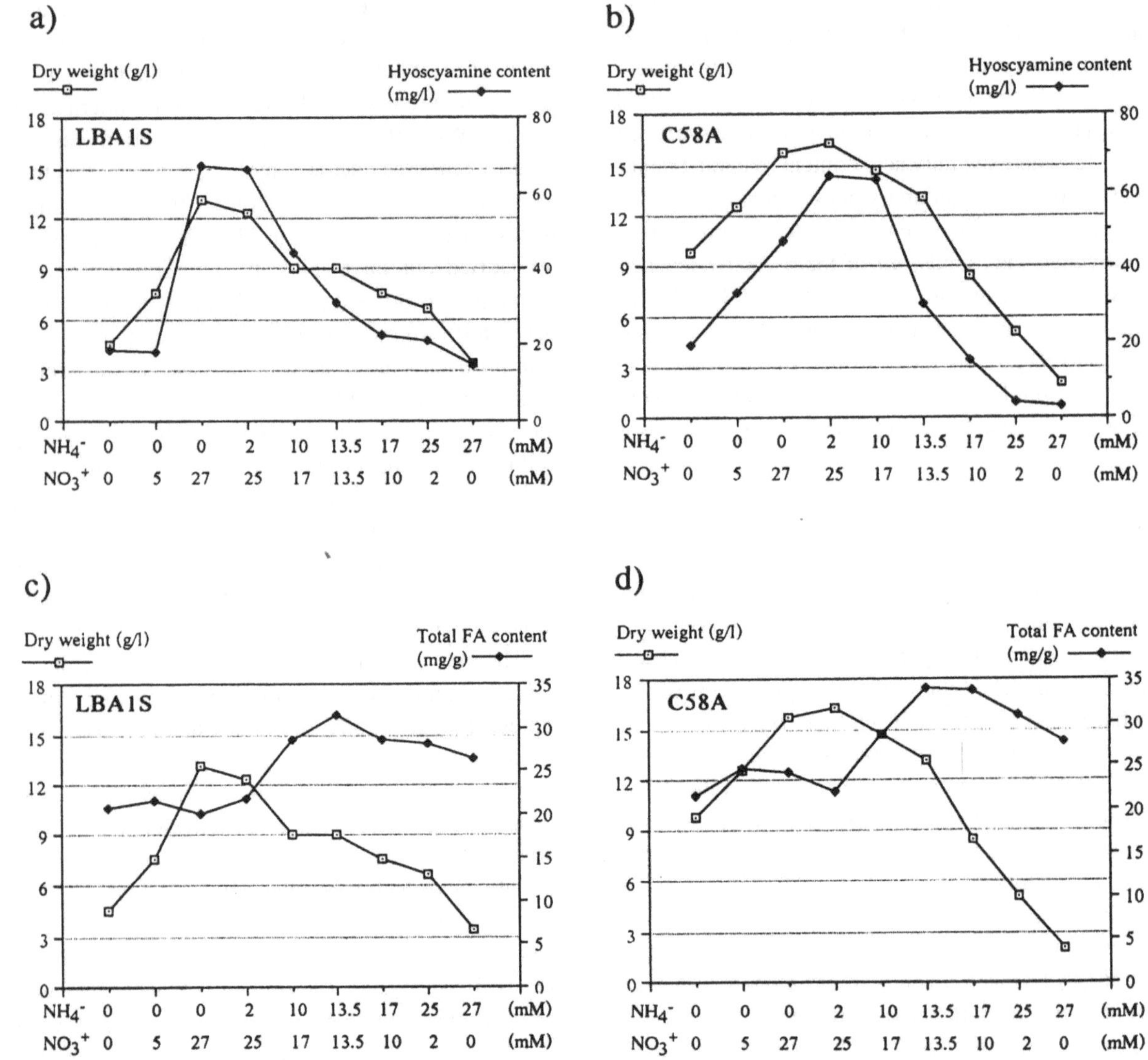

Fig. 3. Effect of ammonium-nitrate ratio on the growth and hyoscyamine accumulation in hairy root clones LBA1S (a) and C58A (b). The same effect on the total fatty acid content of the clones LBA1S and C58A is shown in panels (c) and (d), respectively.

better tolerated these changes. However, C58A was more sensitive to the increase in total nitrogen.

Total fatty acid formation in the transformed root cultures varied from 0.5 to 3.4% of the dry weight of the root material. Their synthesis was active in the exponential growth phase when the roots are rich in actively dividing meristematic cells. The main lipid classes in the transformed root cultures were phospholipids and triglycerides, and the major three individual fatty acids were linoleic, palmitic and linolenic acid. The amount of free fatty acids was small. There were differences in the total amount of lipids and modifications in their relative ratios during the time course of growth, as well as when different nutrients were applied. However, primary metabolism was less influenced by such changes than secondary metabolism.

Transformed roots are generally not so easily manipulated by changing culture conditions as cell suspension cultures, as has been shown in this study, due to their differentiated nature and high degree of genetic stability. However, there are great differences between the individual hairy root clones, and are hence not as

Table 4. Effect of nitrogen content on two hairy root clones, LBA1S and C58A, of *Hyoscyamus muticus*. Ammonium-nitrate ratio was constant (2 : 25).

	Clone LBA1S		Clone C58A	
Total nitrogen (mM)	Growth ($g\,l^{-1}$)	Hyoscyamine content ($mg\,l^{-1}$)	Growth ($g\,l^{-1}$)	Hyoscyamine content ($mg\,l^{-1}$)
0	0.88	0.47	1.34	0.35
1	7.47	18.48	8.60	6.39
10	15.08	75.62	15.20	48.10
27	15.81	73.90	15.47	67.35
50	16.13	81.55	15.49	53.35
60	16.68	83.16	15.99	53.96
75	16.79	70.10	16.62	35.43
100	17.10	79.31	18.26	30.75

uniform as has often been speculated. The *Agrobacterium* strain used for the transformation also certainly has a great influence (Vanhala et al. 1994). Even though transformed roots of *H. muticus* are stable, relatively well growing and produce hyoscyamine in a wide range of media, the requirements for certain essential nutrients, *e.g.* nitrogen and carbon, can vary considerably from clone to clone. Moreover, the alterations of the carbon-nitrogen ratio may have a marked influence on growth and alkaloid production. This therefore makes optimization work difficult because the nutritional conditions have to be optimized separately for each individual clone.

Acknowledgements

We wish to thank Tuulikki Seppänen-Laakso for advice and discussions on fatty acid determinations and Terhi Varjonen for excellent technical assistance. This work was supported by grant nr. 3753/3011/91 from the Finnish Academy (to R.H).

References

Barz W, Herzbeck H, Hüsemann W, Schneiders G & Mangold HK (1980) Alkaloids and lipids of heterotrophic, photomixotrophic and photoautotrophic cell suspension cultures of *Peganum harmala*. Planta Med. 40: 137–148

Christen P, Aoki T & Shimomura K (1992) Characteristics of growth and tropane alkaloid production in *Hyoscyamus albus* hairy roots transformed with *Agrobacterium rhizogenes* A4. Plant Cell Rep. 11: 597–600

Fowler MW (1982) Substrate utilization by plant cell cultures. J. Chem. Tech. Biotechnol. 32: 338–346

Fried J (1991) Lipids. In: Sherma J & Fried B (Eds) Handbook of Thin-Layer Chromatography, Vol 55 (pp 593–623), Marcel Dekker Inc, New York, Basel, Hong Kong

Gamborg OL, Miller RA & Ojima K (1968) Nutrient requirements of suspension cultures of soybean root cells. Exp. Cell Res. 50: 151–158

Hüsemann W (1985) Photoautotrophic growth of cells in culture. In: Vasil IK (Ed) Cell culture and somatic cell genetics of plants, Vol 2 (pp 213–252), Academic Press Inc, Orlando

Jung KH, Kwak SS, Kim SW, Lee H, Choi CY & Liu JR (1992) Improvement of the catharanthine productivity in hairy root cultures of *Catharanthus roseus* by using monosaccharides as a carbon source. Biotechnol. Lett. 14: 695–700

Kurz WGW & Constabel F (1985) Aspects affecting biosynthesis and biotransformation of secondary metabolites in plant cell cultures. CRC Crit. Rev. Biotechnol. 2: 105–118

Matsuzaki T, Koiwai A, Nagao T, Sato F & Yamada Y (1984) Lipid compositions of photomixotrophic green calluses and chlorophyll deficient leaves of tobacco. Agric. Biol. Chem. 48: 1699–1706

MacCarthy JJ & Stumpf PK (1980) Fatty acid composition and biosynthesis in cell suspension cultures of *Glycine max* (L.) Merr., *Catharanthus roseus* G. Don and *Nicotiana tabacum* L. Planta 147: 384–388

Neidleman SL (1987) Effects of temperature on lipid unsaturation. Biotech. Gen. Eng. Rev. 5: 245–268

Nguyen C, Bourgaud F, Forlot P & Guckert A (1992) Establishment of hairy root cultures of *Psoralea* species. Plant Cell Rep. 11: 424–427

Nigra HM, Alvarez MA & Giuletti AM (1990) Effect of carbon and nitrogen sources on growth and solasodine production in batch suspension cultures of *Solanum eleagnifolium* Cav. Plant Cell Tiss. Org. Cult. 21: 55–60

Nikolova P, Moo-Young M & Legge RL (1991) Application of process mass spectroscopy to the detection of metabolic changes in plant tissue culture. Plant Cell Tiss. Org. Cult. 25: 219–224

Oksman-Caldentey K-M, Vuorela H, Strauss A & Hiltunen R (1987) Variation in tropane alkaloid content of *Hyoscyamus muticus* plants and cell culture clones. Planta Med. 53: 349–354

Oksman-Caldentey K-M, Parkkinen O, Joki E & Hiltunen R (1989) Increased production of tropane alkaloids by conventional and transformed root cultures of *Hyoscyamus muticus*. Planta Med. 55: 107

Oksman-Caldentey K-M, Kivelä O & Hiltunen R (1991) Spontaneous shoot organogenesis and plant regeneration from hairy root cultures of *Hyoscyamus muticus*. Plant Sci. 78: 129–136

Panda AK, Mishra S & Bisaria VS (1992) Alkaloid production by plant cell suspension cultures of *Holarrhena antidysenterica*: I. Effect of major nutrients. Biotechol. Bioeng. 39: 1043–1051

Payne J, Hamill JD, Robins RJ & Rhodes MJC (1987) Production of hyoscyamine by 'hairy root' cultures of *Datura stramonium*. Planta Med. 53, 474–478

Radwan SS, Mangold HK, Barz W & Hüsemann W (1979) Lipids in plant tissue cultures. VIII: Reversible changes in the composition of lipids and their constituent fatty acids in response to alternate shifts in the mode of carbon supply. Chem. Phys. Lipids 25: 101–109

Seppänen-Laakso T, Laakso I & Hiltunen R (1990) Simultaneous analysis of bound and free fatty acids in human plasma by quantitative gas chromatography. Acta Pharm. Fenn. 99: 109–117

Sevón N, Hiltunen R & Oksman-Caldentey K-M (1992) Chitosan increases hyoscyamine content in hairy root cultures of *Hyoscyamus muticus*. Pharm. Pharmacol. Lett. 2: 96–99

Taya M, Yoyama A, Kondo O & Kobayashi T (1989) Growth characteristics of plant hairy roots and their cultures in bioreactors. J. Chem. Eng. Jpn. 22: 84–89

Toivonen L, Laakso S & Rosenqvist H (1992) The effect of temperature on hairy root cultures of *Catharanthus roseus*: Growth, indole alkaloid accumulation and membrane lipid composition. Plant Cell Rep. 11: 395–399

Uozumi N, Kohketsu K, Kondo O, Honda H & Kobayashi T (1991) Fed-batch culture of hairy roots using fructose as a carbon source. J. Ferm. Bioeng. 72: 457–460

Vanhala L, Hiltunen R & Oksman-Caldentey K-M (1994) Virulence of different *Agrobacterium* strains on hairy root formation of *Hyoscyamus muticus*. Plant Cell Rep. (in press)

Weber N & Mangold HK (1988) Lipids. In: Constabel F & Vasil IK (Eds) Cell culture and somatic cell genetics of plants, Vol 5 (pp 509–534), Academic Press Inc, San Diego

Plant Cell, Tissue and Organ Culture **38**: 273–279, 1994.

Catharanthine and ajmalicine synthesis in *Catharanthus roseus* hairy root cultures

Medium optimization and elicitation

F. Vázquez-Flota, O. Moreno-Valenzuela, M.L. Miranda-Ham, J. Coello-Coello & V.M. Loyola-Vargas*

*Dept. Biotecnología, División de Biología Vegetal, Centro de Investigación Científica de Yucatán, Apdo. Postal 87, 97310 Cordemex, Yucatán, México (*requests for offprints)*

Key words: Ajmalicine, alkaloids, catharanthine, *Catharanthus roseus*, hairy roots.

Abstract

Two year old, transformed root cultures of *Catharanthus roseus* accumulate ajmalicine and catharanthine (0.57 and 0.36 mg g^{-1} DW, or 7.0 and 3.0 mg l^{-1}, respectively). Changes in the concentration of the medium components, as well as the addition of hydrolytic enzymes and biotic elicitors, were used as strategies to increase these alkaloid yields. Regarding the components of the medium, the results obtained, when sucrose was raised from 3 to 4.5%, are noteworthy. The nitrogen source induced differential responses in the individual alkaloid yields. No net change in the alkaloid content was observed either with changes in the concentration of vitamins or macro- and micronutrients. Though the root culture only shows a limited response to elicitors, *Aspergillus* treatment and the use of macerozyme increased the accumulation of ajmalicine selectively, while the addition of methyl jasmonate increased the yield of both alkaloids.

Abbreviations: MeJa – methyl jasmonate, mU – milliunits

Introduction

Catharanthus roseus produces a wide spectrum of indole alkaloids when cultured in vitro, both in cell suspension or in transformed root cultures (Parr et al. 1988; Toivonen et al. 1989, 1990; Brillanceau et al. 1989). A common approach used in order to obtain higher yields of these molecules is the two stage process in which cultures are grown to maximum biomass in a growth medium and then are transferred to a production medium that limits growth, but stimulates alkaloid biosynthesis (Zenk et al. 1977). However, the development of a single medium that could combine both purposes would be advantageous from a commercial point of view. For *C. roseus* cell suspension cultures, different production media have been formulated (Morris 1986; Smith et al. 1987), and although there is a report of an optimized medium for transformed roots (Toivonen et al. 1991), it is known that the effect on alkaloid accumulation can vary greatly among different lines depending on their nutritional status.

On the other hand, plant cell suspension cultures have been shown to accumulate alkaloids as a response to different stimulators of abiotic or biotic origin (DiCosmo & Misawa 1985; Eilert et al. 1986; Kurz et al. 1987; Godoy-Hernández & Loyola-Vargas 1991). A novel compound, jasmonic acid, has been pointed out as a signal transducer in the elicitation process (Gundlach et al. 1992). A side benefit of the induction by elicitors of the accumulation of alkaloids is that the activities of the enzymes involved in alkaloid biosynthesis would be concomitantly altered and as Stöckigt (1980) pointed out, the need for "cultures capable of synthesizing large quantities of these alkaloids" will be fulfilled.

In the present study, different components of Gamborg's B5 medium have been modified, and different elicitors have been used in order to obtain the appropriate environment for a *C. roseus* transformed root culture to produce high alkaloid yields.

Materials and methods

Plant material

Hairy root R/J1 line was obtained through genetic transformation of seedling roots of *Catharanthus roseus* with *Agrobacterium rhizogenes* strain 1855, bearing plasmid pBI.121.1. The root line was one year old at the beginning of the experiments and has been maintained through subculturing every 14 days in half strength Gamborg's B5 medium (B5/2, Gamborg et al. 1968) supplemented with 3% sucrose. The root cultures were kept on orbital shakers at 100 rpm, 25 °C and in the dark (Ciau-Uitz et al. 1994).

Modification of Gamborg's medium

In order to develop a medium that would lead to higher alkaloid yields, the original mineral and organic composition of B5/2 medium was modified. Regarding the major salts, $CaCl_2.2H_2O$, $MgSO_4.7H_2O$ and NaH_2PO_4 were assayed in concentrations, which ranged from total absence to a 5-fold increase of calcium and magnesium and a 16-fold increase of phosphate. The nitrogen source was also varied, both in concentration as in the nitrate/ammonium ratio. The total nitrogen concentrations assayed ranged from 3.1 to 37 mM (nitrate from 3.1 to 25 mM and ammonium from 0 to 12 mM). Among the minor salts, $CuSO_4.5H_2O$, $MnSO_4.H_20$ and $ZnSO_4.7H_2O$ were evaluated in concentrations, which ranged from zero to a 10-fold increase. The modified organic components were nicotinic acid, thiamine, pyridoxine and myo-inositol in concentrations ranging from 0 to a 16-fold increase over those present in B5/2. Since sucrose has always been the carbon source of choice, a concentration range from 2 to 6% was assayed. Maltose, lactose, glucose, galactose and fructose were also tested.

Elicitation experiments

Hydrolytic enzymes. Five different concentrations were employed for chitinase, macerozyme and cellulase (chitinase, 0.625, 1.0, 2.5, 5.0 and 10.0 mU; macerozyme, 0.005, 0.05, 0.1, 0.5 and 1.0% and cellulase, 0.1, 0.25, 0.5, 1.0 and 2.0%). The enzymes were dissolved in distilled water and filter-sterilized prior to application on day 20 of the culture cycle. The roots were harvested 48 h later and freeze-dried.

Methyl jasmonate. Five treatments with MeJa were applied to the roots on day 30 of the culture cycle: 0.01, 0.1, 1.0, 10 and 100 μM. Root tissues were harvested after 72 h and immediately freeze-dried. An aqueous solution of MeJa (kindly donated by Dr. C. Nessler) was diluted with ethanol to the desired concentration and applied as a filter-sterilized solution to the flasks.

Fungal homogenates. As a result of preliminary experiments, three fungi species were chosen for the biotic elicitation experiments: *Aspergillus* spp., *Trichoderma viride* and *Trichoderma reseii* and a yeast, *Rhodotorula marina*. The fungal strains were maintained in potato-dextrose-agar at 4 °C in the dark. Spores were used to inoculate Erlenmeyer flasks, containing 100 ml of potato-dextrose medium and were cultured in a shaking bath in the dark at 37 °C and 200 rpm. The mycelia were collected after 7 days by filtration through cheesecloth and homogenized in distilled water with a Polytron. The homogenates were then autoclaved. Concentrations of fungal homogenates equivalent to 0.6, 1.2 and 1.8 mg glucose/ml, were used. Total soluble sugars were determined as described in (Dubois et al. 1956). Transformed roots were inoculated on day 20 of their culture cycle and harvested after 72 h.

For the elicitation experiments the initial inoculum was 0.5 g fresh weight in 250 ml Erlenmeyer flasks, containing 100 ml of B5/2 medium.

Growth determinations

Growth, as fresh and dry weight, was evaluated in triplicate, after 14 and 28 days of culture for each nutritional condition and after the incubation time specified for each elicitation experiment.

Alkaloid quantification

Alkaloids were extracted as reported by Monforte-González et al. (1992). Ajmalicine and catharanthine were quantified by TLC densitometry using as solvent system chloroform:acetone (8:2) (Monforte-González et al. 1992).

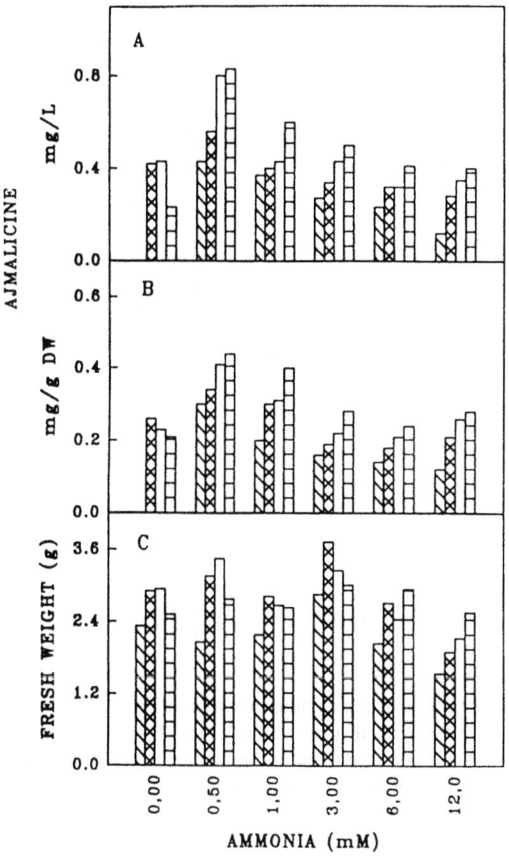

Fig. 1. Ajmalicine yield (A), ajmalicine accumulation (B) and fresh weight (C) of transformed root cultures of *C. roseus* cultured under different combinations of NO_3 and NH_4 for 14 days. The standard deviation was less than 10%. 3.12 mM NO_3 (▨), 6.25 mM NO_3 (▩), 12.5 mM NO_3 (▯), and 25.0 mM NO_3 (▤).

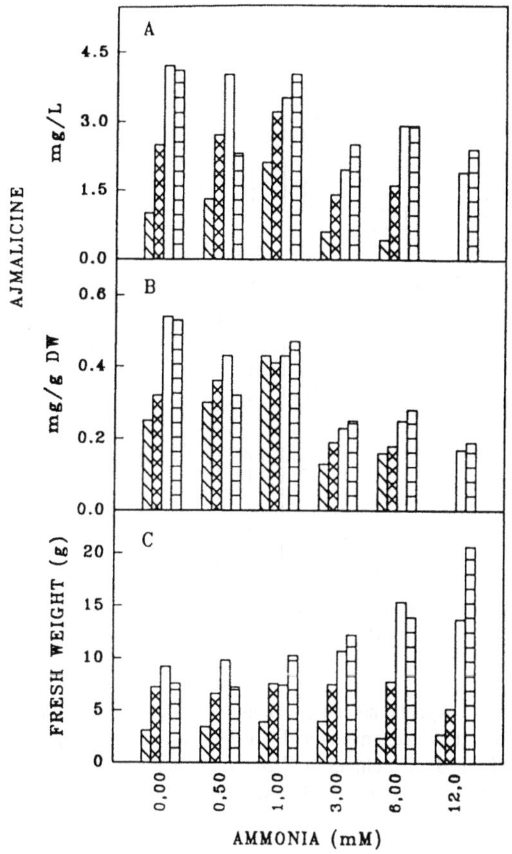

Fig. 2. Ajmalicine yield (A), ajmalicine accumulation (B) and fresh weight (C) of transformed root cultures of *C. roseus* cultured under different combinations of NO_3 and NH_4 for 28 days. The standard deviation was less than 10%. 3.12 mM NO_3 (▨), 6.25 mM NO_3 (▩), 12.5 mM NO_3 (▯), and 25.0 mM NO_3 (▤).

Results and discussion

Two different strategies were tested in order to obtain higher alkaloid yields. The first one involved the modification of the medium composition and the second one the utilization of different elicitation treatments, such as the addition of fungal homogenates, methyl jasmonate and hydrolytic enzymes.

Medium composition

Growth was significantly promoted only when roots were cultured for 28 days in media containing 2.0 mM phosphate (4-fold increase over the original concentration) or with a nitrate:ammonium concentration of 25:12 mM (2/1 ratio). In both cases, there was almost a 2-fold increase in fresh and dry weight; nevertheless, their volumetric alkaloid yields were similar or lower than those obtained in B5/2 (12.5 mM NO_3/0.5 mM NH_4) (Fig. 2). Alkaloid accumulation was promoted in those roots cultured in the absence of phosphate for 14 days, but little growth was observed, thus a very low volumetric yield resulted. Experiments in which roots were first cultured in a complete medium for biomass production and then transferred for a certain period of time without phosphate were not successful (data not shown). Apparently, this lack of response was due to a high internal pool of phosphate (Vázquez-Flota et al. 1992), and may also have been due to the fact that high phosphate levels in the medium are inhibitory for alkaloid accumulation (Knobloch & Berlin 1980).

The results obtained with different nitrogen levels are shown in Fig. 1 and 2. In none of the cases the volumetric alkaloid yield was significantly higher than that found in B5/2 (12.5 mM NO_3/0.5 mM NH_4). There were no important changes in the catharanthine accumulation pattern, whether the roots were cultured for 14 or 28 days (data not shown).

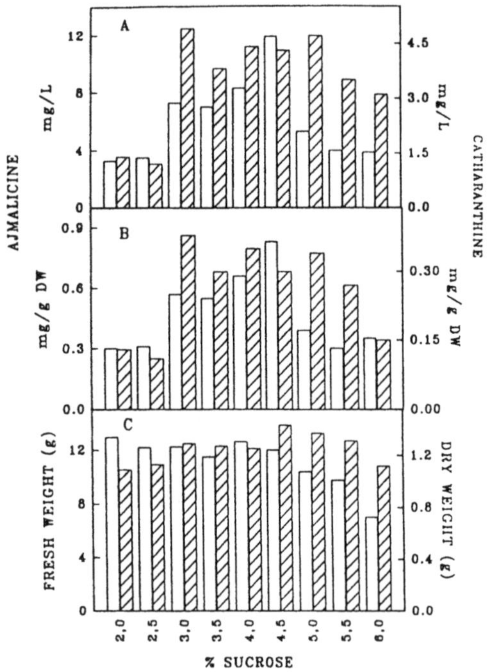

Fig. 3. Ajmalicine and catharanthine yield (A), ajmalicine and catharanthine accumulation (B) and fresh and dry weight (C) of transformed root cultures of *C. roseus* cultured under different concentrations of sucrose for 28 days. The standard deviation was less than 10%.

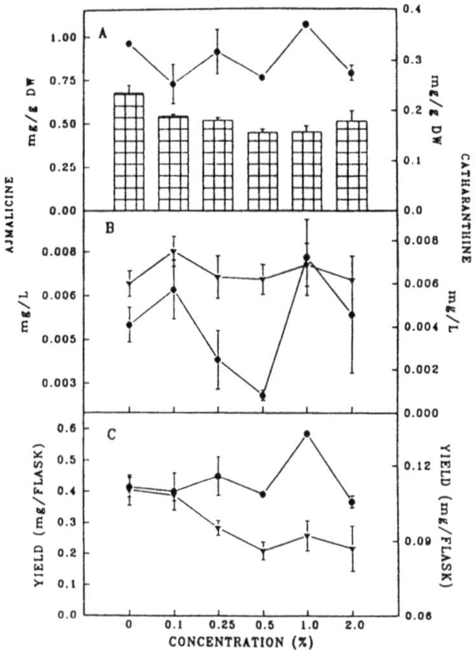

Fig. 5. Effect of the addition of different concentrations of cellulase on the accumulation of individual alkaloids. A, ajmalicine (●) and catharanthine (bars) in the tissues; B, ajmalicine (●) and catharanthine (▼) in the culture medium; C, ajmalicine (●) and catharanthine (▼) yields.

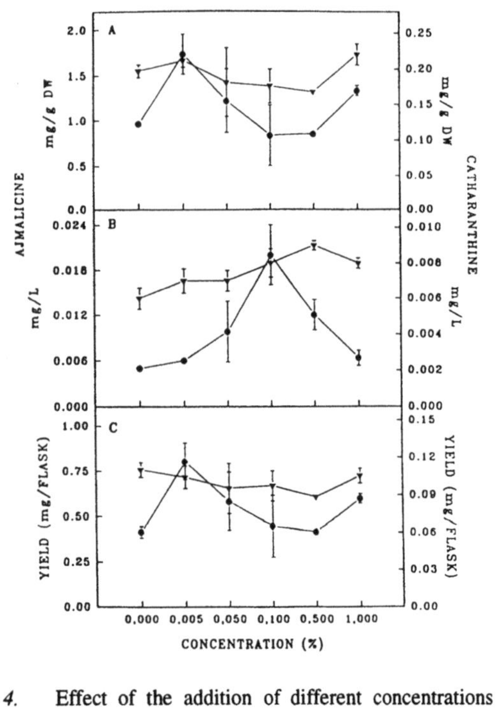

Fig. 4. Effect of the addition of different concentrations of macerozyme on the accumulation of individual alkaloids. A, ajmalicine (●) and catharanthine (▼) in the tissues; B, ajmalicine (●) and catharanthine (▼) in the culture medium; C, ajmalicine (●) and catharanthine (▼) yields.

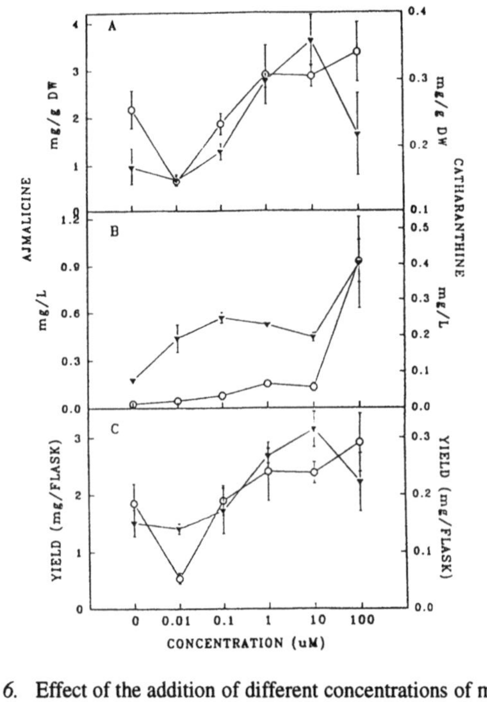

Fig. 6. Effect of the addition of different concentrations of methyl jasmonate on the accumulation of individual alkaloids. A, ajmalicine (▼) and catharanthine (○) in the tissues; B, ajmalicine (▼) and catharanthine (○) in the culture medium; C, ajmalicine (▼) and catharanthine (○) yields.

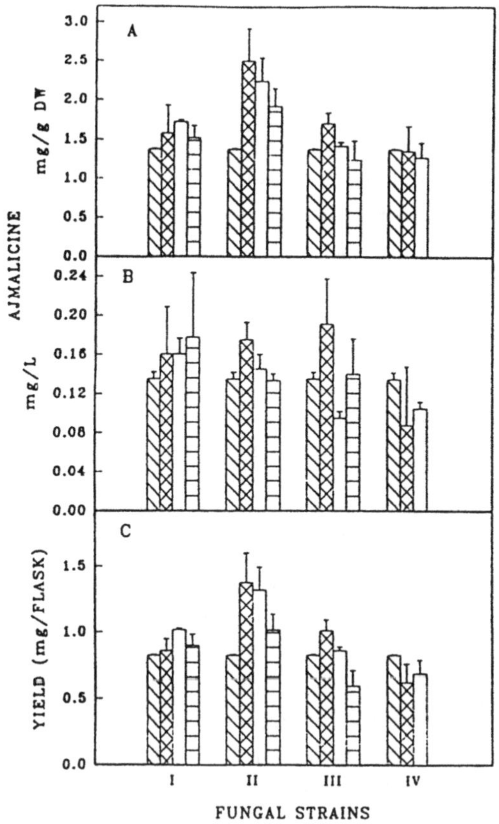

Fig. 7. Effect of the addition of fungal homogenates in different concentrations (mg glucose/ml) on ajmalicine accumulation. A, ajmalicine content in the tissues; B, ajmalicine content in the culture medium; C, yield. I, *Rhodotorula marina*; II, *Aspergillus* spp.; III, *Trichoderma viride*; IV, *Trichoderma reseii*. (▧) Control; (▩) 0.6 mg glucose/ml; (□) 1.2 mg glucose/ml; (▤) 1.8 mg glucose/ml.

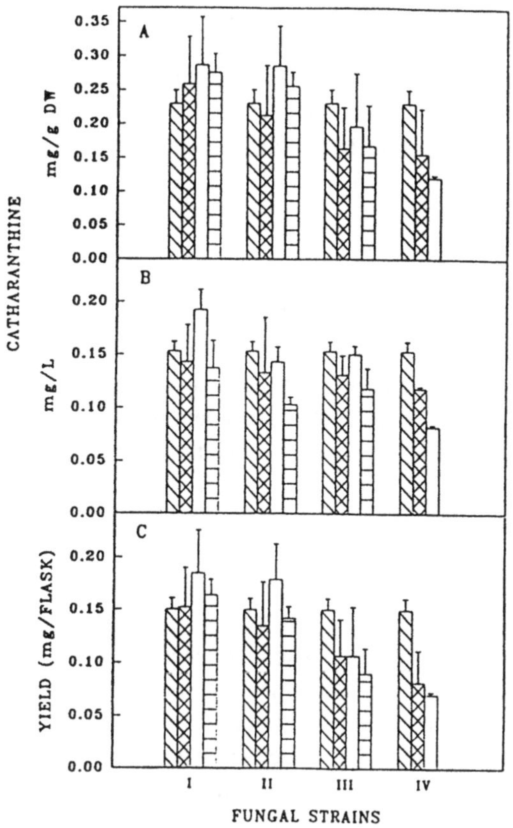

Fig. 8. Effect of the addition of fungal homogenates in diferent concentrations (mg glucose/ml) on catharanthine accumulation. A, ajmalicine content in the tissues; B, ajmalicine content in the culture medium; C, yield. I, *Rhodotorula marina*; II, *Aspergillus* spp.; III, *Trichoderma viride*; IV, *Trichoderma reseii*. (▧) Control; (▩) 0.6 mg glucose/ml; (□) 1.2 mg glucose/ml; (▤) 1.8 mg glucose/ml.

Carbon sources, other than sucrose, modified both growth and alkaloid accumulation, but in none of them these parameters were higher than those observed in the transformed roots cultured in media supplemented with 3% sucrose. The same behavior could be observed in roots cultured for 14 or 28 days. A second culture cycle with carbon sources in which similar growth as in 3% sucrose was noticed (6% glucose, galactose and fructose or 3% glucose with either 3% galactose or 3% fructose), did not increase the alkaloid yield (data not shown). These data contrast from those reported by Jung et al. (1992), in which there was a two-fold increase in catharanthine accumulation, when the carbon source was changed from 3% sucrose to 3% fructose.

Although no detectable increase was observed in the volumetric alkaloid yield in experiments with media supplemented with 2 to 6% sucrose after 14 days of culture (data not shown), there was a 45% increase in ajmalicine yield in roots cultured for 28 days with 4.5% sucrose. The increase in ajmalicine was due to a net increase in its accumulation, since growth was not significantly affected (Fig. 3). This result contrasts with the report that in an optimized medium just growth, but not alkaloid accumulation, was affected (Toivonen et al. 1991).

Elicitation

Hydrolytic enzymes

The addition of certain concentrations of chitinase, macerozyme and cellulase increased the liberation of alkaloids into the media. Ajmalicine liberation was highest when 0.1% macerozyme or 1.0% cellulase was used (Fig. 4 and 5). An increase in the excretion of secondary metabolites as a response to elicitation has been observed in other in vitro systems (Brooks et al. 1986; Threlfall & Whitehead 1988; Vögeli & Chappell 1988). The liberation of alkaloids into the media may be a defense mechanism of plant cells to inhibit growth of pathogens. Catharanthine levels found in the tissues and in the media did not show significant changes when the hydrolytic enzymes were employed (Fig. 4A, 4B, 5A and 5B).

Methyl jasmonate

Both, ajmalicine and catharanthine, were excreted to the medium, depending of the concentration employed. Maximal excretion took place with 100 μM MeJa (Fig. 6). In a similar fashion, the accumulation of these alkaloids in the tissue is also dependent on the concentration. At 10 μM, ajmalicine content was four-fold with respect to the untreated roots (Fig. 6A). The observed dependence of the alkaloid accumulation on the MeJa dose has been documented for other systems (Gundlach et al. 1992).

It has been suggested recently that MeJa diffuses to the cytoplasm, where an esterase hydrolyzes it to jasmonic acid, which has been proposed to be acting as a second messenger in the defense mechanism against pathogens (Farmer & Ryan 1990; Gundlach et al. 1992). The differential response observed with different concentrations of MeJa could be in direct relationship with the quantity that esterases are able to hydrolyze.

Fungi

When the excretion of alkaloids to the medium was evaluated, no response could be observed with any of the strains employed in this work. This fact contrasts markedly with the observations of Nef et al. (1991), who reported that low concentrations of the fungal homogenates induced the liberation of 90% of the total ajmalicine into the medium, while serpentine accumulated preferentially in the tissue.

Only the use of *Aspergillus* homogenates produced an increase in both the accumulation and yield of ajmalicine, whereas, no response was observed with the other fungi and yeast (Fig. 7).

The accumulation of secondary metabolites depends on the dose employed and on the strain, as has been reported for other systems (Kombrink & Hahlbrock 1986; Chappell & Nable 1987; Flores et al. 1988; Collinge & Brodelius 1989; Mukundan & Hjortso 1990a,b). The response of plant cells and tissues to elicitation with fungal homogenates is in direct relation with the composition of the cell wall of the fungi, where their different components are the true elicitors. Some types of cellulose-glucans are better elicitors than chitin-glucan (Kombrink & Hahlbrock 1986). For this work, the oligosaccharides liberated from the cell wall of *Aspergillus* were the best elicitors, even though their identity is still unknown. Nevertheless, the induction response seems to be specific for ajmalicine, since catharanthine did not increase its concentration in the medium or in the tissues (Fig. 8).

Conclusions

Our data for the alkaloid volumetric yield in the B5/2 medium with 4.5% sucrose (15 mg l^{-1} of ajmalicine plus catharanthine) are at the same level as those reported by Toivonen et al. (1991), i.e. 13 mg l^{-1} of the same alkaloids. These results are far from the maximum catharanthine production reported by Jung et al. (1992), who determined 40 mg l^{-1} when fructose was employed as the carbon source.

Regarding the elicitation experiments, it can be seen that the effect of some of these elicitors is specific for the production of an alkaloid in particular, ajmalicine in this case, which contributes at a high extent to the total alkaloid yield. We are currently working on the determination of some of the enzymatic activities involved in the synthesis of ajmalicine under the culture conditions where the optimal sucrose concentration and the best elicitors are employed together.

Acknowledgements

This work was supported by the National Council for Science and Technology (CONACyT, México), Grant No. 0429N. The authors wish to acknowledge the technical assistance of QBB Marina Navarrete-Loeza and the revision of the English version of the manuscript by Dr. Ingrid Olmsted.

References

Brillanceau M-H, David C & Tempé J (1989) Genetic transformation of *Catharanthus roseus* G. Don by *Agrobacterium rhizogenes*. Plant Cell Rep. 8: 63–66

Brooks CJW, Watson DG & Freer IM (1986) Elicitation of capsidiol accumulation in suspended callus cultures of *Capsicum annuum*. Phytochemistry 25: 1089–1092

Chappell J & Nable R (1987) Induction of sesquiterpenoid biosynthesis in tobacco cell suspension cultures by fungal elicitor. Plant Physiol. 85: 469–473

Ciau-Uitz R, Miranda-Ham ML, Coello-Coello J, Chí B, Pacheco LM & Loyola-Vargas VM (1994) Indole alkaloid production by transformed and non-transformed root cultures of *Catharanthus roseus*. In vitro Cell. Develop. Biol. Plant, in press

Collinge M & Brodelius P (1989) Dynamics of benzophenanthridine alkaloid production in suspension cultures of *Eschscholtzia californica* after treatment with a yeast elicitor. Phytochemistry 28: 1101–1104

DiCosmo F & Misawa M (1985) Eliciting secondary metabolism in plant cell cultures. Trends Biotechnol. 3: 318–322

Dubois M, Gilles KA, Hamilton JK, Rebers PA & Smith F (1956) Colorimetric method for determination of sugars and related substances. Anal. Chem. 28: 350–356

Eilert U, Constabel F & Kurz WGW (1986) Elicitor-stimulation of monoterpene indole alkaloid formation in suspension cultures of *Catharanthus roseus*. J. Plant Physiol. 126: 11–22

Farmer EE & Ryan CA (1990) Interplant communication: airbone methyl jasmonate induces synthesis of proteinase inhibitors in plant leaves. Proc. Natl. Acad. Sci. USA 87: 7713–7716

Flores HE, Pickard JJ & Signs M (1988) Elicitation of polyacetylene production in hairy root cultures of Asteraceae. Plant Physiol. 86: 108s Abstract

Gamborg OL, Miller RA & Ojima K (1968) Nutrient requirements of suspension cultures of soybean root cells. Exp. Cell Res. 50: 151–158

Godoy-Hernández G & Loyola-Vargas VM (1991) Effect of fungal homogenate, enzyme inhibitors and osmotic stress on alkaloid content of *Catharanthus roseus* cell suspension cultures. Plant Cell Rep. 10: 537–540

Gundlach H, Müller MJ, Kutchan TM & Zenk MH (1992) Jasmonic acid is a signal transducer in elicitor-induced plant cell cultures. Proc. Natl. Acad. Sci. USA 89: 2389–2393

Jung KH, Kwak SS, Kim SW, Lee H, Choi CY & Liu JR (1992) Improvement of catharanthine productivity in hairy root cultures of *C.roseus* by using monosaccharides as carbon source. Biotechnol. Lett. 14: 695–700

Knobloch K-H & Berlin J (1980) Influence of medium composition on the formation of secondary compounds in cell suspension cultures of *Catharanthus roseus* (L.) G. Don. Z. Naturforsch. 35c: 551–556

Kombrink E & Hahlbrock K (1986) Responses of cultured parsley cells to elicitors from phytopathogenic fungi. Timing and dose dependency of elicitor-induced reactions. Plant Physiol. 81: 216–221

Kurz WGW, Constabel F, Eilert U & Tyler RT (1987) Elicitor treatment: a method for metabolite production by plant cell cultures in vitro. In: Breimer DD & Speiser P (Ed) Topics in Pharmaceutical Sciences 1987, (pp 283–290). Elsevier Science Publishers

Monforte-González M, Ayora-Talavera T, Maldonado-Mendoza IE & Loyola-Vargas VM (1992) Quantitative analysis of serpentine and ajmalicine in plant tissues of *Catharanthus roseus* and hyoscyamine and scopolamine in root tissues of *Datura stramonium* by densitometry in thin layer chromatography. Phytochem. Anal. 3: 117–121

Morris P (1986) Regulation of product synthesis in cell cultures of *Catharanthus roseus*. II. Comparison of production media. Planta Med. 52: 121–126

Mukundan U & Hjortso MA (1990a) Thiophene accumulation in hairy roots of *Tagetes patula* in response to fungal elicitors. Biotechnol. Lett. 12: 609–614

Mukundan U & Hjortso MA (1990b) Effect of fungal elicitor on thiophene production in hairy root cultures of *Tagetes patula*. Appl. Microbiol. Biotechnol. 33: 145–147

Nef C, Rio B & Chrestin H (1991) Induction of catharanthine synthesis and stimulation of major indole alkaloids production by *Catharanthus roseus* cells under non-growth-altering treatment with *Pythium vexans* extracts. Plant Cell Rep. 10: 26–29

Parr AJ, Peerless ACJ, Hamill JD, Walton NJ, Robins RJ & Rhodes MJC (1988) Alkaloid production by transformed root cultures of *Catharanthus roseus*. Plant Cell Rep. 7: 309–312

Smith JI, Quesnel A, Smart NJ, Misawa M & Kurz WGW (1987) The development of a single-stage growth and indole alkaloid production medium for *Catharanthus roseus* (L.) G. Don suspension cultures. Enzyme Microbiol. Technol. 9: 466–469

Stöckigt J (1980) The biosynthesis of heteroyohimbine-type alkaloids. In: Phillipson JD & Zenk MH (Ed) Indole and biogenetically related alkaloids, (pp 113–141). Academic Press, London

Threlfall DR & Whitehead IM (1988) The use of metal ions to induce the formation of secondary products in plant tissue culture. In: Robins RJ & Rhodes MJC (Ed) Manipulating Secondary Metabolism in Culture, (pp 51–56). Cambridge University Press, Cambridge

Toivonen L, Balsevich J & Kurz WGW (1989) Indole alkaloid production by hairy root cultures of *Catharanthus roseus*. Plant Cell Tissue Organ Cult. 18: 79–93

Toivonen L, Ojala M & Kauppinen V (1990) Indole alkaloid production by hairy root cultures of *Catharanthus roseus*: growth kinetics and fermentation. Biotechnol. Lett. 12: 519–524

Toivonen L, Ojala M & Kauppinen V (1991) Studies on the optimization of growth and indole alkaloid production by hairy root cultures of *Catharanthus roseus*. Biotechnol. Bioeng. 37: 673–680

Vázquez-Flota F, Coello J & Loyola-Vargas VM (1992) Growth kinetics and alkaloid production in hairy root cultures of *Catharanthus roseus*. Plant Physiol. 99: 49s (Abstract)

Vögeli U & Chappell J (1988) Induction of sesquiterpene cyclase and suppression of squalene synthetase activities in plant cell cultures treated with fungal elicitor. Plant Physiol. 88: 1291–1296

Zenk MH, El-Shagi H, Arens H, Stöckigt J, Weiler EW & Deus B (1977) Formation of the indole alkaloids serpentine and ajmalicine in cell suspension cultures of *Catharanthus roseus*. In: Barz W, Reinhard E & Zenk MH (Ed) Plant Tissue Culture and its Bio-technological Application, (pp 27–43). Springer-Verlag, Berlin

Plant Cell, Tissue and Organ Culture **38:** 281–287, 1994.

A novel 2-oxoglutarate-dependent dioxygenase involved in vindoline biosynthesis: characterization, purification and kinetic properties

Emidio De Carolis & Vincenzo De Luca
Institut de Recherche en Biologie Végétale, Département de Sciences Biologiques, Université de Montréal, Montréal, Québec, Canada H1X 2B2

Key words: *Catharanthus roseus*, enzymology, 2-oxoglutarate-dependent dioxygenase, vindoline

Abstract

The enzyme, desacetoxyvindoline 4-hydroxylase, was purified to apparent homogeneity from *Catharanthus roseus* by ammonium sulfate precipitation and successive chromatography on Sephadex G-100, green 19-agarose, hydroxylapatite, α-kg sepharose and Mono Q. The 4-hydroxylase was characterized by its strict specificity for position 4 of desacetoxyvindoline suggesting it to catalyze the second to last step in vindoline biosynthesis. The molecular mass of the native and denatured 4-hydroxylase was 45 kDa and 44.7 kDa, respectively, suggesting that the native enzyme is a monomer. Two-dimensional isoelectric focusing under denaturing conditions resolved the purified 4-hydroxylase into three charge isoforms of pIs 4.6, 4.7 and 4.8. The purified 4-hydroxylase exhibited no requirement for divalent cations, but inactive enzyme was reactivated in a time-dependent manner by incubation with ferrous ions. The enzyme was not inhibited by EDTA or SH-group reagents at concentrations up to 10 mM. The mechanism of action of desacetoxyvindoline 4-hydroxylase was investigated. The results of substrate interaction kinetics and product inhibition studies suggest an Ordered Ter Ter mechanism where α-kg is the first substrate to bind followed by the binding of O_2 and desacetoxyvindoline. Their K_m values for α-kg, O_2 and desacetoxyvindoline are 45 µM, 45 µM and 0.03 µM, respectively. The first product to be released was deacetylvindoline followed by CO_2 and succinate, respectively.

Abbreviations: α-kg – α-ketoglutarate or 2-oxoglutarate, NMT – N-methyltransferase, SAM – S-adenosyl-L-methionine, TLC – thin layer chromatography, VBL – vinblastine, VCR – vincristine

Introduction

Alkaloids are widely distributed throughout the plant kingdom. It has been estimated that 15–30% (Robinson 1981) of plants produce alkaloids and that over 25% of these are derived from tryptophan. The family Apocynaceae which consists of the genera *Rauwolfia*, *Catharanthus* and *Aspidosperma* are a very rich source of tryptophan-derived indole alkaloids. The indole alkaloids of this family have been extensively investigated mostly because of their physiological effects on man and their use in pharmacy. The *Vinca* alkaloids represent a class of natural drugs derived from the periwinkle plant, *Catharanthus roseus*. Two commercially important bis-indole alkaloids, vinblastine (VBL) and vincristine (VCR), are known to accumulate in the aerial parts of *C. roseus*. These dimeric indole alkaloids have been extensively investigated because of their usefulness in treating certain neoplasms.

In contrast with the extensive studies on the structural elucidation of indole alkaloids in *C. roseus* (Robinson 1981) which led to knowledge of the biosynthetic pathway of VBL and VCR, less is known of the enzymology of this pathway (De Luca 1993). This situation may be explained by the inherent difficulties in the preparation of specifically substituted indole alkaloid substrates and reaction products that are required to conduct enzymatic studies. The recent advances in our understanding of the late stages of vindoline biosynthesis (Balsevich et al. 1986) has facilitated the synthesis of appropriate substrates for the development of enzyme assays.

Since VBL and VCR accumulate in low amounts, a considerable amount of research has been devoted to study the production of these alkaloids by cell and tissue culture methods (De Luca & Kurz 1988). Unfortunately, this approach has not yet succeeded to produce bis-indole alkaloids since cell cultures are unable to synthesize vindoline, one of the monomeric precursors of VBL and VCR. In order to study enzymes involved in the late stages of vindoline biosynthesis, we decided to use intact *C. roseus* plants which produce and accumulate vindoline.

The biosynthesis of vindoline from tabersonine involves three hydroxylations, one *O*-methylation, one *N*-methylation and an *O*-acetylation (Fig. 1). Recently, a number of these enzymes have been purified and characterized from *C. roseus*. The third last step in vindoline biosynthesis is catalyzed by *N*-1 desmethyldesacetoxyvindoline (16-methoxy-2,3-dihydro-3-hydroxytabersonine) *N*-methyltransferase which has been partially purified (Dethier & De Luca 1993). The last step in vindoline biosynthesis is catalyzed by deacetylvindoline 4-*O*-acetyltransferase (DAT) which has been purified to homogeneity (Power et al. 1990). The enzyme which catalyzes the second last step in vindoline biosynthesis requires indole alkaloid substrate, α-kg, ascorbate, ferrous ions and molecular oxygen for activity and thus is classified as a 2-oxoglutarate-dependent dioxygenase (De Carolis et al. 1990).

The fact that hydroxylation at position 4 is critical for the enzymatic synthesis of vindoline and that desacetoxyvindoline 4-hydroxylase is absent in cell cultures prompted us to develop a protocol for the purification to homogeneity of desacetoxyvindoline 4-hydroxylase. Having developed a purification procedure which yields a highly purified enzyme preparation as well as having a direct enzyme assay which is simple, fast and accurate we attempted to elucidate the kinetic mechanism of desacetoxyvindoline 4-hydroxylase.

Results and discussion

Synthesis of N(1)-[$^{14}CH_3$] desacetoxyvindoline

The assay for desacetoxyvindoline 4-hydroxylase is based on the availability of the substrate, *N*(1)-[$^{14}CH_3$] desacetoxyvindoline (De Carolis et al. 1990). Briefly, crude desalted extracts containing desmethyldesacetoxyvindoline NMT activity from young leaves of *C. roseus* were used to synthesize *N*(1)-[$^{14}CH_3$] desacetoxyvindoline using desmethyldesacetoxyvindoline as the methyl acceptor and *S*-adenosyl-*L*-[$^{14}CH_3$]methionine (SAM) as the methyl donor (Fig. 1). The NMT assay mixture contained 24.5 μM of desmethyldesacetoxyvindoline, 25 nmol labelled SAM (containing 2,750,000 dpm), and 2 mg of protein exhibiting NMT activity (in 100 mM Tris-HCl [pH 8.0]) in a final volume of 320 μl. The enzyme reaction was started by addition of protein and the mixture was incubated for 80 min at 30 °C. The reaction was stopped by addition of 100 μl of 1 N NaOH. Desmethyldesacetoxyvindoline and *N*(1)-[$^{14}CH_3$] desacetoxyvindoline were extracted in 3 × 150 μl of ethyl acetate. The alkaloids were separated by preparative TLC (silica) using 10% methanol in ethyl acetate as the solvent system. The band corresponding to *N*(1)-[$^{14}CH_3$] desacetoxyvindoline was identified, scraped and extracted from the silica powder with methanol. A single product corresponding to *N*(1)-[$^{14}CH_3$]desacetoxyvindoline was obtained with a 36% yield.

Desacetoxyvindoline 4-hydroxylase assay

Desacetoxyvindoline 4-hydroxylase was assayed by measuring the formation of *N*(1)-[$^{14}CH_3$]deacetylvindoline (Fig. 1). The assay mixture contained 0.56 nmol of labelled alkaloid substrate (containing 44,600 DPM), 10 mM α-kg, 7.5 mM ascorbate and 10 μM Fe^{2+} and up to 150 μg of protein in a total volume of 200 μl. The 4-hydroxylase catalyzes the incorporation of one atom of molecular oxygen into position 4 of the alkaloid moiety while the second is incorporated into the keto group of α-kg. The labile intermediate is then decarboxylated leading to the liberation of CO_2 and the formation of succinate. The assay was started by the addition of enzyme and the mixture incubated at 30 °C for 15 min. The reaction was stopped by the addition of 100 μl of 1 M NaOH and the aqueous phase was extracted for indole alkaloids with ethyl acetate. The organic phase was recovered and after evaporation to dryness, the substrate (R_f 0.52) and product (R_f 0.22) were separated by TLC (silica) using 10% methanol in ethyl acetate as the solvent. After chromatography *N*(1)-[$^{14}CH_3$]deacetylvindoline was isolated from the silica support and the radioactivity counted using a scintillation counter.

Figure 2 shows a photograph of an autoradiogram of the chromatographed reaction product of desacetoxyvindoline 4-hydroxylase activity. The lanes represent fractions eluting from a gel filtration column. The autoradiogram shows that hydroxylase activity elutes

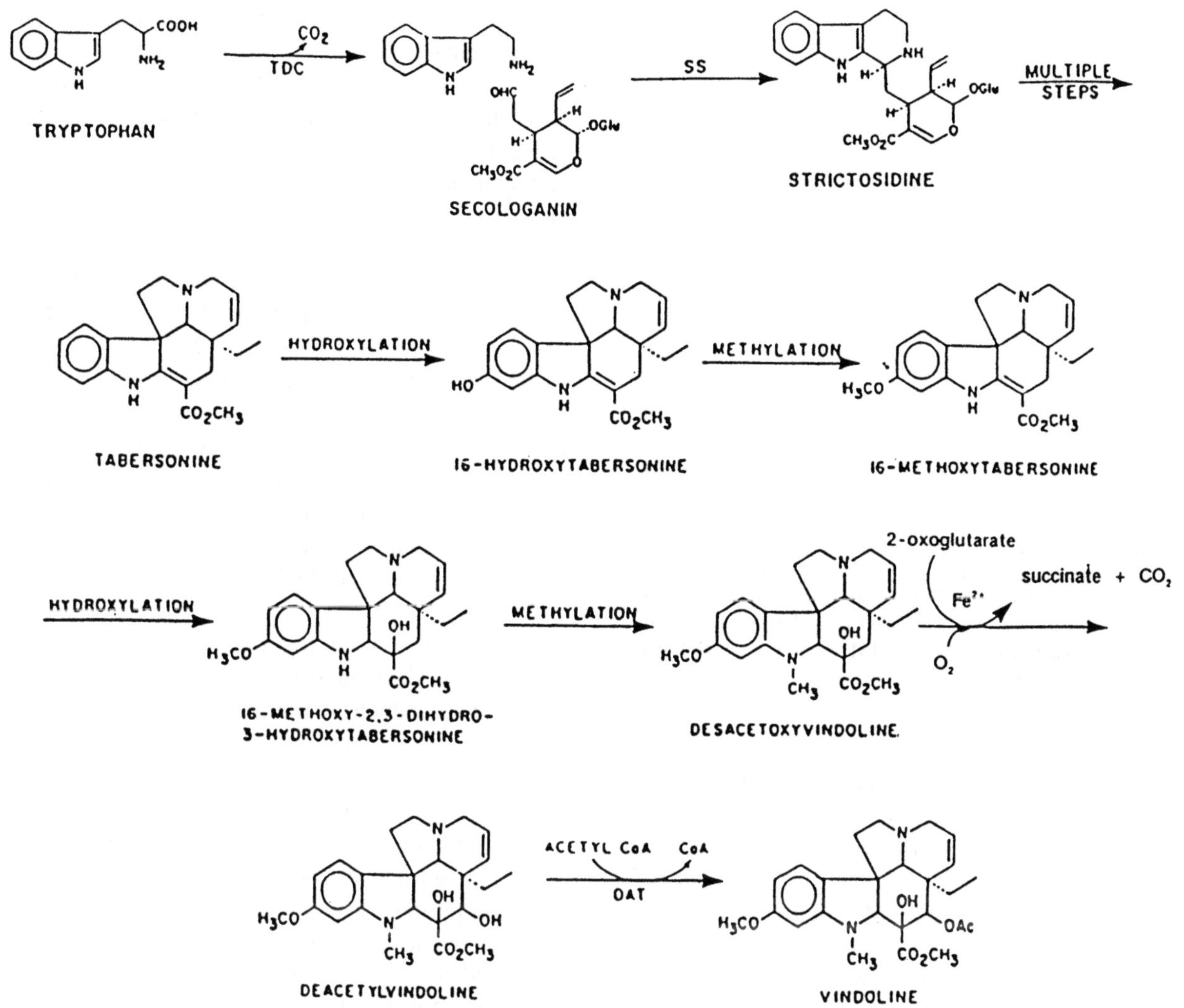

Fig. 1. The biosynthetic pathway from tryptophan to vindoline.

in lanes F to I and that maximal activity occurs in lane H. The efficiency of this assay rendered it suitable for routine use in protein purification on different columns, or in conducting enzyme kinetic studies.

Desacetoxyvindoline 4-hydroxylase is a 2-oxoglutarate dependent dioxygenase

When a protein preparation exhibiting 4-hydroxylase activity was incubated with *N*(1)-[$^{14}CH_3$]desacetoxy-vindoline, no reaction product corresponding to *N*(1)-[$^{14}CH_3$]deacetylvindoline was observed. Addition of ascorbic acid or ferrous sulfate or both to the reaction mixture did not result in any product formation. However, the addition of α-kg produced a radiolabeled product of lower R_f value corresponding to *N*(1)-[$^{14}CH_3$]deacetylvindoline. Addition of ascorbic acid and ferrous sulfate enhanced 4-hydroxylase activity. Removal of molecular oxygen from the reaction mixture containing α-kg resulted in complete inhibition of 4-hydroxylase activity. These results establish that the enzyme is a 2-oxoglutarate-dependent dioxygenase (De Carolis et al. 1990).

Purification of desacetoxyvindoline 4-hydroxylase

As shown previously, a cell free extract of *C. roseus* catalyzed the 4-hydroxylation of *N*(1)-[$^{14}CH_3$]desacetoxyvindoline. Based on these results we used the radiolabeled indole alkaloid as substrate

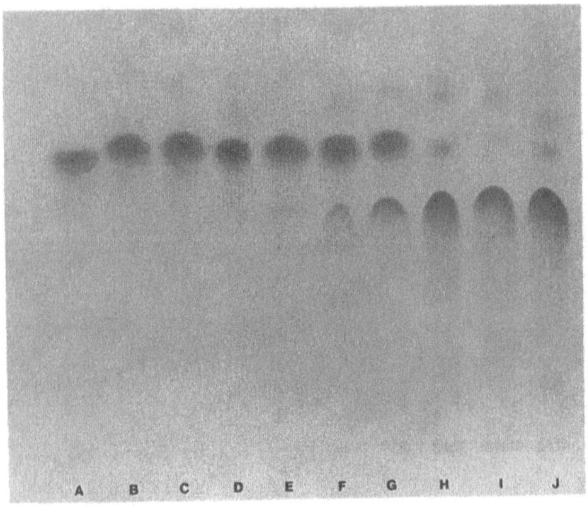

Fig. 2. Photography of an autoradiogram of the chromatographed reaction product of the 4-hydroxylase using the direct assay mixture. Each lane represents fractions eluting from a gel filtration column. In the absence of 4-hydroxylase activity (lanes A–E) only the substrate is present. However, in the presence of increasing 4-hydroxylase activity (lanes F–J), the substrate is converted to product, N(1)-[$^{14}CH_3$]deacetylvindoline. Optimal 4-hydroxylase activity was observed in lane I. Methanol-ethyl acetate was used to develop the silica TLC plate.

to detect 4-hydroxylase activity during enzyme purification (De Carolis & De Luca 1993).

The protein extract of *C. roseus* was fractionated with ammonium sulfate and the protein which precipitated between 35 and 75% saturation was chromatographed by gel filtration on a Sephadex G-100 column. The 4-hydroxylase eluted as a discrete peak with an apparent molecular weight of 45 kDA.

The active fractions were pooled and applied on a green 19-agarose dye-affinity column. The bound protein was eluted with a linear 0.0 to 1.0 M NaCl gradient. The 4-hydroxylase eluted between 0.49 and 0.68 M NaCl and resulted in the elimination of a considerable amount of protein and pigmentation.

The green 19-agarose purified fractions exhibiting 4-hydroxylase activity were chromatographed on a hydroxylapatite column. The bound 4-hydroxylase was eluted with a linear gradient of 10 to 200 mM sodium phosphate. The active 4-hydroxylase eluted at 165 mM phosphate.

Phosphate was removed from the fractions exhibiting 4-hydroxylase activity by chromatography on Sephadex G-25 and the preparation was applied to an α-kg sepharose cosubstrate affinity column. The column was extensively washed and the bound enzyme was selectively eluted with a linear α-kg gradient of 0.0 to 50.0 mM. The 4-hydroxylase eluted at 29.5 mM α-kg. Other bound proteins were nonselectively eluted from the α-kg sepharose column by applying a linear salt gradient of 0.0 to 2.0 M NaCl.

Fractions containing 4-hydroxylase activity which eluted from the α-kg sepharose still contained minor contaminants. As a result, the enzyme was further purified by high performance ion-exchange chromatography on a Mono Q HR 5/5 column. The protein was applied to the column and was eluted with a linear salt gradient of 0.0 to 0.2 M NaCl. The enzyme eluting at 105 mM NaCl was free from other contaminating proteins (De Carolis & De Luca 1993).

This five step purification protocol enriched the enzyme over 2000-fold with a recovery of 1.6% (Table 1) and the specific activity at the Mono Q stage was 86.15 pkat/mg for the 4-hydroxylase. The results obtained with gel filtration chromatography and SDS-PAGE suggested that the native enzyme exists as a monomeric protein of 44.7 kDA (De Carolis & De Luca 1993).

The strategy used to purify the 4-hydroxylase involved the development of a complementary five step protocol which minimized the time required to purify this labile enzyme and diminished the probability for protein denaturation, modification and degradation. However, in several purifications, the typical SDS-PAGE protein profile at the Mono Q step showed a second 40.2 kDa protein which coeluted with the purified 4-hydroxylase (Fig. 3). To examine the similarities between the 4-hydroxylase (44.7 kDa) and the 40.2 kDa both proteins were subjected to protease digestion by the procedure of Cleveland et al. (1977). Proteolysis by trypsin gave identical peptide patterns for the 4-hydroxylase and the 40.2 kDa protein as observed by SDS-PAGE (Fig. 4, lanes A,B). Trypsin, which cleaves preferentially to the carboxy side of basic amino acids gave peptide fragments of 25.8 kDa, 19.2 kDa, 16.6 kDa and 13.6 kDa. Endoproteinase, which cleaves specifically at the carboxyl end of lysine also yielded identical peptide fragments of 25.0 kDa and 20.0 kDa and a minor band of 14.4 kDa for the two proteins (Fig. 4 lanes E,F).

An important factor responsible for the purification of the *C. roseus* 4-hydroxylase was the reactivation and subsequent stabilization of enzyme activity observed when incubating it with ferrous ions. Initially, we found that the enzyme was inactivated after several steps of purification. Studies in our laboratory have shown that Fe^{2+} had little effect on the activity of

Table 1. Purification of desacetoxyvindoline 4-hydroxylase

Step	Total protein mg	Specific activity pkatal/mg	Total activity pkatal	Purification x-old	Recovery %
Crude	1955	0.043	83.5	–	100
Sephadex G-100	100	1.46	146.6	34	176
Green 19-agarose	4.3	15.85	67.4	371	81
Hydroxyapatite	1.4	24.57	34.3	575	41
α-kg Sepharose	0.188	56.83	10.7	1331	12.8
Mono-Q	0.015	86.15	1.3	2018	1.6

One katal of desacetoxyvindoline 4-hydroxylase is defined as the amount of enzyme that catalyzes the conversion of one mole of substrate per second using the direct assay method.

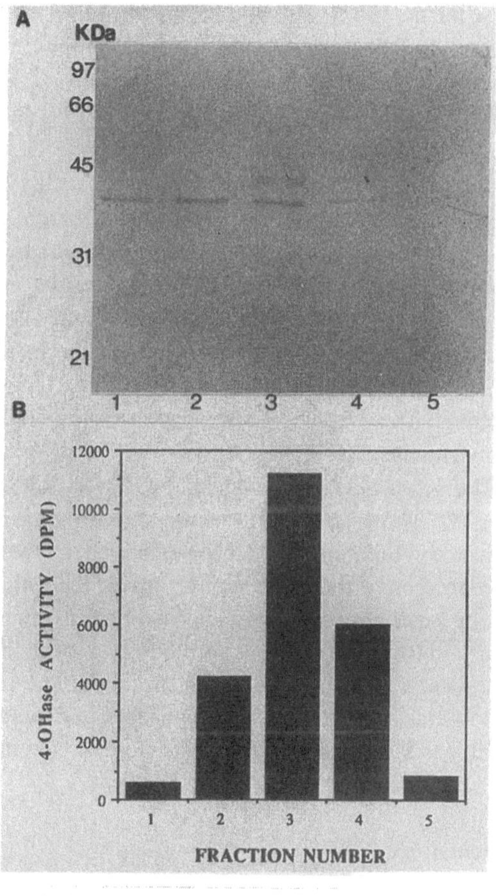

Fig. 3. SDS-PAGE of fractions collected during Mono Q chromatography step. (*A*), protein profile of a Mono Q whereby the 40.2 kDA protein co-eluted with the 4-hydroxylase. (*B*), the 4-hydroxylase activity profile of the Mono Q illustrating the 44.7 kDA band peaked in staining intensity in the fraction containing peak activity. The molecular mass markers are indicated on the left in kDA.

partially purified enzyme (De Carolis et al. 1990), but experiments with the highly purified preparation have

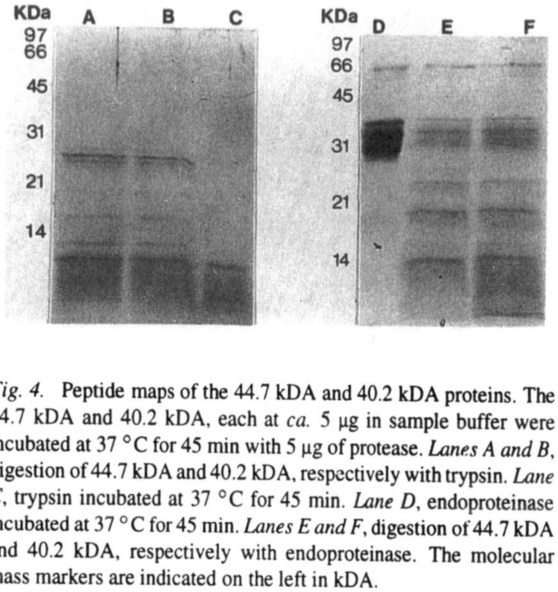

Fig. 4. Peptide maps of the 44.7 kDA and 40.2 kDA proteins. The 44.7 kDA and 40.2 kDA, each at *ca.* 5 μg in sample buffer were incubated at 37 °C for 45 min with 5 μg of protease. *Lanes A and B*, digestion of 44.7 kDA and 40.2 kDA, respectively with trypsin. *Lane C*, trypsin incubated at 37 °C for 45 min. *Lane D*, endoproteinase incubated at 37 °C for 45 min. *Lanes E and F*, digestion of 44.7 kDA and 40.2 kDA, respectively with endoproteinase. The molecular mass markers are indicated on the left in kDA.

shown an absolute requirement for Fe^{2+} (De Carolis & De Luca 1993). Addition of ascorbate to the purified 4-hydroxylase did not reactivate the enzyme, whereas addition of Fe^{2+} increased the enzyme activity by 3.8-fold. Furthermore, reactivation of the enzyme by Fe^{2+} was time-dependent. This suggests that the enzyme loses Fe^{2+} during purification.

Substrate specificity

The desacetoxyvindoline 4-hydroxylase exhibited a strict specificity for position 4 of various indole alkaloid substrates. The most efficient substrates were in the following descending order: [1] > [3] > [2] > [4, 5] (Table 2). Substrates [2] and [3] which lacked the *N*-methyl and both the *N*-methyl and 3-hydroxyl group, respectively were poor substrates. Substrate [4] and

Table 2. Substrate specificity of *C. roseus* 4-hydroxylase.

	Substrate[a]	Relative activity
[1]	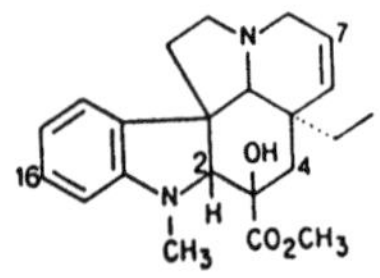	100
[2]	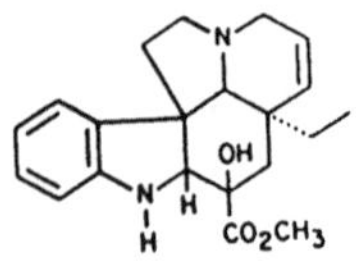	0.6
[3]	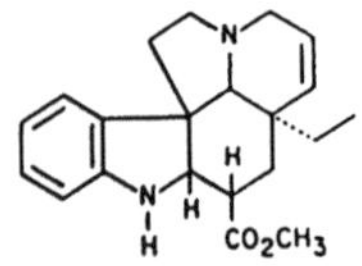	20
[4]	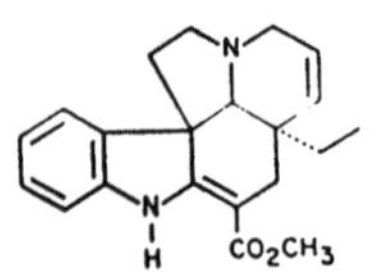	0
[5]	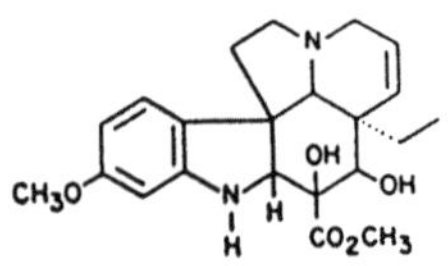	0

[a] Substrate specificity studies of the 4-hydroxylase were conducted using the standard assay for dioxygenases (Rhoads & Udenfriend 1968). The assay contained 9.2 μM [1-^{14}C]α-kg, 5 μM of unlabelled alkaloid, 7.5 μM ascorbate, 0.5 mg catalase, and up to 0.3 mg of protein in a final volume of 1 ml. The release of $^{14}CO_2$ from [1-^{14}C]α-kg is characteristic of 2-oxoglutarate-dependent dioxygenase activity.
[1] 2,3-dihydro-3-hydroxy-*N*(1)-methyltabersonine
[2] 2,3-dihydro-3-hydroxytabersonine
[3] 2,3-dihydrotabersonine
[4] tabersonine
[5] 16-*O*-methyl-2,3-dihydro-3,4-dihydroxytabersonine

[5] exhibited no 4-hydroxylase activity. These results are consistent with the natural occurrence of vindoline in young leaves of *C. roseus* and suggest that desacetoxyvindoline 4-hydroxylase is involved in the second last step in vindoline biosynthesis.

Kinetic analysis

The reaction mechanism and kinetic parameters for desacetoxyvindoline 4-hydroxylase have been elucidated. The intersecting initial velocity patterns observed for all possible substrates (α-kg, O_2 and desacetoxyvindoline) (De Carolis & De Luca 1993) suggest that the binding of these substrates to the 4-hydroxylase occurs by a sequential mechanism (Cleland 1967). The order of binding of O_2 and desacetoxyvindoline was determined by studying the effect on the initial velocity plots at limiting and saturating O_2 concentrations. When O_2 was limiting and desacetoxyvindoline was the fixed substrate the initial velocity pattern gave intersecting lines. However, at saturating O_2 concentrations a plot of α-kg as the variable substrate at changing fixed concentrations of desacetoxyvindoline yielded parallel lines. These differences in initial velocity patterns suggest that O_2 is the second substrate to bind followed by desacetoxyvindoline (Cleland 1967). The first product to be released is deacetylvindoline since noncompetitive inhibition was observed with all three substrates (De Carolis & De Luca 1993). The third product to be released should give competitive inhibition with respect to α-kg and noncompetitive inhibition with O_2 and desacetoxyvindoline. Succinate was the only product which gave this set of patterns and consequently should be the third product to be released. The present results would suggest that α-kg is the first substrate to bind followed by O_2 and desacetoxyvindoline while deacetylvindoline is the first product released followed by CO_2 and succinate. The data is consistent with an Ordered Ter Ter mechanism.

The K_m values for α-kg and O_2 were identical (45 μM). The high affinity of the 4-hydroxylase for desacetoxyvindoline (0.03 μM) may reflect the low concentration of these metabolites inside the cell.

The kinetic mechanism proposed here is in agreement with those of mammalian (Myllyla et al. 1977, Puistola et al. 1980) as well as bacterial (Holme 1975) 2-oxoglutarate-dependent dioxygenases and suggests that this is a general feature of this class of enzyme.

Conclusions

The development of the purification protocol of desacetoxyvindoline 4-hydroxylase has supplied the necessary tools for the molecular cloning of the 4-hydroxylase gene. This clone will be used to study the regulation of the 4-hydroxylase activity in *C. roseus* in response development-specific cues and to environmental stimuli such as light. In addition, the availability of this clone will permit overexpression of this gene in the intact plant and in tissue cultures.

Acknowledgements

This work has been supported by grants from the Natural Sciences and Engineering Research Council of Canada (NSERC) and Les Fonds Pour La Formation De Chercheurs et l'Aide à La Recherche, Québec.

References

Balsevich J, De Luca V & Kurz WGW (1986) Altered alkaloid pattern in dark grown seedlings of *Catharanthus roseus*. The isolation and characterization of 4-desacetoxyvindoline: a novel indole alkaloid and proposed precursor of vindoline. Heterocycles 24: 2415–2421

Cleland WW (1967) Steady state kinetics. In: Boyer PD (Ed) The Enzymes. Vol II (pp 1–61) Academic Press, New York

Cleveland DW, Fisher SG, Kirscher MW & Laemmli UK (1977) Peptide mapping by limited proteolysis in sodium dodecyl sulfate and analysis of gel electrophoresis. J. Biol. Chem. 252: 1102–1106

De Carolis E, Chan F, Balsevich J & De Luca V (1990) Isolation and characterization of a 2-oxoglutarate-dependent dioxygenase involved in the second last step in vindoline biosynthesis. Plant Physiol. 94: 1323–1329

De Carolis E & De Luca V (1993) Purification, characterization and kinetic analysis of a 2-oxoglutarate-dependent dioxygenase involved in vindoline biosynthesis from *Catharanthus roseus*. J. Biol. Chem. 268: 5504–5511

De Luca V (1993) Enzymology of indole alkaloid biosynthesis. In: Lea PJ (Ed) Methods in Plant Biochemistry, Vol 9 (pp 345–368), Academic Press, New York

De Luca V & Kurz WGW (1988) Monoterpene indole alkaloids *(Catharanthus alkaloids)*. In: Constable F & Vasil IK (Eds) Cell Culture and Somatic Genetics of Plants, Vol 5 (pp 385–401). Academic Press, San Diego

Dethier M & De Luca V (1993) Partial purification of an *N*-methyltransferase involved in vindoline biosynthesis in *Catharanthus roseus*. Phytochemistry 32: 673–678

Holme E (1975) A kinetic study of thymine 7-hydroxylase from *Neurospora crassa* Biochemistry 14: 4999–5003

Myllyla R, Tuderman L & Kivirikko KI (1977) Mechanism of prolyl hydroxylase reaction: kinetic analysis of the reaction sequence. Eur. J. Biochem. 80: 349–357

Power R, Kurz WGW & De Luca V (1990) Purification and characterization of acetylcoenzyme A: deacetyvindoline 4-) acetyltransferase from *Catharanthus roseus*. Arch. Biochem. Biophys. 279: 370–376

Puistola U, Turpeenniemi-Hujanen TM, Myllyla R & Kivirikko KI (1980) Studies on the lysyl hydroxylase reaction: Inhibition kinetics and the reaction mechanism. Biochim. Biophys. Acta 611: 51–60

Rhoads RE & Udenfriend S (1968) Decarboxylation of α-ketoglutarate coupled to collagen proline hydroxylase. Proc. Natl. Acad. Sci. 60: 1473–1478

Robinson T (1981) General theories of alkaloid biosynthesis. In: Kleinzeller A, Springer GF & Wittmann HG (Eds) The biochemistry of alkaloids: Molecular biology biochemistry and biophysics. Vol II (pp 12–136). Springer-Verlag, Berlin

Plant Cell, Tissue and Organ Culture **38:** 289–297, 1994.

Are tissue cultures of *Peganum harmala* a useful model system for studying how to manipulate the formation of secondary metabolites?

J. Berlin[1], C. Rügenhagen[1], I.N. Kuzovkina[2], L.F. Fecker[3] & F. Sasse[1]
[1]*GBF – Gesellschaft fur Biotechnologische Forschung m.b.H., Mascheroder Weg 1, D-38124 Braunschweig, Germany;* [2]*Timirjazev Institute of Plant Physiology, Russian Academy of Science, Moscow, Russia;* [3]*BBA – Biologische Bundesanstalt für Land – und Forstwirtschaft, D-38104 Braunschweig, Germany.*

Key words: β-carboline alkaloids, genetic manipulation, hairy root culture, *Peganum harmala*, serotonin, suspension culture

Abstract

This article reviews our present knowledge on the formation of tryptophan derived secondary metabolites in tissue cultures of *Peganum harmala*. With the presence of β-carboline alkaloids and serotonin, *P. harmala* contains two rather simple, interrelated biosynthetic pathways. The long term disadvantage of low and unstable productivity of *P. harmala* suspension culture has recently been overcome by establishing highly productive hairy root cultures. The first β-carboline alkaloid biosynthetic enzymes, specific for the O-methylation of harmalol and harmol as well as for the oxidation of harmaline to harmine, have been detected in these cultures, and they should thus provide a suitable source for studying the yet unknown initial two enzymatic steps of β-carboline alkaloid biosynthesis. Seedlings of *P. harmala* have also been successfully transformed with constructed strains of *Agrobacterium*, as demonstrated by the overexpression of a tryptophan decarboxylase gene from *Catharanthus roseus* in cultures of *P. harmala*. In such transgenic cultures a large overproduction of serotonin was observed. The relative simplicity of these pathways and the rather easy handling of the cultures could make *P. harmala* a useful and attractive model system for studying the interaction, regulation and manipulation of secondary pathways in cultured cells.

Abbreviations: TDC – tryptophan decarboxylase, *tdc* – gene of tryptophan decarboxylase

Introduction

Our research goal over the last few years has been to look for methods by which the formation of secondary metabolites can be improved in cultured plant cells and to analyze which biochemical changes occurred when production was stimulated. One of the culture systems we have worked at was *Peganum harmala*. This plant contains a number of harmane-type β-carboline alkaloids. Indeed, such simple structures suggested that tissue cultures of *P. harmala* would be a useful model system to learn more about how the formation of secondary products can be stimulated and to see whether some general conclusions might be drawn from the biochemical comparison of low and high producing cell cultures. Two other groups had previously reported that callus cultures of *P. harmala* are able to synthesize β-carboline alkaloids. While Reinhard et al. (1968) found only harmine in their tissue culture, Nettleship & Slaytor (1974a) detected the whole spectrum of harmane alkaloids, known as constituents of the plant, and serotonin in root forming callus cultures. Nettleship & Slaytor (1974a) were one of the first to analyze the impact of the culture medium components (e.g. phytohormones, phosphate, nitrogen and carbohydrate source, minor salts, vitamins) on alkaloid formation. Thus, the conditions for further progress seemed to be rather encouraging, especially as screening and selection were new promising tools that had just been introduced into the field when we started our programme in 1979. However, it was soon realized that we could not obtain other or better results than those reported by Nettleship & Slaytor (1974a). The summary of our first findings should thus help to define the narrow bounds by which alkaloid levels can be altered by conventional techniques such as screening, selection, production media and elicitation, if morphological differentiation is required for good expression of a pathway.

Results and discussion

Screening for β-carboline alkaloid overproducing cell lines

It had been suggested by Zenk et al. (1977) that a correlation exists between product levels of the plant and the cultures derived therefrom. Although this had already been disputed by Roller (1978), we compared the alkaloid levels of a large number of seedlings and their cultures. By chance, the callus culture derived from the seedling with the highest β-carboline alkaloid content, had initially also the best alkaloid content (Sasse et al. 1982a). However, the overall correlation of alkaloid levels of plants and their cultures was poor (Sasse et al. 1982a). Even worse, the different yields of the various cultures represented only a momentary picture. After several passages on solid or liquid medium most cultures had quite similar low levels of alkaloids or had even lost the ability to synthesize these compounds (Berlin & Sasse 1988). Sometimes freshly initiated, seemingly undifferentiated callus cultures formed short, star-shaped roots when transferred to liquid medium and such liquid cultures produced more alkaloids and serotonin than the callus cultures (Berlin 1988). The extent of alkaloid and serotonin formation depended clearly upon the extent of organogenesis (i.e. root initiation), as had already been pointed out by Nettleship & Slaytor (1974a). Despite all efforts, by changing phytohormone concentrations and composition, we were not able to establish stable untransformed root cultures of *P. harmala*, as they all changed eventually in fine cell suspension cultures. In conclusion, screening of plants for highest alkaloid levels for establishing the best producing cultures of *P. harmala* will only be useful if the different capacity in alkaloid production is proved to have a genetic rather than a physiological basis. However, even then one may not see differences between genetically different low and high producing plants if the expression of β-carboline alkaloid and serotonin biosynthesis is under the major control of the morphological potential of the cultured cell.

The β-carboline alkaloids of *P. harmala* show typical fluorescences under u.v.-light (harmine, harmol, ruine = blue; harmalol = yellow; harmaline = pale grey-blue). Therefore cell aggregates were screened for colonies showing high fluorescence. Repeated screenings (Yamamoto et al. 1982) for several months (10 and more cycles) indeed helped to enrich aggregates accumulating harmine or harmalol (Sasse et al. 1982a; Berlin 1988). However, these highly accumulating cells showed either poor growth or returned to the low product levels of the parent cultures when they started to grow. Thus, we were not able to isolate a true variant line with permanently increased alkaloid levels. The reason for this failure must be attributed to the fact that we screened for alkaloid accumulating cells whose superior capability for product formation resulted from their higher tendency for morphological differentiation rather than from an alkaloid specific alteration of regulatory controls in alkaloid biosynthesis. The screening for alkaloid-accumulating colonies was probably a screening for the morphologically differentiating cells, which was shown to be an unstable trait. Analytical screening of cultures is not recommended for compounds whose biosynthesis is only compatible with a temporary state of cell differentiation (Berlin & Sasse 1985). Consequently, extensive screenings of *P. harmala* at the plant or the culture levels for obtaining alkaloid overproducing suspension cultures is no longer justified.

Effect of production media and elicitation on the levels of β-carboline alkaloid and serotonin

Nettleship and Slaytor (1974a) had already shown that growth-limiting conditions (removal of 2,4-dichlorophenoxyacetic acid (2,4-D) and/or phosphate) enhanced β-carboline alkaloid and serotonin formation. A more detailed study by Sasse et al. (1982a) confirmed this and showed that calcium and magnesium ions were required for optimal production of the alkaloids and serotonin. During the first growth period on most production media the growth inhibitory effect (20%) was neglectable since specific product levels were increased 2–20-fold depending upon the initial levels of the line under investigation (Sasse et al. 1982a). The alkaloid levels of a high producing line were thus stimulated from 0.75% to 1.6%, while in a low producing line the specific yield increased from 0.02% to 0.4%. This result indicated that the effect of the production media may depend upon the competence of the culture for alkaloid and serotonin formation which indeed depended upon the ability to re-differentiate. This meant also that any production media would only show good results as long as the cells had not lost their capacity for organogenesis. Exactly that was found when the originally highest producing line (0.7 and 1.6% alkaloids on growth and production medium, respectively) gradually changed into a fine non-producing suspension culture. Transfer of such cells to any of the production media did not affect alkaloid and serotonin levels (Berlin & Sasse 1988).

Sometimes, formation of secondary products can be induced by biotic and abiotic elicitors (DiCosmo & Misawa 1985). Media and cell extracts of ca. 50 known phytopathogenic and newly isolated microorganisms were tested for inducing fluorescent β-carboline alkaloids and serotonin in cell suspension cultures (Berlin & Sasse 1988). When growth inhibiting concentrations of elicitors were used, serotonin levels were sometimes strongly enhanced up to 1.5% dry mass (Berlin & Sasse 1988) while alkaloid levels were not or only poorly stimulated. The effect of the elicitors on serotonin biosynthesis in *P. harmala* must be seen as unspecific growth-inhibitory stress induction. The different induction effects of production media and elicitors on serotonin levels seem to result from the different extent of growth inhibition. While growth completely ceased in serotonin-producing elicited cells, growth was only reduced by 20-50% during the first period in the production medium.

The observations described here are not special for secondary products of *P. harmala* but are true for many other systems, in which product formation requires some kind of morphological differentiation. Due to the growth and production instabilities of the initial cultures and their tendency to change into non- or low producing undifferentiated cultures, not only the application of analytical screenings but also of production media and elicitation must be regarded of limited value.

Biochemical reasons for the inability to form serotonin and β-carbolines in undifferentiated cells

As none of the proposed alkaloid specific enzymatic steps was known, tryptophan decarboxylase (TDC) activity, which should be involved in the biosynthesis of both pathways, was used to detect a correlation between enzyme activity and product levels. Freshly initiated callus and suspension cultures with high product levels usually had distinctly higher TDC activity than lines containing low product levels (Sasse et al. 1982b). Increases of TDC activity were noted shortly before increased product accumulation occurred. When cells were transferred from growth medium to a production medium TDC activity and product formation was strongly enhanced (Sasse et al. 1982b). Thus, it was justified in claiming a good correlation of TDC activity with alkaloid and serotonin formation.

When cultures, initially able to re-differentiate, changed into rapidly growing, undifferentiated suspension cultures with low product levels, TDC activity decreased concomitantly until it was no longer measurable. Most cultures having retained the capacity to form at least at little serotonin and alkaloids showed a high sub-culture induction peak of 20–150 pkat/mg protein during the first 24 h. However, after this time TDC activity was no longer measurable in these cultures. It was therefore concluded that TDC activity levels must exert a regulatory role in β-carboline alkaloid and serotonin biosynthesis. Indeed it was clearly shown that the lack of TDC activity is the reason why undifferentiated cell suspension cultures of *P. harmala* are unable to biosynthesize serotonin de novo. Cell cultures unable to synthesize serotonin de-novo transformed added tryptamine, but not added tryptophan, efficiently to serotonin (Sasse et al. 1982b, 1987) which shows that the second and final step of serotonin biosynthesis, the tryptamine-5-hydroxylase, remained well expressed in non-producing cell cultures. Courtois et al. (1988) confirmed the excellent capacity of *P. harmala* suspension cultures for the biotransformation of tryptamine to serotonin. As TDC activity is the rate-limiting step, one should be able to restore serotonin biosynthesis in non-producing undifferentiated cells if one is able to restore TDC activity in these cells. The two ways how this was achieved will be described below.

In the case of β-carbolines it was more difficult to envisage that TDC activity played a major regulatory role in their biosynthesis. Assuming that the proposed biosynthetic pathway with tryptamine as intermediate is correct, it was evident that the loss of TDC activity had also a negative effect on alkaloid levels. However, as feeding of tryptamine did not restore β-carboline alkaloid formation (Sasse et al. 1982b), it could not be claimed that tryptamine supply was a limiting factor. On the other hand, if the biosynthesis of the β-carbolines was strictly channeled and did not involve free tryptamine (which might explain why tryptamine was found to be a much poorer precursor than tryptophan), fed tryptamine might not have been able to reach this pathway. Thus, it was impossible to estimate beforehand whether the loss of TDC activity was, as in the case of serotonin, the only or main reason for the loss of alkaloid production or whether other enzymes were also turned off in undifferentiated cells. The analysis of cell cultures expressing TDC activity in undifferentiated cells could give an answer to this question. Another possibility was to analyze the regulation of β-carboline alkaloids in stable and productive hairy root cultures of *P. harmala.*

Hairy root cultures of Peganum harmala

Hairy root cultures of *P. harmala* were first established by Kuzovkina et al. (1990) by transformation with *Agrobacterium rhizogenes* strain A4. Later it was shown that other wild type strains and constructed strains carrying a binary vector with genes coding for neomycin phosphotransferase II and β-glucuronidase as selection and marker enzymes were also able to efficiently transform seedlings of *P. harmala* (Berlin et al. 1992). Transformed root cultures are easily distinguished from the star-shaped untransformed root cultures. In the presence of ammonium ions (MS-medium) transformed root cultures exist as short, thick, individual roots (Berlin et al. 1992). If nitrate is the main nitrogen source and ammonium ions are low (B5 medium) long, thin, interlacing roots appear which can only be transferred by using tweezers. It was shown that ammonium ions had not only an effect on the appearance of the roots but they also had a rather negative effect on the formation of β-carboline alkaloids (Berlin et al. 1992). In phytohormone-free B5 medium the best root cultures accumulated ca. 1.5–2% β-carboline alkaloids, while the biomass increased 5–10-fold within 3 weeks (Berlin et al. 1992). Such high specific alkaloid levels had also been found occasionally in some untransformed root-forming callus cultures (Berlin 1988). The advantage of the hairy root cultures must thus be seen in their stability and the better growth and higher production rates rather than in better specific yields. All transformed root cultures contained harmine as main alkaloid accounting for 50-70% of all β-carbolines (Berlin et al. 1992). Harmalol was often the second most abundant alkaloid, while the levels of harmol and ruine (harmine-8-O-β-glucoside) varied substantially. Harmaline was a minor alkaloid, only visible due to its intensive fluorescence. Another yellow fluorescent minor band below harmalol might represent dihydroruine (Nettleship & Slaytor 1974a). The specific levels of the alkaloids did not change distinctly during the growth cycle (Berlin et al. 1993a). This was quite different from the accumulation curves of serotonin which was also present in all root cultures. When the root cultures were transferred to fresh medium, a sub-culture induction peak of TDC activity was observed (Berlin et al. 1993a) which declined to low constitutive activity within the next 24-48 h. The specific serotonin content always showed a maximum shortly after the TDC induction peak, confirming the close interrelationship between TDC activity and serotonin levels. Efficient serotonin biosynthesis occurred in the root cultures only during the first days of the culture period. Thus serotonin levels were usually at least 4-fold higher during the first 5 days of culture period than at the end. The decrease of specific serotonin levels was due to dilution by growth as well as to degradation (Fig. 1). It seems possible that some serotonin is oxidized (darkening of the roots) and some is released into the medium in which serotonin is unstable (oxidation). High degradation of serotonin might counteract efforts to enhance TDC activity in root cultures by genetic transformation.

The main question, however, was whether the stably producing root cultures would be a good system for gaining more insight into the biosynthesis of β-carboline alkaloids. As there are still questions regarding the stage at which modifications occur in A-ring and the time of C-ring closure, feeding experiments were performed. Feeding of L-tryptophan to hairy root cultures of *P. harmala* did not significantly affect the total levels of β-carbolines and serotonin (Berlin et al. 1993a). Tryptamine feeding enhanced the levels of serotonin but not of the β-carbolines. The high tryptamine-5-hydroxylase activity had remained well expressed in the root cultures. When tryptophan was fed as radioactive labelled tracer for up to 120 h, just 1-1.5% of the added radioactivity was found in the major alkaloid harmine. Alkaloids were even more poorly labelled when tryptamine was fed as tracer. While it had been expected that labelled tryptamine was efficiently converted to serotonin, it was surprising that 10-15 times more label from tryptophan was found in serotonin than in the alkaloids. The ratio of β-carboline alkaloids to serotonin in these experiments was 2:1. Thus, feeding of the proposed precursors tryptophan and tryptamine to hairy root cultures did not give additional clues to the enzyme sequence of β-carboline alkaloid biosynthesis. It was also checked whether labelled pyruvate was incorporated into the alkaloids. When uptake and labelling of alkaloids was

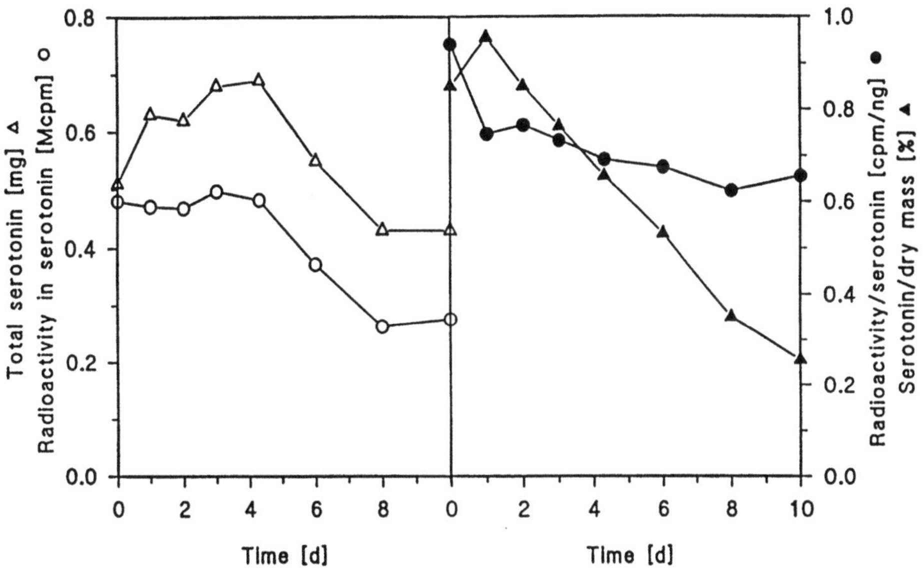

Fig. 1. Fate of labelled serotonin in a hairy root culture. A 3-day-old root culture was labelled with ^{14}C-tryptamine for 24 h. The labelled roots were washed and distributed to 16 flasks with 10 ml conditioned medium. Samples taken over a period of 10 days were analyzed for dry mass, total and specific serotonin content, and for total and specific radioactivity in serotonin. Some de-novo synthesis occurred during the first 24 h in the new flask. After that time a steady decline of specific serotonin levels occurred, which was initially due to dilution (biomass production exceeded de-novo synthesis of serotonin). The decline in total serotonin and of total and specific radioactivity in serotonin indicates that some serotonin was degraded.

followed over a period of 4 days in a one-week-old root culture no label was found in the alkaloids. This was in strong contrast to high incorporation of pyruvate into harmine when it was added to the roots of intact plantlets (Stolle & Gröger 1968). Thus, the roots of intact plants were a much better source for tracer experiments than the hairy root cultures. Whether this is due to the fact that serotonin was not detected in roots of soil grown plantlets (Kuzovkina, unpublished) and thus that tryptamine-5-hydroxylase activity is low or lacking in such roots, is not yet clear.

Feeding of the alkaloids showed the conversion of the dihydro-β-carboline alkaloids to their aromatic forms. Thus harmalol and harmaline feeding enhanced the levels of harmol and harmine, respectively (Berlin et al. 1993a). No evidence was obtained by the feeding experiment for the methylation of harmol to harmine or harmalol to harmaline (Berlin et al 1993a). In conclusion, it has to be understood that the β-carboline alkaloid biosynthetic pathway is extremely difficult to reach by the use of external feeding precursors. β-Carboline alkaloid biosynthesis of *P. harmala* seems to be a well channeled pathway reminiscent of the channeled biosynthesis of cyanogenic glucosides (Conn 1979). Thus, the intermediates of the biosynthetic sequence have to be detected by in vitro enzyme assays. That the root cultures are a good enzyme source for this has most recently been shown by the detection of enzymes catalyzing specifically the O-methylation of harmalol and harmol (see Fig. 2), but not of 6-hydroxytryptamine, serotonin or 5-hydroxytryptophan. It is not yet clear whether one or two enzymes are involved in the O-methylation of the hydroxylated harmane alkaloids. In the same extract methylated harmaline was converted to harmine (Fig. 2). An enzymatic conversion of harmalol to harmol has not yet been detected. Trials to detect enzyme activities hydroxylating tryptamine in the 5-and 6-position or forming the β-carboline structure are under way. Figure 3 summarizes our present knowledge on the biosynthesis of the two tryptophan-derived secondary pathways in *P. harmala* tissue cultures.

Impact of TDC activity on serotonin and β-carbolines

It was mentioned above that a close interrelationship between TDC activity and serotonin levels was found, while the impact of TDC on β-carboline levels was minor and not clear. As TDC activity was the rate-limiting step of serotonin biosynthesis and its loss the reason for the inability of the suspension cultures to form this compound, the establishment of cell cultures

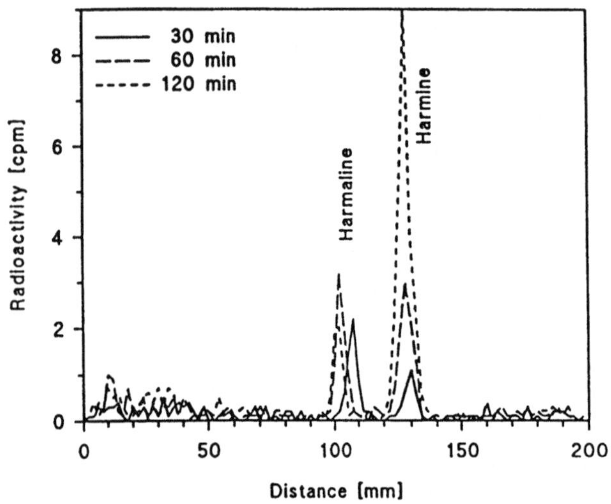

Fig. 2. Sequential conversion of harmalol to harmaline and harmine in crude enzyme extracts of root cultures. A partially purified enzyme extract (including 80% ammonium sulfate precipitation, Sephadex- PD 10-filtration) were used. Assay: [^{14}C-CH_3]-S-adenosylmethionine 15 mM, Harmalol 3 mM, enzyme extract 50-100 μg protein in 125 μl PO_4-puffer pH 8.0; the reaction mixtures were extracted with 250 μl ethylacetate, aliquots were chromatographed on silica gel ($CHCl_3$:MeOH:NH_4OH (25%) 45:5:0.5). The distribution of radioactivity (measured by a TLC-radioactivity scanner) in harmaline and harmine showed that harmalol was first methylated to harmaline which was then oxidized to harmine.

expressing reasonable TDC activity was studied. Two ways seemed to be possible for achieving this goal

- biochemical selection for TDC containing cell lines and
- constitutive overexpression of TDC by genetic transformation.

The basis for the biochemical selection approach was the finding of Sasse et al. (1983) that some toxic tryptophan-analogues (i.e. 4- methyltryptophan) are substrates of TDC and can be detoxified to the corresponding less toxic (> 100-fold) amines. Thus, cell lines containing TDC activity are capable of detoxifying the selective compound and have thus a better chance for survival. Among several 4-methyltryptophan-tolerant clones isolated from a cell suspension culture having lost its capacity to synthesize serotonin and lacking measurable TDC activity, some contained high TDC activity and serotonin levels of up to 2% dry mass (Berlin et al. 1987). The sub-culture TDC induction peak of 120-150 pkat/mg protein decreased rather slowly in these selected lines, so that after 5–7 days TDC activities of 20-40 pkat/mg protein were still measured. Some of these lines have now been grown with or without further selection for more than 6 years and they still show high TDC activity and serotonin levels (Berlin et al. 1993a).

While it was possible to restore serotonin biosynthesis in morphologically undifferentiated, fine suspension cultures by selection for TDC containing cells, it was not possible to restore β-carboline alkaloid biosynthesis in these cells (Berlin et al. 1987). This showed that during dedifferentiation not only TDC but also the activites of other enzymes involved in β-carboline alkaloid biosynthesis were turned off.

In the genetic engineering approach we introduced a cDNA clone coding for the TDC of *Catharanthus roseus* into *P. harmala* callus and root cells (Berlin et al. 1993b). The gene had been cloned between the CaMV 355 promoter and terminator in the pBin19 derived plant vector pBDH5 by Goddijn (1992). Sterile grown seedlings were transformed with *Agrobacterium tumefaciens* strains carrying the *tdc* gene either in a pTi or pRi background (Berlin et al. 1993b). Among the many kanamycin-resistant callus and root cultures, several were found expressing constitutively TDC activity of 30-40 pkat/mg protein over the whole growth period. In transformed control cultures constitutive activities of 0-5 pkat/mg protein were found. Transformed suspension cultures overexpressing TDC activity had serotonin levels of 1-2%, while in control cultures 0.01-0.1% were found. The *tdc*-transgenic root cultures also contained serotonin levels of 1-2%. However, the percent increase was not as high as root culture controls accumulated also between 0.3-0.6% serotonin. Overall, the serotonin levels of the biochemically selected lines and the genetically engineered lines compared quite well. It seemed to be difficult to make the cells produce more than 2% serotonin without additional measures. Indeed, it was shown that the rate-limiting step of serotonin biosynthesis was shifted in the *tdc*-transgenic lines from TDC activity to substrate supply (Berlin et al. 1993b). When tryptophan was fed to TDC overexpressing transgenic lines it was efficiently decarboxylated and immediately hydroxylated to serotonin. Thus, in contrast to what was found for wild type hairy root cultures, tryptophan feeding enhanced the levels of serotonin dramatically, so that specific levels of 5-8% dry mass were readily found in suspension and root cultures (Berlin et al. 1993b). This result showed that the metabolic effect of removing a rate-limiting biosynthetic step does not depend only upon the absolute increase of the corresponding enzyme activity, but also upon the difference between the first and the second rate-limiting step. If, as in the

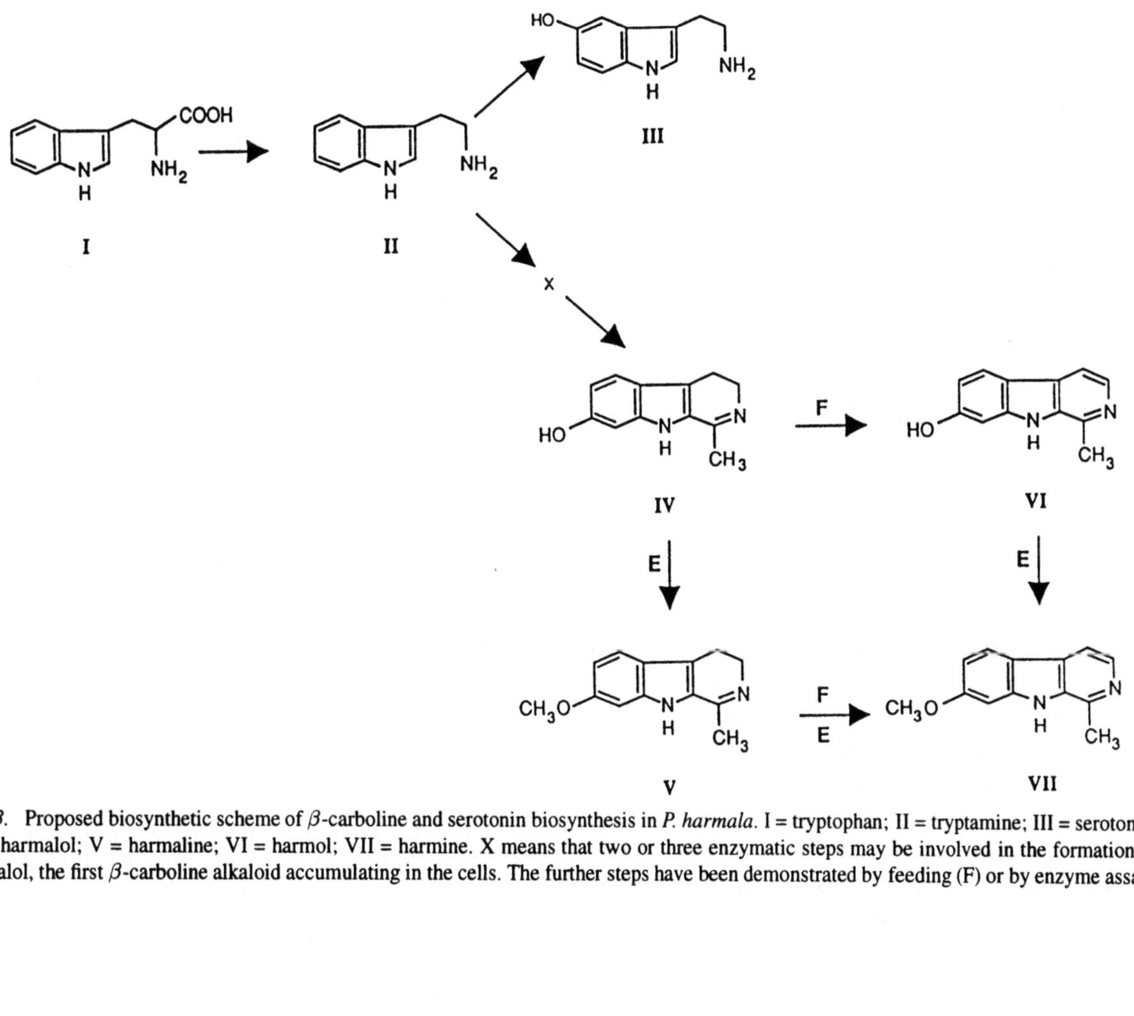

Fig. 3. Proposed biosynthetic scheme of β-carboline and serotonin biosynthesis in *P. harmala*. I = tryptophan; II = tryptamine; III = serotonin; IV = harmalol; V = harmaline; VI = harmol; VII = harmine. X means that two or three enzymatic steps may be involved in the formation of harmalol, the first β-carboline alkaloid accumulating in the cells. The further steps have been demonstrated by feeding (F) or by enzyme assays (E).

case of serotonin biosynthesis, the second rate-limiting step readily becomes so effective, only specific levels of 1-2% may be possible as the enzyme activity can not be saturated. To overcome the new rate-limiting step, tryptophan supply, is a challenging task. It should be possible to select from the *tdc*-transgenic lines tryptophan-overproducing clones containing an anthranilate synthase with lowered feedback control (Widholm 1972).

The levels of β-carboline alkaloids were not significantly altered by the constitutive overexpression of the foreign *tdc*-gene in *P. harmala* cultures (Berlin et al. 1993b). This result was predicted as we could not find any evidence that TDC activity and thus tryptamine supply was rate-limiting in β-carboline alkaloid biosynthesis. Why should tryptamine overproduced by the engineered TDC activity cause a different metabolic effect than tryptamine overproduced by the 4-methyltryptophan-tolerant cell lines with its high TDC activity?

The most interesting question imposed by *P. harmala* is thus: How do the cells manage to channel tryptophan into the evidently strictly separated pathways of serotonin and β-carboline alkaloid biosynthesis? Firstly, one can presently not exclude the possibility that tryptamine, and thus decarboxylation of tryptophan, is not included in β-carboline alkaloid biosynthesis. The zero or very poor incorporation of tryptamine into β-carbolines (Nettleship & Slaytor 1974b; Berlin et al. 1993) would support this. On the other hand, the incorporation of double labelled tryptamine into the alkaloids of the roots of intact plant (Stolle & Gröger 1968) and the missing evidence that 6-hydroxytryptophan could be an intermediate in the β-carboline alkaloid biosynthesis still favors the assumption that decarboxylation of tryptophan is the first step for both biosynthetic pathways. If this assumption is correct, one has to postulate the presence of two TDCs located in two different compartments. The TDCs have not necessarily to be different in their amino acid sequence but may only differ in their targeting signal. One could imagine that tryptamine fed and tryptamine synthesized by a TDC located in the cytoplasm only comes into contact with the tryptamine-5-hydroxylase but not with tryptamine-6-hydroxylase, probably the first enzyme specific in β-carboline alkaloid biosynthesis. It is known that the TDC of *C. roseus* is located in the cytoplasm (DeLuca & Cutler 1987). The TDC of 4- methyltryptophan-tolerant cell lines, as well as stress-induced TDC, could thus also be located in the cytoplasm of *P. harmala* as was the TDC of the transgenic lines (Berlin et al. 1993b). It is of great interest to see whether targeting of the TDC to other cellular compartments, e.g. the compartment of alkaloid biosynthesis would give other metabolic responses.

Thus, beside the detection and characterization of the two missing proposed enzymes of β-carboline alkaloid biosynthesis (tryptamine-6-hydroxylase and the β-carboline alkaloid forming enzyme), it is of particular interest to identify the cellular compartment in which the biosynthesis of the alkaloids occurs. As only 5 or 6 enzymes are involved in the formation of harmine and only 2 in the synthesis of serotonin, it should be possible to study all questions regarding the interrelationship of the two pathways in detail. As in addition, *P. harmala* is a plant species easily accessible to genetic transformation, it is expected that this simple system is an ideal tool for learning more about how the productivity of target pathways can be altered in the desired way.

Acknowledgement

We would like to express special thanks to Drs. J.H.C. Hoge and O.J.M. Goddijn, Plant Molecular Sciences, Leiden University, for having provided the *tdc*-clone for the transgenic research.

References

Berlin J (1988) On the formation of secondary metabolites in plant cell cultures – some general observations and some experimental approaches. In: Production of secondary metabolites by plant cell cultures (pp 89-98). APRIA, Paris

Berlin J & Sasse F (1985) Selection and screening techniques for plant cell cultures. Adv. Biochem. Eng. 31: 99-132

Berlin J & Sasse F (1988) β-Carbolines and indole alkylamines. In: Constabel F & Vasil IK (Eds) Cell Culture and Somatic Cell Genetics of Plants, Vol 5(357-369). Academis Press, Inc., Orlando, Florida

Berlin J, Mollenschott C, Sasse F, Witte L, Piehl HG & Büntemeyer H (1987) Restoration of serotonin biosynthesis in cell suspension cultures of *Peganum harmala* by selection for 4-methyltryptophan-tolerant cell lines. J. Plant Physiol. 131: 225-236

Berlin J, Kuzovkina IN, Rügenhagen C, Fecker L, Commandeur U & Wray V(1992) Hairy root cultures of *Peganum harmala*-II. Characterization of cell lines and effect of culture conditions on the accumulation of β-carboline alkaloids and serotonin. Z. Naturforsch. 47c: 222-230

Berlin J, Rügenhagen C, Greidziak N, Kuzovkina IN, Witte L & Wray V(1993a) Biosynthesis of serotonin and β-carboline alkaloids in hairy root cultures of *Peganum harmala*. Phytochemistry 33: 593–597

Berlin J, Rügenhagen C, Dietze P, Fecker LF, Goddijn OJM & Hoge JHC (1993b) Increased production of serotonin by suspension and root cultures of *Peganum harmala* transformed with a tryptophan decarboxylase cDNA clone of *Catharanthus roseus*. Transgenic Res. 2: 336–344

Conn EE (1979) Biosynthesis of cyanogenic glucosides. Naturwissenschaften 66: 28-34

Courtois D, Yvernel D, Florin B & Petiard V(1988) Conversion of tryptamine to serotonin by cell supension cultures of *Peganum harmala*. Phytochemistry 27: 3137-3141

DeLuca V & Cutler AJ (1987) Subcellular localization of enzymes involved in indole alkaloid biosynthesis in *Catharanthus roseus*. Plant Physiol. 85: 1099-1102

DiCosmo F & Misawa M (1985) Eliciting secondary metabolism in cultured cells. Trends Biotechnol. 3: 318-322

Goddijn OJM (1992) Regulation of terpenoid indole alkaloid biosynthesis in *Catharanthus roseus*: The tryptophan decarboxylase gene. Ph.D.-thesis, Leiden University, Leiden, The Netherlands

Kuzovkina IN, Gohar A & Alterman IE (1990) Production of β-carboline alkaloids in transformed cultures of *Peganum harmala* L. Z. Naturforsch. 45c: 727-728

Nettleship L & Slaytor M (1974a) Adaptation of *Peganum harmala* callus to alkaloid production. J. Exp. Bot. 25: 1114-1123

Nettleship L & Slaytor M (1974b) Limitations of feeding experiments in studying alkaloid biosynthesis in *Peganum harmala* callus cultures. Phytochemistry 13: 735-742

Reinhard E, Corduan G & Volk OH (1968) Nachweis von Harmin in Gewebekulturen von *Peganum harmala*. Phytochemistry 7: 503-504

Roller U (1978) Selection of plants and tissue cultures of *Catharanthus roseus* with high content of serpentine and ajmalicine. In: Alfermann AW & Reinhard E. (Eds) Production of Natural Compounds by Cell Culture Methods (pp 95-108). GSF, München

Sasse F, Heckenberg U & Berlin J (1982a) Accumulation of β-carboline alkaloids and serotonin by cell cultures of *Peganum harmala* L. I. Correlation between plants and cell cultures and influence of medium. Plant Physiol. 69: 400-404

Sasse F, Heckenberg U & Berlin J (1982b) Accumulation of β-carboline alkaloids and serotonin by cell cultures of *Peganum harmala* II. Interrelationship between accumulation of serotonin and activities of related enzymes. Z. Pflanzenphysiol. 105: 315-322

Sasse F, Buchholz M & Berlin J (1983) Site of action of growth inhibitory tryptophan analogues in *Catharanthus roseus* cell suspension cultures. Z. Naturforsch. 38c: 910-915

Sasse F, Witte L & Berlin J (1987) Biotransformation of tryptamine to serotonin by cell suspension cultures of *Peganum harmala*. Planta Medica 53: 354-359

Stolle K & Gröger D (1968) Untersuchungen zur Biosynthese des Harmins. Arch. Pharm. 301: 561-571

Widholm JM (1972) Cultured *Nicotiana tabacum* cells with an altered anthranilate synthetase which is less sensitive to feed-back inhibition. Biochem. Biophys. Acta 261: 52-58

Yamamoto Y, Mizuguchi R & Yamada Y (1982) Selection of a high and stable pigment- producing strain in cultured *Euphorbia millii* cells. Theor. Appl. Genet. 61: 113-116

Zenk MH, El-Shagi H, Arens H, Stöckigt J, Weiler EW & Deus B(1977) Formation of indole alkaloids serpentine and ajamalicine in cell suspension cultures of *Catharanthus roseus*. In: Barz W, Reinhard E & Zenk MH (Eds) Plant Tissue Culture and its Bio-Technological Application (pp 27-43). Springer Verlag, Berlin

Plant Cell, Tissue and Organ Culture **38:** 299–305, 1994.

Breakdown of indole alkaloids in suspension cultures of *Tabernaemontana divaricata* and *Catharanthus roseus*

Jan Schripsema, Denise Dagnino, Rosana I. Dos Santos & Robert Verpoorte
Leiden/Amsterdam Center for Drug Research, Division of Pharmacognosy, Center for Bio-Pharmaceutical Sciences, Leiden University, Gorlaeus Laboratories, P.O. Box 9502, 2300 RA Leiden, The Netherlands

Key words: Breakdown, indole alkaloids, suspension culture, *Catharanthus, Tabernaemontana*

Abstract

The relative importance of breakdown on the accumulation of indole alkaloids has been determined in suspension cultures of *Tabernaemontana divaricata* and *Catharanthus roseus* by the feeding of stable isotope labelled alkaloids. In all cultures a considerable amount of the alkaloid biosynthesized was broken down. The breakdown was found to be dependent on the culture period and the half-life was in the order of several days. The breakdown could not explain the difference between producing and non-producing cultures. Further it was determined that in both cultures the breakdown was due to both biotic and abiotic factors.

Introduction

Breakdown of secondary metabolites is a relatively little investigated subject (Becker 1987). Nevertheless it has to be realized that accumulation of secondary metabolites is the result of production and breakdown. Absence of accumulation can be explained by absence of production, but also the possibility should be considered that it might be due to fast breakdown. Only in a limited number of systems has breakdown been investigated and it has been stated that secondary metabolites are usually not end products of metabolism but show a high degree of turnover (Wink 1987). In early studies, reviewed by Robinson (1974), the occurrence of turnover was concluded from the variation in the amounts of a compound during the development of certain plants, e.g. *Catharanthus roseus* and *Conium maculatum*, and from strong diurnal fluctuations in alkaloid content, as occurs in e.g. *Papaver somniferum*, and further from several more detailed studies using isotopically labelled molecules, which enabled the estimation of turnover rates. A large variation was found in the turnover rates of alkaloids, ranging from several hours to days (Robinson 1974). For plant cell cultures even less is known about the relative importance of breakdown on secondary metabolite accumulation.

Tabernaemontana divaricata and *Catharanthus roseus* are plants belonging to the family Apocynaceae and both accumulate considerable amounts of indole alkaloids. In the present study suspension cultures of *T. divaricata* and of *C. roseus*, which also produce indole alkaloids, were used. The main aims were to determine the importance of the factor breakdown in the accumulation of indole alkaloids, and whether it is a factor which has a significant effect on the accumulation. In the case that breakdown is important further questions would then be what is the cause and what are the possibilities to decrease the breakdown. It also should be considered that breakdown can be biotic (catabolism) or abiotic (chemical degradation).

From the suspension cultures of both species studied, different lines of subculturing are present which seem to differ only in their indole alkaloid accumulation. A comparison of these lines might point to the cause of this difference. Can it be that in a non-accumulating line, this is caused by a high breakdown, in such way that immediately after production the alkaloids are broken down again, avoiding accumulation?

A very suitable method to measure the extent of breakdown and biosynthesis is stable isotope labelling. If one has a pool of isotopically labelled alkaloid then one can determine both biosynthesis and breakdown at the same time: biosynthesis can be estimated by the dilution of label and the change in the total amount of compound present; breakdown can be estimated by the decrease of the amount of labelled compound.

Materials and methods

Cell suspension cultures

The cell suspension cultures from *Tabernaemontana divaricata* (L.) R.Br. ex Roem. et Schult. were derived from a culture which had been growing for one year on a modified MS medium (Schripsema & Verpoorte 1992: strain N). It was then, about 3 years before the present experiments, subcultured on MS medium (Murashige & Skoog 1962) without growth regulators.

The cell suspension cultures from *Catharanthus roseus* (L.) G. Don. were derived from a culture growing on LS medium (Linsmaier & Skoog 1965). Like *T. divaricata* it was adapted to MS medium without growth regulators, about 3 years before the present experiments.

Techniques

The GC-MS analysis of the indole alkaloids has been reported before (Dagnino et al. 1991). The HPLC method for the indole alkaloids was also reported before (Smith 1984). A μBondapak phenyl column was used, and as eluent sodiumphosphate (0.05M), acetonitrile and 2-methoxyethanol (80:15:5) brought to pH 3.5–3.9, depending on the experiment, with H_3PO_4 85%.

Results and discussion

Description of the experimental system

In the *T. divaricata* cultures the main alkaloid produced is O-acetylvallesamine (Fig. 1). It is a monoterpenoid indole alkaloid. It differs from most of the other indole alkaloids by the fact that during the biosynthesis of this compound the carbon next to the aliphatic nitrogen in the tryptamine portion is lost.

The alkaloid spectrum produced by the *T. divaricata* suspension culture used in the present experiments is quite simple. Besides the main alkaloid O-acetylvallesamine, usually more than 90%, vallesamine and voaphylline are found.

For the breakdown experiment labelled alkaloid was needed. This was obtained by growing the cultures on medium in which both ammonium and nitrate (the only nitrogen sources) were labelled with ^{15}N. GC-MS analysis of the O-acetylvallesamine which was obtained from the biomass and medium after a culture period revealed that the alkaloid was for 52% double, 34% single, and 14% not labelled. Measurement of the ^1H-NMR spectrum confirmed this high labelling percentage of O-acetylvallesamine. Instead of the single NH signal from the indole, there was only a small signal of the proton on ^{14}N and a large doublet of the proton on ^{15}N. Integration of those signals showed that 76% of the indole nitrogen was ^{15}N labelled. The labelling percentage of the other nitrogen of the molecule could not be derived from the ^1H-NMR spectrum (Schripsema et al. 1992).

In contrast to the cultures of *T. divaricata*, the *C. roseus* cultures produce a complex mixture of alkaloids. Numerous signals can be seen in the HPLC chromatogram. The main alkaloid produced is ajmalicine (Fig. 1).

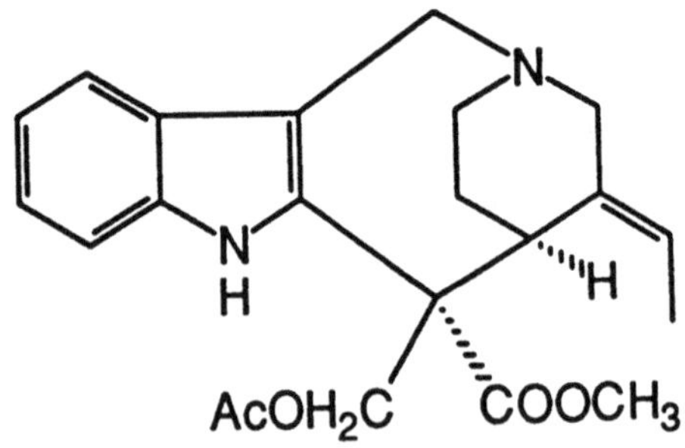

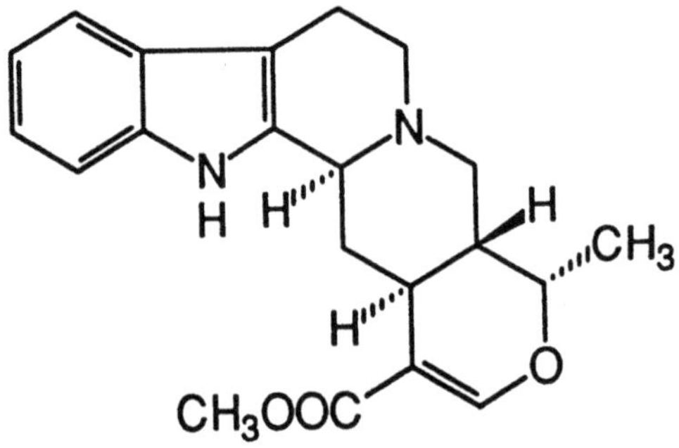

Fig. 1. Structural formulas of O-Acetylvallesamine (top) and Ajmalicine (bottom).

Because ajmalicine is commercially available and because isolation of labelled ajmalicine from the cultures would be more complex, a chemical synthesis of labelled compound was chosen. A suitable method in which the methyl of the carboxymethyl is exchanged was described by Auriola et al. (1991). The methyl was exchanged for deuterated methyl and the ajmalicine obtained in this way was triply labelled (> 98%).

HPLC was always used for the quantitative analysis of the alkaloids in the experiments. With GC-MS the labelling percentages of the individual alkaloids were determined. The combination of these data enables the estimation of both biosynthesis and breakdown. A prerequisite for the calculations is that there should be only one pool or only rapid exchanging pools of indole alkaloid. This was confirmed since both O-acetylvallesamine and ajmalicine were found to give a rapid distribution over the medium and the biomass after addition of the labelled alkaloid to the medium (Dagnino et al. 1993a, Dos Santos et al. 1994). Also during all the time courses similar labelling percentages in the medium and biomass were found indicating a constant interchange between intracellular and extracellular alkaloid.

How much does breakdown influence alkaloid accumulation?

In a first experiment [^{15}N]-O-acetylvallesamine was added to a *T. divaricata* culture at several moments during the batch culture cycle (Dagnino et al. 1993a). The amount added was in all cases kept as low as 20% of the amount already present in the medium to avoid any possible effects of the addition on growth or alkaloid production.

In Fig. 2 the time courses for dry and fresh weight of the culture and also the dissimilation curves are shown. The addition did not have any effect on these time courses. The determination of dissimilation curves (Schripsema et al. 1990) was found to be very useful. Simply weighing each flask during an experiment gives for each individual flask after subtraction of evaporation, a dissimilation curve which is correlated to the dry weight and the sugar consumption of the culture. Differences between flasks can be easily observed. E.g. because of differences in the size of the inoculum, flasks can be earlier or later at a certain stage compared to other flasks.

This experiment was performed in 2-l shake-flasks containing half a liter of medium, so that during the experiments samples could be taken from one flask

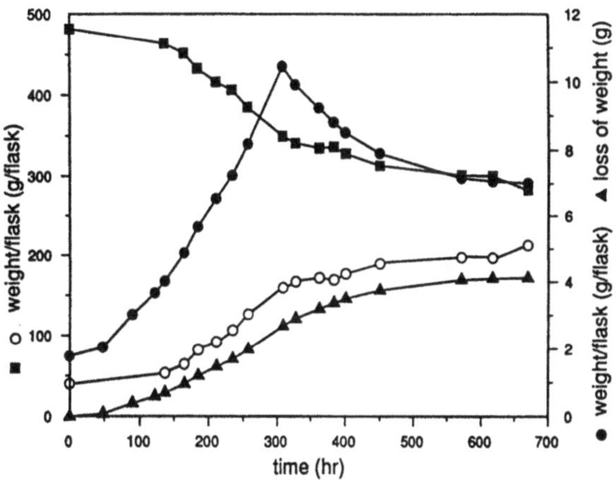

Fig. 2. Time course of dry weight, fresh weight and dissimilation for the *Tabernaemontana divaricata* suspension culture. ▲: Dissimilation, ○: Fresh weight, ●: Dry weight, ■: Medium content.

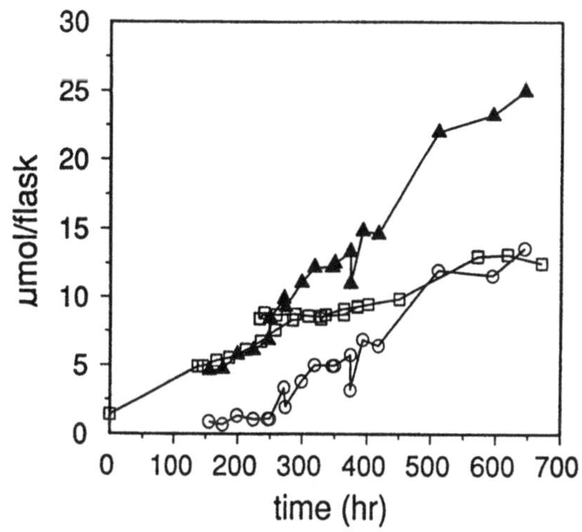

Fig. 3. Accumulation, biosynthesis and breakdown of O-acetylvallesamine in the *Tabernaemontana divaricata* suspension culture. The biosynthesis was calculated from the dilution of [^{15}N]-O-Acetylvallesamine. The breakdown from the decrease. □: Accumulation, ▲: Biosynthesis, ○: Breakdown.

so as to avoid variation between flasks. In Fig. 3 the calculated curves for biosynthesis and breakdown are shown. At the end of the culture period around 50% of the produced O-acetylvallesamine had been broken down. It further appears that the breakdown largely occurred in the beginning of the stationary phase.

In a first experiment with a low accumulating *C. roseus* suspension culture, ajmalicine was added to the culture, in much higher amounts than what would be accumulated by the culture. The growth of the culture was not influenced by the ajmalicine addition. The half-life of ajmalicine in this culture was found to be in the order of 5–10 days (Dos Santos et al. 1994).

Can breakdown explain the difference between high and low accumulating cell lines?

Both from *C. roseus* as from *T. divaricata* cultures separate lines of subculturing have evolved which, although they have always been maintained in exactly the same way, they were grown on the same medium and they have been subcultured in parallel at the same moment, with time differences were observed between the separate lines of subculturing. For *T. divaricata* two cultures are now maintained, one relatively high producing (about 50 mg l^{-1} of the main alkaloid) while the other line hardly produces (less than 1 mg l^{-1}). Both cell lines were compared in their biosynthetic and breakdown capabilities (Dagnino et al. 1993b). In this experiment the relative importance of chemical degradation as compared to the catabolism was also investigated.

Though a relatively large amount of O-acetylvallesamine or voaphylline was added to the cultures in this experiment, this did not seem to influence the biosynthesis as can be seen by comparing the time courses of accumulation of the other terpenoid indole alkaloids after O-acetylvallesamine addition and voaphylline addition with those of the control cultures (Fig. 4).

The normal time courses of accumulation of indole alkaloids in the cultures are shown in Fig. 4A. The non-accumulating line shows a higher accumulation of tryptamine and no accumulation of O-acetylvallesamine or voaphylline.

After feeding [^{15}N]-O-acetylvallesamine (Fig. 4B) the level of O-acetylvallesamine stays more or less constant in the accumulating line while it decreases in the non accumulating line. The production of voaphylline is not influenced while the tryptamine accumulation is lower than in the control.

Interestingly O-acetylvallesamine which was just kept in culture medium without cells showed a similar decrease of O-acetylvallesamine as the non producing culture, indicating that a large amount of the breakdown might be non-biotic.

After feeding of voaphylline (Fig. 4C) similar results were seen. No influence on terpenoid indole alkaloid accumulation, but a decrease of tryptamine after alkaloid feeding. The fact that both after feeding of O-acetylvallesamine as voaphylline the alkaloid accumulation was not influenced can be seen as an indication that no end product inhibition occurs somewhere in the pathway or it should be before tryptamine.

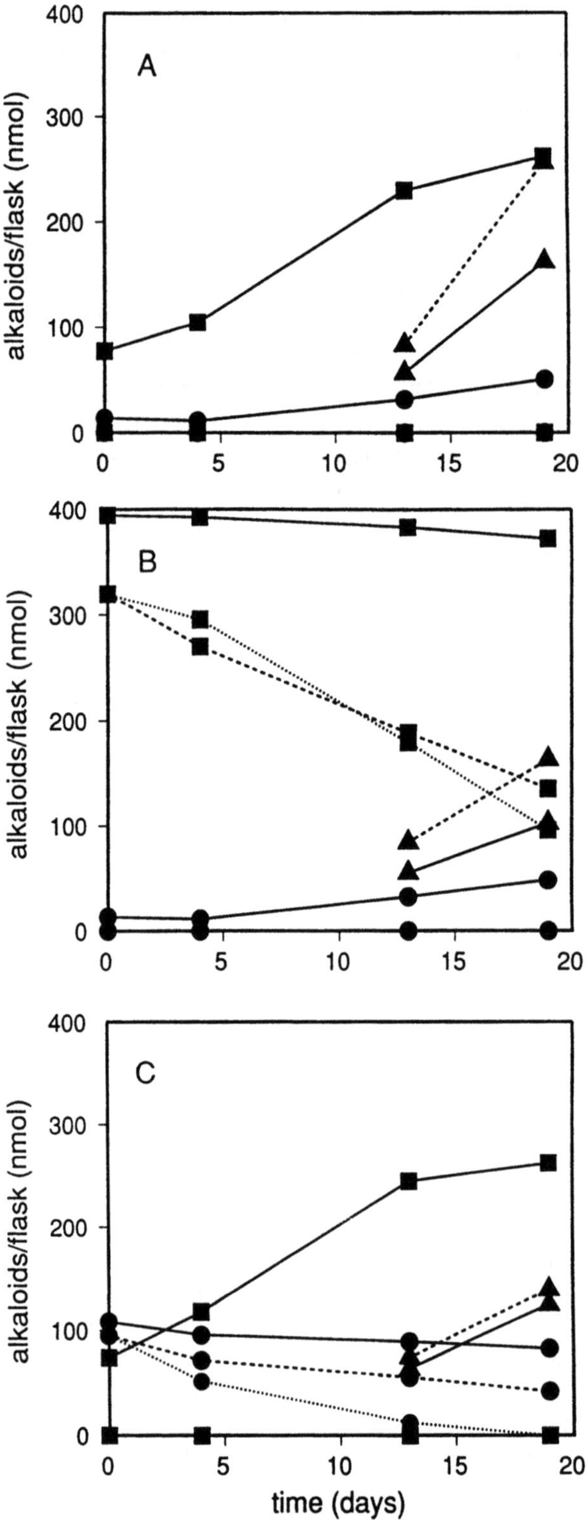

Fig. 4. Time courses of indole alkaloids in the cultures of *Tabernaemontana divaricata*. ■: O-Acetylvallesamine, ●: Voaphylline, ▲: Tryptamine, — : Accumulating line, - - - : Non-accumulating line, ... : Culture medium in the absence of cells. A: Control cultures. B: After feeding of [^{15}N]-O-acetylvallesamine. C: After feeding of [^{15}N]-Voaphylline.

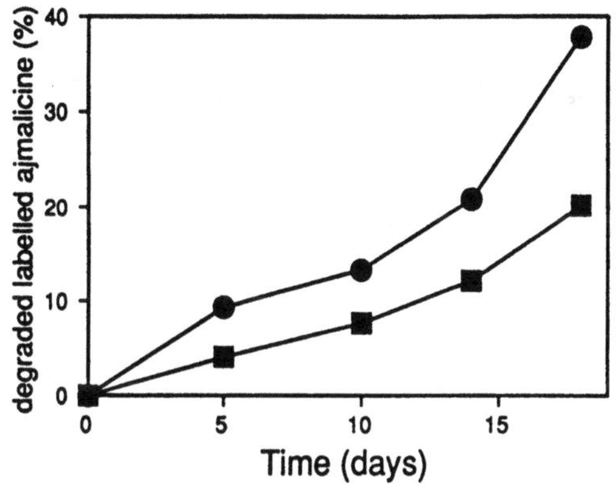

Fig. 5. The breakdown of ajmalicine in the low-accumulating *Catharantus roseus* suspension culture. ●: In the cell culture, ■: In culture medium without cells.

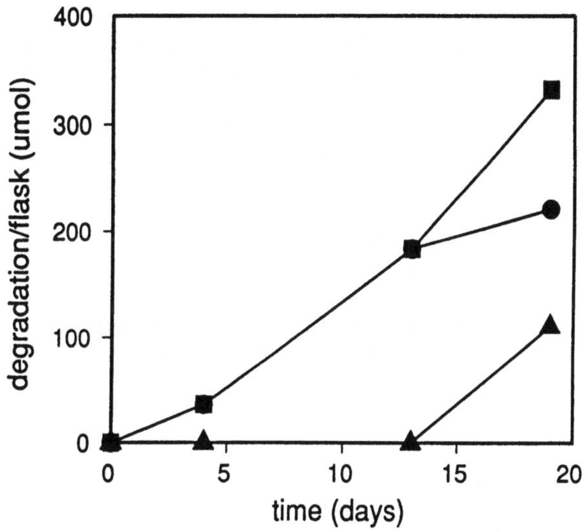

Fig. 6. Estimation of the role of the chemical instability of O-acetylvallesamine and its catabolism on the total amount of O-acetylvallesamine broken down in a suspension culture of *Tabernaemontana divaricata* to which [^{15}N]-O-acetylvallesamine was added. ■: Total amount broken down, ●: Chemical instability, ▲: Catabolism.

The biosynthesis in the experiment was clearly demonstrated by the dilution of label (Table 1). In the accumulating line to which O-acetylvallesamine had been added the labelled compound was diluted with a factor 2.3, while in the other line it only diluted by a factor 1.1.

The breakdown in both lines was similar while the production was clearly different. This showed beyond doubt that the difference in accumulation was caused by a difference in production.

With the *C. roseus* culture similar experiments were done, with a low (± 50 μg ajmalicine per g DW) and a high accumulating line (± 1.2 mg ajmalicine per g DW). In these experiments relatively low amounts of ajmalicine were added. It was found that about 40% of the produced alkaloid (ajmalicine) was broken down during the culture period in the low accumulating line (Fig. 5). The breakdown mainly occurred in the beginning of the stationary phase. It was found that over the same time period ajmalicine in normal culture medium maintained under the same circumstances only without cells was also degraded, showing that a large part of degradation might be abiotic (around 50%). In the high accumulating line comparable results were obtained. Over the culture period about 50% of the ajmalicine produced was broken down while this was also for about 50% due to abiotic breakdown.

The extent and main period of the degradation was similar in both cell lines, when expressed as percentage of the ajmalicine accumulated. It was concluded that as for *T. divaricata* the differences in accumulation were due to differences in the biosynthesis.

Is the cause of alkaloid breakdown biotic or abiotic?

To distinguish between biotic and abiotic breakdown in the *T. divaricata* cultures the decomposition in used culture medium separated from the cells and left under normal culture conditions was compared with the decomposition in the presence of the cells (Table 2). O-acetylvallesamine was more stable in used culture medium than in fresh medium (on which no cells had grown). Only a part of the total amount of breakdown which occurred could be explained by the chemical decomposition in used culture medium. The combination of the data gave Fig. 6 which shows that especially in the first part of the culture period the chemical instability is dominating while in the stationary phase biotransformation is more important. Due to the fact that in a normal culture period the accumulation mainly occurs at the end of the culture period, normally the biotransformation is more important.

Also in the *C. roseus* culture chemical degradation was observed but contrary to the results with *T. divaricata* ajmalicine was not more stable in used culture medium. In the *C. roseus* culture the enzymes in the culture medium were also removed but for the stability of ajmalicine this did not give any difference. However for tabersonine a lower degradation

Table 1. Dilution of ^{15}N labelled alkaloids added and the amount of non-labelled compound accumulated during 19 days of culture. The dilution was calculated by dividing the labelling percentage at time 0 by the final labelling percentage measured.

Labelled alkaloid added	Dilution of labelled compound		Amount of non-labelled compound (nmol/flask)	
	a. line[a]	non-a.line[b]	a. line[a]	non-a. line[b]
O-acetylvallesamine	2.3	1.1	242	13
Voaphylline	1.7	1.1	39	3

[a] Terpenoid indole alkaloid accumulating cell line.
[b] Terpenoid indole alkaloid non-accumulating cell line.

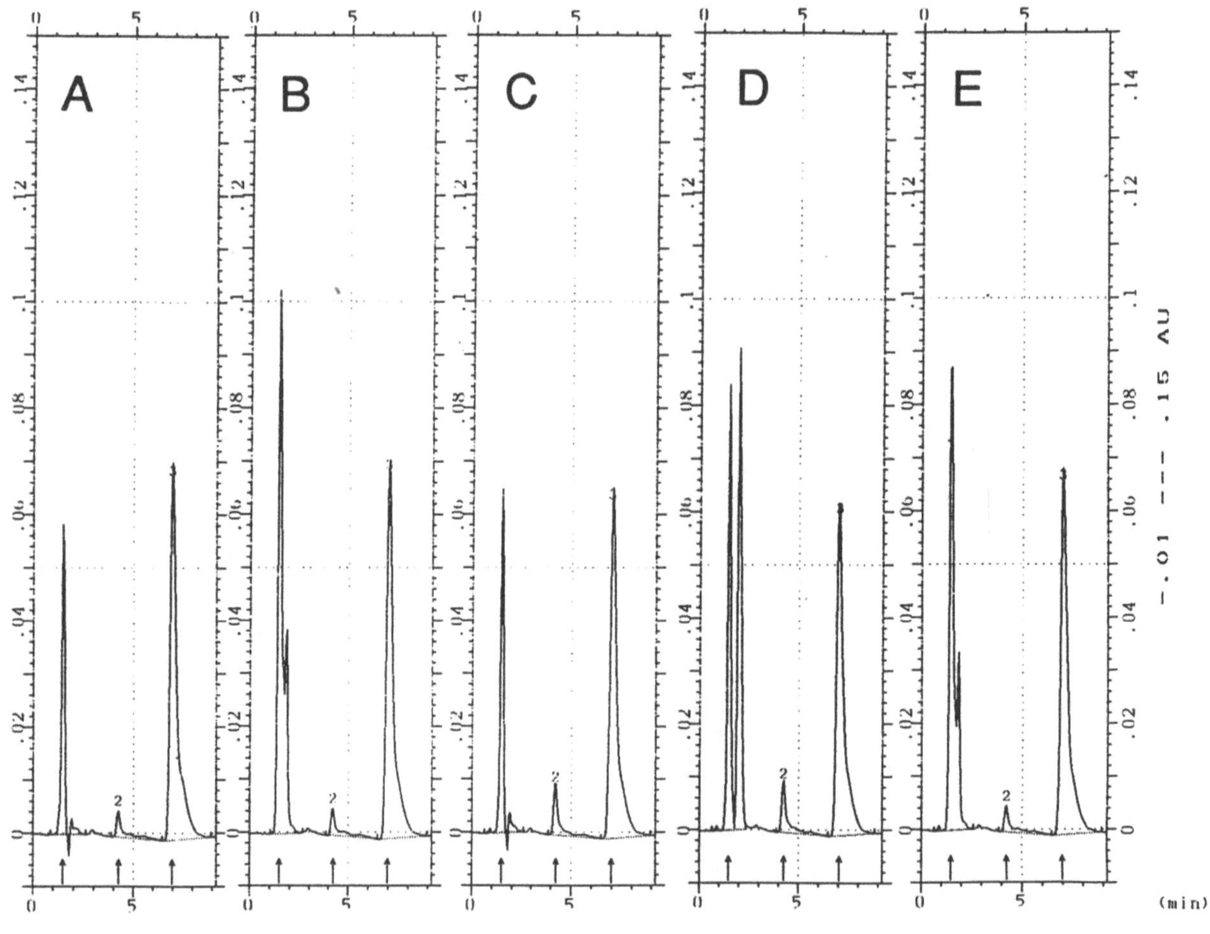

Fig. 7. Chromatograms obtained after different incubations. O-acetylvallesamine was incubated in bis-trispropane buffer pH 6.50 (50 mM), at 30 °C for 1 hour with the following additions: A: none, B: + FAD, C: + crude enzyme extract, D: + FAD + crude enzyme extract, E: + FAD + denatured crude enzyme extract.

was observed after removal of the enzymes by ultrafiltration.

Which enzymes are involved in the biotic breakdown?

Some preliminary experiments were done to investigate the types of reactions occurring during the biotransformation of O-acetylvallesamine. Two types of conversion could be observed with a crude

Table 2. An attempt to distinguish between chemical and biological transformation by comparing the rate of O-acetylvallesamine transformation in fresh culture medium, in used culture medium in the absence of cells and the transformation occurring in the presence of cells.

Time interval	% of O-acetylvallesamine transformed during the time interval				
	In the culture		Without cells		In fresh medium
	a. line[a]	non-a. line[b]	a. line[a]	non-a. line[b]	
4-13 days	38	36	45	34	40
13-19 days	39	31	10	9	41

[a] Terpenoid indole alkaloid accumulating cell line.
[b] Terpenoid indole alkaloid non-accumulating cell line.

enzyme preparation. One was the deacetylation of O-acetylvallesamine to vallesamine. This reaction occurred without the addition of cofactors. The other conversion needed FAD as cofactor, indicating the oxidative nature of the reaction (Fig. 7). At the moment the products which are being formed in this enzymatic reaction and also the products formed by the chemical instability are under investigation.

Conclusions

In suspension cultures of *Tabernaemontana divaricata* and *Catharanthus roseus* a considerable amount of the alkaloid, which is biosynthesized is broken down. The breakdown is independent on the culture period and the half-life was found to be in the order of several days. The difference between producing and non-producing cultures cannot be explained by the breakdown. Both biotic and abiotic factors play a role in the breakdown.

Further studies will focus on the products being formed in the breakdown reactions and the possibilities to influence these reactions.

Acknowledgement

We thank BIO–RIO for the grant received by D. D. and CNPq, Brazil for the grants received by D. D. and R. I. S.

References

Auriola S, Naaranlahti T & Lapinjoki SP (1991) Synthesis of [methyl-^{2}H]-labelled ajmalicine, yohimbine, tabersonine and catharanthine. J. Labelled Compd. Radiopharm. 29: 117–121

Becker H (1987) Regulation of secondary metabolism in plant cell cultures. Plant Tiss. Cell Cult. 199–212

Dagnino D, Schripsema J, Peltenburg A, Verpoorte R & Teunis K (1991) Capillary gas chromatographic analysis of indole alkaloids: investigation of the indole alkaloids present in *Tabernaemontana divaricata* cell suspension culture. J. Nat. Prod. 54: 1558–1563

Dagnino D, Schripsema J & Verpoorte R (1993a) Alkaloid metabolism in *Tabernaemontana divaricata* cell suspension cultures. Phytochemistry 32: 325–329

Dagnino D, Schripsema J & Verpoorte R (1993b) Comparison of terpenoid indole alkaloid production and degradation in two cell lines of *Tabernaemontana divaricata*. Plant Cell Rep. 13: 95–98

Dos Santos RI, Schripsema J & Verpoorte R (1994) Ajmalicine metabolism in *Catharanthus roseus* suspension cultures. Phytochemistry, 35: 677–681

Linsmaier EM & Skoog F (1965) Organic growth & factor requirements of tobacco tissue cultures. Physiol. Plant. 18: 100–127

Murashige T & Skoog F (1962) A revised medium for rapid growth and bio-assays with tobacco tissue cultures. Physiol. Plant. 15: 473–497

Robinson T (1974) Metabolism and function of alkaloids in plants. Science 184: 430–435

Schripsema J, Meijer AH, Van Iren F, Ten Hoopen HJG & Verpoorte R (1990) Dissimilation curves as a simple method for the characterization of growth of plant cell suspension cultures. Plant Cell Tiss. Org. Cult. 22: 55–64

Schripsema J & Verpoorte R (1992) Search for factors related to the indole alkaloid production in cell suspension cultures of *Tabernaemontana divaricata*. Planta Med. 58: 245–249

Schripsema J, Peltenburg-Looman A, Erkelens C & Verpoorte R (1992) Nitrogen metabolism in cultures of *Tabernaemontana divaricata*. Phytochemistry 30: 3951–3954

Smith E (1984) Analysis of *Cinchona* alkaloids by high-performance liquid chromatography. J. Chromatogr. 299: 233–244

Wink M (1987) Physiology of the accumulation of secondary metabolites with special reference to alkaloids. In: Constable F & Vasil IK (Eds) Cell Culture and Somatic Cell Genetics of Plants. Vol 4: Cell Culture in Phytochemistry (pp 17–42). Academic Press, San Diego

Plant Cell, Tissue and Organ Culture **38**: 307–319, 1994.

The cell culture medium – a functional extracellular compartment of suspension-cultured cells

Michael Wink
Institut für Pharmazeutische Biologie, Universität Heidelberg, Im Neuenheimer Feld 364, D-69120 Heidelberg, Germany

Key words: Extracellular space, lytic compartment, secretion, sink-source system, spent medium

Abstract

The spent medium of suspension-cultured cells of *Lupinus polyphyllus* was analyzed by capillary GC and GC-MS and shown to contain ethanol (up to 160 mmol l^{-1}), organic acids (lactate, benzoate, succinate, fumarate, malate), amino acids (main components: alanine, glycine, serine, aspartate, ornithine, glutamate), and quinolizidine alkaloids (lupanine and an uncharacterized malonylderivative). In addition, cells obviously secrete polysaccharides and enzymes (acid phosphatase, phosphodiesterase, DNAse, esterase, α-mannosidase, α-galactosidase, β-glucosidase, lipase, protease and peroxidase) into the medium. Typically these enzymes are localized in the vacuole of intact cells. Cytosolic enzymes, such as glutamate dehydrogenase and malate dehydrogenase were retained by the cells. Peroxidase is overexpressed in suspension-cultured lupin cells but only one basic isoenzyme is secreted, whereas the others are retained in the vacuole. In lupin leaves this isoenzyme is sequestered in the vacuole, implying that secretion is selective and needs a change in the sorting signals of the peroxidase protein. The cell culture medium shares many features of the vacuole. We assume therefore that the medium functions as a lytic compartment. In addition it provides a sink-source system for nutrients and metabolites.

Abbreviations: ADH – alcohol dehydrogenase, FW – fresh weight, GC – gas chromatography, GC-MS – gas chromatography – mass spectrometry, POD – peroxidase, QA – quinolizidine alkaloids

Introduction

It is sometimes assumed that the cell culture medium of suspension-cultured cells only functions as a source of nutrients for the growing cells. However, in vitro cultured bacterial, fungal and animal cells actively secrete a number of metabolites and proteins into the culture medium. Thus, the medium also functions as an external storage compartment to some degree.

A number of in vitro cultured cells need the addition of "conditioned medium" for growth, implying that the spent culture medium must contain some useful metabolites. There is evidence in the literature that the growth medium of plant cells sequesters a number of secondary compounds, peptides and polysaccharides (reviews Barz et al. 1990; Guern et al. 1987; Wink 1985a), and, in addition, remarkably high activities of various hydrolytic and oxidative enzymes (reviews Barz & Köster 1981; Ohlsen et al. 1969; Wink 1984a,b, 1985a) have been detected in the spent medium.

We have proposed previously that the spent culture medium functions as a lytic compartment of in vitro cultured plant cells (Wink 1984b). In this study, we have analyzed the culture medium of cell suspension cultures of *Lupinus polyphyllus* in more detail and come to the conclusion that the medium functions as an extracellular storage compartment for ions and metabolites and, at the same time, as a lytic compartment.

Materials and methods

Cell and tissue cultures

Cell suspension cultures of *Lupinus polyphyllus* were maintained in a modified MS medium (50 ml/flask) and were transferred to fresh medium every 10 to 14 days (Wink et al. 1980, 1983, 1992). The spent medium was prepared by filtration.

Root cultures of *Hyoscyamus albus, H. muticus, H. niger, H. pusillus, Atropa belladonna, Datura stramonium* and *Nicotiana tabacum* were cultivated in liquid medium as described by Sauerwein et al. (1992).

Analysis of the spent medium

Organic acids, sugars, amino acids and ethanol
The spent medium was subjected to ion exchange chromatography (Amberlite IRA-400 and Amberlite IR-120), resulting in a fraction containing organic acids and sugars and another one with amino acids and alkaloids. After freeze drying the organic acids and sugars were derivatized by N-methyl-N-(trimethylsilyl)trifluoroacetamide (MSTFA), the amino acids by trifluoroacetic anhydride (TFA) and isobutyric acid. Organic acids and amino acids were separated by capillary GC using a DB-1 (J & W Scientific) column. GC-MS was performed as described in Wink et al. (1983) and Wink & Witte (1984). Ethanol was determined enzymatically according to Wink (1985a).

Enzyme assays
Hydrolase and nuclease activities were determined photometrically (405 nm) using p-nitrophenyl-based substrates (Wink 1985a). Peroxidase was assayed photometrically (470 nm) with guaiacol and H_2O_2 as substrates. Protease activity was recorded by incubating the medium with azocoll and measuring the liberated chromophore at 520 nm (Wink 1985a).

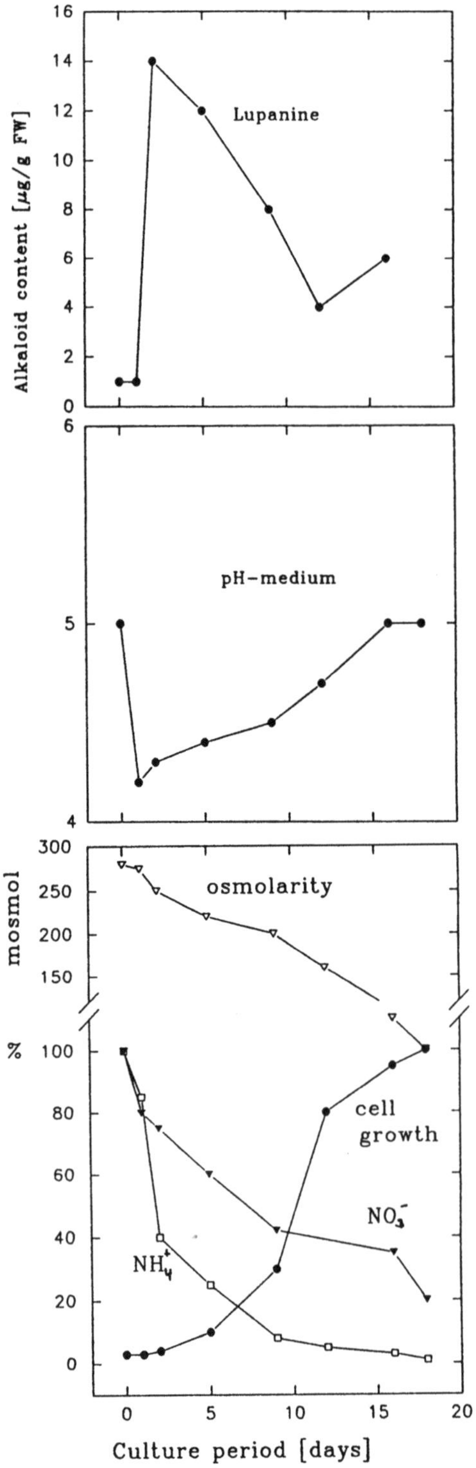

Fig. 1. Growth (FW), nutrient uptake and alkaloid formation in cell suspension cultures of *Lupinus polyphyllus* (after Wink 1984a, Wink & Hartmann 1982b).

Results and discussion

Cell suspension cultures of *Lupinus polyphyllus* form small aggregates and show typical growth kinetics of suspension-cultured cells. There is an initial lag phase after transfer of the cells into fresh medium lasting approximately 3 days, followed by exponential growth

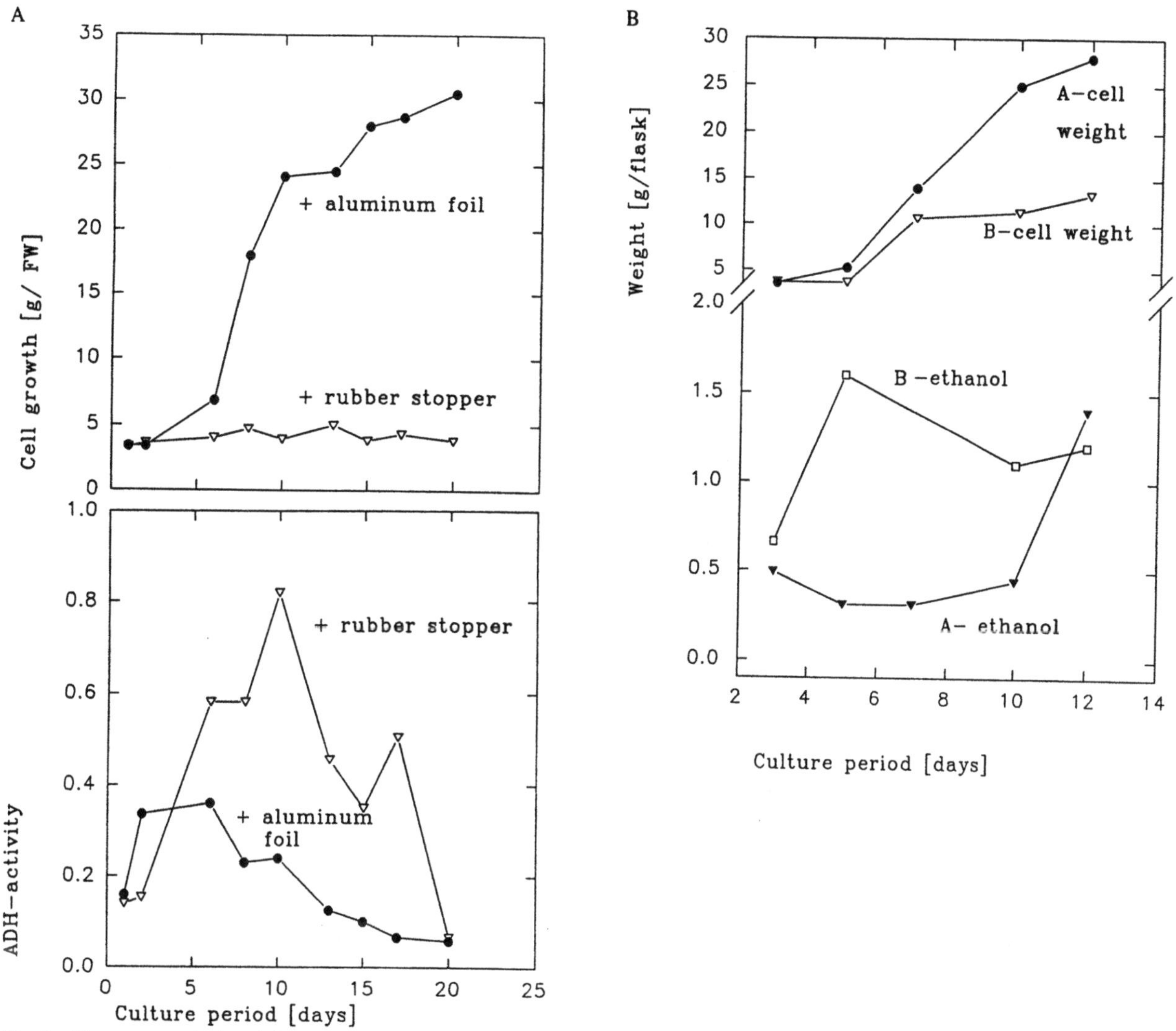

Fig. 2. Time-course of ethanol formation, ADH activity and growth in cell suspension cultures of *Lupinus polyphyllus*. (*A*) ADH activity in relation to oxygen supply. Control Erlenmeyer flasks were closed by aluminum foil allowing exchange of O_2. When flasks were sealed with a rubber stopper, oxygen was limited and ADH became induced. (*B*) Ethanol formation (in %) in control (A) and sealed (B) (by rubber stopper) flasks. Note the increase of ethanol in control cultures in stationary phase.

for 5 to 6 days. The culture, which has formed a thick suspension, then enters the stationary phase (Fig. 1).

The cell culture medium serves as a nutrient source during the initial stage since the carbon source sucrose, the nitrogen source ammonium nitrate and other ions are readily taken up (Fig. 1). As a consequence osmolarity and conductance of the medium decrease during cell growth. In the following we investigated the possible role of the cell culture medium as an external sink for suspension-cultured cells.

Hydrogen ions

The uptake of sucrose from the medium is accompanied by an initial drop of the hydrogen ion concentration of the medium (Fig. 1), thus protons are actively secreted into the medium. This process is thought to be mediated by a proton pumping ATPase at the cytoplasma membrane.

Ethanol

When cells reach a certain density their oxygen supply becomes limiting. As a consequence alcohol dehydro-

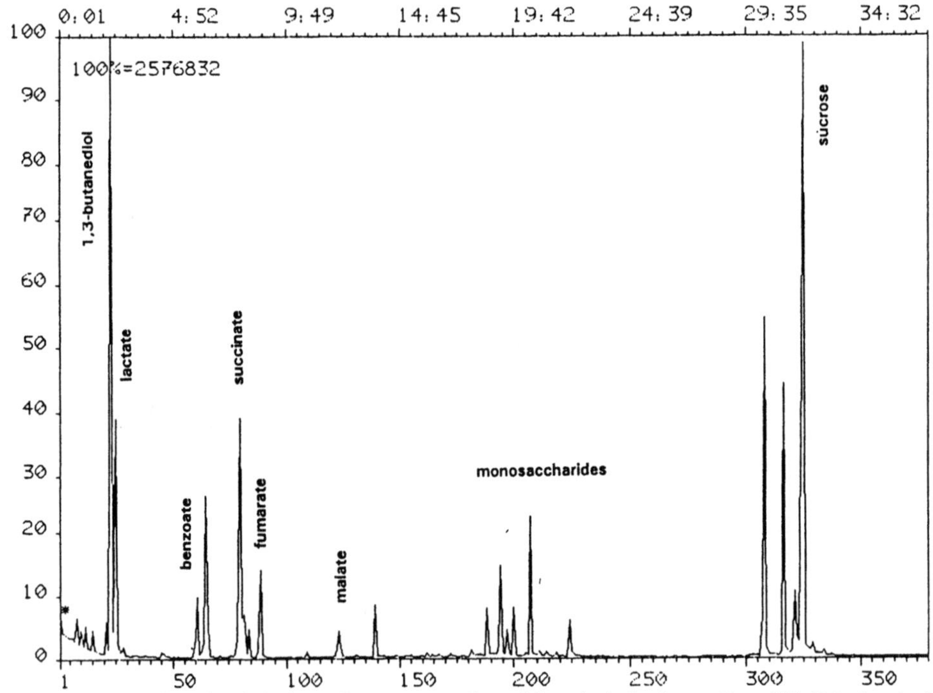

Fig. 3. Separation of organic acids and carbohydrates from spent medium of *L. polyphyllus* by capillary GC. Detection by GC-MS as a total ion chromatogram (TIC). Mono- and disaccharides were not further characterized.

Table 1. Occurrence of ethanol in cell suspension cultures of lupins (after Wink 1985a).

Species	Ethanol (mmol l^{-1})
Lupinus polyphyllus	26–160
Lupinus luteus	11
Lupinus mutabilis	31–42
Lupinus hartwegii	46–145

genase (ADH) is induced, which seems to be typical for this enzyme. As a result the enhanced ADH activity leads to the production of ethanol (Fig. 2B), of which more than 50% is released into the culture medium. When the O_2-exchange is limited by closing the Erlenmeyer flasks with a rubber stopper, ADH induction and ethanol formation are immediately enhanced (Fig. 2 A,B) whereas cell growth is substantially reduced. Ethanol contents of suspension cultures of lupins were up to 160 mmol l^{-1} (Table 1) (Wink 1985a).

Organic acids

Whereas the original medium does not contain organic acids, the analysis of the spent medium by capillary GC and GC-MS clearly shows the presence of a number of organic acids (Fig. 3). Succinate is most abundant followed by lactate, fumarate, benzoate and malate. Succinate, fumarate and malate are intermediates of the citric acid cycle, which are often stored in the vacuole, and lactate is a product of anaerobic metabolism. Lactate formation is probably correlated with anaerobic conditions in densely grown cultures and thus with the ethanol production mentioned before.

Amino acids

In addition to organic acids GC and GC-MS analysis revealed the presence of many amino acids (Fig. 4): Most abundant are alanine, glycine, serine, aspartate, ornithine and glutamic acid, whereas others are present in smaller amounts. We assume that these amino acids, which are normally stored in the vacuole, were also actively secreted into the culture medium. It might be objected that these amino acids result from the proteolytic degradation of medium-released proteins by extracellular proteases. Although this possibility cannot be totally excluded, the presence of substantial amounts of the non-protein amino acid ornithine (Fig. 4) points into the direction of secretion.

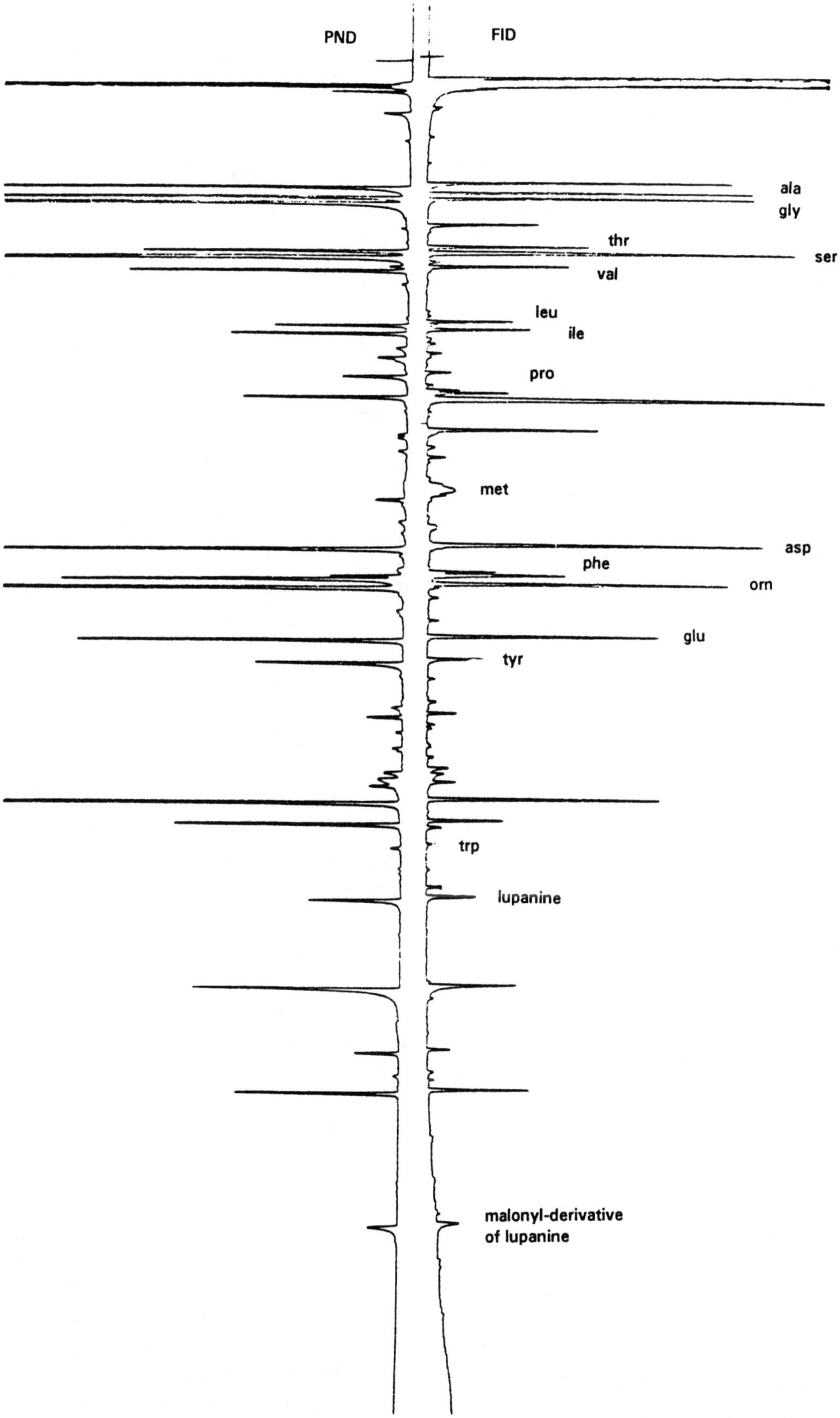

Fig. 4. Separation of amino acids from spent medium of *L. polyphyllus* by capillary GC. Detection by flame ionization (FID) and nitrogen-specific (PND) detectors. Identification by GC-MS. Note the presence of a malonyl derivative of lupanine, which has not been found in intact lupins before.

Table 2. Enzyme composition of vacuoles and cell culture medium of suspension-cultured cells of *Lupinus polyphyllus* (after Wink 1984b, 1985a).

Enzymes	Vacuole	Cell suspension culture	
		Cells	Medium
Acid phosphatase	+	58%	42%
Phosphodiesterase	+	53%	47%
DNAse	+	50%	50%
Esterase	+	57%	43%
α-Mannosidase	+	38%	62%
α-Galactosidase	+	37%	63%
β-Glucosidase	+	52%	48%
Lipase	+	60%	40%
Protease	+	62%	38%
Peroxidase	+	35%	65%
Glutamate dehydrogenase	–	98%	2%
Malate dehydrogenase	–	91%	9%
actual pH	2.3–6.5	7	4–6

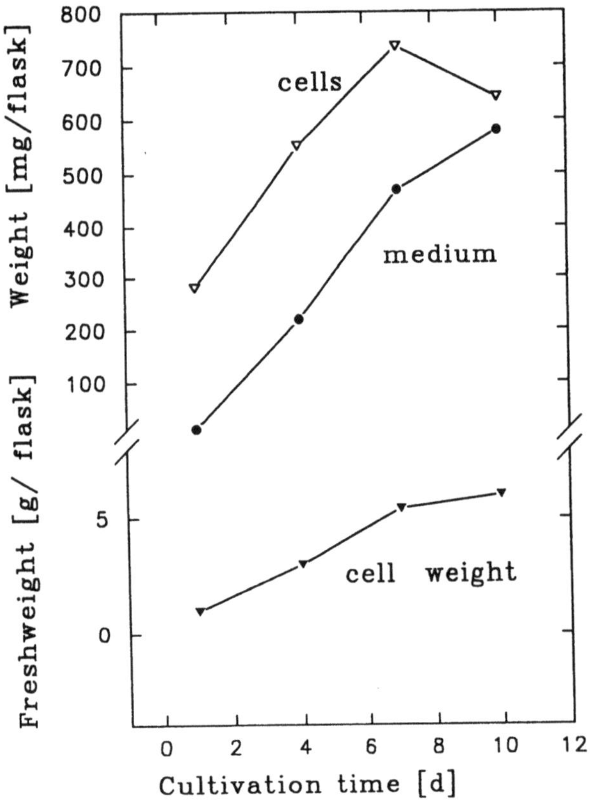

Fig. 5. Accumulation of soluble polysaccharides in cells and growthmedium of *L.polyphyllus*.

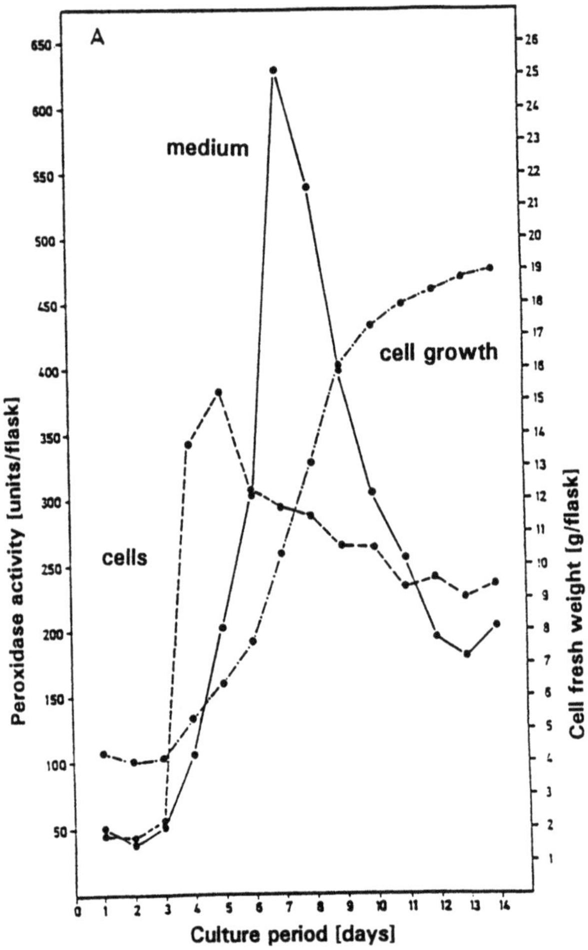

Fig. 6. Time-course of peroxidase activities in cells and medium of *L. polyphyllus* (after Perrey et al. 1989).

Polysaccharides

Suspension-cultured cells of *L. polyphyllus* excrete polysaccharides into the medium (Fig. 5) in a growth-correlated fashion. At the end of the culture period about 50% of the soluble polysaccharides can be recovered from the medium.

Enzymes

We have shown before that the medium of lupin cells contains a rich set of enzymatic activities (Wink 1984b, 1985a) (Table 2), mostly hydrolytic enzymes (acid phosphatase, DNAse, esterase, protease, α-mannosidase, α-galactosidase and β-glucosidase) and peroxidase (POD). Whereas typical cytosolic enzymes (such as glutamate dehydrogenase and malate dehydrogenase) are retained by the cells, vacuolar enzymes (Boller 1982), as those mentioned in Table 2, seem

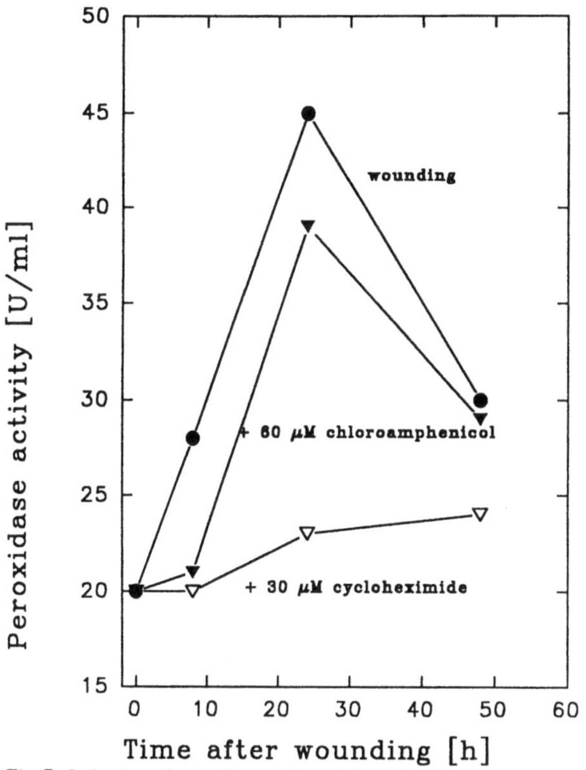

Fig. 7. Induction of peroxidase activity in lupin leaflets by wounding (C. Bartosch & Wink, unpublished). Inhibition by cycloheximide indicates that induction involves the *de novo* synthesis of peroxidase enzymes. The inactivity of chloroamphenicol shows that POD is encoded by nuclear and not by mt or cpDNA.

pI	Leaf complete	Leaf vacuole	Suspension culture apoplast	Suspension culture cells	Suspension culture vacuole	Suspension culture medium
10.7	—	—				—
10.5	—	—		—	—	
10.3	—	—				
9.6	—	—		—	—	
9.3	—	—		—	—	
8.5	—	—	—	—	—	
8.0	—					
7.3	—		—			
7.0	—					
6.7	—					
4.0	—					
3.5	—					

Fig. 8. Schematic representation of the localisation of peroxidase isozymes in plants and cell cultures of *Lupinus polyphyllus* (after Perrey et al. 1989).

to be actively secreted. Between 42 and 65% of the enzymic activity was recovered from the medium. Some of these enzymes are also associated with cell walls and may be released into the medium via this route.

In the case of peroxidase the situation is more complex: Peroxidase provides one of the highest enzymatic activities found in our lupin cultures (Fig. 6). As compared to intact plants its activity is strongly enhanced under in vitro conditions. In this context it should be recalled that peroxidase is an inducible enzyme (in leaves of *L.polyphyllus* POD can be rapidly induced by wounding; Fig. 7). We suggest therefore that the stress conditions which prevail in suspension cultures (osmotic and shear stress) have induced POD enzymes. Also other stress proteins such as heat shock proteins (HSP 70) and osmotins are strongly induced under in vitro conditions (Perrey & Wink 1991; Wink et al. 1992). We have previously analyzed the POD isozymes of leaves and suspension-cultured cells by isoelectric focussing (Fig. 8) (Perrey et al. 1989). In leaves, there are 5 dominant basic isozymes, all of them are located in the vacuole. In suspension-cultured cells we can identify 4 of these isozymes in the cells, and also here they are localized in the vacuole. The most basic isozyme is not present intracellularly but in the medium. Thus the secretion of POD isozymes must be a selective process and cannot be the result of cell breakage or dissociation from the cell wall. In analogy to other enzymes which are secreted by plant cells, it can be assumed that the specificity lies in the presence or absence of respective signal peptides (Chrispeels 1991; Wink 1993). In order to prove this assumption we have started to clone and identify the respective molecular signals of lupin POD. As a first step we have selected a partial cDNA clone which encodes lupin POD (Perrey et al. 1991). This clone should help us to select full length clones of the other vacuolar isozymes and to compare it with that encoding the medium residing isozyme.

The release of hydrolytic enzymes into the culture medium is not restricted to suspension-cultured cells. We have analyzed enzymatic activities in spent medium of root cultures of *Hyoscyamus, Atropa, Datura* and *Nicotiana* (Fig. 9) (Sauerwein & Wink, unpublished). Peroxidase and protease were generally

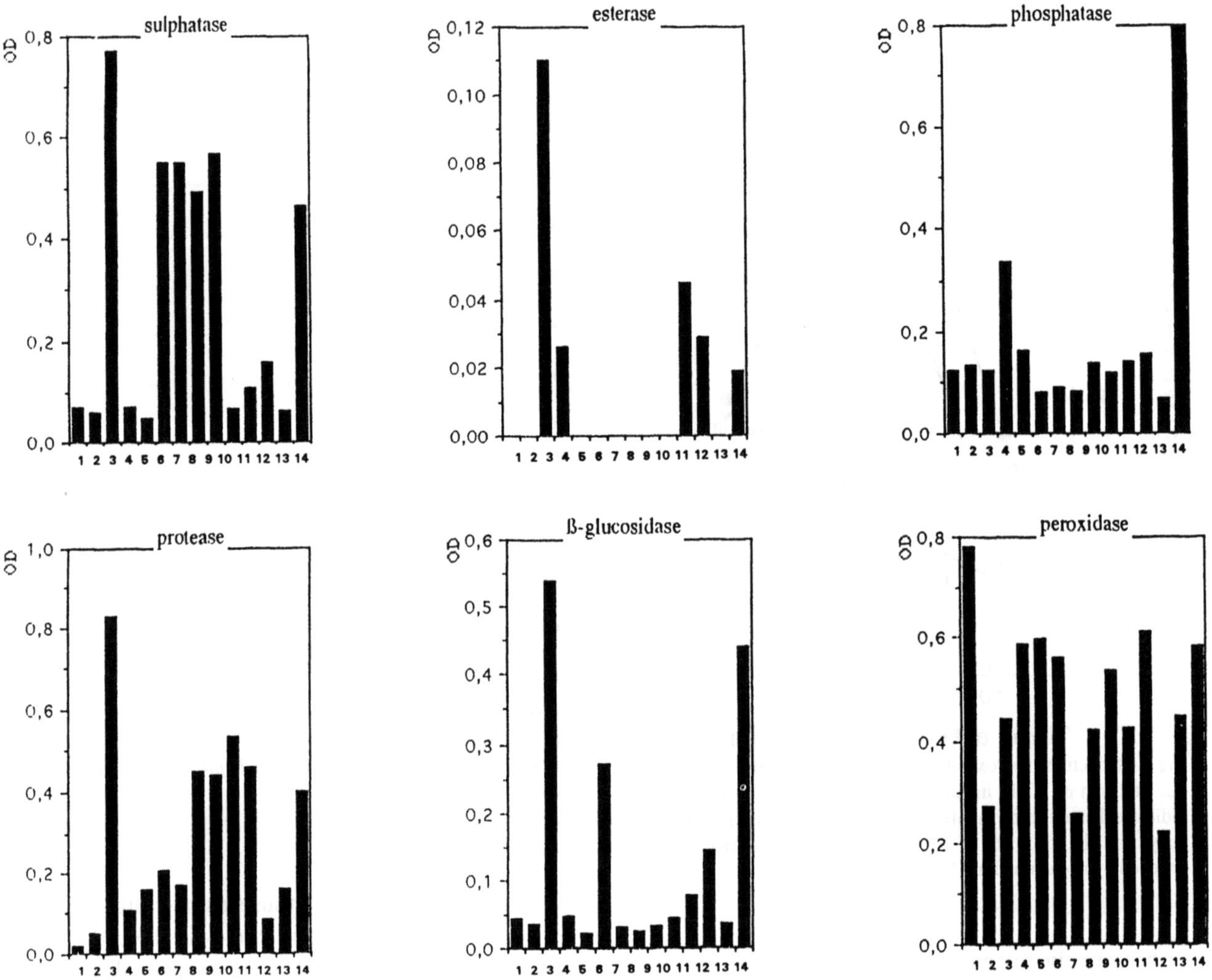

Fig. 9. Comparison of the activities of extracellular enzymes in spent medium of suspension-cultured cells of *L.polyphyllus* (No 14) with that of root cultures of various Solanaceae (M. Wink & M. Sauerwein, unpublished). 1: *Hyoscyamus muticus*, 2: *H. aureus*, 3: *H. niger*, 4: *H. pusillus*, 5,6: *H. albus*, 7–9: *H. albus* transformed by *Agrobacterium*, 10: *Atropa belladonna*, 11,12: *Datura stramonium*, 13: *Nicotiana tabacum*, 14: *Lupinus polyphyllus*.

present in high activities, compared to spent medium of *L. polyphyllus*, whereas phosphatase, esterase, sulphatase, and β-glucosidase were generally much less active.

Quinolizidine alkaloids

Quinolizidine alkaloids represent the main natural products of lupins which function as chemical defence compounds against herbivores and to a minor degree against microorganisms and competing plants (reviews: Wink 1984a, 1985b, 1987a–d, 1988, 1989, 1992, 1993a,b). Several questions concerning their biosynthesis, storage, transport and degradation have been worked out already (Table 3).

Taking these data into account we tried to answer the problem why lupin cell cultures (both callus and suspension cultures) only produce small amounts of alkaloids as compared to the intact plant. Lupanine was always the main alkaloid irrespective whether the intact plant accumulated a different major alkaloid (Wink & Hartmann 1980, 1985; Wink et al. 1983). We have concluded that the pathway leading to lupanine is the basic pathway of quinolizidine alkaloid biosynthesis and that the other alkaloids are derived from it. Thus, in cell cultures only this basic alkaloid pathway seems to be expressed (Wink & Hartmann 1980; Wink et al. 1983; Wink 1987b).

In the intact plant, QA formation takes place in the chloroplast and is light-dependent (Wink & Hartmann 1982c). It could be shown that light-grown green, chlorophyll-rich cell cultures produced more lupanine than colourless heterotrophic cultures (Wink & Hartmann 1980, 1982a,b). If kept under dark and light regime, alkaloid formation in cell cultures was also light-dependent and showed a diurnal cycle (Wink & Hartmann 1982b) similar to the situation in the plant (Wink & Witte 1984).

As compared to the intact plant the absolute amount of lupanine formed was reduced by one to two orders of magnitude (Wink et al. 1983; Wink 1987a–c). It could be determined experimentally that the overall activity of the enzymes of QA biosynthesis was reduced (Wink 1984a, 1987a–c). But alkaloid storage and degradation were additional critical factors. In the plant, QA which are formed in the leaf, are translocated via the phloem all over the plant (Wink & Witte 1984, 1991). Main sites of alkaloid accumulation are epidermal and subepidermal cells of stems and leaves (Wink 1984a; Wink et al. 1984). The subcellular site of QA storage is the vacuole (Mende & Wink 1987) and QA are transported across the tonoplast by a carrier-mediated proton-antiport system (Mende & Wink 1987). Thus storage and transport are gene-encoded processes and governed by gene regulation.

We have therefore analyzed the capacity of suspension-cultured cells to take up quinolizidine alkaloids. Sparteine is resorbed in a growth-dependent fashion (Fig. 10); uptake was highest during active growth and reduced during the stationary phase. Uptake was time-, pH and temperature-dependent (Fig. 11) (Mende 1987). These kinetics provide experimental evidence that sparteine uptake is not achieved by free diffusion but that it is a carrier-mediated process (Mende & Wink 1987; Mende 1987). We have assayed the uptake activity in suspension-cultured cells (when it was highest) and compared it to that of epidermal cells (Fig. 12). It is evident that the uptake capacity of suspension-cultured cells is substantially lower (Wink & Mende 1987).

When alkaloids were included into the culture medium, they were taken up by the cells within 48 h (Wink 1985c). The alkaloid level inside the cells decreased rapidly and after 3–7 days alkaloids were no longer detectable. In case of ^{14}C-labelled alkaloids we could trap $^{14}CO_2$ concomitantly. In addition, lupin cells could be maintained on media without a nitrogen source if the alkaloid sparteine was added as a sole N-source (Wink & Witte 1985). We assumed, that cultured cells rapidly degrade and metabolize sparteine and utilize its nitrogen. This degradation activity seems to be markedly enhanced in cultured cells and only

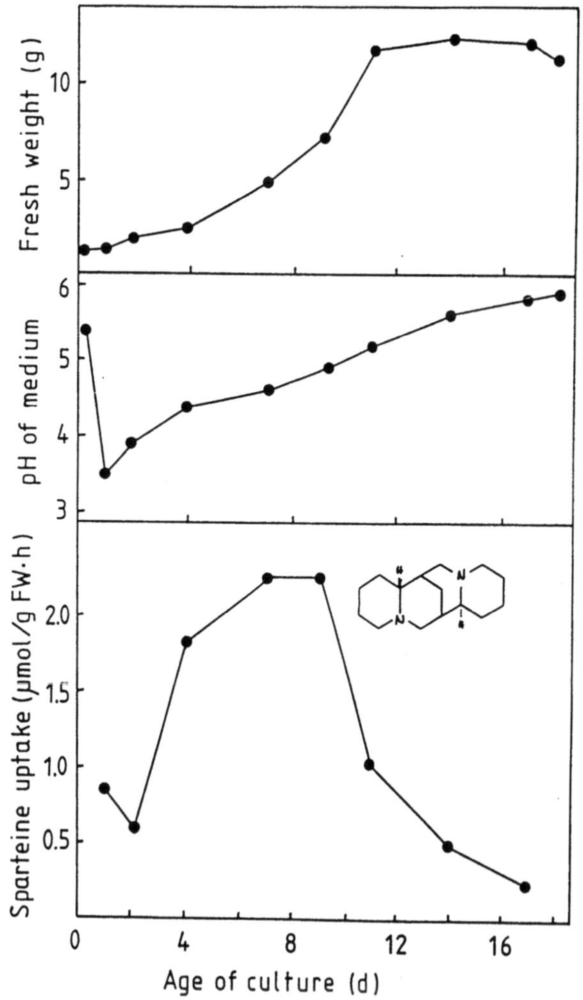

Fig. 10. Uptake of sparteine by cell suspension cultures of *L. polyphyllus* in relation to age of culture (after Mende 1987).

comparable to the situation of germinating seeds: here QA are stored as a nitrogen source which is utilized during germination and early growth of the seedling (Wink & Witte 1985).

So far, we had concluded that the biosynthetic and storage capacities of lupin alkaloids are reduced in lupin cell culture and concomitantly that alkaloid degradation is enhanced. But where is the site of degradation? Lupanine is regularly released into the culture medium to some degree (Fig. 4), which was seen in induction experiments (Wink 1985c) and during diurnal cycles. The enzymes (especially peroxidase) which are present in the culture medium were shown to be able to degrade the alkaloid molecules (Wink 1984a,b, 1985a,c). We suggest that peroxidative degradation of secondary compounds in the culture medium is not an exception but occurs more regularly (reviews: Barz & Köster 1981; Barz et al. 1990). In the differenti-

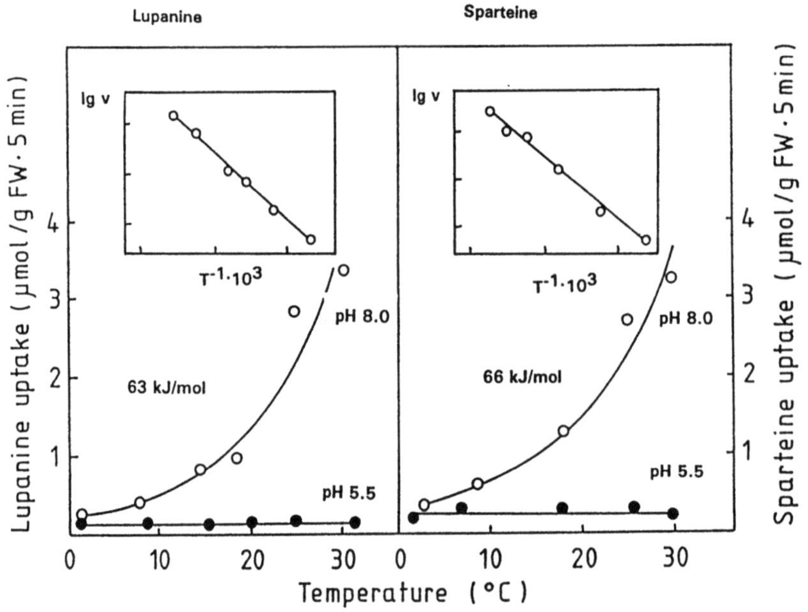

Fig. 11. Temperature-dependence of sparteine and lupanine uptake by suspension-cultured cells of *L.polyphyllus* (after Mende 1987).

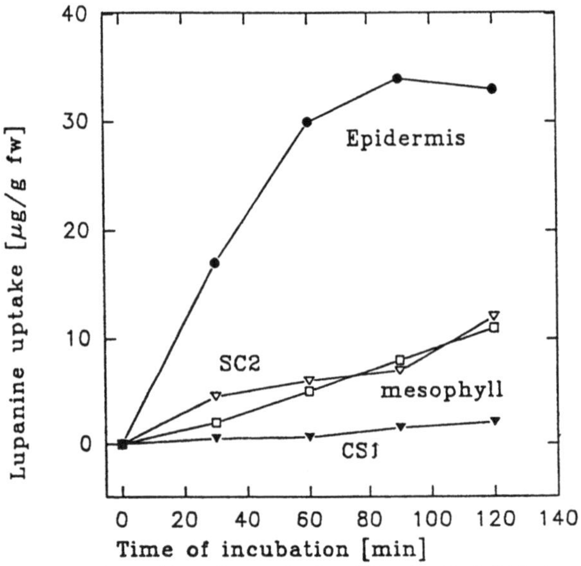

Fig. 12. Rate of lupanine uptake in epidermal cells (which store QA predominantly), mesophyll cells and suspension-cultured cells (after Wink & Mende 1987).

ated plant, biosynthesis, transport, accumulation and degradation are regulated in space and time resulting in alkaloid concentrations which are present at the right place and time to ward off attacks by herbivores and microorganisms (Wink 1987d, 1988, 1992, 1993a). The importance of differentiation is demonstrated by the finding that already shoot cultures (but not root cultures) are able to produce 10–50 times more alkaloid than undifferentiated suspension cultures (Wink 1987a, 1989, 1990).

Conclusions

The cell culture medium serves as a source in the first instance since it provides all the necessary nutrients and ions. During growth and aging of the cultures it also functions as an external sink. Since the volume of medium is relatively big as compared to the intracellular sink (e.g., vacuoles) relatively large amounts of metabolites (organic acids, amino acids, ethanol and ions), enzymes and polysaccharides can be sequestered here (Guern et al. 1987). In contrast to the differentiated plant which has a variety of spatially separated intracellular sink tissues but only limited space in the apoplast, the medium of suspension-cultured cells provides a general and spacious "pool" which promotes the secretion of metabolites, ions and macromolecular constituents (Fig. 13).

The secretion of enzymes and metabolites into an extracellular space is a common phenomenon in bacterial, fungal and animal cells. Is the situation of suspension-cultured cells exceptional for plants? First of all, plant cells secrete those metabolites which are necessary for the cellwall formation into the extracellular space (apoplast), which includes corresponding enzymes. In addition, plants release enzymes into the

Table 3. Overview of the biochemistry of quinolizidine alkaloids (QA) (after Wink 1984a,b, 1985a–c, 1987a–d, 1990, 1992a,b, 1993b).

OCCURRENCE OF QA:
Fabaceae: *Lupinus, Cytisus, Genista, Laburnum, Thermopsis, Baptisia, Sophora* etc.

BIOSYNTHESIS:
Sequence: Lysine → cadaverine → lupanine
Enzymes involved: Lysine decarboxylase, oxosparteine synthase
Alkaloid synthesis only in green tissue
Localisation of alkaloid biosynthesis in leaf chloroplast
Regulation by light (pH, thioredoxin, precursor availability)
Diurnal fluctuation of enzymes and alkaloid contents

ACCUMULATION:
All parts of a plant accumulate alkaloids
Vacuole functions as intracellular storage compartment
Epidermal/subepidermal tissues main site of alkaloid storage in leaves and stems
Storage in seeds

TRANSPORT:
Long distance transport in phloem
Passage across tonoplast with aid of an alkaloid transporter; Mg- ATP and K^+ necessary for uptake
Also carrier-mediated transport across plasmalemma

DEGRADATION:
Diurnal degradation in all organs
During germination and growth of the seedling (Mobilization of alkaloidal nitrogen)

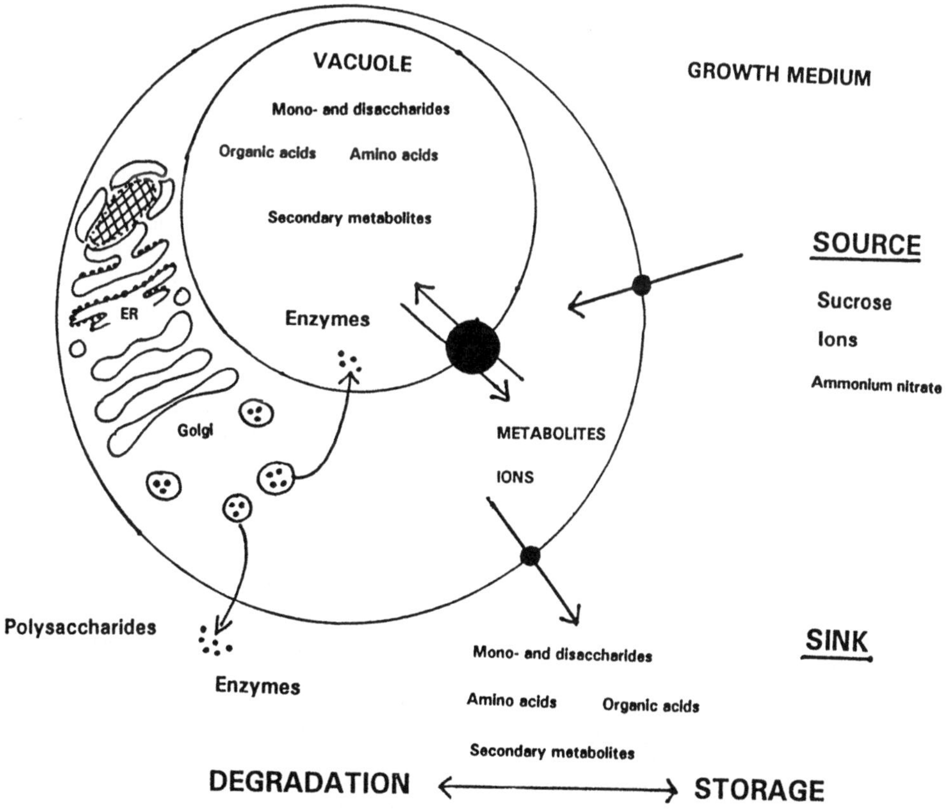

Fig. 13. Schematic illustration of the 'sink-source' system of in vitro cultured cells.

extracellular space when attacked by microbes: The secretion of chitinase, glucanase and other enzymes has been studied in this context. Since suspension-cultured cells can be stressed by the culture conditions (Wink et al. 1992), the secretion of high amounts of peroxidase and hydrolytic enzymes in the medium may be an analogous response.

Furthermore, the medium composition of organic acids, amino acids and enzymes is very similar to that of vacuoles. Thus it is possible that the medium functions as a big, extracellular vacuole. Besides its function as a storage and osmotic compartment (review Wink 1993c), the vacuole provides a lytic environment analogous to that of lysosomes in animal cells (Marty et al. 1980; Matile 1975, 1978; Boller & Wiemken 1986; Willenbrink 1987; Wink 1993c). Since lytic processes can be encountered in the medium, it can be regarded as an external lytic compartment (Wink 1984b; Barz et al. 1990).

Summarizing these findings, the medium of suspension cultured cells functions as a nutrient source on one hand and as a storage ('sink') and lytic compartment on the other hand (Fig. 13) and thus complements the activities of the vacuole (Wink 1993c). Therefore, a detailed analysis of the medium should not be neglected when studying primary and secondary metabolism of in vitro cultured cells and tissues since it is an essential part of a culture system.

Acknowledgements

Work from our laboratory was supported by grants of the Deutsche Forschungsgemeinschaft. I would like to thank my coworkers, Drs. P. Mende, M.T. Hauser, R. Perrey, M. Sauerwein for cooperation and Mrs C. Theuring, U. Schade, M. Weyerer and U. Kahl for technical assistance. Special thanks are due to Dr. L. Witte (Braunschweig) for performing the GC-MS measurements.

References

Barz W, Beimen A, Dräger B, Jaques U, Otto Ch, Süper E & Upmeier B (1990) Turnover and storage of secondary metabolites in cell culture. In: Charlwood BV, Rhodes MJC (Eds) Secondary Products from Plant Tissue Culture (pp 79–102). Clarendon Press, Oxford

Barz W & Köster J (1981) Turnover and degradation of secondary products. In: Conn EE (Ed) The Biochemistry of Plants. Vol. 7. Secondary plant products (pp 35–84). Academic Press, Orlando

Boller T (1982) Enzymatic equipment of plant vacuoles. Physiol. Veget. 20: 247–257

Boller T & Wiemken A (1986) Dynamics of vacuolar compartmentation. Annu. Rev. Plant Physiol. 37:137–164

Chrispeels MJ (1991) Sorting of proteins in the secretory system. Annual Rev. Plant Physiol. Plant Mol. Biol. 42: 21–53

Guern J, Renaudin JP & Brown SC (1987) The compartmentation of secondary metabolites in plant cell cultures. In: Constabel F & Vasil I (Ed) Cell Culture and Somatic Cell Genetics of Plants, Vol. 4: Cell Culture in Phytochemistry (pp 43–76). Academic Press, New York

Marty F, Branton D & Leigh RA (1980) Plant vacuoles. In: Tolbert NE (Ed) The Biochemistry of Plants: A Comprehensive Treatise, Vol 1 (pp 625–658). Macmillan Press, New York

Matile P (1975) The lytic compartment of plant cells. Springer Verlag, Berlin

Matile P (1978) Biochemistry and function of vacuoles. Annual Rev. Plant Physiol. 29: 193–213

Mende P (1987) Untersuchungen zur Aufnahme von Chinolizidinalkaloiden in Zellsuspensionskulturen, Epidermen, Protoplasten und Vakuolen von *Lupinus polyphyllus*. Dissertation, TU Braunschweig

Mende P & Wink M (1987) Uptake of the quinolizidine alkaloid lupanine by protoplasts and vacuoles of *Lupinus polyphyllus* cell suspension cultures. J. Plant Physiol. 129:229–242

Ohlsen AC, Evans JJ, Frederick DP & Jansen E (1969) Plant suspension culture media macromolecules — Pectic substances, protein and peroxidase. Plant Physiol. 44: 1594–1600

Perrey R & Wink M (1991) Constitutive expression and molecular characterization of a cDNA clone encoding a partial HSP70 gene in cell suspension cultures of *Lupinus polyphyllus*. J. Plant Physiol. 137: 744–748

Perrey R, Hauser M-T, & Wink M (1989) Cellular and subcellular localization of peroxidase isoenzymes in plants and cell suspension cultures from *Lupinus polyphyllus*. Z. Naturforsch. 44c: 931–936

Perrey R, Warskulat U & Wink M (1991) Molecular cloning of a *Lupinus polyphyllus* cDNA encoding a basic peroxidase isoenzyme of cell suspension cultures. J. Plant Physiol. 137: 537–540

Sauerwein M, Wink M & Shimomura K (1992) Influence of light and phytohormones on alkaloid production in transformed root cultures of *Hyoscyamus albus*. J. Plant Physiol. 140: 147–152

Willenbrink J (1987) Die pflanzliche Vacuole als Speicher. Naturwissenschaften 74: 22–29

Wink M (1984a) Stoffwechsel und Funktion der Chinolizidinalkaloide in Pflanzen und pflanzlichen Zellkulturen. Habilitationthesis, Technische Universität Braunschweig

Wink M (1984b) Evidence for an extracellular lytic compartment of plant cell suspension cultures: The cell culture medium. Naturwissenschaften 71: 635.

Wink M (1985a) Composition of the spent culture medium. Time course of ethanol formation and the excretion of hydrolytic enzymes into the culture medium of suspension-cultured cells of *Lupinus polyphyllus*. J. Plant Physiol. 121: 287–293

Wink M (1985b) Chemische Verteidigung der Lupinen: Zur biologischen Bedeutung der Chinolizidinalkaloide. Plant Syst. Evol. 150: 65–81

Wink M (1985c) Metabolism of quinolizidine alkaloids in plants and cell suspension cultures: Induction and degradation. In: Neumann K-H, Barz W & Reinhard E (Eds) Primary and Secondary Metabolism of Plants and Cell Cultures (pp 107–116). Springer Verlag, Berlin

Wink M (1987a) Physiology of the accumulation of secondary metabolites with special reference to alkaloids. In: Constabel F

& Vasil I (Ed) Cell Culture and Somatic Cell Genetics of Plants, Vol. 4: Cell Culture in Phytochemistry (pp 17–41). Academic Press, New York

Wink M (1987b) Quinolizidine alkaloids: Biochemistry, metabolism, and function in plants and cell suspension cultures. Planta Med. 53: 509–514

Wink M (1987c) Why do lupin cell cultures fail to produce alkaloids in large quantities? Plant Cell Tissue Organ Cult. 8: 103–111

Wink M (1987d) Chemical ecology of quinolizidine alkaloids. ACS Symposium. Series 330: 524–533

Wink M (1988) Plant breeding: Importance of plant secondary metabolites for protection against pathogens and herbivores. Theor. Appl. Gen. 75: 225–233

Wink M (1989) Induction of alkaloid formation in plant cell cultures. In: Kotyk A, Skoda J, Paces V & Kostka V (Eds) Highlights of Modern Biochemistry (pp 1183–1192). Elsevier

Wink M (1990) Physiology of secondary product formation in plants. In: Charlwood BV, Rhodes MJC (Eds) Secondary Products from Plant Tissue Culture (pp 23–41). Clarendon Press, Oxford

Wink M (1992) The Role of Quinolizidine Alkaloids in Plant Insect Interactions. In: Bernays EA (Ed) Insect-Plant Interactions, Vol. IV (pp 133–169). CRC-Press, Boca Raton

Wink M (1993a) Quinolizidine alkaloids. In: Methods in Plant Biochemistry, Vol. 8, Alkaloids and Sulphur Compounds (pp 197–239). Academic Press, London

Wink M (1993b) Allelochemical properties and the raison d'etre of alkaloids. In: Cordell G (Ed) The Alkaloids, Vol. 43 (pp 1–118). Academic Press, Orlando

Wink M (1993c) The plant vacuole – a multifunctional compartment. J. Exp. Bot. 44 Suppl: 231–246

Wink M & Hartmann T (1980) Production of quinolizidine alkaloids by photomixotrophic cell suspension cultures: Biochemical and biogenetic aspects. Planta Med. 40: 149–155

Wink M & Hartmann T (1982a) Physiological and biochemical aspects of quinolizidine alkaloid formation in cell suspension cultures. In: Fujiwara A (Ed) Plant Tissue Culture 1982 (pp 333–334). IAPTC, Tokyo

Wink M & Hartmann T (1982b) Diurnal fluctuation of quinolizidine alkaloid accumulation in legume plants and photomixotrophic cell suspension cultures. Z. Naturforsch. 37c: 369–375

Wink M & Hartmann T (1982c) Localization of the enzymes of quinolizidine alkaloid biosynthesis in leaf chloroplast of *Lupinus polyphyllus*. Plant Physiol. 70: 74–77

Wink M & Hartmann T (1985) Enzymology of quinolizidine alkaloid biosynthesis. In: Zalewski RI & Skolik JJ (Eds) Natural Products Chemistry 1984 (pp 511–520). Elsevier, Amsterdam

Wink M & Mende P (1987) Uptake of lupanine by alkaloid-storing epidermal cells of *Lupinus polyphyllus*. Planta Med. 53: 465–469

Wink M & Witte L (1984) Turnover and transport of quinolizidine alkaloids: Diurnal variation of lupanine in the phloem sap, leaves and fruits of *Lupinus albus* L. Planta 161: 519–524

Wink M & Witte L (1985) Quinolizidine alkaloids as nitrogen source for lupin seedlings and cell suspension cultures. Z. Naturforsch. 40c: 767–775

Wink M & Witte L (1991) Storage of quinolizidine alkaloids in *Macrosiphum albifrons* and *Aphis genistae* (Homoptera: Aphididae). Entomol. Gener. 15: 237–254

Wink M, Witte L, Schiebel HM & Hartmann T (1980) Alkaloid pattern of cell suspension cultures and differentiated plants of *Lupinus polyphyllus*. Planta Med. 38: 238–245

Wink M, Witte L, Hartmann T, Theuring C & Volz V (1983) Accumulation of quinolizidine alkaloids in plants and cell suspension cultures: Genera *Lupinus, Cytisus, Baptisia, Genista, Laburnum, and Sophora.* Planta Med. 48: 253–257

Wink M, Heinen HJ, Vogt H & Schiebel HM (1984) Cellular localization of quinolizidine alkaloids by laser desorption mass spectrometry (LAMMA 1000). Plant Cell Rep. 3: 230–233

Wink M, Perrey R, Schneider M, Warskulat U, von Borstel K & Mende P (1992) Alkaloid metabolism and gene expression in cell suspension cultures of *Lupinus polyphyllus* and *L. hartwegii*. In: Oono et al. (Eds) Plant Tissue Culture and Gene Manipulation for Breeding and Formation of Phytochemicals (pp 101–119). National Institute of Agrobiological Resources, Tsukuba, Japan

Plant Cell, Tissue and Organ Culture **38:** 321–326, 1994.

Secondary metabolites in hairy root cultures of *Leontopodium alpinum* Cass. (Edelweiss)

Ingrid Hook
Department of Pharmacognosy, School of Pharmacy, Trinity College, 18 Shrewsbury Road, Dublin 4, Ireland

Key words: Anthocyanins, chlorogenic acid, Compositae, Edelweiss, essential oil, hairy roots

Abstract

The formation of 5 hairy root lines of *Leontopodium alpinum* was induced by infection of sterile plants with *Agrobacterium rhizogenes*. The transformed roots were grown as batch cultures in a phytohormone-free modified Murashige & Skoog medium. A time-course experiment with the most productive line showed that a culture period of 6 weeks was optimum for biomass production yielding a 70-fold increase in fresh weight. A 70% enhancement of anthocyanin formation could be induced by addition of benzyladenine (to a final concentration of 0.5 mg l^{-1}) to the culture medium 14 days before harvest. The presence in the cultures of chlorogenic acid as well as other hydroxycinnamic acid esters was confirmed by TLC. An essential oil (ca. 0.6%) was separated from the hairy roots by steam distillation, a high variability in oil yield being observed between the different lines. GC analyses showed the oils to be complex mixtures of > 30 compounds, with 2 of these consistently representing ca. 60% of the oils. The essential oils isolated from hairy roots were found to be qualitatively similar to the natural root oil, although quantitative differences in oil components were apparent. Oil yields could be increased by growing roots in the absence of light.

Abbreviations: MS – Murashige & Skoog, BAP – benzylaminopurine

Introduction

Leontopodium alpinum Cass. (Edelweiss; Compositae; Inuleae) is a well-known plant indigenous to the alpine regions of Europe but few researchers have examined the plant for secondary metabolites. Compounds which have been identified in aerial parts include hydrocarbons (Bicci et al. 1975), flavonoids (Tira et al. 1970), sterols and hydroxycinnamic acid esters (Hennessy et al. 1989). We have reported the presence of dicaffeoylquinic and chlorogenic acids, sterols (Hennessy et al. 1989) and an essential oil (Comey et al. 1992a) in roots from cultivated plants. As roots from wild edelweiss are difficult to obtain because of the protected status of the plant and their separation from cultivated plants is inefficient due to their fibrous nature, there is insufficient root material available for phytochemical evaluation. Hairy root cultures of a number of dicotyledonous plants have been established in recent years and found to produce the same secondary metabolites as natural roots (Payne et al. 1987; Parr & Hamill 1987; Sauerwein & Shimomura 1991; Hu & Alfermann 1993; Trotin et al. 1993). We initiated the development and optimized production of hairy roots of *L. alpinum*, in an attempt to produce an alternative source of root material in amounts sufficient for phytochemical investigations.

Materials and methods

Plant material and hairy root induction

Commercially available seeds of *L. alpinum* were used as source material. These were surface sterilized, germinated aseptically and the resulting plantlets micropropagated according to the published protocol (Hook 1993). For the induction of 'hairy' roots sterile plants were inoculated with a 36-hour activated form of *Agrobacterium rhizogenes* (Strain 9402; kanamycin resistant). After 7 weeks those roots showing negative geotropism were tranferred to an agar-solidified phytohormone-free medium (HR) containing ampicillin to destroy residual *A. rhizogenes*. 5 lines of hairy roots were established between September and December 1989 (Hook 1993).

Batch culture of hairy roots

Roots transformed by infection with *A. rhizogenes* were routinely cultured in conical flasks containing HR medium (150 ml in a 250 ml flask, 600 ml in a 1 l flask or 3000 ml in a 5 l flask). This was a modified Murashige & Skoog medium (1962) and consisted of MS salts, thiamine HCl (0.4 mg l^{-1}), mesoinositol (100 mg l^{-1}) and sucrose (30 g l^{-1}). Cultures were grown at 25 °C (± 2°), on an orbital shaker (90 rpm), under fluorescent lights (Super Gro 65–80 W) and a photoperiod of 18 h light/6 h dark. In those experiments where roots were grown in the absence of light, cultures were kept at identical conditions but entirely surrounded by aluminium foil. All experiments were carried out in triplicate.

Chlorogenic acid extraction and detection

A hairy root sample (HRa) was refluxed with methanol, filtered and the extract examined by TLC. A pure standard of chlorogenic acid was used as reference. TLC was performed on precoated silica gel plates ($60F_{254}$), developed in ethyl acetate-formic acid-glacial acetic acid-water (100:11:11:27) and visualized under UV_{254} and with ferric chloride reagent (Wagner et al. 1984).

Anthocyanin extraction and determination

Pigmentation of cultures was determined after homogenization and extraction with methanol containing 1% HCl. After filtration, absorbances were measured at 525 nm. Yields of anthocyanin were calculated from a calibration curve constructed from the purified pigment isolated from *L. alpinum* suspension cultures. Experiments were carried out in duplicate.

Essential oil determination

Freshly harvested roots were dried (< 40°). 15 g were crushed, mixed with water (300 ml) and subjected to steam distillation for 5 h using an Apparatus for the Determination of Essential Oils in Vegetable Drugs (Anonymous 1980). The volume was recorded and the separated oil used for chromatographic evaluation.

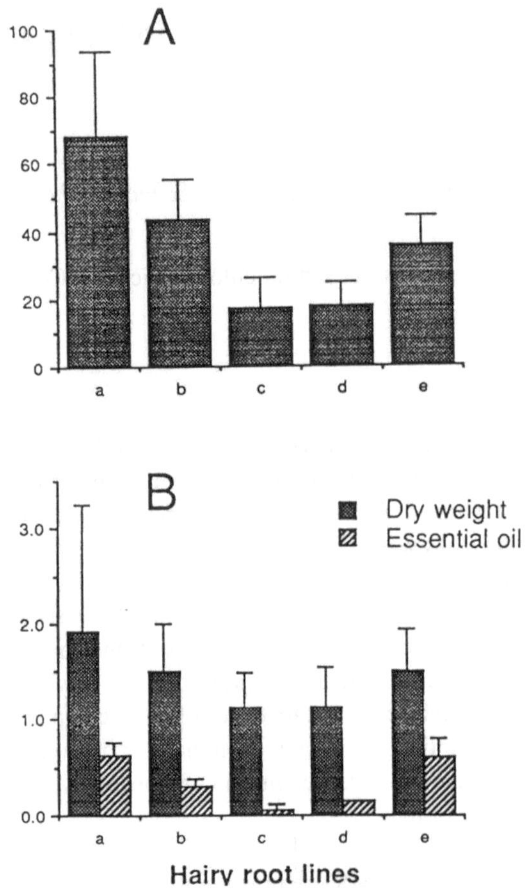

Fig. 1. Growth Index (*A*) and yields of dry weight and essential oil (*B*) for different lines of hairy root cultures of *Leontopodium alpinum* (April '91 – March '93). The culture period was 6 weeks. Cultures were grown in 250 ml flasks containing 150 ml of medium.

Gas Chromatography of Essential Oil

Analyses were carried out on a GC-FID instrument fitted with a glass column (2 m × 3 mm i.d.) packed with 15% Carbowax 20 M on Gas Chrom. W.A.W. (80–100 mesh), operated at 120–200 °C (2° min^{-1} rise), with a nitrogen flow of 30 ml min^{-1}. Percentage relative abundances of individual compounds were calculated by internal normalisation protocols.

Results

Growth characteristics of 'hairy' roots

All root lines separated displayed different growth characteristics: Line HRa grew as grey-beige coloured,

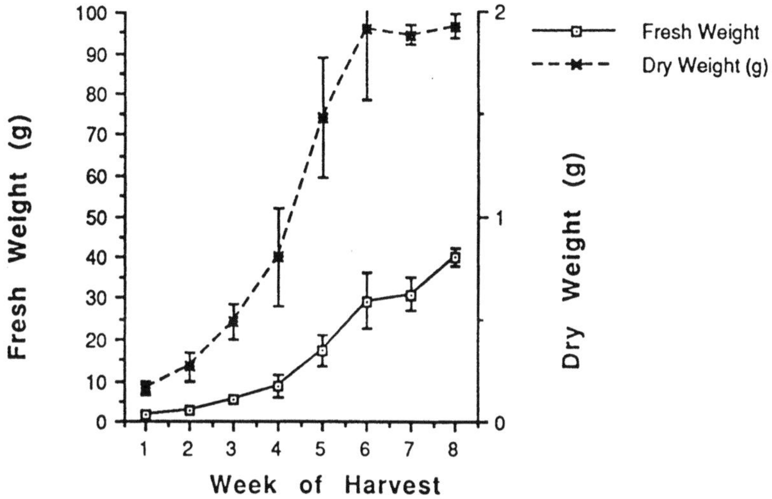

Fig. 2. Time course experiment with line HRa. The hairy root culture was grown in 250 ml flasks containing 150 ml of medium.

thin roots showing many laterals and dense outgrowths of root hairs; line HRb was beige coloured, very thin, with few laterals or root hairs; line HRc had poor root growth, with a tendency to callus formation and subsequently died; line HRd developed as green-beige, short friable roots, with dense root hairs; line HRe resembled line HRa. Proof of genetic transformation was not based on opine detection, but on their ability to grow on a kanamycin-containing HR medium, their active proliferation, negative geotropismm and abundant lateral branching (Jaziri et al. 1988). Growth indices (defined as fresh wt of roots at harvest ÷ fresh wt of inoculum at subculture) were distinctly different with line HRa showing distinct superiority on a continuous basis over a 2-year period (Fig. 1). Because of this it was used for more detailed examinations of growth parameters. Results of a time-course experiment carried out over 8 weeks (Fig. 2) showed that biomass dry wt did not increase after 6 weeks and the percentage weekly increase in dry wt decreased dramatically after week 5. The colour of the culture medium also visibly darkened at around week 6 and the pH increased from 4–4.5 to 7 after 8 weeks, suggesting cell lysis and metabolite decomposition. It was concluded that the optimum culture period for *L. alpinum* hairy roots (HRa) was 6 weeks.

Presence of hydroxycinnamic acid esters

TLC of the methanolic extract of freshly dried roots (HRa) indicated the presence of the same hydroxycinnamic acid esters as were found in natural roots of cultivated plants (Hennessy et al. 1989), i.e. chlorogenic acid (Rf 0.41) and dicaffeoylquinic acids (Rf 0.56 and 0.85) (Wagner et al. 1984).

Enhancement of anthocyanin formation

With suspension cultures of *L. alpinum* we had found that addition of BAP (to a final concentration of 0.5 mg BAP l^{-1}) 14 days prior to harvest resulted in a ca. 5-fold increase in anthocyanin yields (Comey et al. 1992b). When this protocol was applied to hairy roots which showed occasional pigmentation, the anthocyanin content could be increased by up to 70%.

Essential oil content

Normal roots of cultivated *L. alpinum* plants have been shown to produce up to 2.0% of a pale brown essential oil (Comey et al. 1992a). A much lower percentage has been isolated by steam distillation from the hairy roots. Fig. 1B shows that the different root lines gave varying yields, with line HRa again being most productive (0.62%). GC analyses indicated the oil to be a complex mixture of more than 30 compounds, 20 of which were present in concentrations up to 1.0% and 2 consistently represented nearly 60% of the sample. A preliminary GC-MS analysis in conjunction with a data library search indicated many of the constituents to be various sesquiterpenes. In this respect *L. alpinum* resembles the taxonomically related *Inula racemosa*, which also produces an essential oil consisting of mainly sesquiterpenes (Bokadia et al. 1986). The compounds in edelweiss root oil could not be identified by GC-MS alone and are currently the subject of

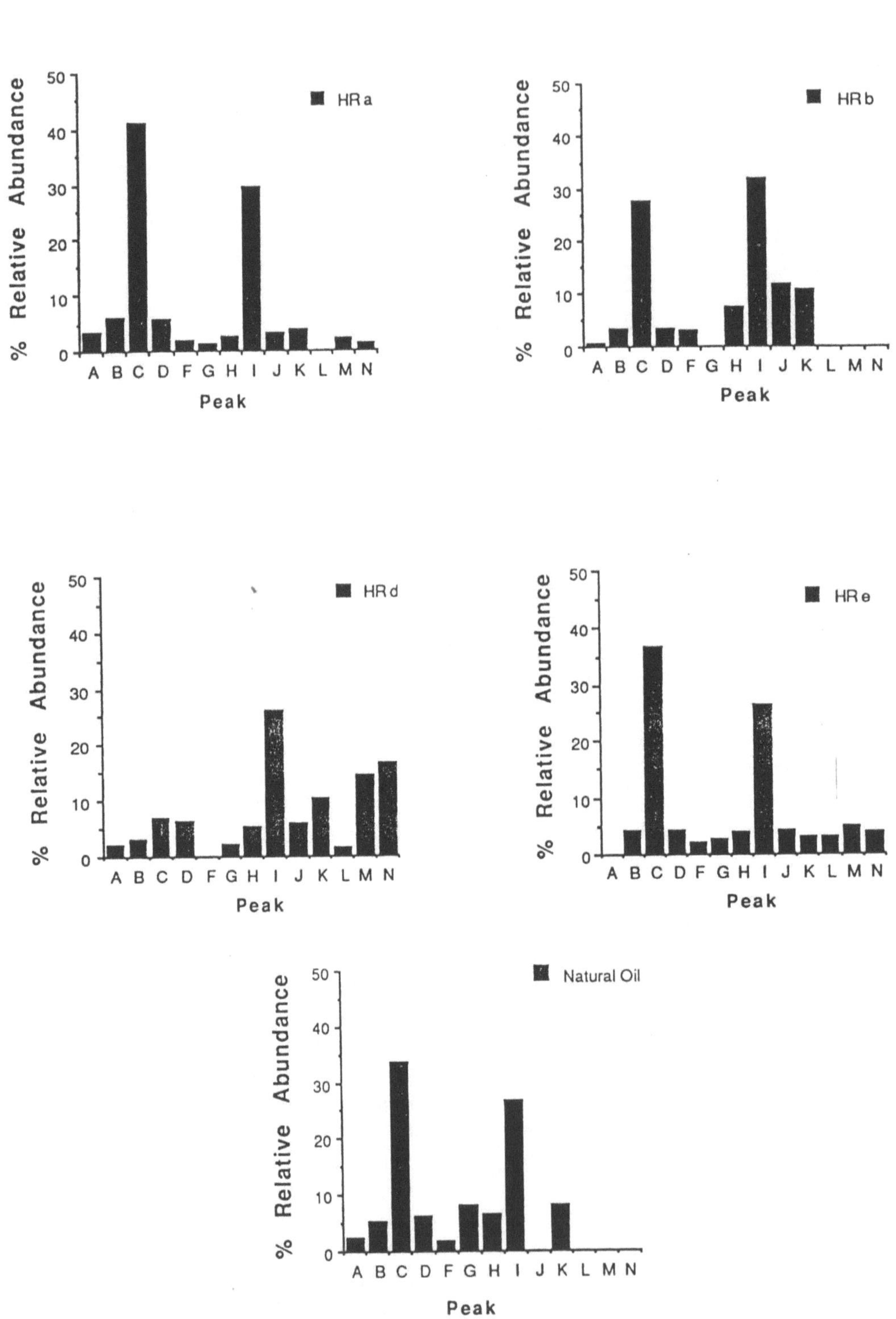

Fig. 3. Comparison of gas chromatographic profiles of essential oils distilled from natural, soil-grown roots (Comey et al. 1992a) and hairy roots cultured under light, in 250 ml flasks containing 150 ml medium for 6 weeks. Only compounds are shown with relative abundances > 1%.

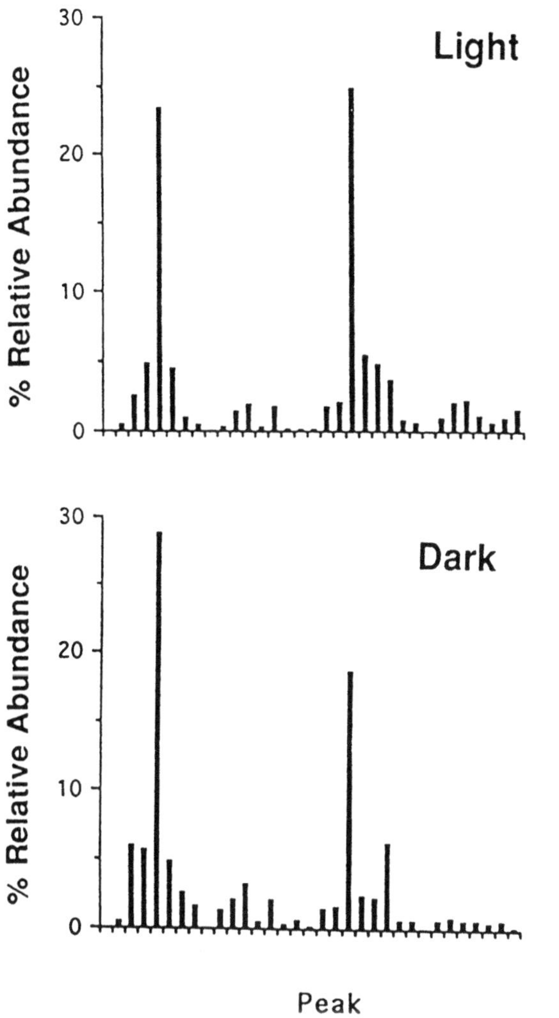

Fig. 4. Comparison of gas chromatographic profiles of essential oils distilled from hairy roots (line HRa) grown under light or dark conditions, using 1000 ml flasks containing 600 ml medium. All detectable oil components are shown.

a full phytochemical investigation. GC analyses did however indicate that the hairy root-derived oil was qualitatively similar to that isolated from natural roots, although quantitative differences were apparent (Fig. 3).

Enhancement of essential oil production

The presence of light during the culture period was found to significantly affect hairy root growth (HRa) and oil production. The growth index of roots cultured in the light was always ca. twice that of roots grown in the dark, while the percentage essential oil formed in dark-grown roots was always nearly twice that of light grown samples. GC analyses have shown that quantitatively the two oils show major differences, with oil from dark grown roots showing significantly higher concentrations of the lower-boiling components (Fig. 4).

Discussion

The production by suspension cultures of secondary metabolites typical of the intact plant has proved disappointing, especially with regard to essential oils (Mulder-Krieger et al. 1988). However 'hairy' roots developed in response to a genetic transformation by strains of *Agrobacterium rhizogenes* are increasingly being studied as sources of plant secondary metabolites (Hashimoto & Yamada 1991). Their development is currently restricted to dicotyledonous plants and for the production of metabolites normally present in natural roots of the intact plant (Saito et al. 1992). No reports to date have been published regarding any essential oil production in hairy roots. We have found that such roots developed on *L. alpinum* form an oil qualitatively similar to that produced by intact plants. Although the percentage yield is lower than from natural roots, because of the high growth index associated with some root lines, these cultures will form a useful alternative source of root material for phytochemical investigations. It has also been found possible to enhance oil production by growing roots in the absence of light. Investigations are ongoing with regard to other aspects of essential oil formation, as well as anthocyanin and chlorogenic acid production.

References

Anonymous (1980) British Pharmacopoeia, Vol II, Appendix XI. E. HMSO, London

Bicci C, Nano GM & Tira S (1975) *n*-Paraffin components of some Gnaphalieae. Planta Med. 28: 389–391

Bokadia MM, MacLeod AJ, Mehta SC, Mehta BK & Patel H (1986) The essential oil of *Inula racemosa*. Phytochemistry 25: 2887–2888

Comey N, Hook I & Sheridan H (1992a) Essential oil from normal and hairy roots of *Leontopodium alpinum*. Proceedings of 23rd International Symposium on Essential Oils, Ayr, Scotland

Comey N, Hook I & Sheridan H (1992b) Enhancement of anthocyanin production in cell cultures and hairy roots of *Leontopodium alpinum*. Planta Med. 58, Suppl. 1: A 605

Hashimoto T & Yamada Y (1991) Organ Culture and Manipulation. In: Hostettmann K (Ed) Methods in Plant Biochemistry, Vol 6. Assays for Bioactivity (pp 323–350). Academic Press, London

Hennessy D, Hook I, Sheridan H & McGee A (1989) Hydroxycinnamic acid esters from cell suspension cultures and plants of *Leontopodium alpinum*. Phytochemistry 28: 489–490

Hook I (1993) *Leontopodium alpinum* Cass. (Edelweiss): In vitro culture and production of secondary metabolites (Ch. XV). In: Bajaj YPS (Ed) Biotechnology in Agriculture and Forestry, Vol 21. Medicinal and Aromatic Plants IV (pp 217–232). Springer-Verlag, Heidelberg

Hu ZB & Alfermann AW (1993) Diterpenoid production in hairy root cultures of *Salvia miltirrhiza*. Phytochemistry 32: 699–703

Jaziri M, Legros M, Homes J & Vanhaelen M (1988) Tropine alkaloids production by hairy root cultures of *Datura stramonium* and *Hyoscyamus niger*. Phytochemistry 27: 419–420

Mulder-Krieger Th, Verpoorte R, Svendsen AB & Scheffer JJC (1988) Production of essential oils and flavours in plant cell and tissue cultures. A review. Plant Cell Tissue Organ Cult. 13: 85–154

Murashige T & Skoog F (1962) A revised medium for rapid growth and bioassays with tobacco tissue cultures. Physiol. Plant. 15: 473–497

Parr AJ & Hamill JD (1987) Relationship between *Agrobacterium rhizogenes* transformed hairy roots and intact, infected *Nicotiana* plants. Phytochemistry 26: 3241–3245

Payne J, Hamill JD, Robbins RJ & Rhodes MJC (1987) Production of hyoscyamine by 'hairy' root cultures of *Datura stramonium*. Planta Med. 53: 474–478

Saito K, Yamazaki M & Murakoshi I (1992) Transgenic medicinal plants: *Agrobacterium* mediated foreign gene transfer and production of secondary metabolites. J. Nat. Prod. 55: 149–162

Sauerwein M & Shimomura K (1991) Alkaloid production in hairy roots of *Hyoscyamus albus* transformed with *Agrobacterium rhizogenes*. Phytochemistry 30: 3277–3280

Tira S, Galeffi C & Di Modica G (1970) Flavonoids of Gnaphalieae: *Leontopodium alpinum*. Experientia 26: 1192

Trotin F, Moumou Y & Vasseur J (1993) Flavanol production by *Fagopyrum esculentum* hairy and normal root cultures. Phytochemistry 32: 929–931

Wagner H, Bladt S & Zgainski EM (1984) Plant Drug Analysis. Springer-Verlag, Berlin

Plant Cell, Tissue and Organ Culture **38**: 327–335, 1994.

Glycosylation in cardenolide biosynthesis

Christoph Theurer, Hans-Joachim Treumann, Thomas Faust, Ursula May & Wolfgang Kreis*
Pharmazeutisches Institut, Eberhard-Karls-Universität Tübingen, Auf der Morgenstelle 8, D-72076 Tübingen, Germany (requests for offprints)*

Key words: biotransformation, cardiac glycosides, β-glucosidase, glucosyltransferase, *Digitalis lanata*, radiolabelled cardiac glycosides.

Abstract

The glycosylation and deglycosylation of cardiac glycosides was investigated using cell suspension cultures and shoot cultures, both established from *Digitalis lanata* EHRH. plants, as well as isolated enzymes. Shoots were capable of glucosylating digitoxigenin, evatromonoside, digiproside, glucodigitoxigenin and digitoxin. Suspension cultured *Digitalis* cells glucosylated all the substrates mentioned but digiproside, whereas the UDP-glucose-dependent cardenolide glucosyltransferase isolated from that source did not accept digitoxigenin and digiproside as substrates. It is concluded that at least three different glucosyltransferases are involved in cardiac glycoside formation in *Digitalis*. Similar experiments carried out with glucosylated cardenolides which were administered to cultured cells, shoots and a cardenolide β-glucosidase isolated from young leaves revealed that at least two different glucosidases occur in *Digitalis lanata*, albeit in different tissues or during different phases of development. The biotransformation of glucoevatromonoside was investigated using unlabelled compound and [^{14}C-glucose]-glucoevatromonoside synthesized enzymatically. After 7 d of incubation almost no radioactivity could be recovered from the cardenolide fraction, indicating that the terminal glucose of glucoevatromonoside was now incorporated into volatile, hydrophilic and insoluble compounds. Since, on the other hand, large amounts of cardenolides were found in the experiments with unlabelled glucoevatromonoside it is assumed that steady state or pool size regulation is achieved by the coordinated action of a cardenolide glucosidase and a glucosyltransferase.

Abbreviations: Acdox – D-acetyldigitoxose, dgen – digoxigenin, dox – D-digitoxose, dten – digitoxigenin, dtl – D-digitalose, fuc – D-fucose, gten – gitoxigenin, qun – D-quinovose, CGH – cardenolide 16′-O-glucohydrolase, DFT – UDP-fucose:digitoxigenin 3-O-fucosyltransferase, DGT – UDP-glucose:Digitoxin 16′-O-glucosyltransferase, DQT – UDP-quinovose:digitoxigenin 3-O- quinovosyltransferase

Introduction

Cardiac glycosides are plant steroids composed of either a C_{23} (cardenolides) or a C_{24} (bufadienolides) genin and an oligosaccharide chain of variable length. These compounds are widely used in the treatment of certain forms of cardiac insufficiency. The leaves of *Digitalis lanata* (Scrophulariaceae), which are the most important source of cardiac glycosides, contain about 80 different cardenolide-type glycosides built from 5 different genins and 10 different sugars. The sugars are connected to position 3β of the genin forming an unbranched side chain of up to 5 links, including unusual deoxysugars, like D-digitoxose, D-digitalose and D-fucose (see Wichtl et al. 1987, for a review). Most of the *Digitalis* cardenolides have a terminal glucose; these compounds are called primary glycosides. Cardenolides without a terminal glucose have been termed secondary glycosides (see Fig. 1 for cardenolide structures).

The biosynthesis of cardenolides is thought to pass through cholesterol, pregnenolone, progesterone and pregnane intermediates. Condensation with a C_2 unit finally leads to the formation of digitoxigenin, which in turn serves as the precursor for the various types of cardiac glycosides (Luckner 1990). Almost nothing is known about the biosynthesis of the oligosaccharide side chain, especially with regard to the formation of

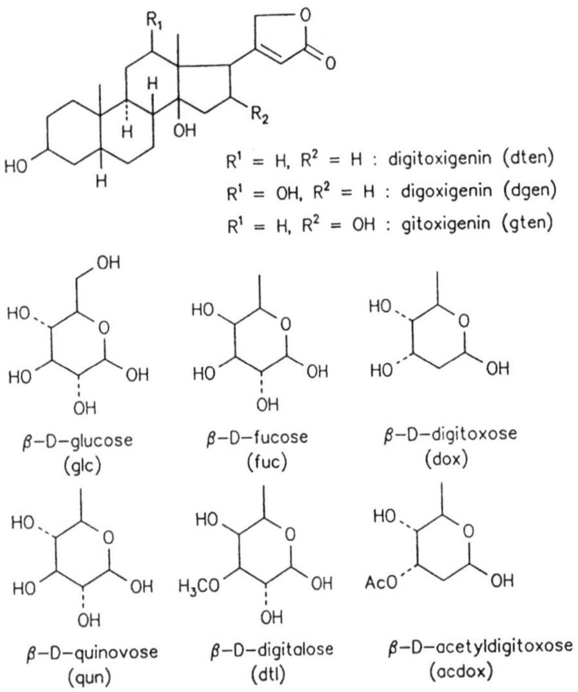

Fig. 1. Chemical structures of the cardenolide genins and sugars occurring in the cardiac glycosides mentioned in the text. Abbreviations for the respective compounds are given in parentheses.

the various deoxy sugars. According to one hypothesis these rare sugars are attached to the respective genins or "growing" glycosides by the action of specific glycosyltransferases. It may also be, however, that more common sugar nucleotides, such as UDP-D-glucose or UDP-D-galactose, are involved and that these rare sugars are only formed at the glycoside stage. A combination of both ways may also come into play (Fig. 2).

Plant cell suspension cultures are regarded as appropriate tools to study the biosynthesis of secondary plant products, provided that the pathway of interest is realized at least in part in the suspended cells. Cell cultures of various cardenolide-producing species have been established and employed to study cardenolide biosynthesis (see Reinhard & Alfermann 1980, for a review). In any case, undifferentiated suspension cultures were not able to produce cardenolides. However, cardenolide genins were modified when fed exogenously. Oxidation and epimerization of the 3β-hydroxyl and 5β-hydroxylation and glucosylation of the 3-hydroxyl seem to be quite common reactions. Cultured *Digitalis* cells have also been used to modify cardenolide mono-, di-, tri-, and tetrasaccharides. Side-chain glucosylation/deglucosylation and acetylation/deacetylation as well as steroid 12β- and 16β-hydroxylation have been reported. Enzymatic studies in the field of cardenolide glucosylation/deglucosylation have led to the isolation of a sterol glucosyltransferase (Yoshikawa & Furuya 1979) which also glucosylated digitoxigenin to some extent, a UDP-glucose-dependent digitoxin 16'-O-glucosyltransferase (Kreis et al. 1986) and a cardenolide-specific β-glucosidase (Kreis & May 1990). The cardenolide-specific enzymes have not yet been characterized with respect to their substrate preferences since purified enzymes have not been available.

In the current study we used cell suspension cultures, shoot cultures and purified enzymes of crude protein extracts of various sources to examine the glycosylation and deglycosylation of cardenolides. The main focus was on the glycosylation of digitoxigenin, which is supposed to be a central intermediate in cardenolide biosynthesis.

Material and methods

Shoot cultures

Shoot cultures were initiated from axillary buds of individual *Digitalis lanata* EHRH. (Scrophulariaceae) plants as described in previous papers (Schöner & Reinhard 1986, Stuhlemmer et al. 1993) and established as continuous cultures in liquid medium. In the biotransformation studies presented here the shoot culture strain D (Stuhlemmer et al. 1993) was used.

Suspension cultures

Cells from line K 1 OHD were subcultured every 10.5 days in a modified MS medium (Murashige & Skoog 1962) as described earlier (Kreis & Reinhard 1985).

Enzyme assays

Cardenolide 16'-O-glucohydrolase (CGH) and UDP-glucose:digitoxin 16'-O-glucosyltransferase (DGT) were assayed as described previously (Kreis & May 1990, Kreis et al. 1993). When partially purified enzyme preparations were used, the incubation time and protein concentration were adapted in such way that at least 70% of the cardenolide substrate was left unconverted. Digitoxigenin 3β-O-fucosyltransferase (DFT) and digitoxigenin 3β-O-quinovosyltransferase

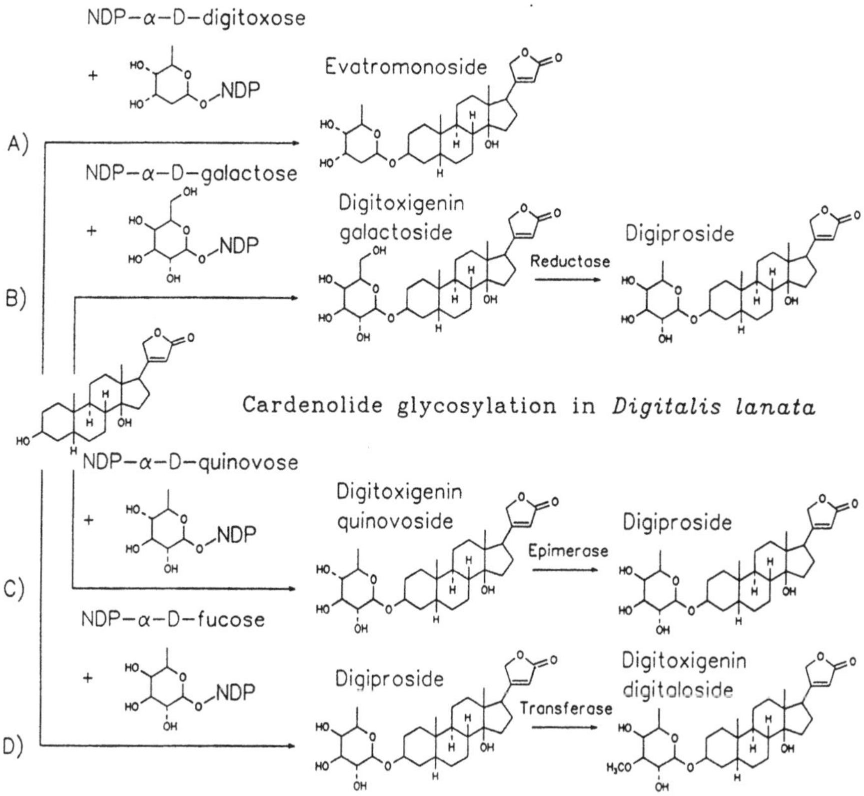

Fig. 2. Possible first biosynthetic steps leading to the formation of cardiac monosaccharides comprising rare (di)deoxy sugars. (A) Specific glycosyltransferases may exist using unusual sugar nucleotides to form the respective cardenolide glycosides. (B) More abundant sugar nucleotides, such as UDP-α-D-galactose, may be used to form intermediate glycosides which are then modified by oxidoreductases or other enzymes. Combinations of the two pathways are possible, involving modifications such as epimerisation (C) or group transfer (D). Digitoxigenin β-D-galactoside is a hypothetical intermediate not yet isolated. All of the other cardenolides depicted have been reported to occur in *Digitalis lanata*. In addition, their corresponding primary glycosides, i.e., disaccharides with a terminal glucose moiety, are known to occur.

(DQT) were assayed as described by Faust et al. (1992b).

Enzyme purification

CGH and DGT were purified following the protocols developed recently (May & Kreis, manuscript in preparation; Kreis & Westrich, manuscript in preparation). In order to analyze the substrate specificity of the CGH the enzyme was isolated from young *Digitalis lanata* leaves and purified by subsequent chromatography on Phenylsepharose CL-4B (hydrophobic interaction), CM sepharose CL-6B (cation exchange) and DEAE-Sephacel (anion exchange). The DGT was isolated from the cell culture line K 1 OHD (Kreis et al. 1986) and partially purified by ammonium sulphate fractionation, gel filtration on Sephacryl S-300 and ion exchange chromatography on Mono Q before its substrate specificity was tested.

With a view toward separating the DFT from the UDP-fucose 4′-epimerase the crude protein extract was fractionated with solid $(NH_4)_2SO_4$. Precipitated protein (30–45% $(NH_4)_2SO_4$ fraction) was collected by centrifugation and dissolved in buffer. The protein concentrations were determined according to Bradford (1976) with BSA as the standard protein.

Biotransformation experiments

The biotransformation experiments were carried out in 500 ml Erlenmeyer flasks containing 50 ml shoot culture medium (Stuhlemmer et al. 1993) inoculated with 15 g shoots (wet mass) or in 300 ml Erlenmeyer flasks containing 30 ml suspension culture medium (Kreis & Reinhard 1985) inoculated with 2 g cells (wet mass). The respective cardenolide substrates were added on day 1 or 3 of a culture cycle. At the end of an experiment tissue and spent medium were separated by filtering through paper. An aliquot of the medium was diluted with an equal volume of methanol, the precipitate spun down and the clear supernatant subjected to further analysis. Suspension-cultured cells were extracted and the extracts processed exactly as described earlier (Kreis & Reinhard 1988); for cardenolide extraction from cultured shoots the method of Hoelz et al. (1992) was adapted.

Cardenolide analysis and identification

Cardenolide-containing extracts were analyzed by HPLC according to Wichtl et al. (1982) with the modifications introduced by Schöner & Reinhard (1986) on a combination of two Shandon ODS 5 μm columns (pre-column 8 mm i.d. × 40 mm; main column 8 mm i.d. × 250 mm) with an LC 1084B equipped with a 79850 LC terminal (Hewlett Packard, Waldbronn, Germany). A step gradient of 84% acetonitrile (solvent B) and double-distilled water (solvent A) was applied: start (25% B), 5 min (25% B), 10 min (30% B), 15 min (30% B), 25 min (35% B), 30 min (40% B), 35 min (50% B), 50 min (65% B), 52 min (95% B), 58 min (95% B), 60 min (25% B). In addition, the extracts were analyzed by TLC as described by Stuhlemmer et al. (1993).

Samples containing mixtures of glucodigifucoside and glucodigiquinovoside were analyzed by micellary electrokinetic chromatography. A modular capillary electrophoresis instrument (Grom, Herrenberg, Germany) and a Chromatopac C-R6A (Shimadzu, Kyoto, Japan) for data processing were used. Details of this method have been published elsewhere (Gaus et al. 1993).

The identity of the cardenolides was checked by direct chromatographic comparison with authentic samples using HPLC and TLC. UV spectra were recorded on-line during HPLC analysis using an LC 1090 (Hewlett-Packard, Waldbronn, Germany) equipped with a diode array detector.

Chemicals

Digiproside was synthesized from digitoxigenin and 2,3,4-tri-O-acetyl-α-D-fucopyranosyl bromide as described previously (Faust et al. 1992b). Lanatoside A, digitalinum verum, glucoverodoxin, odorobioside G and β-methyldigitoxin, which was used as the internal cardenolide standard, were from Boehringer Mannheim (Mannheim, Germany). Digitoxin, evatromonoside and lanatoside C were obtained from Carl Roth (Karlsruhe, Germany). Glucodigifucoside was isolated from *Digitalis lanata* leaves (Theurer, unpublished) and glucoevatromonoside from leaves of *Digitalis heywoodii* P. and M. SILVA (Kreis, unpublished). A sample of authentic glucodigifucoside was generously provided by Prof. Wichtl, Marburg.

Radiolabelled glucoevatromonoside

Labelled glucoevatromonoside was synthesized enzymatically using a partially purified digitoxin glucosyltransferase (see above) and [^{14}C-glucose]-UDPG and evatromonoside as the substrates. The product was purified by column chromatography on RP-8 silica gel (Merck, Darmstadt). The purity of the synthesized glucoevatromonoside was confirmed by means of a Radio TLC analyzer (Berthold, Wildbad, Germany). Details on the procedure have been described elsewhere (Theurer et al. 1992).

Results

Suspension-cultured cells of the genus *Digitalis* fail to produce cardenolides. One way to trigger cardenolide production is to induce morphological differentiation in morphogenic or embryogenic cell lines. However, these systems only produce small amounts of a cardenolide mixture whose composition does not represent the pattern found in the mature plant. Shoot cultures, on the other hand, accumulate cardenolides also seen in the leaf. About 10 individual cardenolides were identified, including glucodigifucoside (dten-fuc-glc), digitalinum verum (gten-dtl-glc) and the lanatosides A (dten-dox-dox-acdox-glc) and C (dgen-dox-dox-acdox-glc) (Stuhlemmer et al. 1993). These cardenolides include (di)deoxysugars such as D-digitoxose, D-digitalose and D-fucose (see Fig. 1, for cardenolide structures). Therefore, we decided to use axenic shoots to study cardenolide glycosylation/deglycosylation and compared the results obtained with those obtained in experiments with non-producing cell suspension cultures and partially purified enzymes.

Cardenolide glucosylation

Digitoxigenin (dten), glucodigitoxigenin (dten-glc), evatromonoside (dten-dox), digiproside (dten-fuc) and digitoxin (dten-dox-dox-dox) were glucosylated by cultured shoots to digitoxigenin glucoside, glucodigitoxigeninglucoside, glucoevatromonoside, glucodigifucoside and purpureaglycoside A, respectively (Fig.

Cardenolide glucosylation in *Digitalis lanata*

Cell suspension	DGT	Substrate	Shoots
Yes	No	Digitoxigenin	Yes
Yes	Yes	Evatromonoside	Yes
Yes	Yes	Digitoxin	Yes
No	No	Digiproside	Yes
Yes	n.d.	Glucodigitoxigenin	Yes

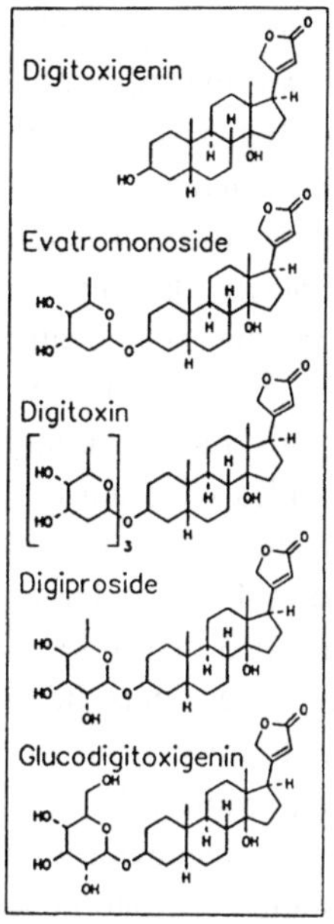

Fig. 3. Cardenolide glucosylation in *Digitalis lanata*. The respective cardenolide substrates were administered to cell cultures, shoot cultures and a cardenolide specific glucosyltransferase (DGT) isolated from the *Digitalis lanata* cell suspension culture K 1 OHD. N.d.= not determined.

Cardenolide deglucosylation in *Digitalis lanata*

Cell suspension	Substrate	Shoots	CGH
n.d.	Glucodigitoxigenin	n.d.	n.d.
Yes	Glucoevatromonoside	Yes	No
No	Purpureaglycoside A	No	Yes
No	Glucodigifucoside	No	No
n.d.	Glucodigiquinovoside	n.d.	Yes

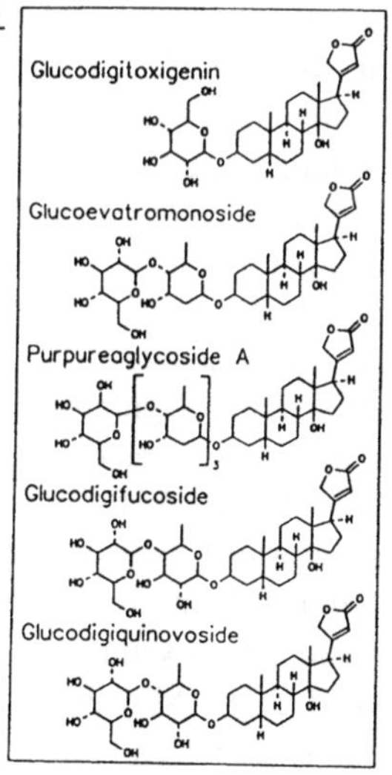

Fig. 4. Cardenolide deglucosylation in *Digitalis lanata*. The respective cardenolide substrates were administered to cell cultures, shoot cultures and a cardenolide specific β-glucosidase (CGH) isolated from young *Digitalis lanata* leaves. N.d.= not determined.

3). Cell suspension cultures were able to glucosylate all the cardenolide substrates mentioned but digiproside. The partially purified UDP-glucose digitoxin 16'-O-glucosyltransferase accepted digitoxin and evatromonoside as cardenolide substrates, but only UDP-glucose served as a glycosyl donor; other sugar nucleotides tested were UDP-D-galactose, UDP-D-fucose, GDP-D-glucose and ADP-D-glucose. We conclude that at least three different cardenolide glucosyltransferases exist in *Digitalis lanata*. Two of them are present in non-producing suspension cultured cells, whereas a third one is only present in cardenolide-producing tissue. It is supposed that the enzyme glucosylating cardenolide genins is a sterol glucosyltransferase similar to the one reported by Yoshikawa & Furuya (1979). They isolated the enzyme from *Digitalis purpurea* cell cultures and found that it has a weak glucosylation activity towards digitoxigenin. The DGT described by Kreis et al. (1986) seems to be specific for digitoxosides, the best substrates being monodigitoxosides such as evatromonoside.

The glucosylated tridigitoxosides, e.g. lanatoside A and purpureaglycoside A, are compounds known to be stored in the vacuoles of various *Digitalis* tissues (Kreis & Reinhard 1987, Hoelz et al. 1992, Kreis et al. 1993). In similar experiments with glucoevatromonoside it could be demonstrated that this compound also accumulates in the vacuoles (Kreis & Treumann, unpublished). Since monodigitoxosides are thought to be intermediates in the formation of bis- and tridigitoxosides, the question arises whether stored glucoevatromonoside can be remetabolized from its storage site and channelled back into cardenolide metabolism. Therefore, we became interested in enzymes hydrolyzing cardiac glycosides with a terminal glucose to their corresponding glucose-free derivatives.

Cardenolide deglucosylation

After harvest or during the technical isolation of cardenolides a variable part of the primary cardiac glycosides present in *Digitalis* leaves is degraded, yielding the so-called "secondary glycosides", i.e., glycosides without a terminal glucose. When *Digitalis* leaf powders containing large amounts of lanatosides, purpureaglycosides or glucoevatromonoside were wetted with water and then kept for several hours in a humid atmosphere the cardenolides mentioned were hydrolysed and increasing levels of the corresponding secondary glycosides were found. Glucodigifucoside, a major cardenolide of *Digitalis lanata* during all stages of leaf development, was not hydrolysed under these conditions (Fig. 4).

As early as 1935, Stoll and co-workers (Stoll et al. 1935) reported on enzyme activities in *Digitalis* leaves capable of hydrolysing primary glycosides to their corresponding secondary forms. At that time the enzymes could not be solubilized from the leaf material and were therefore called "desmoenzymes". Since then, a cardenolide specific glucohydrolase termed cardenolide 16′-O-glucohydrolase (CGH) has been solubilized from the leaves of various *Digitalis* species and the enzyme from *Digitalis lanata* purified about 500-fold (May & Kreis, manuscript in preparation). A partially purified enzyme was used to test its substrate preferences. Only cardenolide substrates will be considered here. The best substrate was lanatoside C, but purpureaglycoside A and glucodigiquinovoside were hydrolysed as well (Fig. 4). The CGH seems not to be identical with the unspecific β-glucosidase reported to be present in *Digitalis* cell cultures and leaves (Kreis et al. 1986, Hoelz et al. 1992). This assumption was made because CGH-free cell cultures contained high levels of a β-glucosidase hydrolysing *p*-dinitrophenyl β-D-glucoside.

Primary cardenolides are rarely deglucosylated when applied exogenously, even in tissues with high CGH activity. This observation was made not only with cell suspension cultures but also with shoot cultures. However, cardenolide deglucosylation could be provoked when the tissue under consideration was treated with potassium cyanide prior to the biotransformation experiment. Under these conditions lanatoside C was hydrolysed by cultured *Digitalis grandiflora* Mill. cells (Kreis et al. 1993). These results can be explained if we assume that glucosylated cardenolides are channelled to a hypothetical carrier at the tonoplast, avoiding any contact with the CGH, which is membrane-bound but not associated with the tonoplast. In the presence of cyanide the active uptake of primary glycosides is completely inhibited and they may hence come into contact with the CGH. Neither shoot cultures nor phytohormone-habituated cell suspension cultures of *Digitalis lanata* were capable of hydrolysing lanatoside C in the presence of cyanide, whereas glucoevatromonoside was rapidly deglucosylated in the respective experiments. Phytohormone-habituated cell suspension cultures of *Digitalis lanata* completely lack CGH and hence were not able to cleave lanatoside C. Low CGH activity was detected in shoot cultures (Kreis & May 1990) but was obviously too weak to produce a visible effect in biotransformation experiments under uptake-inhibitory conditions. Interestingly, glucoevatromonoside was hardly hydrolysed by the isolated CGH.

When glucoevatromonoside was fed to cultured shoots it was taken up very rapidly and eventually biotransformed to glucodigoxigeninmonodigitoxoside, which was the major cardenolide at the end of the experiment (Fig. 5). Similar biotransformation experiments were carried out with suspension cultures where glucoevatromonoside was taken up by cells and isolated protoplasts (data not shown). Recently, it was demonstrated that glucoevatromonoside is hydrolysed in *Digitalis* shoot cultures to evatromonoside in the presence of cyanide (Christmann et al. 1993). Since tetrasaccharides were not deglucosylated under these conditions and because only weak CGH activity could be detected in shoots and none at all in suspended cells it was concluded that more than one cardenolide specific β-glucosidase must be present.

In order to investigate whether the glucose moiety of glucoevatromonoside is turned over we synthesized [^{14}C-glucose]-glucoevatromonoside. Optimal incubation times and substrate concentrations for the enzymatic synthesis of [^{14}C-glucose]-glucoevatromonoside from evatromonoside and [^{14}C-glucose]-UDP-glucose were determined in preliminary experiments not documented here (Theurer et al. 1992).

The radioactive substrate was taken up by the cells and radioactive glucodigoxigenin monodigitoxoside together with untransformed substrate could be detected after 2 d of incubation. At the end of the experiment almost no radioactivity was found in the cardenolide fraction, indicating that the labelled glucose was cleaved off and replaced by unlabelled glucose due to the coordinated action of a cardenolide glucosidase and a glucosyltransferase.

To summarize, it is concluded that at least two different cardenolide specific β-glucosidases are present in *Digitalis*. The β-glucosidases detected so far seem to hydrolyse glycosidic bonds formed between β-D-glucose and the equatorial 4-hydroxyl of the next sugar. They probably differ in their chain-length specificity. Glucoevatromonoside, for example, is rarely hydrolysed by the purified CGH (May & Kreis, manuscript in preparation) and is supposed to be a good substrate for a second cardenolide-specific glucosidase which has already been detected *in vitro* but has not yet been purified or characterized.

Discussion

The glycosylation of digitoxigenin and the cardiac glycosides has been investigated extensively using cell

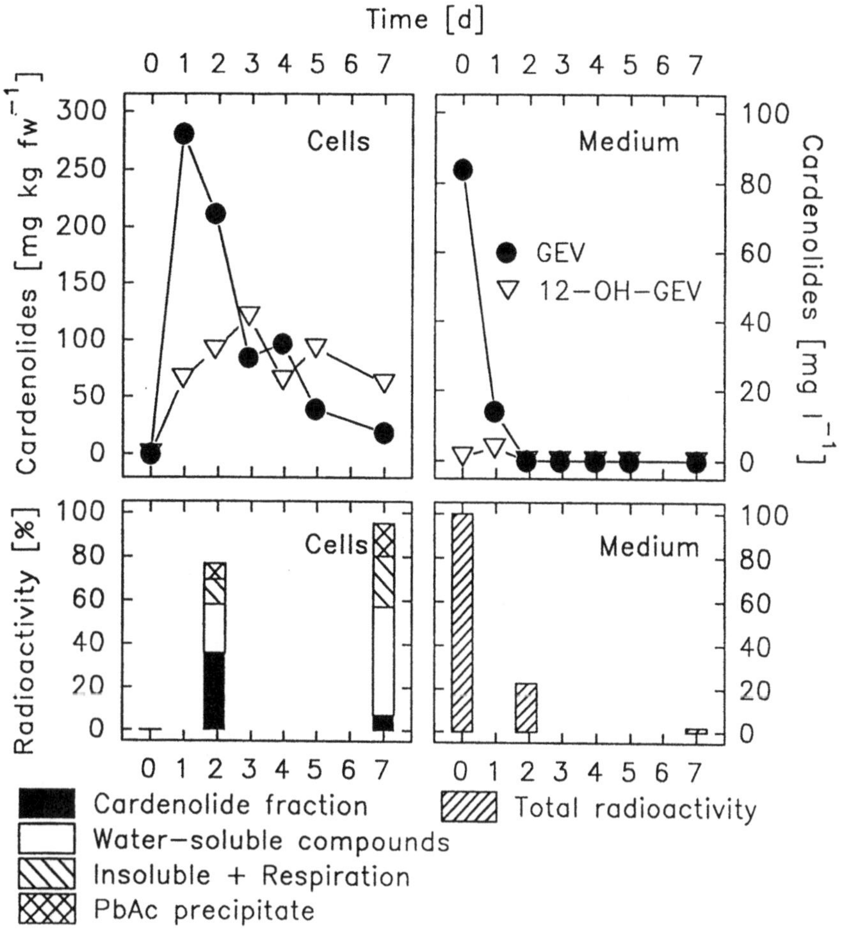

Fig. 5. Biotransformation of glucoevatromonoside by shoot cultures of *Digitalis lanata*. The time course of accumulation and biotransformation of unlabelled glucoevatromonoside is shown in the upper part of the figure. The lower part summarizes the results from experiments with [^{14}C-glucose]-glucoevatromonoside. Not all of the radioactivity administered could be found in the ethanol extract of the cells or in the bathing medium. This loss was attributed to the formation of insoluble products and loss due to respiration ("Insoluble + Respiration"). Radioactivity was also found in the lead acetate precipitate ("PbAc precipitate") and in the hydrophilic waste phase ("Water-soluble compounds") produced during extract purification.

suspension cultures or root cultures of various plant species (see the reviews by Reinhard & Alfermann 1980, Suga & Hirata 1990, Kreis 1993). Suspension cultures of *Nerium oleander* were shown to form digitoxigenin β-D-glucoside and 3-epidigitoxigenin β-D-glucoside (Paper & Franz 1990). Similar results have been obtained with cell suspension cultures of various *Digitalis* (see Reinhard & Alfermann 1980, for a review) and *Strophanthus* species (Kawaguchi et al. 1988, 1989, 1991).

The combination of the biosynthetic potentials of unrelated plant species and the formation of novel cardiac glycosides by biotransformation was achieved by Furuya's group (Kawaguchi et al. 1990), who administered digitoxigenin to hairy root cultures of *Panax ginseng*. Two new glycosides, namely 3-epidigitoxigenin β-D-gentiobioside and digitoxigenin β-D-sophoroside, were isolated. In this way the *Digitalis*-borne aglycone digitoxigenin was combined with sugar side chains known from the ginseng saponins.

When digitoxigenin was fed to shoot cultures of *Digitalis lanata* several cardenolide mono- and disaccharides were formed, including glucodigitoxigenin, glucodigitoxigeninglucoside (dten-glc-glc), glucodigifucoside (dten-fuc-glc), odorobioside G (dten-dtl-glc) and several of the corresponding 12β-hydroxy derivatives (Faust et al. 1992a). This was the first report on the fucosylation and digitalosylation of digitoxigenin in biotransformation experiments. However, no digitoxosylation of the cardenolide genin was observed. Since the cultivated shoots were demonstrated to form digitoxosides endogenously (Stuhlemmer et al. 1993) and since several pregnane precursors were shown to boost the amount of cardenolides containing digitoxose (Kreis et al. 1991) it was concluded that fucosylation (including digitalosylation), glucosylation and quinovosylation but not digitoxosylation can occur at the cardenolide genin stage.

Any investigations of the enzymatic steps in cardenolide glycoside formation are limited by the unavailability of putative co-substrates of the respective glycosyltransferases. In an attempt to elucidate the fucosylation of digitoxigenin on the enzyme level we recently synthesized UDP-α-D-fucose by condensing α-D-fucopyranosyl phosphate and commercial uridine 5′-monophosphomorpholidate (Faust et al. 1992b). A partially purified enzyme from young leaves of *Digitalis lanata* EHRH. has been shown to catalyze the transfer of D-fucose from synthetic UDP-α-D-fucose to cardenolide genins such as digitoxigenin. In these experiments digitoxigeninquinovoside was formed as a side product and the question was whether the epimerisation of fucose occurs at the cardenolide level or at the sugar nucleotide level. Using partially purified enzyme preparations obtained from young *Digitalis lanata* leaves, it could be demonstrated that UDP-D-fucose was epimerized to UDP-D-quinovose and that this sugar nucleotide was used for cardenolide formation (Faust & Kreis, unpublished).

The results presented here and the observations made in previous studies led us to the conclusion that the formation and conversion of cardiac glycosides is part of a very complex, highly structured metabolic grid. A detailed knowledge of the properties and substrate preferences of the different enzymes involved in glycoside formation and hydrolysis will help us to understand the cardenolide pathway, its regulation and the biosynthetic relationship among the individual cardenolides. Our recent results indicate that cardenolide glycosylation, especially digitoxosylation, may take place at an early pregnane stage of cardenolide biosynthesis. Due to this modification the pregnane molecule may be tagged so as to be recognized and channelled into the cardenolide pathway. The occurrence of so-called digitanols, i.e., pregnane glycosides with structural similarities to cardiac glycosides, supports this idea. The formation of cardenolide fucosides (and digitalosides) may be part of a shunt pathway not used in the formation of cardenolide tetrasaccharides, which are sometimes regarded as the final products of the cardenolide pathway. According to this hypothesis, glucodigifucoside, a major cardenolide in *Digitalis lanata* leaves, may be regarded as the end product of this shunt pathway. Work on the isolation and purification of further cardenolide glycosylating and deglycosylating enzymes is in progress.

Acknowledgement

Our research was financially supported by a grant from the Deutsche Forschungsgemeinschaft.

References

Bradford MM (1976) A rapid and sensitive method for the quantitation of microgram quantities of protein utilizing the principle of protein-dye binding. Anal. Biochem. 72: 248–254

Christmann J, Kreis W, Reinhard E (1993) Uptake, transport and storage of cardenolides in foxglove. Cardenolide sinks and occurrence of cardenolides in the sieve tubes of *Digitalis lanata*. Botanica Acta, 106: 419–427

Faust T, Haussmann W, Kreis W, Reinhard E, Stuhlemmer U & Theurer C (1992a) Fucosylation and digitoxosylation in cardenolide biosynthesis. 2nd Dutch-German symposium on plant cell cultures, Münster, book of abstracts, p. 39

Faust T, Theurer C, Eger K & Kreis W (1992b) Studies on the biosynthesis of cardiac glycosides. Fucosyltransferase activity in leaves of *Digitalis lanata*. Planta Med. (Suppl.) 58: 672

Gaus H-J, Treumann H-J, Kreis W & Bayer E (1993) Separation of cardiac glycosides by micellary electrokinetic capillary electrophoresis. J. Chromatogr., 635: 319–322

Hoelz H, Kreis W, Haug B & Reinhard E (1992) Storage of cardiac glycosides in vacuoles of *Digitalis lanata* mesophyll cells. Phytochemistry 31: 1167–1171

Kawaguchi K, Hirotani M & Furuya T (1988) Biotransformation of digitoxigenin by cell suspension cultures of *Strophanthus amboensis*. Phytochemistry 27: 3475–3479

Kawaguchi K, Hirotani M & Furuya T (1989) Biotransformation of digitoxigenin by cell suspension cultures of *Strophanthus intermedius*. Phytochemistry 28: 1093–1097

Kawaguchi K, Hirotani M, Yoshikawa T & Furuya T (1990) Biotransformation of digitoxigenin by ginseng hairy root cultures. Phytochemistry 29: 837–843

Kawaguchi K, Hirotani M & Furuya T (1991) Biotransformation of digitoxigenin by cell suspension cultures of *Strophanthus divaricatus*. Phytochemistry 30: 1503–1506

Kreis W (1994) Biotransformations. In: Mavituna F (ed) Plant Secondary metabolites via bioreactor culture. Kluwer Academic Publishers, Dordrecht, in press

Kreis W & May U (1990) Cardenolide glucosyltransferases and glucohydrolases in leaves and cell cultures of three *Digitalis* (Scrophulariaceae) species. J. Plant Physiol. 136: 247–252

Kreis W & Reinhard E (1985) Rapid isolation of vacuoles from suspension-cultured *Digitalis lanata* cells. J. Plant Physiol. 121: 385–390

Kreis W & Reinhard E (1987) Selective uptake and vacuolar storage of primary cardiac glycosides by suspension-cultured *Digitalis lanata* cells. J. Plant Physiol. 128: 311–326

Kreis W & Reinhard E (1988) 12β-hydroxylation of digitoxin by suspension-cultured *Digitalis lanata* cells. Production of deacetyllanatoside C using a two-stage culture method. Planta Med. 54: 95–100

Kreis W, May U & Reinhard E (1986) UDP-glucose:digitoxin 16'-O-glucosyltransferase from suspension-cultured *Digitalis lanata* cells. Plant Cell Rep. 5: 442–445

Kreis W, Eisenbeiss M, Haussmann W, Stuhlemmer U & Reinhard E (1991) Cardenolide formation in *Digitalis* shoot cultures. Workshop "Biochemistry of Plant Terpenoid Biosynthesis", Leiden, book of abstracts

Kreis W, Hoelz H, Sutor R & Reinhard E (1993) Cellular organization of cardenolide biotransformation in *Digitalis grandiflora*. Planta, 191: 246–251

Luckner M (1990) Secondary metabolism in microorganisms, plants and animals, 3rd edition, Gustav Fischer, Jena

Murashige T & Skoog, F (1962) A revised medium for rapid growth and bioassays with tobacco tissue cultures. Physiol. Plant. 15: 473–497

Paper D & Franz G (1990) Biotransformation of 5βH-pregnan-3βol-20-one and cardenolides by cell suspension cultures of *Nerium oleander* L. Plant Cell Rep. 8: 651–655

Reinhard E & Alfermann AW (1980) Biotransformations by plant cell cultures. Adv. Biochem. Engin. 16: 49–83

Schöner S & Reinhard E (1986) Long-term cultivation of *Digitalis lanata* clones propagated *in vitro*: cardenolide content of the regenerated plants. Planta Med. 52: 478–481

Stoll A, Hoffmann A & Kreis W (1935) Über glucosidspaltende Enzyme der *Digitalis*-Blätter. Hoppe-Seyler's Z. Physiol. Chem. 235: 249–264

Stuhlemmer U, Kreis W, Eisenbeiss M & Reinhard E (1993) Cardiac glycosides in partly submerged shoots of *Digitalis lanata*. Planta Med., 59: 539–545

Suga T & Hirata T (1990) Biotransformation of exogenous substrates by plant cell cultures. Phytochemistry 29: 2393–2406

Theurer C, Treumann A, Eisenbeiss M & Kreis W (1992) Studies on the biosynthesis of cardiac glycosides. Enzymatic synthesis of [^{14}C-glucose]-glucoevatromonoside. Planta Med. (Suppl.) 58: 658

Wichtl M, Bühl G & Huesmann K (1987) Fingerhut. *Digitalis* L. – bekannte und weniger bekannte Vertreter einer wichtigen Arzneipflanzengattung. Deutsche Apotheker Ztg. 127: 2391–2400

Wichtl M, Mangkudidjojo M & Wichtl-Bleier W (1982) Hochleistungs-Flüssigkeitschromatographische Analyse von *Digitalis*-Blattextrakten. I. Qualitative Analyse. J. Chromatogr. 234: 503–508

Yoshikawa T & Furuya T (1979) Purification and properties of sterol:UDPG glucosyltransferase in cell culture of *Digitalis purpurea*. Phytochemistry 18: 239–241

Plant Cell, Tissue and Organ Culture **38**: 337–344, 1994.

Enzymes in cardenolide-accumulating shoot cultures of *Digitalis purpurea* L.

Hanns Ulrich Seitz & Dorothea Elisabeth Gärtner
Botanical Institute, University of Tübingen, Auf der Morgenstelle 1, D-72076 Tübingen, Germany

Key words: Cardenolides, *Digitalis purpurea*, 3β-hydroxysteroid-5β-oxidoreductase, progesterone 5β-reductase, shoot cultures

Abstract

In contrast to undifferentiated cell suspension cultures of *Digitalis lanata*, photomixotrophic shoot cultures of *Digitalis purpurea* accumulate cardiac glycosides in substantial concentrations. They are used to investigate enzymes of the cardenolide pathway. All cardenolides are 5β-configurated. The progesterone 5β-reductase and the 3β-hydroxysteroid-5β-oxidoreductase are present in shoot cultures but not in undifferentiated cell cultures. These enzymes provide precursors for cardenolides, whereas the presence of the progesterone 5α-reductase, also present in shoot cultures, is discussed with regard to its role in phytosterol biosynthesis and may be attributed to the general steroid pathway. The progesterone 5α-reductase had an activity maximum during the early growth period seven days after onset of cultivation, whereas the corresponding progesterone 5β-reductase activity was highest on day 11. The maximum cardenolide accumulation was after 24 days. The enzyme activities present in crude extracts from shoot cultures were characterized with regard to their requirements for NADPH and NADH, pH-optimum, temperature optimum, affinity to their substrates and their localization in the cell. The progesterone 5β-reductase was purified 769-fold.

Abbreviations: DW – dry weight, FW – fresh weight, PVP – polyvinylpyrrolidone

Introduction

Cardiac glycosides of *Digitalis* species are widely used drugs against heart diseases and many efforts have been made to produce these secondary metabolites using the cell culture technology. However, undifferentiated cell cultures of *Digitalis* species show only a very limited capacity for cardiac glycoside accumulation (Lui & Staba 1979; Rücker et al. 1976; Corchete et al. 1990). Most frequently, no accumulation of cardenolides by cell cultures was observed (Graves & Smith 1967; Hirotani & Furuya 1977). But cardiac glycoside accumulation did coincide with shoot and green plantlet formation (Diettrich et al. 1990), although its biosynthesis does not seem to be directly linked to the chlorophyll content of the tissue (Hagimori et al. 1982). Due to our lack of knowledge about the enzyme systems involved in cardenolide biosynthesis the pathway is still hypothetical. In its present form it is mainly based on feeding experiments with putative precursors and intermediates. At present four enzymic reactions have been described (Fig. 1). The 'cholesterol side-chain cleaving enzyme' has been detected in protein extracts from leaves of *Digitalis purpurea.* It catalyses the transformation of cholesterol to pregnenolone (Pilgrim 1972). The Δ5-3β-hydroxysteroid dehydrogenase/ Δ5-Δ4-ketosteroid isomerase (3β-HSD) forms progesterone from pregnenolone and has been detected in extracts from suspension-cultured cells and intact leaves of *Digitalis lanata* (Seidel et al. 1990). These enzymes may also be part of the primary metabolism leading to other sterols which may be integrated into cell membranes. This part of the biosynthesis may be designated as general steroid pathway. Since all cardenolides are 5β-configurated there is a requirement for a stereospecific enzyme forming a 5β-configurated steroid skeleton as the starting point of the cardenolide pathway. The first stereospecific enzyme forming 5β-pregnane-3,20-dione from progesterone was described recently in our laboratory (Gärtner et al. 1990). It was detected in young leaves of *Digitalis purpurea.* In addition to the 5β-derivatives the corresponding

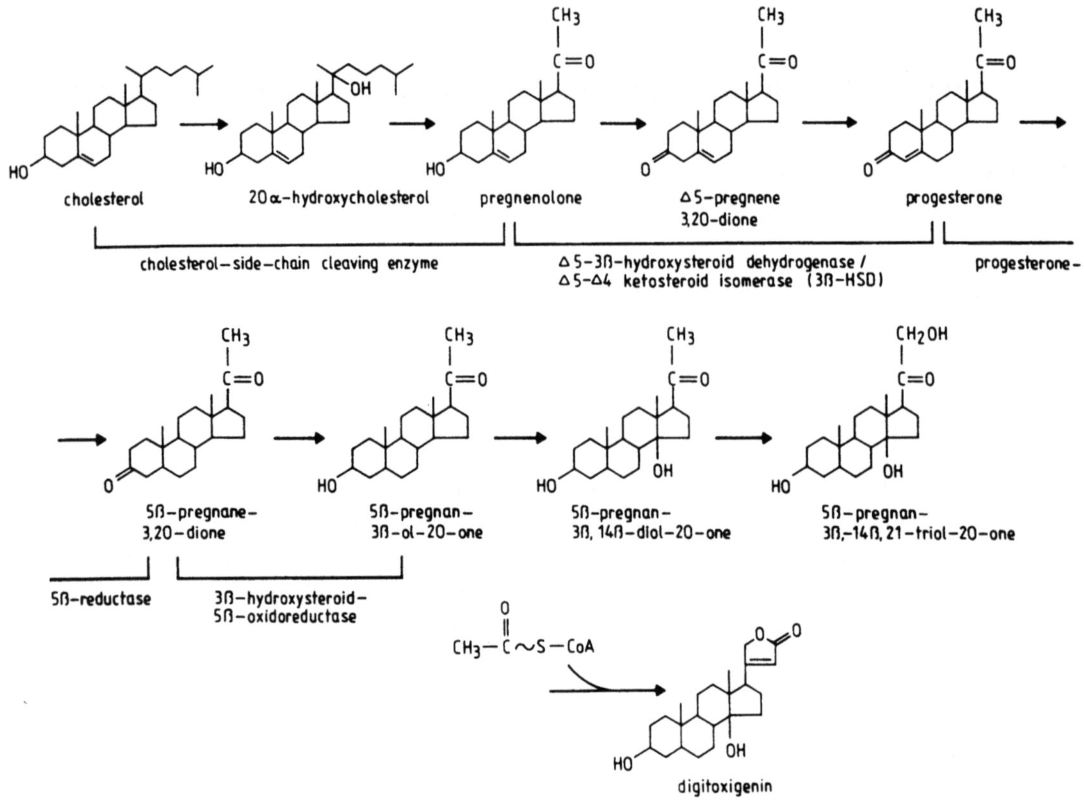

Fig. 1. Probable biosynthetic route and enzymic steps of formation of cardenolides in *Digitalis*-species.

5α-pregnane-3,20-dione was formed. This activity has already been detected in cell-suspension cultures of *Digitalis lanata* (Wendroth & Seitz 1990). The role of the latter activity with regard to cardenolide biosynthesis is questionable since all cardenolides are 5β-configurated and no cardenolides were formed when 5α-pregnane-3,20-dione was fed *in vivo* (Tschesche et al. 1970).

In order to elucidate the enzymic steps leading to cardenolides a cardenolide-forming system is of great importance. Therefore an axenic, mixotrophic shoot culture was established. In the present communication we describe the occurrence of cardenolides and the progesterone 5β-reductase in shoot cultures of *Digitalis purpurea*. In addition to the progesterone 5β-reductase a 3β-hydroxysteroid-5β-oxidoreductase is described. Both enzymes are characterized and their role in cardenolide biosynthesis is discussed (Fig. 2). With regard to regulatory steps in the secondary metabolism there is a need for purification of the enzymes.

Materials and methods

Shoot cultures of digitalis purpurea

Shoot cultures were established and cultivated as described elsewhere (Gärtner & Seitz 1993).

Protein extractions and enzyme assays

The enzyme proteins were extracted and assayed as described previously. Progesterone 5α-reductase activity was determined as described by Wendroth & Seitz (1990). Progesterone 5β-reductase was extracted and assayed after Gärtner et al. (1990). The methods for assaying 3β-hydroxysteroid-5α-oxidoreductase and 3β-hydroxysteroid-5β-oxidoreductase have been described by Warneck & Seitz (1990) and Gärtner & Seitz (1993) respectively.

Product identification and quantification

The compounds present in the extracts from enzyme assays were analysed by gas chromatography as previously described (Gärtner & Seitz 1993).

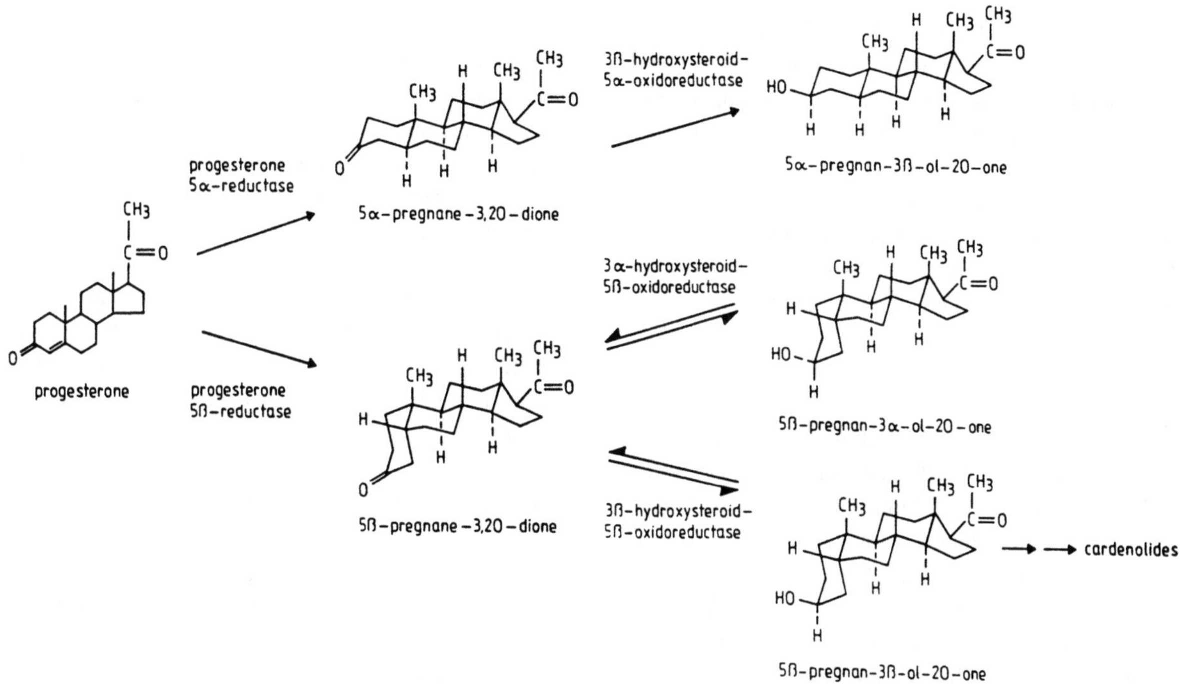

Fig. 2. Biosynthetic pathways of progesterone-derived pregnanes and the respective enzymic steps.

Isolation, identification and quantitative analysis of cardenolides

The aglycones of cardiac glycosides were prepared from dried leaves by hydrolysis with trifluoroacetic acid and analysed by HPLC. For details see Gärtner & Seitz (1993).

Purification procedures

DEAE-Sephacel. After adding 20% (v/v) glycerol to the supernatant of a 100,000 g centrifugation it was applied to a DEAE-Sephacel column (20 × 2.6 cm) pre-equilibrated with buffer A (25 mM TrisHCl, 0.25 M sucrose, 2 mM EDTA, 1 mM DDT, pH 7.5). The proteins were eluted with a linear NaCl gradient ranging from 50–500 mM in buffer A. Fractions of 3.7 ml were collected and assayed for conductivity, total protein and catalytic activities.

Blue-Sepharose Cl-6B. Active fractions from the DEAE column (45–50 ml) were combined and concentrated (Centricon 30, Amicon). The concentrated solution was brought to 20% glycerol and layered onto a Blue-Sepharose Cl-6B column (10 × 1 cm), pre-equilibrated with buffer B (50 mM TrisHCl, 0.25 M sucrose, 2 mM EDTA, 1 mM DTT, pH 7.5 and eluted with a linear gradient (0–1 M NaCl in buffer B) followed by 1.5 M NaCl (in buffer B). Fractions of 2.6 ml were collected and assayed for total protein and catalytic activities.

2′,5′-ADP-Sepharose 4B. Active fractions from the preceeding column were again concentrated by ultrafiltration and redissolved in buffer C (10 mM KP_i pH 7.4 containing 2 mM EDTA, 1 mM DTT and 20% glycerol) and applied to a column packed with 2′,5′ADP-Sepharose 4B (3.5 × 2.5 cm), pre-equilibrated with buffer C. The proteins were eluted with buffer C (3 h) and a linear gradient of 2′-AMP ranging from 0 to 6 mM (1 h). The fractions (2.6 ml) were assayed for total protein and catalytic activity.

Non-dissociating PAGE. Active fractions from the 2′,5′-ADP-Sepharose 4B separation were concentrated (Centricon 30) and diluted with buffer D (66 mM TrisHCl, 10% glycerol, 0.5% DTE, pH 8.6). The discontinuous non-dissociating PAGE was performed on a 8% separating gel (pH 9.2) and 5% stacking gel (pH 8.6) with 50 mM Tris at pH 8.4 containing 380 mM glycine. After electrophoresis one lane was silver stained according to Ansorg (1985). The other part of the gel was chopped into 5 mm slices. These slices were homogenized in buffer E (100 mM TrisHCl, 0.25

M sucrose, 2 mM EDTA, 4 mM DTT, pH 7.5 and vigorously shaken for 12 h at 4°C. This extract was then centrifuged for 15 min at 16,000 g. The supernatant was assayed for total protein and enzyme activity.

SDS-PAGE. The purity of active fractions from the different purification steps was assayed by SDS-PAGE according to Laemmli (1970) using 10% separating and 5% stacking gels and silver staining.

Analytical methods

Protein was determined according to Bradford (1976) using BSA as a standard. The glucose concentration in the medium was determined according to Dubois et al. (1956) with glucose as a standard. Chlorophyll was extracted with 100% acetone. The concentration was determined according to Lichtenthaler (1987). The nitrate concentration in the medium was monitored with the assay from Boehringer (Mannheim, FRG).

Results

Characterization of the shoot cultures

Axenic, mixotrophic shoot cultures of *Digitalis purpurea* are ideal sources for the isolation of enzymes involved in cardenolide metabolism. Compared to soil-grown seedlings this system is much better to handle. To obtain information on optimum growth conditions these shoot cultures were characterized with regard to several parameters. The results are summarized in Fig. 3. The fresh weight was monitored during 35 days. A lag phase of about 7 d was followed by a growth phase between day 10 and 21. After that no growth was observed. The fresh weight increased by a factor of 10 during 21 d.

The mixotrophic shoot cultures use glucose as a carbon source and nitrate as the major nitrogen source. During the first seven days after the onset of cultivation the consumption of glucose was very low as was the increase in fresh weight. After seven days the glucose content of the medium decreased continuously reaching a final concentration of 24% after 31 days. The nitrate content decreased rapidly, but at the end of the cultivation period this decrease slowed down parallel to the deceleration in growth. After 31 d only 3.3% of the initial concentration remained in the medium indicating that the nitrate could be rate limiting for growth.

The chlorophyll content of the plantlets increased continuously with time. The amount of chlorophyll a increased more rapidly than that of chlorophyll b, leading to changes in the ratio of chlorophyll a/b. During the phase of maximum growth the ratio was 3:1, possibly representing the period of maximum photosynthetic activity. At the end of the cultivation period the ratio was 1:6.

The amount of soluble protein measured in the supernatant of a 100,000 g centrifugation was greatest in the middle of the growth phase (14 d). The changes in the microsomal protein concentrations were less pronounced.

Characterization of the stereospecific enzymes

In extracts from shoot cultures of *Digitalis purpurea* five enzyme activities are present which are related to the sterol or cardenolide biosynthesis. As previously described cell cultures of *Digitalis lanata* contain progesterone 5α-reductase and this enzyme is also present in shoot cultures of *Digitalis purpurea*. This enzyme is ER-bound and catalyses the reduction of progesterone to 5α-pregnane-3,20-dione. NADPH is utilized as the co-substrate with an apparent K_m-value of 130 μM. The K_m-value for progesterone is 30 μM. The activity of this enzyme yields an A/B trans ring junction.

The corresponding progesterone 5β-reductase forms 5β-pregnane-3,20-dione which is cis-configurated with regard to ring A and B as in the case of cardenolides. The reaction is irreversible and in contrary to the progesterone 5α-reductase the enzyme is located in the soluble fraction of the cell. The enzyme requires NADPH as a co-substrate (K_m = 6 μM). The apparent K_m-value for progesterone is very similar (22 μM).

By the following reduction step at C-3 also stereoisomeres can be produced. The formation of 5α-pregnan-3β-ol-20-one from 5α-pregnane-3,20-dione is catalyzed by the 3β-hydroxysteroid-5α-oxidoreductase. In cell cultures of *Digitalis lanata* this enzyme is located up to 66 % in the soluble part of the cell. It requires also NADPH (K_m = 120 μM) but it accepts also NADH. The apparent K_m-value with 5α-pregnane-3,20-dione as the substrate is 20 μM. By reducing the analogous β-derivative two stereoisomeres can be formed which are on the one hand 5β-pregnan-3α-ol-20-one, on the other hand 5β-pregnan-3β-ol-20-one. The formation of a 3α-configurated product is catalyzed by the 3α-hydroxysteroid-5β-oxidoreductase located to 97% in the cytoplasm. The

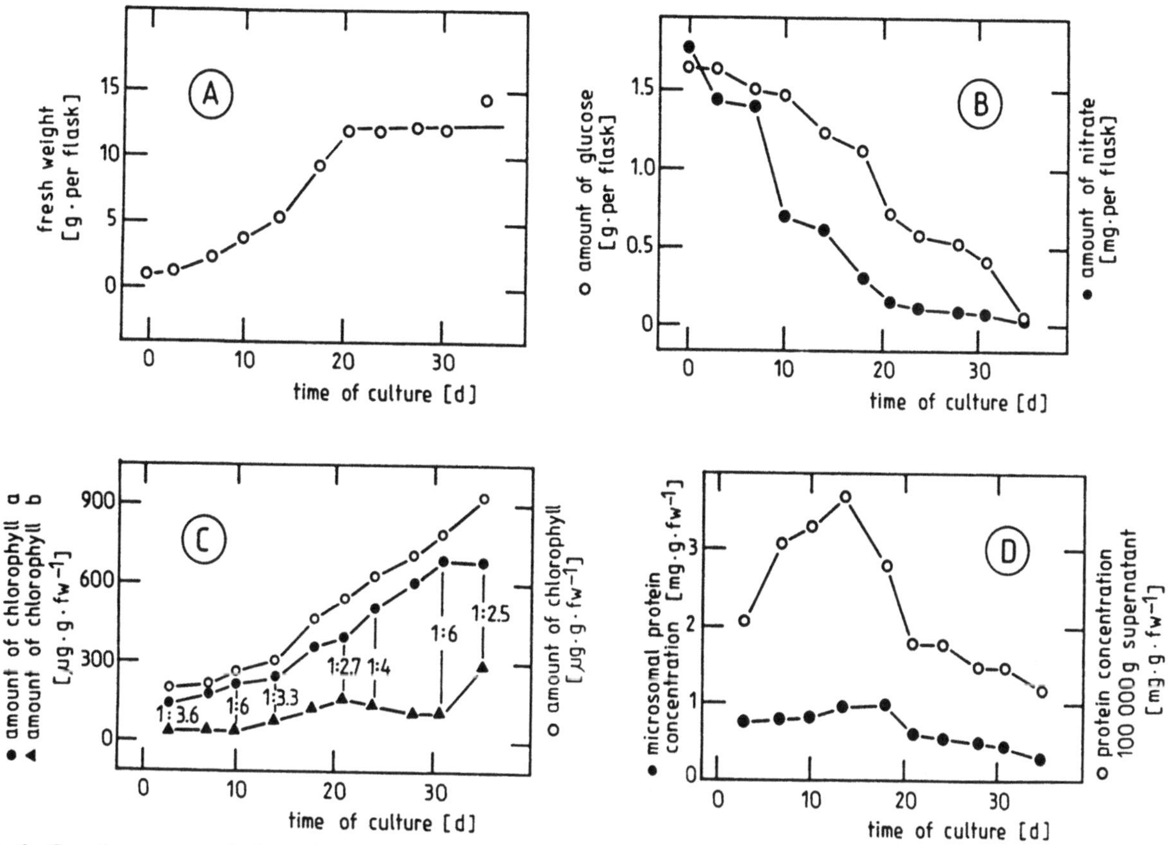

Fig. 3. Growth parameters of mixotrophic, aseptically grown shoot cultures of *Digitalis purpurea*. The shoots are grown in 300 ml Erlenmeyer flasks in 50 ml of a modified MS-medium. A. Fresh weight (g per flask). B. Glucose content in the medium (left scale: g per flask) o—o and nitrate content in the medium (right scale: mg per flask) •—•. C. Amount of total chlorophyll o—o, chlorophyll a •—• and b ▲—▲ (µg g^{-1} FW). D. Amount of soluble o—o and microsomal •—• protein (mg g^{-1} FW).

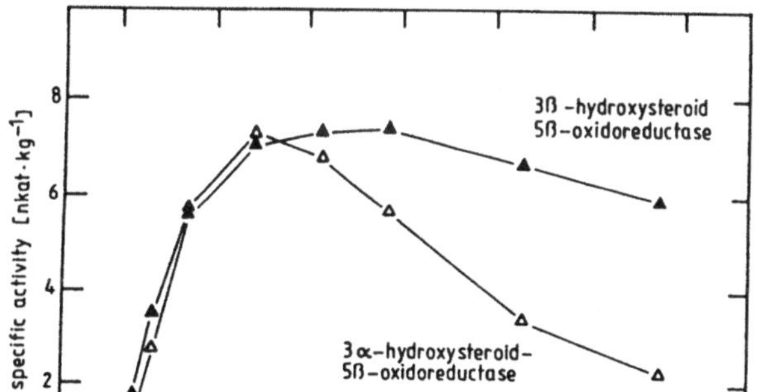

Fig. 4. Effect of substrate concentrations on the catalytic activities of 3α-hydroxysteroid-5β-oxidoreductase △—△ and 3β-hydroxysteroid-5β-oxidoreductase ▲—▲.

co-substrate for this reversible enzymic conversion is NADH with a K_m-value of 400 µM whereas the affinity for NADPH was much higher (K_m = 910 µM). The microsomal activity was only present with NADPH at a K_m-value of 1.33 mM. This activity is inhibited at higher concentrations of the substrate 5β-pregnane-3,20-dione (Fig. 4). The apparent K_m for the reverse reaction with 5β-pregnan-3α-ol-20-one as a substrate is 235 µM.

The 3β-hydroxysteroid-5β-oxidoreductase catalyzes the reversible reduction to 5β-pregnan-3β-ol-20-one. About 90% of the activity are found to be soluble. The activity in the soluble fraction as well as the minor part in the microsomes is dependent on NADPH (K_m = 300 µM). The affinity to NADH is much lower (1.39 mM) for the extract from the soluble part and 730 µM for the microsomal activity. With regard to the substrate 5β-pregnane-3,20-dione the affinity is in the same order of magnitude (K_m = 17.1 µM) as for most of the other enzymes described here. In contrary to the above-mentioned case no substrate inhibition has been observed. The K_m-value for the oxidation is 31.7 µM. In view of the fact that all cardenolides are

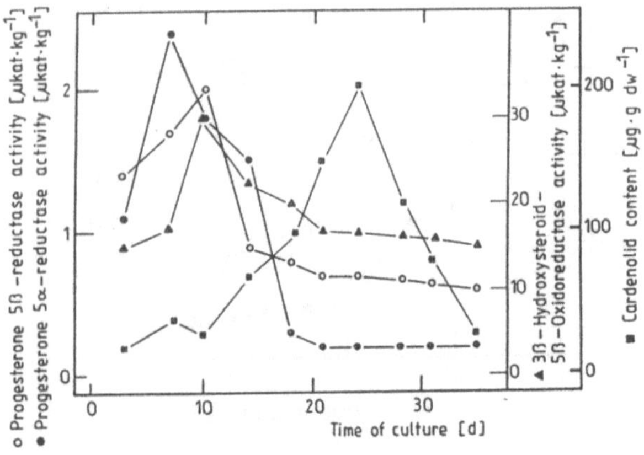

Fig. 5. Time course of cardenolide accumulation ■—■ and the specific activities of progesterone 5β-reductase o—o (left scale), progesterone 5α-reductase •—• (left scale), and 3β-hydroxysteroid-5β-oxidoreductase ▲—▲ (right scale).

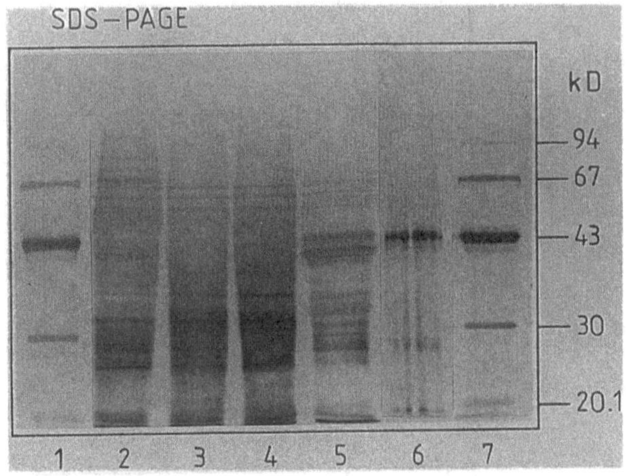

Fig. 6. SDS-PAGE of active fractions from different stages of purification on a 10% separating gel according to Laemmli (1970). The gels were silver stained according to Ansorg (1985). Lane 1 + 7 = low molecular weight standard; Lane 2 = 100,000 g supernatant; Lane 3 = DEAE-Sephacel; Lane 4 = Blue-Sepharose Cl-6B; Lane 5 = 2′5′-ADP-Sepharose 4B; Lane 6 = active fractions from non-dissociating polyacylamide gel electrophoresis.

β-configurated at C-3 and C-5 this is the second stereospecific enzyme of the putative biosynthetic pathway.

Activity changes of progesterone 5α-reductase, progesterone 5β-reductase, 3β-hydroxysteroid-5β-oxidoreductase and cardenolide content

The progesterone 5α-reductase is a microsomal-bound enzyme (Wendroth & Seitz 1990). Its activity increases markedly with a maximum at 7 d followed by a steep decline (Fig. 5). After 18 days no extractable activity is observed. The corresponding progesterone 5β-reductase is a soluble enzyme (Gärtner et al. 1990) exhibiting a maximum activity on day 11, after which it decreases, levelling off after 18 d. The 3β-hydroxysteroid-5β-oxidoreductase shows a very similar time course. The activity of the latter enzyme is 10 times higher than that of the progesterone 5β-reductase. Such a ratio has been also reported for animal systems (Morikawa & Tamaoki 1990; Penning et al. 1984).

In *Digitalis purpurea* three aglycones are accumulated, namely digitoxigenine, gitoxigenine and gitaloxigenine. Their time courses of accumulation are very similar. Maximum accumulation is reached after 24 days and then it declines rapidly.The major aglycone is digitoxigenin.

Purification of progesterone 5β-reductase

The purification procedure for the progesterone 5β-reductase from shoot cultures of *Digitalis purpurea* is summarized in Table 1. It includes three chromatographic steps and a separation on a non-dissociating gel. A substantial purification of 460-fold was achieved by affinity chromatography on 2′,5′-ADP-Sepharose 4B.

From a non-dissociating polyacrylamide gel the enzyme protein could be eluted by homogenizing the gel matrix and incubating the slices in assay buffer. This step resulted in a 769-fold purification. The molecular mass of the native protein was determined by gel chromatography on Sephadex G-200 to be 280 kDa.

The purification steps were checked by SDS-PAGE. After non-dissociating PAGE one band is present on the SDS-gel exhibiting a molecular mass of 43 kDa (Fig. 6). Preliminary attempts to microsequence this polypeptide failed, since the aminoterminus is masked. Now experiments are underway to apply the microsequencing technique to fragments of the protein after proteolytic digestion.

Discussion

Cardenolides have the genin as a complex nucleus in common and thus the enzymes of the putative biosynthetic pathway are of special interest. It is widely accepted that cholesterol, pregnenolone and progesterone are part of this pathway but these compounds

Table 1. Purification of progesterone 5β-reductase from shoot cultures of *Digitalis purpurea*.

Purification step	Total enzyme activity (pkat)	Specific activity (µkat kg $^{-1}$)	Purification (-fold)	Overall yield (%)
cytosol	75	0.5	–	–
DEAE-sephacel	72	7.7	15	96
Blue-sepharose CL-6B	70	27.8	56	93
2′,5′-ADP-sepharose 4B	12	230.0	460	16
PAGE	2	384.4	769	3

are also utilized for the synthesis of plant sterols. This section of the pathway may be designated as general steroid metabolism. The question arises now, which enzyme initiates the specific steps leading to cardenolides.

The reduction of the double bond between C-4 and C-5 of progesterone results in two isomeres with 5α- and 5β-configuration. The progesterone 5α-reductase is microsomal, whereas the progesterone 5β-reductase is soluble. Both enzymes are NADPH-dependent. A quite similar situation was found in the rat liver (McGuire & Tomkins 1959). We conclude that this enzymic step is very specific.

The reduction of the 3-keto group to a hydroxyl gives rise to stereoisomeres. By the action of the 3β-hydroxysteroid-5α-oxidoreductase 5α-pregnan-3β-ol-20-one is formed. The 5β-pregnane-3,20-dione can either be reduced by the 3α-hydroxysteroid-5β-oxidoreductase to 5β-pregnan-3α-ol-20-one or by the 3β-hydroxysteroid-5β-oxidoreductase to 5β-pregnan-3β-ol-20-one. These three enzymes are present in the soluble and the microsomal fraction. The reaction is reversible and NADPH as well as NADH are accepted as co-substrates. The K_m-values for the reverse reaction are much higher than for the reduction, thus indicating that the reduction is the preferred process. Very similar properties were observed in animal systems (Campbell & Karavolas 1990; Morikawa & Tamaoki 1990; Penning et al. 1984). These results strongly indicate that the second reduction steps are less specific.

Leaf discs of *Digitalis lanata* incorporate 5β-pregnane-3,20-dione and 5β-pregnan-3β-ol-20-one into cardenolides, which are without any exception β-configurated at C-3 and C-5; an interconversion of both compounds has been also observed (Tschesche et al. 1970). In callus and cell cultures of *Digitalis purpurea* the conversion of 5β-pregnane-3,20-dione to 5β-pregnan-3β-ol-20-one as well as to 5β-pregnan-3α-ol-20-one has been described by Hirotani & Furuya (1977). Feeding 5β-pregnan-3β-ol-20-one to these cultures resulted in the appearance of 5β-pregnane-3,20-dione, 5β-pregnan-3α-ol-20-one and 5β-pregnan-3β-ol-20β-diol. This may be explained by the presence of an isomerase which transforms 5β-pregnan-3β-ol-20-one to the corresponding 3α-isomer or by the oxidation of 5β-pregnan-3β-ol-20-one to 5β-pregnane-3,20-dione. We assume that these two conversions are catalyzed by separate enzymes, since the 3α-hydroxysteroid-5β-oxidoreductase prefers NADH whereas NADPH is the superior co-substrate for the 3β-hydroxysteroid-5β-oxidoreductase. They differ also with regard to substrate inhibition; higher substrate concentrations are only inhibitory to the 3α-hydroxysteroid-5β-oxidoreductase. Moreover, the activities are eluted at different salt concentrations from a DEAE-Sephacel column.

Feeding experiments and enzymological data indicate that the 'cholesterol side-chain cleaving enzyme' and the 3β-HSD may be part of the general steroid pathway providing constituents of biomembranes. These enzymes convert a variety of different substrates and can also be found in plant species which do not accumulate cardenolides.

The reduction of progesterone seems to be the initial step in cardenolide biosynthesis, since feeding cholesterol (Hagimori et al. 1983) and pregnenolone (Stuhlemmer, personal communication) did not stimulate an increase in the cardenolide content, whereas progesterone led to an increase of 180% (Hagimori et al. 1983). Feeding 5β-pregnane-3,20-dione stimulated cardenolide biosynthesis by 200–300% (Stuhlemmer, personal communication), whereas the corresponding 5α-pregnane-3,20-dione did not change the cardenolide content.

In order to initiate studies on the regulation of cardenolide biosynthesis progesterone 5β-reductase was purified about 770-fold. Its molecular mass was determined by SDS-PAGE to be 43 kDa. The native protein has a molecular mass of about 280 kDa as determined by gel filtration, thus indicating that it is a homopolymer of several subunits, contrary to the situation described for the cytosolic Δ4–3-ketosteroid 5β-reductase from rat liver with monomers of 37 kDa (Okuda & Okuda 1984).

Conclusions

Mixotrophic shoot cultures from *Digitalis purpurea* are well suited for studying the cardenolide pathway. The initial step in cardenolide biosynthesis is the stereospecific, NADPH-dependent and irreversible reduction of progesterone yielding an A/B cis-ring junction which is characteristic for all cardenolides. The subsequent step involves the reduction of the 3-keto group. This step is less specific.

Acknowledgement

This work was supported by a grant from the Deutsche Forschungsgemeinschaft and a fellowship (D.E.G.) from the Land Baden-Württemberg.

References

Ansorg W (1985) Fast and sensitive detection of protein and DNA bands by treatment with potassium permanganate. J. Biochem. Biophys. Meth. 11: 13–20

Bradford MM (1976) A rapid and sensitive method for the quantification of microgram quantities of protein utilizing the principle of protein-dye-binding. Anal. Biochem. 72: 248–254

Campbell JS & Karavolas HJ (1990) Purification of the NADP:5α-dihydroprogesterone 3α-hydroxysteroid oxidoreductase from female rat pituitary cytosol. J. Steroid Biochem. Mol. Biol. 37: 215–222

Corchete MP, Sanchez JM, Cacho M, Moran M & Fernandez-Tarrago J (1990) Cardenolide content in suspension cell cultures derived from root and leaf callus of *Digitalis thapsi* L. J. Plant Physiol. 137: 196–200

Diettrich B, Mertinat H & Luckner M (1990) Formation of *Digitalis lanata* clone lines by shoot tip culture. Planta Med. 56: 53–58

Dubois M, Gilles KA, Hamilton JK, Rebers PA & Smith F (1956) Colorimetric method for determination of sugars and related substances. Anal. Chem. 28: 350–356

Gärtner DE, Wendroth S & Seitz HU (1990) A stereospecific enzyme of the putative biosynthetic pathway of cardenolides. Characterization of a progesterone 5β-reductase from leaves of *Digitalis purpurea* L. FEBS Lett. 271: 239–242

Gärtner DE & Seitz HU (1993) Enzyme activities in cardenolide-accumulating, mixotrophic shoot cultures of *Digitalis purpurea* L. J. Plant Physiol. 141: 269–275

Graves JMH & Smith WH (1967) Transformation of pregnenolone and progesterone by cultured plant cells. Nature 214: 1248

Hagimori M, Matsumoto T & Obi Y (1982) Studies on the production of *Digitalis* cardenolides by plant tissue culture. II Effect of light and plant growth substances on digitoxin formation by undifferentiated cell and shoot-forming cultures of *Digitalis purpurea* L. grown in liquid media. Plant Physiol. 69: 653–656

Hagimori M, Matsumoto T & Obi Y (1983) Effects of mineral salts, initial pH and precursors on digitoxin formation by shoot-forming cultures of *Digitalis purpurea* L. grown in liquid media. Agric. Biol. Chem. 47: 565–571

Hirotani M & Furuya T (1977) Restoration of cardenolide-synthesis in redifferentiated shoots from callus cultures of *Digitalis purpurea*. Phytochemistry 16: 610–611

Laemmli UK (1970) Cleavage of structural proteins during the assembly of the head of bacteriophage T4. Nature 227: 680–685

Lichtenthaler HK (1987) Chlorophylls and carotinoids: Pigments of photosynthetic biomembranes. In: Packer L & Douce R (Eds) Methods in enzymology, Vol 148 (pp 350–382). Academic Press, San Diego, New York, Berkeley

Lui JHC & Staba EJ (1979) Effects of precursors on serially propagated *Digitalis lanata* leaf and root cultures. Phytochemistry 18: 1913–1916

McGuire JS & Tomkins GM (1959) The multiplicity and specificity of Δ^4-3-ketosteroid hydrogenase (5α). Arch. Biochem. Biophys. 82: 476–477

Morikawa T & Tamaoki BJ (1990) Purification and characterization of 3α-hydroxysteroid dehydrogenase from chicken hepatic cytosol. J. Steroid Biochem. Mol. Biol. 37: 569–574

Okuda A & Okuda K (1984) Purification and characterization of Δ_4-3-ketosteroid 5β-reductase. J. Biol. Chem. 259: 7519–7524

Penning TM, Mukharji I, Barrows S & Talalay P (1984) Purification and properties of a 3α-hydroxysteroid dehydrogenase of rat liver cytosol and its inhibition by anti-inflammatory drugs. Biochem. J. 222: 601–611

Pilgrim H (1972) 'Cholesterol side-chain cleaving enzyme' Aktivität in Keimlingen und *in vitro* kultivierten Geweben von *Digitalis purpurea*. Phytochemistry 11: 1725–1728

Rücker W, Jentzsch K & Wichtl M (1976) Wurzeldifferenzierung und Glykosidbildung bei *in vitro* kultivierten Blattexplantaten von *Digitalis purpurea* L. Z. Pflanzenphysiol. 80: 323–335

Seidel S, Kreis W & Reinhard E (1990) Δ5–3β-Hydroxysteroid dehydrogenase/ Δ5-Δ4-ketosteroid isomerase (3β-HSD), a possible enzyme of cardiac glycoside biosynthesis in cell cultures and plants of *Digitalis lanata* EHRH. Plant Cell Rep. 8: 621–624

Tschesche R, Hombach R, Scholten H & Peters M (1970) Neue Beiträge zur Biogenese der Cardenolide in *Digitalis lanata*. Phytochemistry 9: 1505–1515

Warneck HM & Seitz HU (1990) 3β-Hydroxysteroid oxidoreductase in suspension cultures of *Digitalis lanata* EHRH. Z. Naturforsch. 45c: 963–972

Wendroth S & Seitz HU (1990) Characterisation and localisation of progesterone 5α-reductase from cell cultures of foxglove (*Digitalis lanata* EHRH). Biochem. J. 266: 41–46

Plant Cell, Tissue and Organ Culture **38**: 345–349, 1994.

Enzymes involved in the metabolism of 3-hydroxy-3-methylglutaryl-coenzyme A in *Catharanthus roseus*

Robert van der Heijden[1], Veronika de Boer-Hlupá[1], Robert Verpoorte[1] & Johannis A. Duine[2]
[1]*Division of Pharmacognosy, Leiden/Amsterdam Center for Drug Research, Leiden University, Gorlaeus Laboratories, P.O. Box 9502, NL-2300 RA Leiden, The Netherlands;* [2]*Department of Microbiology and Enzymology, Delft University of Technology, Julianalaan 67, NL-2628 BC Delft, The Netherlands*

Key words: Catharanthus roseus, enzyme, HMG-CoA, HPLC, metabolism, mevalonate

Abstract

3-Hydroxy-3-methylglutaryl-coenzyme A (HMG-CoA) is an important intermediate in various metabolic pathways, e.g. sterol biosynthesis, ketogenesis and leucine catabolism. The reactions and enzymes involved in the metabolism of HMG-CoA are briefly reviewed. These enzymes have been studied in *Catharanthus roseus*, a model system for studies on the regulation of secondary metabolic pathways, particularly those leading to terpenoid-indole alkaloids. By using HPLC, three HMG-CoA catabolizing enzyme activities have been detected in protein extracts from suspension cultured *C. roseus* cells: HMG-CoA lyase, 3′-nucleotidase and (tentatively identified) 3-methylglutaconyl-CoA hydratase (HMG-CoA hydrolyase). The enzymes have been partially purified. HMG-CoA is formed from three molecules of acetyl-CoA, via reactions which are catalyzed by two (as in yeast and animal cells, via intermediacy of acetoacetyl-CoA) or by just one enzyme (as in e.g. radish). It is yet not clear which process occurs in *C. roseus*.

Abbreviations: AACT – acetoacetyl-CoA thiolase, AACT/HMGS – acetoacetyl-COA thiolase/HMG-CoA synthase, CoASH – coenzyme A (reduced form), HMG-CoA – 3-hydroxy-3-methylglutaryl-CoA, MG-CoA – 3-methylglutaconyl-CoA

Introduction

Mevalonate is a building block of a vast array of plant primary and secondary metabolites e.g. phytosterols, terpenoids and terpenoid-indole alkaloids. At present much research is focused on the regulation of the mevalonate biosynthesis in order to, eventually, manipulate the accumulation of mevalonate derived metabolites. An important intermediate in the biosynthesis of mevalonate is 3-hydroxy-3-methylglutaryl-coenzyme A (HMG-CoA). Some reactions and enzymes involved in the biosynthesis of mevalonate and the metabolism of HMG-CoA are presented in Fig. 1.

Mevalonate is formed from HMG-CoA in a reaction catalyzed by the enzyme HMG-CoA reductase. From several animal and plant sources, HMG-CoA reductase (E.C. 1.1.1.34) has been well characterized (Roitelman et al. 1992; Bach 1987; Bach et al. 1990, 1991). The enzyme is a major point of regulation in (chole-)sterol biosynthesis. In animal cells, HMG-CoA reductase activity is regulated at the gene, mRNA and protein level (reviewed by Lutton 1991). Similar processes seem to play a role in plant systems; recently evidence for a protein kinase cascade in higher plants was provided by the isolation and purification of an HMG-CoA reductase kinase from cauliflower inflorescences (Mackintosh et al. 1992).

HMG-CoA is, however, also a substrate for the enzymes HMG-CoA lyase and 3-methylglutaconyl-CoA hydratase. Relatively little is known about these enzymes.

HMG-CoA lyase (E.C. 4.1.3.4) catalyzes the cleavage of (3*S*)-HMG-CoA to acetoacetate and acetyl-CoA, a reaction involved in ketogenesis and leucine catabolism. The enzyme is localized in mitochondria (Clinkenbeard et al. 1975). In plants this activity has been detected in among others *Euphorbia lathyris*

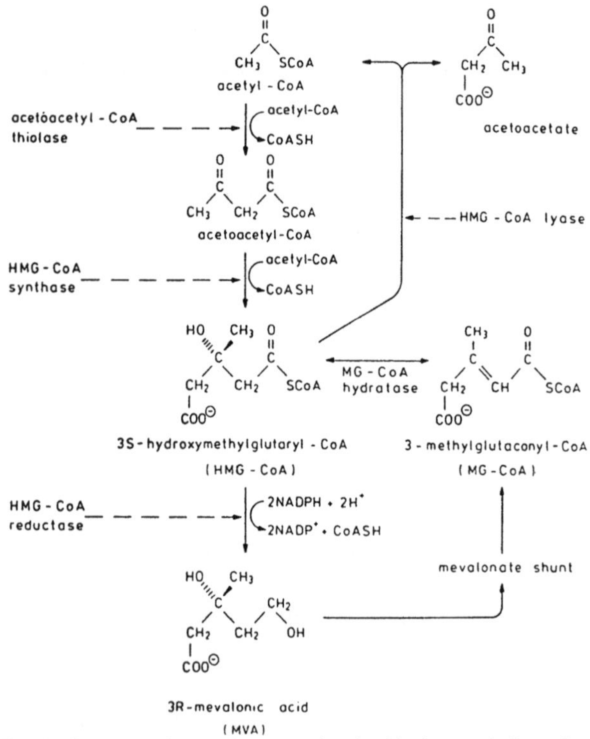

Fig. 1. Some reactions and enzymes involved in the metabolism of HMG-CoA and the biosynthesis of mevalonate.

(Skrukrud et al. 1988) and radish seedlings (Bach et al. 1991). From the latter source it was partially purified and was found to have a molecular mass of 70,000 Da and a pI of 7 (Bach et al. 1991; Weber & Bach 1993).

3-Methylglutaconyl-CoA hydratase (HMG-CoA hydrolyase, E.C. 4.2.1.18) catalyzes the reversible hydration of 3-methylglutaconyl-CoA (MG-CoA) to HMG-CoA, a reaction involved in leucine catabolism and the mevalonate shunt pathway. It has been partially purified from sheep liver (Hilz et al. 1958); to our knowledge no data are available on the plant enzyme.

HMG-CoA is biosynthesized from three molecules of acetyl-CoA. In yeast and animal cells two enzymes are involved in this process: acetoacetyl-CoA thiolase and HMG-CoA synthase (Fig. 1). However, it was reported recently that a single protein (named AACT/HMGS), isolated from radish seedlings, was able to catalyze both reactions involved in HMG-CoA biosynthesis (Bach et al. 1991). AACT/HMGS activity was subsequently detected in maize seedlings (Weber 1992) and tobacco cultures (Gondet et al. 1992). It is, however, not yet generally accepted that this is the only pathway for HMG-CoA biosynthesis in plants.

Acetoacetyl-CoA thiolase (AACT, acetyl-CoA:acetyl-CoA C-transferase, E.C. 2.3.1.9) catalyzes the condensation of two molecules acetyl-CoA to acetoacetyl-CoA and CoASH. In plants AACT activity has been detected in among others *Parthenium argentatum* and *Phaseolus radiatus* (reviewed in Bach et al. 1991). In yeast and animal cells two forms of AACT have been characterized, a cytosolic form which is involved in cholesterol biosynthesis and a mitochondrial form which is involved in ketogenesis. The forms have distinct differences in pI; most cytosolic forms have a pI between 5.1–5.8, while most mitochondrial forms have a pI of about 8.

HMG-CoA synthase (E.C. 4.1.3.5) catalyzes the reaction of acetyl-CoA and acetoacetyl-CoA to form HMG-CoA and CoASH. In plants HMG-CoA synthase activity has been detected in only a few species (reviewed in Bach et al. 1991).

The plant enzyme AACT/HMGS, which catalyzes the formation of HMG-CoA from three molecules of acetyl-CoA, has been purified from radish seedlings. It is a monomeric enzyme with a molecular mass of 55,500 Da and a pH optimum of 8. Maximum in vitro activity is obtained in the presence of (chelated) Fe^{2+} and a quinone (e.g. pyrroloquinoline quinone, PQQ) (Bach et al. 1991; Weber 1992).

Furthermore HMG-CoA is a substrate for various (non-specific) enzymes, such as thioesterases, pyrophosphatases and 3′-nucleotidases, the latter two reacting with the CoA-moiety of the molecule. Thus it was reported for chalcone synthase, an enzyme involved in flavonoid biosynthesis, that by the action of 3′-nucleotidases the substrate malonyl-CoA was converted to an efficient chalcone synthase inhibitor: malonyl-3′-dephospho-CoA (Hinderer & Seitz 1986). It is not clear whether this mechanism plays a regulatory role in vivo; if so, similar processes could occur in HMG-CoA metabolism.

Here we present some results of our research on the regulation of the mevalonate biosynthesis, which is focused on the metabolism of HMG-CoA in *Catharanthus roseus*. The aim of this research is to characterize the regulation mechanisms involved in the channelization of mevalonate into terpenoid (-indole alkaloid) biosynthesis. For this, an HPLC system has been developed to separate the CoA-esters involved in HMG-CoA metabolism (Fig. 1). This system enables the characterization of metabolic processes by direct identification and quantification of substrate consumption and product formation. Three HMG-CoA catabolizing enzyme activities have been detected and subsequently characterized and partially purified. In addition, spectrophotometric assays have been used to determine the activities of some enzymes involved in HMG-CoA metabolism in *C. roseus* plants and cultures.

HPLC-analysis of CoASH and its esters involved in HMG-CoA metabolism; detection of HMG-CoA catabolizing enzymes in *Catharanthus roseus* cultures

Various HPLC methods have been described for the analysis of short-chain acyl-CoA derivatives in (animal) tissue extracts e.g. by gradient elution 17 CoA-derivatives could be separated (King & Reiss 1985). We modified this system to an isocratic one which is able to separate CoASH, acetyl-CoA, acetoacetyl-CoA and HMG-CoA, all of which are involved in HMG-CoA metabolism. The conditions used are as follows: column: Hypersil ODS (250 × 4.6 mm, particle size 5 μm); solvent system: 0.2 M sodium phosphate buffer, pH 5.0, containing 4.5% (v/v) acetonitrile; flow rate: 1.5 ml/min. The compounds were detected by UV at 254 nm or by a photodiode array detector. k′-Values are 2.2, 4.5, 8.2 and 9.1 for CoASH, HMG-CoA, acetyl-CoA and acetoacetyl-CoA respectively. Acetyl-CoA and acetoacetyl-CoA do not show baseline separation. Due to the stability of the CoA-esters at acidic pH, analyses could be facilitated using an autosampler. However, at the pH-values at which enzyme incubations are performed (pH 7–8.2) the CoA-esters hydrolyze, making short incubation times and adequate controls essential (Van der Heijden et al. in preparation). The HPLC system is now used to detect activities of some enzymes involved in HMG-CoA metabolism in *C. roseus* plants and cell cultures.

As HMG-CoA is a substrate to several enzymes, after incubation of HMG-CoA with an enzyme extract from a *C. roseus* suspension culture, several products can be expected. A typical chromatogram obtained from such an experiment is presented in Fig. 2. After optimization of the incubation conditions and by fractionation of the protein extracts the activities described below have been identified.

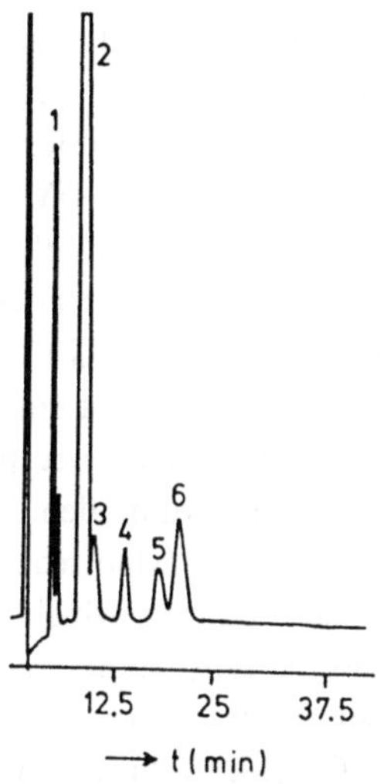

Fig. 2. Chromatogram of an incubation of 3*R/S*-HMG-CoA with an enzyme preparation from a *Catharanthus roseus* suspension culture. Peak identification: (1) CoASH, (2) HMG-CoA, (3) 3′-dephospho-CoASH (sometimes co-eluting with a compound derived from the protein extract), (4) acetyl-CoA, (5) 3′-dephospho-HMG-CoA, (6) MG-CoA. Proteins were extracted from a 5-day-old *C. roseus* suspension culture. 20 g of cells were frozen in liquid nitrogen and powdered in a blender. Powdered biomass was thawed in the presence of 2 g polyvinylpolypyrrolidone and 20 ml extraction buffer (1 M sucrose, 2 mM EDTA, 20 mM 2-mercaptoethanol, 10 uM leupeptin in 0.1 M KH_2PO_4 pH 7.5). Incubations were performed for 10 min at 30 °C in 0.2 M Tris-HCl buffer pH 8.2 containing 0.2 mM EDTA and 20 mM $MgCl_2$ and 0.5 mM 3*R/S*-HMG-CoA. HPLC conditions as mentioned in the text.

HMG-CoA lyase activity in a Catharanthus roseus *suspension culture*

HMG-CoA lyase activity results into the formation of acetyl-CoA and acetoacetate. Acetyl-CoA is eluted with a retention time of 14.1 min (Fig. 2). However, acetoacetate is not detected. After optimization of the incubation conditions etc., the HPLC system is now used for assaying HMG-CoA lyase activity (Van der Heijden et al. in preparation). It was found that HMG-CoA lyase is a soluble enzyme, whose activity is strongly stimulated by relatively high concentrations of Mg^{2+}; optimum activities are found in the presence of 10–30 mM Mg^{2+} and 10 mM dithiothreitol (DTT). The enzyme has been purified about 8 times by ammonium sulfate precipitation and anion-exchange chromatography, respectively. The enzyme is unstable. After anion-exchange chromatography, HMG-CoA lyase is separated from other HMG-CoA catabolizing enzymes (see below). HMG-CoA lyase is specific with respect to the stereochemistry of HMG-CoA (as with the enzyme from other sources, it is likely that only (3*S*)-HMG-CoA is accepted as substrate).

3′-Nucleotidase activity in a Catharanthus roseus *suspension culture*

Another HMG-CoA catabolizing enzyme activity gave a product with a k′-value of 10.6. It was noted that the enzyme was not specific with respect to the *R* or *S* form of HMG-CoA (in all experiments 3*R/S*-HMG-

CoA was used as substrate). The product showed an identical UV spectrum as the substrate. The longer retention time in the HPLC system, the nonspecificity and the identicity of the UV spectra suggested that the enzyme removed the 3′-phosphate group from the adenosine residue of the coenzyme A part of the molecule. Incubation of HMG-CoA with rye grass 3′-nucleotidase (obtained from a commercial source) gave a product which was with respect to retention behaviour and UV spectrum identical to the product from the *C. roseus* enzyme reaction. The activity from the *C. roseus* culture, which is thus identified as 3′-nucleotidase activity, is not specific for HMG-CoA, it also dephosphorylates CoASH, acetyl-CoA and acetoacetyl-CoA. k′-Values in our HPLC system for 3′-dephospho-CoASH, -HMG-CoA, -acetyl-CoA and -acetoacetyl-CoA are 5.6, 10.6, 20.2 and 25.4, respectively. The enzyme is relatively stable (it can be stored for several months at −80 °C), is soluble (it is found in a 105,000 g supernatant) and was partly purified by ammonium sulfate precipitation (precipitates between 40–60% saturation).

3-Methylglutaconyl-CoA hydratase activity in Catharanthus roseus *suspension cultures*

The third HMG-CoA catabolizing activity gave a product with a k′-value in the HPLC system of 12.4 (compound 6 in Fig. 2). The maximum in its UV spectrum was found at 260 nm, for HMG-CoA it is found at 257 nm. Furthermore, in all incubations performed, only a limited amount (less than 20%) of the substrate could be converted to product, suggesting a reaction equilibrium favoring HMG-CoA formation. These data point to 3-methylglutaconyl-CoA hydratase (HMG-CoA hydrolyase) activity, which was further confirmed by using a spectrophotometric assay (Hilz et al. 1958). Direct proof of the identity of the product is now being sought by means of mass spectrometry.

The activity of this soluble and relatively stable enzyme is not stimulated by Mg^{2+} and DTT. This enzyme was purified 42 times by ammonium sulfate precipitation and chromatography over Q-Sepharose, hydroxyapatite and Superose 6 respectively. The activity was separated from HMG-CoA lyase activity by chromatography over Q-Sepharose.

HMG-CoA biosynthesis in Catharanthus roseus

Experiments were also performed on the biosynthesis of HMG-CoA, the formation of which is of particular interest aspect since it may involve either 1 (AACT/HMGS) or 2 (AACT and HMG-CoA synthase) enzymes. It is not clear which mechanism(s) occur(s) in *C. roseus*.

Assuming the involvement of two enzymes, the activities of AACT and HMG-CoA synthase have been determined in *C. roseus* tissues and cultures (Table 1) using spectrophotometric assays (Miziorko 1985). High activities of AACT and HMG-CoA synthase were found in the flower and the stem. The AACT activity was confirmed by HPLC since the substrates acetoacetyl-CoA and CoASH were converted to acetyl-CoA. The activity of HMG-CoA synthase could not be confirmed by HPLC. This could be due to the presence of HMG-CoA catabolizing enzymes. Both AACT and HMG-CoA synthase, which are unstable and inactivated by higher salt concentrations (> 200 mM NaCl), have been partially purified by acetone precipitation, ion-exchange chromatography and gel filtration. However, on all chromatographic media tested (Superose 6, Mono-Q, Blue Sepharose, Red Sepharose and Orange A) no separation of AACT and HMG-CoA synthase activity was obtained.

Table 1. Specific acetoacetyl-CoA thiolase (AACT) and HMG-CoA synthase activities (pkatal/mg protein) in *C. roseus* plants and cultures.

C. roseus	AACT	HMG-CoA synthase
Plant		
root	969	45
flower	1152	134
stem	1587	119
leaf	234	–*
latex	585	72
Seedlings, 8 days old		
light grown	685	65
etiolated	518	30
Hairy roots, 14 days old	518	58
Suspension culture		
7 days, light grown	952	73
12 days, light grown	250	32
10 days, dark grown	534	10

* Low activity, but sensitivity of assay was reduced by the presence of chlorophyll.

Further efforts are now being made to detect AACT/HMGS activity (the enzyme which produces HMG-CoA directly from acetyl-CoA) in *C. roseus.*

Conclusions

HMG-CoA is an intermediate in a series of metabolic processes, such as sterol biosynthesis, ketogenesis and leucine catabolism. In animal systems, these processes have been rather well characterized, but for plant systems little insight into regulatory aspects has been obtained so far.

In *C. roseus* cultures at least four HMG-CoA catabolizing enzyme activities are present: HMG-CoA lyase, MG-CoA hydratase (HMG-CoA hydrolyase), 3′-nucleotidase and HMG-CoA reductase. (No efforts have been made to determine the latter activity. In our experimental set up HMG-CoA reductase activity is expected to be low due to the absence of NADPH.)

The products of these catabolic enzymes are CoA derivatives which can be separated by HPLC from the substrate. The system provides thus an analytical tool by which both substrate consumption and formation of product(s) can be monitored. This is of great importance when crude enzyme preparations are used in studies on HMG-CoA catabolism. Such requirements are usually not met by, for example, radiochemical and spectrophotometric assays.

In vitro the activities compete for HMG-CoA; however, the situation in vivo is likely to be quite different. Localization studies are needed to give an insight into the cellular and subcellular compartmentation of the activities and to understand the role of compartmentation in the regulation of terpenoid biosynthesis (reviewed by Gray 1987) and related pathways of secondary metabolism.

Acknowledgement

The research of Dr. R. van der Heijden has been made possible by a fellowship of the Royal Netherlands Academy of Arts and Sciences.

References

Bach TJ (1987) Synthesis and metabolism of mevalonic acid in plants. Plant Physiol. Biochem. 25: 163–178

Bach TJ, Weber T & Motel A (1990) Some properties of enzymes involved in the biosynthesis and metabolism of 3-hydroxy-3-methylglutaryl-CoA in plants. In: Towers GHN & Stafford HA (Eds) Recent Advances in Phytochemistry, Vol 24, Biochemistry of the Mevalonic Acid Pathway to Terpenoids (pp 1–82). Plenum Press, New York

Bach TJ, Boronat A, Caelles C, Ferrer A, Weber T & Wettstein A (1991) Aspects related to mevalonate biosynthesis in plants. Lipids 26: 637–648

Clinkenbeard KD, Reed WD, Mooney RA & Lane MD (1975) Intracellular localization of the 3-hydroxy-3-methylglutaryl-Coenzyme A cycle enzymes in liver. J. Biol. Chem. 250: 3108–3116

Gondet L, Weber T, Maillot-Vernier P, Benveniste P & Bach TJ (1992) Regulatory role of microsomal 3-hydroxy-3-methylglutaryl-CoA reductase in a tobacco mutant that overproduces sterols. Biochem. Biophys. Res. Commun. 186: 888–893

Gray JC (1987) Control of isoprenoid biosynthesis in higher plants. Adv. Bot. Res. 14: 25–91

Hilz H, Knappe J, Ringelmann E & Lynen F (1958) Methylglutaconase, eine neue Hydratase, die am Stoffwechsel verzweigter Carbonsäuren beteiligt ist. Biochem. Z. 329: 476–489

Hinderer W & Seitz HU (1986) In vitro inhibition of carrot chalcone synthase by 3′-nucleotidase: the role of the 3′-phosphate group of malonyl-coenzyme A in flavonoid biosynthesis. Arch. Biochem. Biophys. 246: 217–224

King TD & Reiss PD (1985) Separation and measurement of short-chain coenzyme A compounds in rat liver by reversed-phase high-performance liquid chromatography. Anal. Biochem. 146: 173–179

Lutton C (1991) Dietary cholesterol, membrane cholesterol and cholesterol synthesis. Biochimie 73: 1327–1334

Mackintosh RW, Davies, SP, Clarke PR, Weekes J, Gillespie JG, Gibb BJ & Hardie DG (1992) Evidence for a protein kinase cascade in higher plants, 3-hydroxy-3-methylglutaryl-CoA reductase kinase. Eur. J. Biochem. 209: 923–931

Miziorko HM (1985) 3-Hydroxy-3-methylglutaryl-CoA synthase from chicken liver. In: Law JH & Rilling HC (Eds) Methods in Enzymology, Vol 110 (pp 19–26). Academic Press, Orlando

Roitelman J, Olender EH, Bar-Nun S, Dunn WA Jr & Simoni RD (1992) Immunological evidence for eight spans in the membrane domain of 3-hydroxy-3-methylglutaryl coenzyme A reductase: implications for enzyme degradation in the endoplasmatic reticulum. J. Cell Biol. 117: 959–973

Skrukrud CL, Taylor SE, Hawkins DR, Nemethy EK & Calvin M (1988) Subcellular fractionation of triterpenoid biosynthesis in *Euphorbia lathyris* latex. Physiol. Plant. 74: 306–316

Weber T (1992) Studies on the synthesis and the metabolism of 3-hydroxy-3-methylglutaryl-coenzyme A in *Raphanus sativus*. Karlsruhe Contributions to Plant Physiology 24: 1–129

Weber T & Bach TJ (1993) Partial purification and characterization of membrane-associated 3-hydroxy-3-methylglutaryl-coenzyme A lyase from radish seedlings. Z. Naturforsch. C 48: 444–450

Plant Cell, Tissue and Organ Culture **38**: 351–356, 1994.

Regulation of 3-hydroxy-3-methylglutaryl-coenzyme A reductase by wounding and methyl jasmonate

Implications for the production of anti-cancer alkaloids

Ignacio E. Maldonado-Mendoza, Ronald J. Burnett, Melina Lòpez-Meyer & Craig L. Nessler*
Department of Biology, Texas A&M University College Station, Texas 77843–3258, USA (requests for offprints)*

Key words: *Camptotheca acuminata*, *Catharanthus roseus*, HMGR, methyl jasmonate, terpenoid indole alkaloids

Abstract

HMGR (3-hydroxy-3-methylglutaryl-coenzyme A reductase; E.C.1.1.1.34) supplies mevalonate for the synthesis of many plant primary and secondary metabolites, including the terpenoid component of indole alkaloids. Suspension cultures of *Camptotheca acuminata* and *Catharanthus roseus*, two species valued for their anticancer indole alkaloids, were treated with the elicitation signal transducer methyl jasmonate (MeJA). RNA gel blot analysis from MeJA treated cultures showed a transient suppression of HMGR mRNA, followed by an induction in HMGR message. Leaf disks from transgenic tobacco plants containing a chimeric *hmg1*::GUS construct were also treated with MeJA and showed a dose dependent suppression of wound-inducible GUS activity. The suppression of the wound response by MeJA was limited to the first 4 h post-wounding, after which time MeJA application had no effect. The results are discussed in relation to the differential regulation of HMGR isogenes in higher plants.

Abbreviations: GUS – β-glucuronidase, *hmg* – gene of hmgr, HMGR – 3-hydroxy-3-methylglutaryl-coenzyme A reductase, JA – jasmonic acid, MeJA – methyl jasmonate, MUG – methylumbelliferyl-β-d-glucuronide, TDC – tryptophan decarboxylase, SDS – sodium dodecyl sulfate, SS – strictosidine synthase

Introduction

In green plants mevalonate is used in the synthesis of a wide variety of primary and secondary metabolites. Mevalonate derivatives are important in photosynthesis (chlorophylls, carotenoids, and plastoquinone), respiration (ubiquinone), and general growth and development (abscisic acid, cytokinins, and gibberellins). Mevalonate is also used to form a number of plant secondary products including rubber, phytoalexins, and terpenoid indole alkaloids.

The branch-point enzyme HMGR catalyzes the final step in the synthesis of mevalonate and shunts HMG-CoA into the isoprenoid pathway. Plant HMGRs are encoded by small gene families which display complex developmental and environmental regulation (Caelles et al. 1989; Learned & Fink 1989; Narita & Gruissem 1989; Yang et al. 1991; Choi et al. 1992; Chye et al. 1992; Genschik et al. 1992).

Our interest in HMGR is based on its role in providing an essential, and perhaps limiting, mevalonate precursor for the synthesis of indole alkaloids (Fig. 1). The indole portion of these alkaloids is provided by the enzymatic decarboxylation of tryptophan to tryptamine, catalyzed by TDC (Fig. 1; Lückner 1984). Strictosidine, a key intermediate in the synthesis of over 1,000 indole alkaloids, is assembled by the condensation of tryptamine with the monoterpene glucoside secologanin by SS (Fig. 1; Stöckigt & Zenk 1977). Several pharmacologically active alkaloids are derived from strictosidine including quinine, strychnine, and the anticancer agents vinblastine, vincristine, and camptothecin (Cordell 1974).

Because of the importance of strictosidine in indole alkaloid synthesis, most molecular studies have focused on the enzymes in this pathway. SS cDNAs have been cloned from *Rauvolfia serpentina* (Kutchan et al. 1988) and *Catharanthus roseus* (McKnight et al.

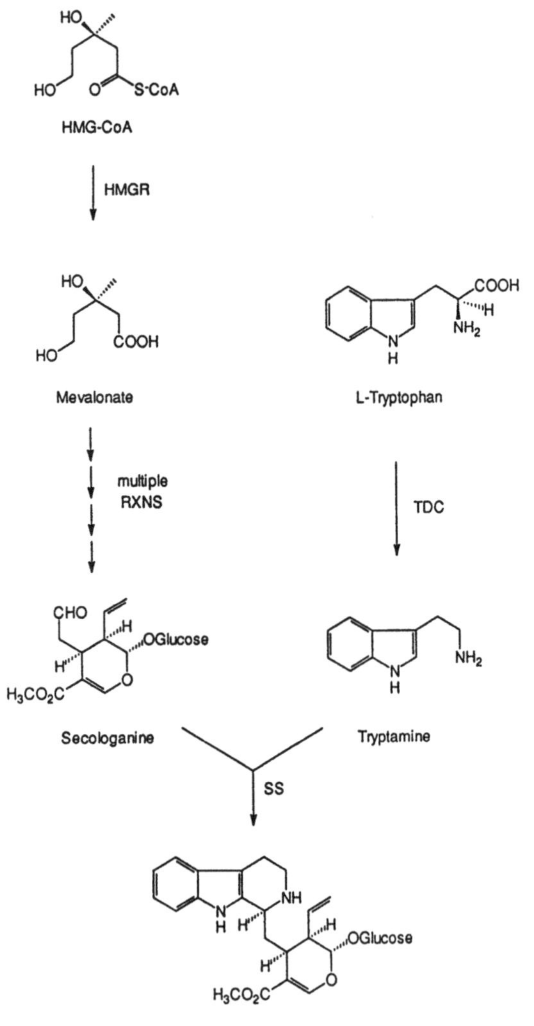

Fig. 1. Outline of Strictosidine Biosynthesis. Strictosidine, an intermediate in indole alkaloid synthesis, is assembled by strictosidine synthase (SS) which couples tryptamine to secologanin. Decarboxylation of tryptophan by tryptophan decarboxylase (TDC) yields the tryptamine whereas secologanin, the monoterpenoid portion of strictosidine, is made from mevalonate through a series of reactions beginning with HMGR.

1990; Pasquali et al. 1992), and a TDC cDNA has been isolated from *C. roseus* (De Luca et al. 1989).

Recently an HMGR cDNA has been cloned from *Catharanthus* (Maldonado-Mendoza et al. 1992), however, little information is available concerning the regulation of HMGR or other genes involved in secologanin biosynthesis in species which synthesize indole alkaloids. Addition of secologanin to *C. roseus* tissue cultures has been found to increase the yield of indole alkaloids, which suggests that terpenoid synthesis may be limiting for *in vitro* alkaloid production (Merillon et al. 1986; 1989). We have therefore begun to explore the molecular regulation of the terpenoid portion of the indole alkaloid pathway.

As a first step, we have cloned and sequenced an HMGR gene from *Camptotheca acuminata* (Burnett et al. submitted), a Chinese tree which is the natural source of the anti-tumor monoindole alkaloid camptothecin. Camptothecin is a specific inhibitor of DNA topoisomerase I, and is therefore toxic to rapidly growing cells.

In this report we examine HMGR expression in suspension cultures of *Catharanthus* and *Camptotheca* treated with methyl jasmonate (MeJA), the elicitation signal transducing phytohormone. In addition we use an HMGR promoter-reporter gene construct to determine the time course of MeJA suppression of HMGR wound induction.

Materials and methods

Plant materials

Camptotheca cell suspension cultures were induced and grown in media described by van Hengel (1992). *Catharanthus roseus* cell suspensions were cultured in the media of Sakato & Misawa (1974). Suspension cultures were grown on a gyratory shaker at 120 r.p.m. under a 16h/8h light/dark cycle. Tobacco plants (*Nicotiana tabacum* cv Xanthi) used for transformations were grown on 1/2 MS media (Murashige & Skoog 1962) in Magenta boxes.

RNA isolation and analysis

Total RNA from suspension cultures was isolated using the procedure of Jones et al. (1985). Poly A^+ RNA was selected by oligo dT cellulose chromatography with the FastTrack mRNA isolation kit (Invitrogen, San Diego, CA).

For RNA gel blot analysis, poly A^+ RNA (2 μg/lane) was denatured, separated on formaldehyde gels and blotted to nylon (MSI). Nucleic acid blots were probed with ^{32}P labeled HMGR probe in 5X SSC, 0.5% SDS, 5X Denhardt's reagent, 100 μg/ml denatured calf thymus DNA, 50% formamide at 42 °C. Final wash conditions were 1X SSC, 0.5% SDS at 68 °C for 1h.

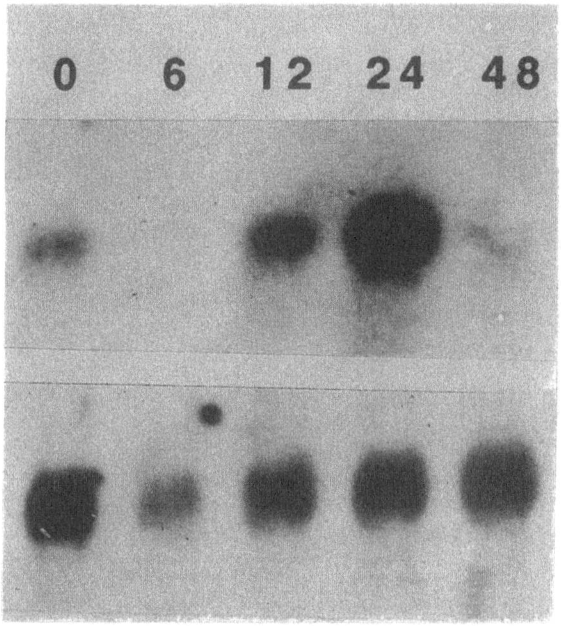

Fig. 2. Effect of MeJA on HMGR expression in *Camptotheca acuminata* and *Catharanthus roseus* suspension cultures. RNA was isolated from *Camptotheca* (A) and *Catharanthus* (B) cell suspension cultures at intervals after addition of 100 μM MeJA to the media. Two micrograms of poly A^+ RNA per lane was separated on 1.2% agarose, blotted, and hybridized with species-specific HMGR probes.

Plasmid construction and plant transformation

An *hmg1*::GUS translational fusion was made by ligating a 1790 bp EcoRI-NcoI fragment from the *hmg1* promoter (GenBank accession number L10390) to a β-glucuronidase reporter gene (pRAJ275; Jefferson 1987) containing the NOS polyadenylation signal at its 3′ end (Fig. 2). The chimeric gene construct was transferred to the binary vector Bin19 (Bevan 1984), electroporated into *Agrobacterium tumefaciens* LBA 4404 and used to transform tobacco leaf disks (Horsch et al. 1985).

Wounding and MeJA treatment

To examine the effect of wounding on *hmg1*::GUS expression in transgenic tobacco plants, 7mm disks were punched from 10–15 cm leaves using a cork borer. Wounded leaf disks were then incubated in 1/2 MS medium in constant light at 25 °C for 24 h. The effect of MeJA concentration on the wound response was assayed by treating leaf disks with 10^{-4}M, 10^{-5}M, 10^{-6}M MeJA for 24 h.

The interaction of the wound response with MeJA treatment was assayed in leaf disks which were treated with 10^{-4}M MeJA at various intervals after wounding (15 min, 30 min and 1, 2, 3, 4, 5, 6, 12, and 18 h). The disks were all harvested at 24 h after wounding and assayed for GUS activity.

GUS assays

Quantitative GUS assays were performed as described by Jefferson et al. (1987). Fluorescence of methylumbelliferone cleaved from MUG was measured in a DNA fluorometer (Hoeffer Scientific, San Francisco, CA) and expressed as nanomoles of MUG per minute per milligram of protein as measured by the method of Bradford (1976).

Results and discussion

MeJA regulated HMGR expression in suspension cultures

Exposure of plant tissue cultures to fungal elicitors is a proven method for increasing the production of some classes of indole alkaloids (Eilert et al. 1987). The results of Gundlach et al. (1992) indicate that jasmonic acid (JA) is a general signal transducer for elicitation in higher plants. Using several well characterized tissue culture systems, these investigators have shown that elicitation caused a transient increase in JA levels that can be correlated with the onset of secondary product synthesis. Furthermore, they also demonstrated that exogenous MeJA could substitute for fungal wall preparations in these cultures to elicit secondary metabolite production and the *de novo* transcription of defense genes such as phenylalanine ammonia lyase.

The regulation of HMGR expression by MeJA in tissue cultures is of particular interest to us because of the role this enzyme plays in providing secologanin for monoterpenoid indole alkaloid synthesis (Fig. 1). In *Catharanthus roseus* mRNAs for TDC and SS, two of the early enzymes in the indole alkaloid pathway (Fig. 1), are induced by elicitation with fungal wall extracts (Roewer et al. 1992). We therefore investigated the effect of MeJA on HMGR expression in *Camptotheca* and *Catharanthus* cell suspension cultures.

Fig. 2 shows gel blots of poly A^+ RNA extracted from *Camptotheca* and *Catharanthus* cell suspension cultures which were treated with 100 μM MeJA and harvested at various times. The blots were hybridized with probes corresponding to conserved HMGR coding regions from each species (Maldonado-Mendoza et al. 1992; Burnett et al. submitted). In both *Camptotheca* and *Catharanthus*, HMGR message transiently decreased after 6 h exposure to MeJA and then increased in abundance over basal levels by 24 h. In *Camptotheca* cultures the steady-state HMGR message had declined by 48 h, whereas in *Catharanthus* the message levels continued to increase.

Multiple copies of HMGR genes have been identified in all higher plant species examined to date. A minimum of two HMGR isogenes are known from *Arabidopsis thaliana* (Caelles et al. 1989; Learned and Fink 1989), tobacco (Genschik et al. 1992), and tomato (Narita & Gruissem 1989), whereas three functional HMGR genes have been reported in potato (Choi et al. 1992) and the rubber tree *Hevea brasiliensis*, (Chye et al. 1992).

DNA gel blots indicate that *Camptotheca* HMGRs also constitute a small family of genes with divergent members (Burnett et al. submitted). The expression of one of the *Camptotheca* HMGR isogenes, *hmg1*, has been studied in transgenic tobacco (Burnett et al. submitted). This gene is wound inducible, but its expression is suppressed by treatment with MeJA. Furthermore, expression of the *hmg1* gene continues to be suppressed for as long as the tissues are exposed to MeJA.

Although we observed a transient drop in the steady-state levels of HMGR message in both *Camptotheca* and *Catharanthus* tissue cultures after 6 h of MeJA treatment, by 24 h message levels had surpassed the levels seen in untreated controls (Fig. 2). Thus it appears that the recovery, and later induction of HMGR expression seen in MeJA treated cultures is the result of transcription from family members distinct from those suppressed by MeJA.

Quantitative regulation of hmg1::GUS expression by MeJA

Previously we have shown that the 5′ region of the *Camptotheca hmg1* promoter is necessary and sufficient to regulate the wound-induced, MeJA-suppressed expression of the β-glucuronidase (GUS) reporter gene in transgenic tobacco (Burnett et al. submitted).

Fig. 3 shows the relative GUS activity in wounded leaf disks from an *hmg1*::GUS plant treated with 0, 10^{-6}, 10^{-5} and 10^{-4} M MeJA for 24 h. In this experiment, wounding induced a 3.5-fold increase in GUS activity relative to controls. Treatment with MeJA, in contrast, showed a dose dependant suppression of wound-inducible GUS activity. It is also important to note that even at the lowest concentration of MeJA tested (10^{-6}M), the levels of GUS activity in treated leaf disks were below those of unwounded controls. This suggests that the HMGR gene expression is transcriptionally silenced by MeJA treatment rather than being simply not turned on in response to wounding.

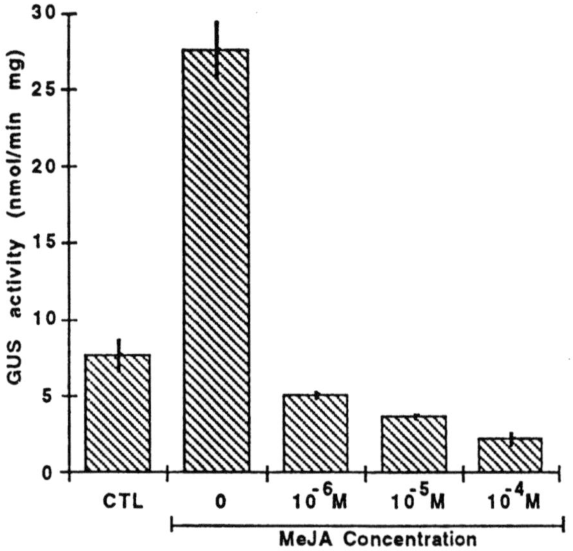

Fig. 3. Effect of wounding and MeJA concentration on the expression of *hmg1*::GUS in transgenic tobacco. The effect of wounding on *hmg1*::GUS expression was assayed in 7 mm leaf disks floated on 1/2 MS medium containing 0, 10^{-6}, 10^{-5}, 10^{-4} M MeJA for 24 h in constant light at 25 °C. Controls, unwounded leaf disks, were assayed immediately after removal from the leaf.

Temporal regulation of hmg1::GUS expression by MeJA

In order to examine the temporal regulation of wound-induced *hmg1*::GUS expression by MeJA, leaf disks were treated with 100 μM MeJA at various time intervals after wounding. The wounded leaf disks were then harvested after 24 h and assayed for GUS activity.

As can be seen in Fig. 4, the relative amount of GUS activity in wounded leaf disks treated with MeJA is comparable to the unwounded controls at time points from 0 to 4 h after wounding. By the fifth hour post-wounding, however, the level of wound-induced GUS activity had climbed to a level similar to the MeJA minus, wounded control. With slight variation, the level of GUS activity remained equally high in disks treated with MeJA at 6, 12, and 18 h after wounding.

These data indicate that the transcriptional/translational wound induction of GUS activity from the *hmg1* promoter has a minimum lag time of 4 h

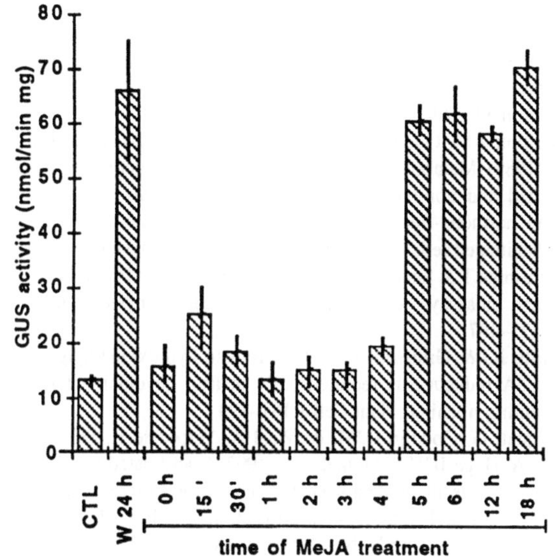

Fig. 4. Temporal analysis of MeJA suppression of the wound-induced expression of *hmg1*::GUS in transgenic tobacco. The timing of MeJA suppression of wound-induced *hmg1*::GUS expression was assayed in 7 mm leaf disks floated on 1/2 MS medium to which 100 μM MeJA was added at 0, 15 min, 30 min, 1, 2, 3, 4, 5, 6, 12, 18 h after wounding. Disks were assayed after 24 h in constant light at 25 °C.

after which MeJA application is no longer effective as a suppressor. Furthermore, the rise in GUS activity to levels approximating wounded controls after 4 h, also suggests that the MeJA factor(s) which silence the *hmg1* promoter cannot displace the positive regulatory factors which accompany wounding.

Conclusions

The common occurrence of small families of HMGR genes in plants may be required by the variety of end-products which use mevalonate as a pre-cursor as well as the diversity of conditions under which these products are made. Different HMGR activities are associated with normal plant growth and development, as well as environmental stresses such as wounding and pathogen challenge (Bostock & Stermer 1989).

The three HMGR isogenes of potato appear to be differentially expressed in response to wounding and elicitation (Yang et al. 1991; Choi et al. 1992). All three isogenes are wound inducible, but only one of them (*hmg1*) is suppressed by pathogen application or elicitation with arachidonic acid (Choi et al. 1992). Unlike *hmg1*, other two potato isogenes (*hmg2*, *hmg3*) show increased expression in response to these stimuli (Choi et al. 1992).

Our data support the existence of differentially regulated HMGR isogenes in both *Catharanthus* and *Camptotheca*. The transient suppression and then induction of HMGR transcripts reported here for suspension cultures treated with MeJA suggests that individual members of the HMGR gene family contribute to the steady-state HMGR message level according to their individual response to environmental cues. It also appears that the *Camptotheca hmg1* isogene (Burnett et al. submitted) has a regulatory paradigm similar to the potato *hmg1* gene (Choi et al. 1992). It will be of interest to determine whether the negative regulatory ligands responsible for MeJA suppression of the *hmg1* genes are the same as, or different than, the ligands which induce expression in other HMGR isogenes.

Acknowledgements

This work was supported by the National Institutes of Health (CA57592–01) and the U.S. Department of Agriculture (90–37262–5375). I.E.M.M. (CONCyT #15834) and M.LM. (CONCyT #58385) were supported by fellowships from the Consejo Nacional de Ciencia y Tecnología de México.

References

Bevan M (1984) Binary *Agrobacterium* vectors for plant cell transformation. Nucl. Acids Res. 12: 8711–8721

Bostock RM & Stermer BA (1989) Perspectives on wound healing in resistance to pathogens. Ann. Rev. Phytopathol. 27: 343–371

Bradford MM (1976) A rapid and sensitive method for the quantitation of microgram quantities of protein utilizing the principle of protein-dye binding. Anal. Biochem. 72: 248–254

Caelles C, Ferrer A, Bacells L, Hegardt FG & Boronat A (1989) Isolation and structural characterization of a cDNA encoding *Arabidopsis thaliana* 3-hydroxyl-3-methylglutaryl coenzyme A reductase. Plant Mol. Biol. 13: 627–638

Choi D, Ward BL & Bostock RM (1992) Differential induction and suppression of potato 3-hydroxy-3-methylglutaryl coenzyme A reductase genes in response to *Phytophthora infestans* and to its elicitor arachidonic acid. Plant Cell 4: 1333–1344

Chye M-L, Tan C-T & Chua N-H (1992) Three genes encode 3-hydroxyl-3-methylglutaryl CoA reductase in *Hevea brasiliensis*: *hmg1* and *hmg3* are differentially expressed. Plant Mol. Biol. 19: 473–484

Cordell GA (1974) The biosynthesis of indole alkaloids. Lloydia 37: 219–298

De Luca V, Marineau C & Brisson N (1989) Molecular cloning and analysis of a cDNA encoding a plant tryptophan decarboxylase: comparison with animal dopa decarboxylases. Proc. Natl. Acad. Sci. USA 86: 2582–2586

Eilert U, De Luca V, Constabel F & Kurz WGW (1987) Elicitor-mediated induction of tryptophan decarboxylase and strictosidine synthase activities in cell suspension cultures of *Catharanthus roseus*. Arch. Biochem. Biophys. 254: 491–497

Genschik P, Criqui M-C, Parmentier Y, Marbach J, Durr A, Fleck J & Jamet E (1992) Isolation and characterization of a cDNA encoding a 3-hydroxy-3-methylglutaryl coenzyme A reductase from *Nicotiana sylvestris*. Plant Mol. Biol. 20: 337–341

Gundlach H, Müller MJ, Kutchan TM & Zenk MH (1992) Jasmonic acid is a signal transducer in elicitor-induced plant cell cultures. Proc. Natl. Acad. Sci. USA 89: 2389–2393

Horsch RB, Fry JE, Hoffmann NL, Eichholtz D, Rogers SG & Fraley RT (1985) A simple and general method for transferring genes into plants. Science 227: 1229–1231

Jefferson RA (1987) Assaying chimeric genes in plants: The GUS gene fusion system. Plant Mol. Biol. Rep. 5: 387–405

Jefferson RA, Kavanagh TA & Bevan MW (1987) GUS fusions: β-Glucuronidase as a sensitive and versatile gene fusion marker in higher plants. EMBO J. 6: 3901–3907

Jones JDG, Dunsmuir P & Bedbrook J (1985) High level expression of introduced chimeric genes in regenerated transformed plants. EMBO J. 4: 2411–2418

Kutchan TM, Hampp N, Lottspeich F, Beyreuther K & Zenk MH (1988) The cDNA clone for strictosidine synthase from *Rauvolfia serpentina*. DNA sequence determination and expression in *Escherichia coli*. FEBS. Lett. 237: 40–44

Learned RM & Fink GR (1989) 3-hydroxy-3-methylglutaryl coenzyme A reductase from *Arabidopsis thaliana* is structurally distinct from the yeast and animal enzyme. Proc. Natl. Acad. Sci. USA 86: 2779–2783

Lückner M (1984) Secondary Metabolism in Microorganisms, Plants and Animals, Ed. 2. Springer-Verlag, New York

Maldonado-Mendoza IE, Burnett RJ & Nessler CL (1992) Nucleotide sequence of a cDNA encoding a 3-hydroxy-3-methylglutaryl coenzyme A reductase from *Catharanthus roseus*. Plant Physiol. 100: 1613–1614

McKnight TD, Rossner CA, Devagupta R, Scott AI & Nessler CL (1990) Nucleotide sequence of a cDNA encoding the vacuolar protein strictosidine synthase from *Catharanthus roseus*. Nucl. Acids Res. 16: 4939

Merillon J-M, Doireau P, Guillot A, Chenieux J-C & Rideau M (1986) Indole alkaloid accumulation and tryptophan decarboxylase activity in *Catharanthus roseus* cells cultured in three different media. Plant Cell Rep. 5: 23–26

Merillon J-M, Ouelhazi L, Doireau P, Chenieux J-C & Rideau M (1989) Metabolic changes and alkaloid production in habituated and non-habituated cells of *Catharanthus roseus* grown in hormone-free medium. Comparing hormone-deprived non-habituated cells with habituated cells. J. Plant Physiol. 134: 54–60

Murashige T & Skoog, F (1962) A revised medium for rapid growth and bioassays with tobacco tissue culture. Physiol. Plant. 15: 473–497

Narita JO & Gruissem W (1989) Tomato hydroxymethylglutaryl-CoA reductase is required early in fruit development but not during ripening. Plant Cell 1:181–190

Pasquali G, Goddijn OJM, de Waal A, Verpoorte R, Schilperoort RA, Hoge JHC & Memelink J (1992) Coordinated regulation of two indole alkaloid biosynthetic genes from *Catharanthus roseus* by auxin and elicitors. Plant Mol. Biol. 18: 1121–1131

Roewer IA, Cloutier N, Nessler CL & De Luca V (1992) Transient induction of tryptophan decarboxylase (TDC) and strictosidine synthase (SS) genes in cell suspension culture of *Catharanthus roseus*. Plant Cell Rep.11: 86–89

Sakato K & Misawa M (1974) Effects of chemical and physical conditions on growth of *Camptotheca acuminata* cell cultures. Agric. Biol. Chem. 38: 491–49

Stöckigt J & Zenk MH (1977) Isovincoside (strictosidine), the key intermediate in the enzymatic formation of indole alkaloids. FEBS Lett. 79: 233–237

van Hengel AJ, Harkes MP, Wichers HJ, Hesselink PGM & Buitelaar RM (1992) Characterization of callus formation and camptothecin production by cell lines of *Camptotheca acuminata*. Plant Cell Tiss. Organ Cult. 28: 11–18

Yang Z, Park H, Lacy GH & Cramer CL (1991) Differential activation of potato 3-hydroxy-3-methylglutaryl coenzyme A reductase genes by wounding and pathogen challenge. Plant Cell 3: 397–405

Index

MIX
Papier aus verantwortungsvollen Quellen
Paper from responsible sources
FSC® C105338

If you have any concerns about our products,
you can contact us on
ProductSafety@springernature.com

In case Publisher is established outside the EU,
the EU authorized representative is:
Springer Nature Customer Service Center GmbH
Europaplatz 3, 69115 Heidelberg, Germany

Printed by Libri Plureos GmbH
in Hamburg, Germany